PIMLICO

606

# CHARLES DARWIN

## THE POWER OF PLACE

Janet Browne is a zoologist and historian of science. She is at present Professor in the History of Biology at the Wellcome Trust Centre for the History of Medicine at University College, London. The first volume of her biography *Charles Darwin: Voyaging* is also published in Pimlico.

**Praise for *Charles Darwin: The Power of Place***

'The second, final volume of her magnificent life of [Darwin]. Much the best biography of Darwin to date, it makes irresistible reading.' Miranda Seymour, Top Five Books of the Year, *Sunday Times*

'Biographies of Charles Darwin are scarcely an endangered species...a few are moderately interesting and fulfilling, but none has offered the promise of this, the second volume of Janet Browne's study of the world's greatest biologist.' Robin McKie, *Observer*

'Magisterial, beautifully written and paced, wise, scholarly and a rivetting read.' Jackie Wullschlager, *Financial Times*

'A marvellous book...This second part of the life stands on its own. Soothing, unhurried and absorbing' Jane Ridley, *Spectator*

'From beginning to end, the book is richly informative and a delight to read.' Michael T. Ghiselin, *Times Literary Supplement*

'Browne's first volume was warmly received when it appeared seven years ago, and the second triumphantly fulfils its promise...[a] remarkable book.' James Secord, *Daily Telegraph*

'One of the most distinguished of all modern biographies.' *Guardian*

'A richly detailed, vivid and definitive portrait with not a word wasted: the best life of Charles Darwin.' *Kirkus Review*

'Browne's subject is monumental, but her writing style is never overburdened by the weight. Rather, her prose is elegant in its clarity of thought, her craftsmanship impeccable in the way it weaves a coherent whole from the innumerable threads of thought, experience and persona that comprised this colossal life.' *Publishers Weekly*

'Monumental and absorbing' J. B. Pick, *Scotsman*

'If you want to take the measure of his greatness, this is the book you should read.' Anthony Daniels, *Sunday Telegraph*

'Browne's triumph in this second of her two volumes is that, while analysing a great scientist, she continues to evoke a flesh-and-blood mortal.' G. S. Rousseau, *Literary Review*

'Out of all the virtues of Janet Browne's outstanding biography, the clinching one is the unassuming manner that respects and conveys the spirit of its subject so well.' Marek Kohn, *Independent*

'A magnificent achievement. Browne combines a clear and sympathetic account of Darwin's later years with a brilliant analysis of the phenomenon that became known as Darwinism.' John Gardiner, *BBC History*

'Janet Browne's meticulously organised and beautifully written two-volume biography...is as racy and exciting as any Victorian novel, and a fitting tribute to a truly great man.' John Banville, *Irish Times*

'Sharing her subject's passion for detail, but presenting her mountain of facts in a buoyant narrative, Browne maps the intricate ecosystem Darwin inhabited.' Jane Gregory, *New Scientist*

# CHARLES DARWIN
# THE POWER OF PLACE

Volume II of a Biography

JANET BROWNE

PIMLICO

For Kit and Evie (again)

Published by Pimlico 2003
2 4 6 8 10 9 7 5 3 1



First published in Great Britain by Jonathan Cape 2002
Pimlico edition 2003

Pimlico
Random House, 20 Vauxhall Bridge Road,
London SW1V 2SA

Random House Australia (Pty) Limited
20 Alfred Street, Milsons Point, Sydney,
New South Wales 2061, Australia

Random House New Zealand Limited
18 Poland Road, Glenfield,
Auckland 10, New Zealand

Random House South Africa (Pty) Limited
Endulini, 5A Jubilee Road, Parktown 2193, South Africa

Random House UK Limited Reg. No. 954009

A CIP catalogue record for this book
is available from the British Library

ISBN 0-7126-6837-3

Papers used by Random House UK Limited are natural,
recyclable products made from wood grown in sustainable forests.
The manufacturing processes conform to the environmental
regulations of the country of origin

Printed and bound in Great Britain by
Clays Ltd, St Ives

# *Contents*

*part one*

# AUTHOR

*chapter*

I

# STORMY WATERS

IF CHARLES DARWIN had spent the first half of his life in the world of Jane Austen, he now stepped forward into the pages of Anthony Trollope.

Victorian Britain seemed to be at peace with itself as political agitation at home and memories of the Crimean War and Indian uprising gave way to relative stability in the late 1850s and early 1860s. Free trade and carboniferous capitalism pushed ahead as the great manufacturing industries of the nation boomed. In the grand houses of London, Viscount Palmerston picked up his silk hat to become prime minister in 1857, followed in short order by Lord Derby in 1858, and then Palmerston again in 1859, while Benjamin Disraeli, William Gladstone, and Richard Bright stalked the wings impatient to transform the face of party politics. Cathedral cities hummed with religious controversy; books and magazines poured from the presses; the newly affluent took tours and holidays; and a whole army of clerks, civil servants, bureaucrats, bankers, and accountants was called into being to administer the fresh commercial horizons that accompanied the emerging empire, as India, China, Canada, South America, and the Antipodes increasingly fell under British economic domination. Steam technology was the hero of society. At that time Britain possessed two-thirds of the world's capacity for cotton factory production and accounted for half the world's output of coal and iron, an unmatched degree of industrial preeminence. The length of railway track snaking across the countryside doubled from 1850 to 1868. Lawnmowers, water-closets, gas lights, iron girders, encaustic tiles, and much, much more were available to those who could afford them. Although Queen Victoria and her ministers were soon to encounter complex foreign affairs in Garibaldi's Italy and painful consequences from the Civil War in the United States of America, the ethos of "improvement" prompted significant developments in domestic housing, health, education, communications, dress, and manners. "The genius of England is universally admitted to be of an eminently enterprising and speculative character," declared the magazine *Once a Week*.[1] Confidence soared. Social boundaries shifted.

Even so, the contradictions at the heart of Victorian life were more obvious than ever. Fraud, filth, overcrowding, poverty, death, and violence were a fact of life in the urban slums. Rural communities had lost in a decade more than 40 percent of the male workforce to industrial, colonial, and military demands and bleakly faced another round of agricultural depression and distress. The nation's religious faith, although never coherent, was fracturing into fervour or dissent. While many from the ruling ranks of society turned a blind eye to these issues, a remarkable array of novelists, statisticians, medical men, radical divines, and social activists were starting to reveal the squalor alongside prosperity and discovering the interesting in the ordinary. In time, parliamentary leaders would open their minds to a second round of political reform in the nineteenth century, egged on by the high sense of purpose, moral earnestness, doctrines of self-help, and appreciation of decorum that characterised the emerging middle classes. From real-life Westminster to imaginary Barchester and back again, Trollope easily captured in his novels this sense of the personal and parochial. But life was not simple even for those whom Lord Salisbury called "persons of substance." These mid-century years were not so much an age of equipoise as framed by social and political contrasts. It was an age of capital, labour, complacency, and faith; at the same time, an age of cities, misery, change, commerce, deference, and doubt.

In among the contrasts stood the unobtrusive figure of Charles Darwin. Supported by a family fortune derived from the Industrial Revolution, Darwin was content to become a thoroughly respectable Victorian gentleman. He put away his *Beagle* shotguns, cast a discerning eye over his investments, and began to participate in the growing sense of national prosperity. He had no need to seek employment. Like many others in his circle he was free to pursue his interests, in his case a magnificent obsession with natural history.

In 1858 he was forty-nine years old, a steady and likable individual, "one of the kindest and truest men that it was ever my good fortune to know," said Thomas Henry Huxley. His scientific status was already secure, although he had not yet revealed his theories about species to anyone other than a few close friends. His *Origin of Species* was yet to come. His personal position was equally secure. He was married and comfortably settled with his wife, Emma, and their children in a country house in the village of Downe, in Kent, near enough to the attractions of the metropolis but a world away from its problems. For years now he had been troubled by continued ill health—"being ill was normal." Yet his home at Down House was the safe harbour he sought for the end of his personal voyage. "Few persons can have lived a more retired life than we have done," he wrote with undisguised pleasure. "My life goes on like clock-work and I am fixed on the spot where I shall end it."[2]

In fact Darwin was far more sociable than his words allowed. In London, his friends were clever and influential, a cosmopolitan mix of university professors, authors, manufacturers, government officials, landowners, and politicians; here and there a baronet or a literary lady or two, a few old comrades from his time on the *Beagle,* and a clutch of intelligent nieces ready to discuss the latest concerts or exhibitions. Whenever he went to town, he sought out the company of his older brother Erasmus, pleasantly fixed in his bachelor ways, and his cousins Fanny and Hensleigh Wedgwood, all living close to each other on the outskirts of Bloomsbury and forming the hub of an extended circle of intermarrying Wedgwoods and Darwins.[3] Darwin had married his cousin Emma Wedgwood in 1839. Other cousin marriages among the clan drew the generations together.

Erasmus hosted dinner parties for him, gossiped, and kept parcels until his arrival. If Darwin was alone he would stay overnight and meet his old friend the geologist Sir Charles Lyell or some other scientific colleague for breakfast—meetings which he valued for keeping in touch and maintaining his intellectual momentum. Otherwise, he would bring Emma and the youngest children up for the pantomime or trips to the dentist. They would stay with Fanny and Hensleigh and see their other relatives visiting from the shires.

His country friends were no less pillars of the community. Darwin welcomed the soothing rhythm of local affairs, always willing to discuss the state of the weather or his poor health with neighbours, organise parish charities, and sympathise with John Innes, the resident vicar, over difficult young curates or problems with the village school. Every so often, a little debate about church doctrine with Innes made his strolls around the country lanes agreeably lively. Innes was just the kind of relaxed clergyman that Darwin himself might once have become if the voyage of the *Beagle* had not intervened. "I do not attack Moses," the naturalist remarked affably to him, "and I think Moses can take care of himself."[4] At Downe Darwin took on duties as a local magistrate, an occupation at the heart of provincial life in which law-abiding, landowning gentlemen like himself imposed fines on poachers or issued licenses for keeping pigs.

This society was reassuringly sedate. Darwin and Emma regularly met Sir John Lubbock, mathematician and fellow of the Royal Society, and his son John, a young naturalist, who lived a few miles away at High Elms. They enjoyed the company of the Bonham Carters, in a neighbouring village, and George Ward Norman, a director of the Bank of England and country gentleman of Downe. Every so often they invited weekend guests from London, sending a horse and carriage to the nearest railway station to pick up visiting groups. Joseph Hooker, the assistant director of Kew Gardens, and Thomas Henry Huxley, the biologist and writer, were particular friends. Somewhat surprisingly for a period in which the prevailing motifs were industrialisation, social movement, urban expansion, and religious dissent, Darwin's parish was utterly secluded, almost a relic of a former age in its social structure and restricted occupational patterns. The long trip to the railway station made it seem further away from modernity than it really was. Downe village was small, no more than five hundred people in the 1861 census, and relatively stable considering its proximity to growing suburban centres like Bromley and Maidstone. The national post office had altered the spelling from Down to Downe late in the 1850s, a change that Darwin resolutely ignored when addressing letters. Twenty years later the population had increased by only fifty. Falling readily into the provincial swing of things, he unashamedly called himself a "Farmer" in Bagshawe's *Directory.* A Wedgwood niece tartly observed after one of these weekend parties, "We have enough dullness in the family & plenty of virtue—a little vice would make a pleasant variety."[5]

Yet underneath the mild exterior, Darwin's mind teemed with ideas—daring and unusual proposals that he hesitated to put before the world. He had balked at disclosing his theories before they were ready, fretting anxiously over his work, doggedly probing every crevice of the evidence, building up a tightly packed argument that he hoped would protect his scheme from at least some of the intense criticism he knew it would provoke. Ever since returning from the *Beagle* voyage in 1836, some twenty-two years before, he had believed that living beings were not created by divine fiat. From that time on, he had sought an alternative explanation that would depend on natural processes rather than on God's direct action.

He had found it in Thomas Malthus's *Essay on the principle of population,* an economic principle of checks and balances that Darwin applied to the survival rates of animals and plants and called "natural selection."

Since then he had focused his energies on documenting the origin of species by these natural means, "slaving away" in private. Only recently, in 1856, Charles Lyell had pressed him to get on and publish, and in that year Darwin began writing a long manuscript intended for future publication. "I have found it quite impossible to publish any preliminary essay or sketch," he confided to another friend, "but I am doing my work as completely as my present materials allow without waiting to perfect them." This unfinished manuscript on natural selection already ran to 250,000 words, comprising eleven chapters of a probable fourteen, a great pile of paper on his study table that he ruefully called his "big book on species." He knew it was his life's work, before which everything else faded into irrelevance.[6]

As Darwin now conceived it, natural selection operated on living beings as if it were a statistical necessity, a law of nature stripped of any divine influences, invincible, predominant, and fierce, relentlessly honing animals, plants, and humans in the struggle for existence. His theories had no room for biblical teachings about Adam and Eve or the Garden of Eden. Organisms either adapted or died. His vision of nature had moved far beyond the cosy notions that fortified most Victorians, views about the perfect adaptation between animals and plants and their environment that, for many, mirrored the social stability they thought they saw around them. "Every class of society accepts with cheerfulness that lot which Providence has assigned to it," Palmerston optimistically declared. He might almost have gone on to include animals and plants. Darwin, on the other hand, saw the natural world as a constant competitive struggle for survival.

Much of the lasting fascination of Darwin's life story surely lies in the relationship between this prolific inner world of the mind and the private and public lives that he created for himself. His power of analysis was outstanding; his creative imagination remarkable. As a biologist, his distinctive gift was to envisage all living beings not only in their relations to one another but also in their relations to the places in which they lived and to the unfolding sequence of time. He would become one of the most famous scientists of his day, a Victorian celebrity whose work even in his own lifetime was regarded as a foundation stone for the modern world, not least for the manner in which he changed the way human beings thought about themselves and their own place in nature. And yet he liked to be a countryman, pottering around his garden. He was an invalid plagued by disorders that probably fed on his intense intellectual activities. He was a husband, father, friend, and employer, as well as a naturalist, author, and thinker. To explore what sort of person he was adds significantly to the evaluation of his part in history—or, putting it round another way, to know something of Darwin and the way he operated explains a good deal that might otherwise be perplexing about the scale of his achievement and the revolution in thought that customarily bears his name. The manner in which his daily life interlocked with his theories and with his public role as the author of *On the Origin of Species* brings to light a long and unexpectedly eventful life story.[7]

The *Origin of Species* was to dominate the second half of his life. Where Darwin had once voyaged on the *Beagle* through new oceans of thought, he now turned his mind towards writing and publishing, towards being an author. The events surrounding the book's publication were exacting enough. Afterwards Darwin would emerge as a remarkable tactician—a man who prefered to remain behind the scenes but a canny and dedicated publicist for all that. The strategic effort that he put into disseminating his views was intense. As has long been clear, the Darwinian revolution was neither completely Darwinian nor completely revolutionary,[8] and there was no steady march towards the publication and approval of his ideas. The recasting of contemporary scientific horizons was hardly carried out by him alone, and what would pass for "Darwinism" was never a monolithic structure. The chronicle of Darwin's mature years was in fact to be the story of how he negotiated the reception of his *Origin of Species,* and this in turn would comprise a web of will-power, strategy, conflict, the loss of friends, disappointment, pleasure, ruthless determination, and great personal exertion set within larger currents of scientific and social transformation—a story at root about the making and validating of new scientific knowledge. The project was immeasurably enhanced by his producing a book at a time when the publishing industry was expanding and review journals were enjoying rapidly diversifying audiences. He benefitted from the public support of his friends, many of whose careers progressively interwove with his own. His book closely meshed with major transformations in nineteenth-century thought and came to symbolise the fresh perspective. What could it have been like to be Darwin during what would come to be called the Darwinian revolution?

Even Darwin himself might have found the question difficult to answer. "I wish I could feel all was deserved by me," he was to say. Indeed, many of the modern images of Darwin—as an invalid, as a methodical country squire, as a solitary genius—while perceptive and accurate enough, fall short of doing full justice to the many sides of his character that would emerge during the last twenty-five years of his life. In retrospect, it is plain that there was much more to Darwin than his theory, and more to the theory than Darwin. His scientific associates almost inevitably met a different figure from the man that his wife and children knew, a man different again from the controversial author that the public encountered when the *Origin of Species* was published or that his servants passed in the hall. Friends and enemies responded to him variously. Darwin responded variously in return. "If I had been a friend of myself, I should have hated me,"[9] he remarked pensively to Huxley at the height of the controversy over the *Origin*.

As might be expected, he not only lived his own life, he lived also in the lives of others. In personal terms the choices he made and the paths he pursued were necessarily of the moment. Every plus had its minus and some of his actions can now be seen to be deeply exploitative. He manipulated his household and daily routines in order to allow the production of his book and the other volumes that were to follow. The impressive tenacity and persistence that he brought to his natural history researches were at times utterly selfish. The dedicated collection of facts that made his writings so powerful rested on an implacable ability to take advantage of the knowledge of others. His illnesses, usually so disruptive and debilitating, as well as his well-known reticence and modesty, served as an effective way to avoid unwanted responsibilities. And despite his reputation for shyness, the people who knew him well recalled his "hearty laugh" and "jovial" manner, his "By Jove!" in letters and conversation. Even though he lost his religious faith and shocked his readers, he was acclaimed as a good man, a benevolent sage, and he was buried in Westminster Abbey. All his best qualities—the qualities that inspired Leslie Stephen to dub him "a noble old hero of science"—were grounded in ordinary longings, faults, and frailties. Darwin was one of the most human of men, surrounded by concentric circles of friends and relatives. In many ways, his biography is in part the biography of Victorian family life—of what it was like to make and live with science.

In all this, there was a special resonance between the man and his domestic setting. By now the Darwins had lived at Down House for sixteen years, a fruitful, relatively placid time during which he and Emma produced eight of their ten children. Two of these children had died in childhood. One was buried in the churchyard of the small flint-walled Anglican church in Downe, the other in Malvern after a disastrous trip to the water-cure. He and Emma had mourned heavily over these losses. But despite the emotional toll, Darwin felt wholly content in Downe. The remaining children were William, the oldest, born in 1839, Henrietta (b. 1843), George (b. 1845), Elizabeth (b. 1847), Francis (b. 1848), Leonard (b. 1850), Horace (b. 1851), and little Charles, the baby, born in 1856. He and his family were an integral part of the fabric of town and country life that characterised the landed classes in Britain during the middle years of the nineteenth century. Without this sense of place, Darwin could hardly have hoped to bring his work on natural selection and the origin of species to completion. His home, and the lifestyle of a country gentleman that he created within it, gave him the peace he needed and time to consider every part of his argument.

Without this sense of place, too, his work would not have taken the singular character that it did. His life and his science were of a piece. The tumble of ideas that had characterised the first half of his existence was giving way to the methodical intensity of documenting and reinforcing his notions. His home and garden were his experimental laboratories, his book-lined study was his manufactory; these were the places where he most liked to be. He discovered that he valued routine—and went to great lengths to create a well-regulated household in which he was left free for the steady construction of facts. More than this, his home and his homelife became an actual part of his intellectual enterprise. Over the years, Darwin bred pigeons, grew pots of seeds in his outhouses, observed bees moving across his flowerbeds, tracked worms in the fields that he saw from his drawing-room window, counted blades of grass in his lawn, watched his infant children in the nursery, and pondered the twists and turns of climbing weeds in his hedges, all the while seeking the detailed evidence of adaptation in living beings that he believed to be the keystone of his project. Although his *Beagle* experiences were still important to him and always carried due weight in his writings, and his particular insight into nature remained undimmed, these home-based researches were the hidden triumph of his theory of evolution. His family setting, his house and garden, the surrounding Kent countryside, and his own sense of himself at the heart of the life he had created and the property he owned provided the finely crafted examples of adaptation in action that lifted his work far out of the ordinary. His thinking path, the path he called the Sandwalk that skirted the edge of a copse at the bottom of the Down House garden, became the private source of his conviction that his theory was true—true, if only he could show it.

Solitude served him well here. But Darwin was not a complete rural recluse. Systematically, he turned his house into the hub of an ever-expanding web of scientific correspondence. Tucked away in his study, day after day, month after month, Darwin wrote letters to a remarkable number and variety of individuals. He relied on these letters for every aspect of his evolutionary endeavour, using them not only to pursue his investigations across the globe but also to give his arguments the international spread and universal application that he and his colleagues regarded as essential footings for any new scientific concept. They were his primary research tool. Furthermore, after the *Origin of Species* was published, he deliberately used his correspondence to propel his ideas into the public domain—the primary means by which he ensured his book was being read and reviewed. His study inside Down House became an intellectual factory, a centre of administration and calculation, in which he churned out requests for information and processed answers, kept himself at the leading edge of contemporary science, and ultimately orchestrated a transformation in Victorian thought.

This took place on the grand scale. Darwin wrote or received some fourteen thousand letters that are still in existence in libraries the world over, and there must have been as many again now lost to posterity.[10] By far the largest number of letters were exchanged with his closest scientific friends—Charles Lyell, Joseph Hooker, Asa Gray, and Thomas Henry Huxley, men who supported and helped him through thick and thin. These friends were prominent naturalists in their own right, each in his way representing major branches of the Victorian natural history sciences and important scientific institutions. Through them Darwin gained access to the machinery of international intellectual endeavour, and while Huxley may have earned most of the subsequent historical plaudits as Darwin's chief defender and publicist, the roles played in Darwin's life by Joseph Hooker at the Royal Botanic Gardens at Kew, by Asa Gray at Harvard University, and by Charles Lyell, an independent gentleman-geologist in London, should not be underestimated. This intimate network proved crucial to Darwin both personally and in the eventual acceptance of evolutionary ideas. Otherwise, the largest group of his correspondents were German-speaking naturalists, more than one hundred different individuals. Darwin wrote to his overseas contacts in old-fashioned, stilted English, apologising quaintly for his lack of languages.

He also hunted down anyone who could help him on specific issues, from civil servants, army officers, diplomats, fur-trappers, horse-breeders, society ladies, Welsh hill-farmers, zookeepers, pigeon-fanciers, gardeners, asylum owners, and kennel hands, through to his own elderly aunts or energetic nieces and nephews. Many of his letters went to residents of far-flung regions—India, Jamaica, New Zealand, Canada, Australia, China, Borneo, the Hawaiian Islands—reflecting the increasing European domination of the globe and rapidly improving channels of communication.[11] There was only one postbag in Downe village, and it looks as if Darwin's daily activities could have filled it alone. In 1851 he spent £20 on "stationery, stamps & newspapers" (nearly £1,000 in modern terms), paying a monthly invoice to James Verrel, the newsagent in Bromley High Street, and a smaller sum to Albert Sales of the George Inn at the crossroads in Downe, the village publican, grocer, and postmaster. By 1877 Darwin's expenditure on postage and stationery had doubled to £53 14s. 7d, a sum roughly equal to his butler's annual salary. Every part of his life was run by letters—and the lives of his family members too. "Everyone obeyed the advice given by a family poet," remarked a granddaughter cordially some fifty years later.

> *Write a letter, write a letter;*
> *Good advice will make us better;*
> *Father, mother, sister, brother,*
> *Let us all advise each other.*[12]

If there was any single factor that characterised the heart of Darwin's scientific undertaking it was this systematic use of correspondence. Darwin made the most of his position as a gentleman and scientific author to obtain what he needed. He was a skilful strategist. The flow of information that he initiated was almost always one-way. Like countless other well-established figures of the period, Darwin regarded his correspondence primarily as a supply system, designed to answer his own wants. "If it would not cause you too much trouble," he would write. "Pray add to your kindness," "I feel that you will think you have fallen on a most troublesome petitioner," "I trust to your kindness to excuse my troubling you." "If any man wants to gain a good opinion of his fellow men, he ought to do what I am doing, pester them with letters," he once said to John Jenner Weir, the ornithologist. "Best & most beloved of men, I supplicate & entreat you to observe one point for me," he cried to Hooker. There was no need for Darwin to doubt the legitimacy of this one-way arrangement. After all, he occupied an assured place in the intellectual elite, at the heart of an expanding scientific and social meritocracy that in turn lay at the hub of one of the most powerful and systematically organised empires known to history. He made vigorous use of these advantages.

People usually did what he asked. From time to time, he would reward his correspondents by forwarding their articles or introducing them to London experts, and perhaps they felt this was return enough. It is clear that he functioned near the top of a hierarchical social structure that facilitated such interactions. Among his closer friends, however, Darwin was unwilling to appear quite so exploitative. There he built rather more of a network of give and take, responding to his colleagues with friendly encouragement and support.

One way or another, these men and women, near and far, contributed materially to his developing project and its subsequent trajectory in the world at large. Darwin's completed theory of evolution ought perhaps to be seen as the interplay between the creative vision residing in a single mind and a mass of information gathered from many different hands, including his own. Whether the man himself might, with hindsight, be characterised either as a hero of science, an observer *par excellence,* a political animal, or a nervous, reclusive revolutionary, his achievements were manifestly the product of a highly efficient Victorian communication system, firmly embedded in what can be called knowledge-producing relationships. With pen and ink and postage stamps he set about constructing what he hoped would be "a considerable revolution in natural history." Alone at his desk, captain of his ship, safely anchored in his country estate on the edge of a tiny village in Kent, he was in turn manager, chief executive, broker, and strategist for a world-wide enterprise. Once, in a passing compulsion, he attached a mirror to the inside of his study window, angled so that he could catch the first glimpse of the postman turning up the drive. It stayed there for the rest of his life.

Such a life obviously depended on the postal system, the preeminent collective enterprise of the Victorian period, and Darwin sensed the splendour of this organisation as readily as Anthony Trollope, who, after novelising the nation before breakfast, would go to his employment in the General Post Office in London. No one would believe the number of letters surging across nineteenth-century Britain, said Rowland Hill, the inventor of the penny postage system. By mid-century, 600 million letters were dispatched every year. Twenty-five thousand delivery men travelled 149,000 miles to distribute these letters, carrying in their sacks a weight of nearly 4,300 tons. Carrier services transported 72 million newspapers, 12 million book parcels, and 7 million money-orders a year, and 68 million letters moved around the capital alone, requiring eleven deliveries a day. Prime ministers, civil servants, and Queen Victoria ran the country with a daily outpouring of well-turned phrases, and countless novelists relied on the prompt arrival and dispatch of letters to carry their plots along, confident that readers were involved in similar processes. The tide of correspondence, wrote Hill, "knew no ebb."[13] Letters became more informal, more up-to-date, more personal, and more frequent, pulling the edges of family and empire together and convincing Britons that their country was one of the most advanced nations in the world.

Now that nearly 150 years have passed since the *Origin of Species* was published, and its place in modern thought is assured, Darwin's life can be explored from the inside looking out, from the domestic, respectable, pastoral setting in which he chose to locate himself to the extensive letter-based connections he created with the world beyond; and from the outside looking in, through the eyes of others, close or far away. His book came to represent the spirit of the age. Although Trollope and Darwin never met, each would have felt completely at home in the other's sphere. Yet if anyone had told Darwin how famous he would become, he would have been very surprised.

## II

Picking up a thin, well-wrapped package one morning in June 1858, Darwin wondered who could be writing to him from Ternate, an island in the Dutch East Indies halfway between Celebes and New Guinea. His web of correspondents already circled the globe. India, Africa, Tasmania, South America—over the years he had gathered contacts in every quarter feeding his insatiable appetite for facts.

He was interested to recognise Alfred Russel Wallace's handwriting on the package. He knew Wallace to be a talented man, full of the intrepid scientific spirit that Darwin most admired, and at that time travelling rough in Indonesia and Malaysia collecting rare natural history specimens. A year or so beforehand Darwin had asked Wallace if he could possibly get the skins of some Malayan poultry for him. He hoped there were unusual details about tropical birds or animals described inside.

But the package contained nothing of the sort. Although Wallace's words were unassuming and polite enough, they had cataclysmic effect. Darwin's life was never the same again.

What the packet enclosed was a short handwritten essay which, line by line, spelled out virtually the same theory of evolution by natural selection that Darwin believed was his alone. Isolated in the jungle for four years, Wallace had independently hit on the same argument as Darwin. All of Darwin's main ideas were repeated. To Darwin's agitated mind these ideas seemed to hang together in Wallace's essay far better than they did in his own unpublished writings. Wallace wrote clearly—so clearly that no one could mistake his meaning. The struggle for survival among animals and plants; competition and extinction; the improvement of domestic races by selection; the divergence of species into different forms: all these were included. Malthus was there. So was Lyell. Wallace demonstrably removed the divine Creator and proposed an entirely natural origin for species. His words indicated that he fully understood the significance of what he was saying. "It is the object of the present paper to show . . . that there is a general principle in nature which will cause many varieties to survive the parent species, and to give rise to successive variations departing further and further from the original type."

> It is evident that, of all the individuals composing the species, those forming the least numerous and most feebly organized variety would suffer first, and, were the pressure severe, must soon become extinct. . . . The superior variety would then alone remain, and on a return to favourable circumstances would rapidly increase in numbers and occupy the place of the extinct species and variety. The variety would now have replaced the species, of which it would be a more perfectly adapted and more highly organized form.[14]

Darwin was stunned. "I never saw a more striking coincidence," he moaned helplessly. "If Wallace had my MS sketch written out in 1842 he could not have made a better short abstract!"[15]

He was well and truly forestalled. It was impossible to pretend otherwise. All his originality was smashed, all his years of hard work suddenly useless. For a moment the news hit him like the death of a child. Then, his mind churned with painful emotions—not anxiety or panic, he confessed afterwards, but much baser feelings of mortification, possessiveness, irritation, and rancour, each flaring up one by one after the first unaccountable, humiliating surprise. Hour after hour they returned, making him cross and edgy. "It is miserable in me to care at all about priority," he complained to Hooker and Lyell that day. "Full of trumpery feelings." These were probably the most lonely hours of his life, facing the knowledge that what mattered to him now was not so much the long-gone moment of discovery but the possession, the ownership, of his theory. Wallace's easy brilliance forced him to confront the focus of his entire working life. Had it all been a waste of time? Those years he had spent labouring over barnacles, the deterioration of his physical health, the endless attention to notes and letters, and the huge manuscript so close to completion? He caught himself wondering truculently if his own letters to Wallace, brief as they had been, might somehow have given the game away. The resemblance of their ideas was startling.

Nevertheless Wallace was obviously acting in good faith. It was evident he had no idea that Darwin was so well advanced on a project so similar to his own, even though they had discussed species and varieties in letters beforehand. In his accompanying note he asked Darwin to pass the essay on to Sir Charles Lyell if it seemed sufficiently interesting. Since Lyell was often instrumental in bringing the work of unknown naturalists into the public eye, this was a reasonable request to make, and Wallace, who had no personal access to prominent scientific figures, manifestly hoped for the kind of friendly introduction that Darwin could provide. Moreover, his essay explicitly drew on Lyell's *Principles of Geology*, especially on Lyell's critical account of Lamarck's theory of transformation and his commentary on the creation, adaptation, and extinction of species. The essay, in short, had been composed for Lyell, not for Darwin. Yet it would have been near-impossible at that period for Wallace to write directly to Lyell. A favourable word from Darwin would help him along. And Wallace knew that Darwin and Lyell were friends—Darwin had told Wallace so in a previous letter. Beyond that, Wallace knew that Darwin's *Journal of Researches* had been warmly dedicated to Lyell.

One letter—and Darwin was shipwrecked. His dilemma was profound, as intense as any in his life, although the course of his action was plain to him. All his moral instincts—his strong sense of duty, his unquestioning acceptance of gentlemanly responsibilities, his pleasant nature, his honourable feelings—told him he must comply with Wallace's request, and, moreover, acknowledge to Lyell that Wallace had got there first. There is no reason to suppose that he hesitated, or at least not for long.[16] A less scrupulous person might perhaps have destroyed the essay and pretended it never arrived. A long journey from the Far East supplied a ready excuse should one be needed. A less candid man might have delayed and delayed, unwilling to concede the point until his own work was published.[17]

Darwin's honour as a gentleman—as he understood it—was at stake.[18] In the competitive scientific world in which he chose to live, publication, originality, and priority made a delicate trio. It was easy to succumb to the temptation to be secretive or be overly quick to publish; and yet the creator of any fresh insight, then as now, must eventually relinquish possession and place his or her ideas in the public domain in order to be given credit for advancing knowledge. The primary spur for Victorians like Darwin was not so much to gain individual power, as it might have been for the politician, nor wealth, as for the businessman, but reputation and professional pride—the need for recognition of the value of one's endeavour by others in the field. Darwin had always believed that science was something more than a race to publish new findings, or at least on those occasions when he heard of controversies between other men he usually discussed them in censorious terms with Hooker. For this reason, he had considered it wise to wait for his work to mature, wanting above all that it should be right, and regarding any personal desire to be ahead of the field as merely secondary, no doubt idealistic and self-deluded, but sincere. Though this could once have been called a deliberate delay, and might well be attributed to Darwin's cautious attitude to the external political situation when he first began thinking about evolution, it is likely that by now he was struggling with his own tendency to be a perfectionist. Deep down below, his vision of science glinted with a wish—in turn debilitating or inspiring—to arrive at the truth.

But with Wallace's letter in his hand he had never felt more vulnerable. It was hard to give up his claim to his life's work, work in which his whole identity was wrapped. Nonetheless at some point that day he decided he must respect Wallace's priority. Perhaps he came to think that he wanted to live an honourable life—one that he could contemplate in old age with equanimity, and in which his deeds were fair and his social behaviour as honest as he could make it.[19] He wrote to Lyell of "justice" and his dread of any possible "dishonour." More than this, he may have found solace in an elevated spirit of resignation. He might not be able to conquer the storm of inner feeling but he was able to control his outward conduct. "I certainly was a little annoyed to lose all priority," he told Lyell, but "resigned myself to my fate."[20]

He dispatched the essay to Lyell as Wallace had asked.

> My dear Lyell,
>
> Some year or so ago, you recommended me to read a paper by Wallace in the Annals, which had interested you & as I was writing to him, I knew this would please him much, so I told him. He has today sent me the enclosed & asked me to forward it to you. It seems to me well worth reading. Your words have come true with a vengeance that I shd. be forestalled. You said this when I explained to you here very briefly my views of "Natural Selection" depending on the struggle for existence. . . . Please return me the MS which he does not say he wishes me to publish; but I shall of course at once write & offer to send to any Journal. So all my originality, whatever it may amount to, will be smashed. Though my Book, if it will ever have any value, will not be deteriorated, as all the labour consists in the application of the theory. I hope you will approve of Wallace's sketch, that I may tell him what you say.[21]

So saying, he gave up his right to his life's work and passed the matter over to his oldest friend.

## III

When Darwin called this event a coincidence he was already building up and retreating behind a protective fence of his own making. He evidently found it far less stressful to characterise the situation as a "striking coincidence" than to contemplate the alternatives—alternatives that implied any number of authors might be racing towards his own personal goal or that his concepts were much less innovative than he thought. He disliked finding out that someone else could conjure up his own private brainwave. He had invested his time, his health, and his happiness in the work, and to lose this intellectual capital overnight was as cruel to him as any financial disaster. Darwin had read widely and carefully for many years, always evaluating his own views against those he met in the writings of other naturalists. He genuinely believed no one else held the same combination of ideas as he did. Even his most gifted scientific friends, it seemed to him, were sometimes puzzled by the thrust of his arguments. "I occasionally sounded not a few naturalists, and never happened to come across a single one who seemed to doubt the permanence of species. Even Lyell and Hooker though they would listen with interest to me, never seemed to agree."[22] As a consequence, Darwin felt that he was the only one who really understood his theory or could see how it might work as a biological explanation, those crystalline glimpses of a new kind of knowledge that sustained him through all reversals and urged him to continue. It was hard to accept that he was not the innovator he imagined he was. Along with everything else, his scientific vanity was badly shaken.

Yet Wallace's letter was really no more of a coincidence than the invitation to travel on the *Beagle* had been. To start with, there were differences between the two theories that Darwin noticed only when he studied Wallace's work more thoroughly later on in the summer. Wallace attended far more than he did to the replacement of a parent species by an offspring variety. Wallace wrote of the way that a group of advantaged individuals, say a variety of pigeon that could fly further in times of food shortage, might in time supplant those birds that possessed less stamina. Group replaced group. His view of nature was thus less concerned with individuals than Darwin's. Second, he declared his belief that there could be no parallel between the natural process of "selection" and what went on under artificial conditions of domestication—a point diametrically opposed to Darwin.[23] Any parallel with breeders "selecting" traits in their animals, Wallace said, was "altogether false." Issues emerging from these differences kept the two naturalists engaged in close debate for the rest of Darwin's life.

Darwin had also long been blind to many of the changing currents around him. If he had been less inwardly focused on his own projects, or less preoccupied with his health and that of his family, he might not have been so hopelessly taken aback. Wallace had scattered suggestive pointers about the way his thoughts were tending in several articles published in London journals during the 1850s and deliberately raised the problem of accurate distinctions between species and varieties in letters to Darwin. Lyell had drawn these signals to Darwin's attention during a weekend visit to Down House in 1856.

Other suggestive pointers were just as plain to see. Evolution, Lyell observed, was hanging tensely in the air. Evolution—or something very like it.

If Darwin had lived in London, as Lyell did, or mixed more frequently with the intellectual avant-garde, he must surely have noticed the general swing of progressive, liberal opinion among a small circle of influential figures. Speculative developmental ideas enjoyed fairly wide currency. Mostly these ideas were loosely based on the concept of an inbuilt advance of mankind and society, ideas that ultimately rested on the ideologies of enlightenment and transformation disseminated by European thinkers of the late eighteenth century and revolutionary period, among them Jean-Baptiste Lamarck and Darwin's own grandfather, Dr. Erasmus Darwin, and that were now revitalised with high Victorian notions of striding forwards. During the 1820s and 1830s, most aspects of this transformist philosophy of nature, including the possibility of changing the nature of the human mind and the structure of society itself, had in Britain come to be associated with scientific rationalism and as often as not the lurking threat of political activism. But by the 1850s, intellectuals were equally liable to embrace the same motifs in the safer form of self-advance, economic progress, and the steady march of civilisation while still taking Lamarck's name as a general catch-all label for any progressive transmutationary ideas.

Living and working in Paris through the revolutionary period, Lamarck had in fact boldly taken the whole of nature for his study. A true child of the enlightenment, he had believed all living beings were subject to rational laws and proposed that species had evolved gradually over eons of time and in interaction with the environment—*transformisme,* he called it. The mechanisms by which this might take place were equally of his time and were based on his opinion that organisms possessed an innate driving force that pushed them forward in the scale of nature. He thought this *sentiment interieur,* when stimulated by the animal's needs, would adapt the animal or plant to its surroundings, as, for example, birds that feed by the shoreline might develop webbed feet. Lamarck's entirely secular scheme lent itself to becoming incorporated into many other developmental systems based on environmentalism, self-help, and human progress, and was the foundation-stone for some exceptional biological work in the second quarter of the nineteenth century by men such as Robert Grant, Darwin's former friend at Edinburgh University.[24] As a young man Lyell had found Lamarck's ideas so dangerous—and so exciting—that he devoted a huge section of his *Principles of Geology* to a blow-by-blow refutation of them.

Now, in 1858, developmental ideas were again exciting the attention of the intelligentsia. The men and women of the *Westminster Review,* for instance, led by the charismatic editor John Chapman and Mary Ann Evans (the novelist George Eliot), were fascinated by the idea of inbuilt natural laws and steady advance in biology and in human society. To them, God played only a minimal role. Their friends Henry Buckle and Herbert Spencer extolled the march of civilisation. In Spencer's writings, this took the form of a law of development that Spencer applied to animals and plants as readily as to politics, economics, technology, and human society. Buckle looked more to the history of nations. He told his readers that the sweep of past history indicated that civilised societies would inevitably replace the less advanced. It did not take too much effort for these men and women to assume that their own British culture had come through this mill to exist in the present day as the most advanced in the world.

Much of Spencer's worldview drew on the same texts and cultural convictions as Darwin's. During the 1850s he had immersed himself in elaborating the framework of a personal philosophy in which all progress was based on the idea of differentiation. In 1852 he published an essay on the "development hypothesis" in which he defended the theory of animal transmutation, followed by a Malthusian essay, "Theory of Population," in the *Westminster Review,* holding that population pressure drove the weakest to the wall. His anti-religious *Principles of Psychology* (1855) followed shortly afterwards, and by the end of the decade he had begun an ambitious, lifelong reevaluation of metaphysics, the first part published as *First Principles* in 1862. Spencer believed that biological and social progress were constitutive parts of one broad evolutionary continuum—they were governed by the same immutable laws and controlled by the same forces of nature. He and Buckle fluently adopted Malthusian views as part of the common context of Victorian political economy, biology, and society.

In the same progressive set, George Henry Lewes, the editor of the forward-looking *Leader,* regular contributor to the *Westminster,* and admirer of Eliot, delved into anatomy and physiology, proposing that human thought was merely a product of the brain's activity rather than a gift from God. Supported by William Benjamin Carpenter, another physiologist, Lewes pushed divine agencies right to the background. The two men rejected conventional natural theology, the system of explanation still entrenched in the old universities and predominant in the general run of science publications, and regarded questions about the origins of natural beings and natural phenomena as relevant and intelligible, perhaps even soluble, without any necessary dependence on the divine.

Above all, the anonymous transmutationary tract *Vestiges of the Natural History of Creation* effortlessly maintained its appeal. *Vestiges* was by now in its tenth, most enlarged, and most popular edition. "Mr. Vestiges," as the author was popularly known (he was really Robert Chambers, an Edinburgh publisher and writer), proposed a vast developmental sequence of change in the natural universe, moving from the very first clouds of gas out of which the earth emerged at the beginning of time, through the successive depositions of geological strata, to the first appearance of life and its progressive changes until mankind appeared from apes. Human beings might change still further, it was suggested, especially through advances in mental constitution. Mr. Vestiges displayed a keen interest in phrenology and mental progress. All this was effected without any obvious divine intervention. Whoever the unknown author might be, he or she put forward ideas about the ascent of mankind in easily comprehensible terms that attracted hard-working members of the bourgeoisie and secular thinkers with reform on their minds, appealed to high-minded social theorists advocating advance, and repelled traditional scientists.[25]

Progress and the march of history were, in short, prominent themes of the age. Linguists and philologists emerged from dusty libraries to draw attention to family lineages in the relationships of words; historians like Buckle spoke of the advance of nations; even reforming Anglican priests like Charles Kingsley could bring his social-realist novel *Alton Locke* (1850) to a climax with the hero's dream of a metamorphosis from jellyfish to man. Harriet Martineau shocked pious readers by confidently proclaiming her religious doubt in *Letters on the Laws of Man's Nature and Development* (1851).

Great currents of change were making their presence felt, in turn political, social, religious, scientific, and economic. Secular thought, and the splintering of the national church, were accelerating, even though congregations were larger than ever before, some people evidently taking fresh impetus from participating in vigorous dissent and others stubbornly adhering to the conventional values of established Anglican doctrine.[26] Inside the great gates of an Oxford college, the Rev. Baden Powell frankly discounted miracles and advised contemporaries to take a "perfectly unbiassed and dispassionate view" of evolution, while John Henry Newman turned toward Catholicism.[27] Middle-class liberals advocated self-help and the development of the individual human mind through the principles of phrenology. Medical men investigated the chemical origins of life, the material causes of thought, and the possibility of spontaneous generation while physical scientists emphasised the existence of natural laws, laws fashioned by the Creator but left by Him to run under their own steam.[28] In Darwin's former circle of acquaintance, his friends Robert Grant, Edward Blyth, and Hewett Cottrell Watson each advocated some form of evolutionary change—Grant most plainly in his comparative anatomy lectures at University College London.[29] Even Lyell was famous for presenting a naturalistic, developmental account of the earth's physical changes in succeeding editions of his *Principles of Geology* and for putting forward a well-publicised programme for abandoning the Bible as any sort of accurate guide to science.[30] Divine agencies rarely appeared in the writings of figures such as these. One by one, Victorian thinkers claimed the right to investigate the world around them without recourse to God's miraculous powers. It was much the same in economic terms. Across the nation at large, expanding economic horizons highlighted the prevalence of cutthroat competition, commercial success, and entrepreneurial adaption to circumstance—natural laws of society, it was increasingly said. For the masses, ideas of development and advance supplied potent rhetoric for working people to climb out of the mire of class and poverty, for social revolution, if needs be. There was plenty of evolutionism around for those who had eyes to see it.

Despite Lyell's warning, Darwin seemingly closed his mind to the possibility that other thinkers might be moving along the same road as he and that any one of them might come up with the same answer. He had the impression that most of his contemporaries were stuck in the creationist mould. There was, for example, the popularity of Hugh Miller's book about fossils, emotively titled *Footprints of the Creator,* and the continued enthusiastic readership for the natural theological panaceas of the Bridgewater Treatises. Then there were Richard Owen's and Louis Agassiz's schemes of archetypes and homological relationships between organisms that depended on a supreme intelligence for their existence. Darwin saw provincial vicars refuting geology as "a lie and delusion of Satan," tracts that he and Lyell read in amused disbelief.[31] He was aware of Philip Henry Gosse's *Omphalos,* published in 1857, in which Gosse declared that God created fossils inside the rocks, and Adam and Eve with belly buttons, expressly to give the appearance of a history that had existed long before the first day of creation—the same ardent desire to reconcile faith and science that Gosse's son Edmund would pity in *Father and Son* decades later, and that goaded Kingsley into announcing that God would scarcely have written on the rocks "one enormous and superfluous lie." Natural history journals were full of the terminology of design and perfection. It was easy for Darwin to believe he that was the outsider, alone except for Lyell, Hooker, and a few others, in holding these new ideas against the conservative tide. He was used to bracing himself for theological opposition, to persuading, to chipping away at the reservations of those people he let into his secret.

He had no particular reason to consider Wallace a potential hazard either. In the starkest sense, Darwin had no real cause to notice him at all during the short time he had corresponded with him, regarding him as a mere collector, a self-financing naturalist obliged to sell unusual specimens to cover his costs, a traveller who would be only too willing to supply information and specimens to a gentleman at home. He could not even remember having met Wallace, although Wallace afterwards said they had once spoken briefly at the British Museum. Wallace possessed no independent means, was not of the landed gentry or the university-educated classes, and, thought Darwin blandly, probably had little occasion for contemplative philosophical endeavour. Darwin knew—or thought he knew—scores of helpful men who supplied gentlemen with valuable bits of information, not exactly as a labourer might supply a master, or a protégé satisfy a patron, but rather as a necessary functionary in a service economy that depended on hierarchies of exchange where knowledge was the commodity and shifting social boundaries provided the channels of opportunity and collaboration that made Victorian science so very distinctive.[32] At this point in Darwin's life, the flow of material was always one way, towards himself. If he thought about Wallace at all, he probably would have regarded him merely as supplying basic data that he—Darwin—would turn into acceptable science. To be sure, he pleasantly acknowledged Wallace's interest in theories of animal distribution. "I am a firm believer that without speculation there is no good and original observation," he told him in 1857. Yet as soon as personal contact was established, he hardly bothered to engage Wallace in serious discussion. All he wanted from Wallace was exotic poultry skins and answers to a few specialised natural history questions.

Much of Darwin's shock apparently hinged on realising he had been thoroughly mistaken about the man. His years of dedicated research at Down House left him unprepared for seeing his hard-won insights reiterated, as he thought, by a nobody from nowhere.

Who was this person making such a leap into the intellectual dark from Ternate?

## IV

Wallace certainly came from the other side of the Victorian cultural divide, a man of great "personal magnetism" and "lofty ideals," said his friend E. B. Poulton. His relatively humble background and progressive social convictions in fact made him a much more probable candidate than Darwin to come up with a radical doctrine like evolution. He was fourteen years younger than Darwin, a likable, mild-mannered man, full of visions for a reformed society, at times endearingly innocent of the ways of the world, full of "boyish joyous exuberance," energetic and endlessly curious about the things around him. He had "an abounding interest . . . in human knowledge in all its phases, especially new ones," continued Poulton.[33] Unconventional ideas almost ran in his veins. Later on he would speak vehemently against vaccination, become embroiled in a debate about life on Mars, defend the reality of séances and spiritualism, and support socialism and the urgent need for land nationalisation, as well as take a lifelong public stand in defence of evolution by natural selection. Even dyed-in-the-wool conservative newspapers like the *Christian Commonwealth* celebrated his reforming ardour.[34] So similar to Darwin in his intellectual grasp of the inner meanings of biology and so dissimilar in almost every other way, he was destined to become Darwin's alter ego, the other man of the evolutionary story. He was to captivate, intrigue, and exasperate Darwin for the rest of his life.

Wallace had none of the opportunities available to gentlemen's sons.[35] His father was a provincial solicitor, at that time a relatively lowly occupation, who slipped into financial disarray and led his family from place to place in search of ever-diminishing employment. They were "old-fashioned religious people belonging to the Church of England," he said. In the end the family survived almost entirely on Wallace's mother's small inheritance. Wallace's father died in 1834. The older brothers left home when they reached thirteen or fourteen to become apprenticed as land surveyors, builders, watchmakers, and draughtsmen: respectable, craft-based professions for young men who needed to make their own way in life with no patrimony. Wallace's sister learned French and became a teacher and then a governess, the only acceptable path for unmarried women of her background. Wallace himself, born in Usk, in Wales, in 1823, the fifth living child of the family, accompanied his widowed mother to live in south-east England and attended Hertford Grammar School before spending six months working in London (aged fourteen) with his older brother John and then assisting another brother, William, a land surveyor, during the Midlands railway boom.

It was an itinerant existence. Wallace tramped the countryside with his brother William for several years and moved in and out of the surveying profession as the need arose. He never really had a base, although he felt affectionate about the town of Neath, near Swansea, where he spent several formative years; and his experiences there, walking the Welsh landscape, gave him a pronounced taste for science and natural history.[36] In 1844, aged twenty-one, he temporarily tried his hand at teaching elementary arithmetic and English in the Collegiate School in Leicester, but he returned to surveying when the next flush of railway development came along. Such a life offered few benefits. With considerable poignancy he afterwards listed his mental and physical deficiencies at this time, regretting his inability to hold a tune or learn a language, his shyness and "want of confidence," his lack of "physical courage" and "love of solitude." When he was old and famous he liked to recollect that his first trip on a railway train was in the section reserved for third class, standing up in an open truck like cattle with other working men. Another "wretched third class carriage" followed by a damp bed in their Bristol lodgings proved fatal to his brother William, who died of congested lungs in 1846. Wallace deplored this unnecessary death, symptomatic of the class divide in British society and buttressed by widespread Malthusian assumptions that working people were plentiful in number, irresponsible, and dispensable. He remembered the few good meals he ate during those early years of adversity. For him, as for the vast majority of the populace, the hungry 1840s were a dreadful ordeal.

Wallace's mental development was therefore grounded in the provincial, industrialising countryside of the mill-hands, weavers, factory inspectors, railway men, itinerant labourers, poor law commissioners, and sanitary engineers of Victorian Britain, the same rural manufacturing worlds brought to life in Mrs. Gaskell's *Mary Barton* and *Cousin Phillis*. He was self-educated in the sciences. From the first, he was a high-minded early socialist, greatly influenced as a young man by listening to lectures in the Hall of Science in London's Tottenham Court Road on Robert Owen's industrial democracy at New Lanark, and devoting himself to learning popular doctrines like phrenology and mesmerism which promised mental and social improvement for all. "I have always looked upon Owen as my first teacher in the philosophy of human nature and my first guide through the labyrinth of social science."

Those lectures by Robert Owen took a prominently anti-Malthusian line. In his *New View of Society* (1813) Owen explained that Malthus's deadly checks were not inevitable. Greater agricultural production was possible, argued Owen, if reforms in the use and ownership of land were made. Mankind could be educated out of distress, and self-governing communities would moralise and civilise the new industrial workers. In essence, Wallace encountered Malthus in a different way from Darwin. This argument for the malleability of the human mind, and its advance under appropriate external conditions, created a lasting impression. He displayed a lifelong curiosity about the workings of the psyche, and in time learned to perform hypnotism on his acquaintances.[37]

As he wandered the country, Wallace attended lectures on these and other topics in public meeting halls and made good use of the reading rooms provided in the Free Libraries and Mechanics' Institutes opening up in towns all over the provinces, soaking up the reforming political views of men like Cobden and Bright from cheap printed pamphlets.[38] He came to the conclusion that "the orthodox religion of the day was degrading and hideous, and that the only true and wholly beneficial religion was that which inculcated the service of humanity, and whose only dogma was the brotherhood of man. Thus was laid the foundation of my religious scepticism."[39]

What little religious belief Wallace possessed "very quickly vanished" under such influences, particularly after he studied a fourpenny booklet summarising David Friedrich Strauss's *Life of Jesus,* one of the texts most likely to sow doubt in the minds of British church-goers in the first half of the nineteenth century. Ever afterwards Wallace entertained advanced, completely secular views on human and social progress. He read George Combe's phrenological and socially prescriptive *Constitution of Man,* Alexander von Humboldt's travels, Lyell's *Principles of Geology,* Malthus's *Essay on Population,* and Darwin's *Journal of Researches*—in short, anything of a scientific, radical, or philosophical nature that he could lay his hands on.[40]

These kinds of activities would not have brought Wallace into contact with high science had it not been for his chance meeting at a Mechanics' Institute library with the man, other than Darwin, who was to affect his life most deeply. Henry Walter Bates was an apprentice hosiery manufacturer in Leicester, roughly Wallace's age, and "an enthusiastic entomologist."[41] Bates introduced him to the outdoor delights of beetle and butterfly collecting (Wallace was already interested in plant collecting), and they began making local excursions together. Their subsequent letters to each other were filled with youthful erudition and ambitious talk about books and travelling together collecting specimens.[42] At first they recognised that this was just talk. Nonetheless, in 1845, they discussed *Vestiges,* which they both read closely, and Wallace began reflecting on the origins of the human race. "I well remember the excitement caused by the publication of the *Vestiges,* and the eagerness and delight with which I read it," he remarked.[43] Men like Wallace were ideal targets for *Vestiges'* heady mix of individual self-education, transmutation, and political reform—a text generated and judged by standards different from those of elite science. Wallace was captivated by the book's theory of self-generated change, and he wholeheartedly committed himself to the idea of transmutation.

Two years later, the friends resolved to cut loose from Britain and earn their living collecting natural history specimens on the River Amazon. This was an adventurous move made possible only by having nothing to lose. What decided them was reading W. H. Edwards's *Voyage up the Amazon,* after which "Bates and myself at once agreed that this was the very place for us to go." Like many other relatively unknown figures in nineteenth-century natural history, they planned to capitalise on the growing commercial opportunities presented by science and sell rare materials to collectors and museums on their return. To this end, they haunted the British Museum and the outer rim of London's scientific societies, asking questions, meeting curators, and learning their clientele's wishes, swiftly followed by the acquisition of a natural history agent, Samuel Stevens of Bloomsbury, who put them under contract and advanced the money for them to go.[44] Stevens served the pair well during their absence. He kept potential purchasers in the British Museum and Entomological Society gently on the boil and published short notices about the travellers' achievements every so often in natural history journals. But money was tight. Most independent collectors needed four or five affluent subscribers lined up before feeling secure enough to set sail and usually took advantage of some form of semi-official support system, as in letters and contacts supplied by patrons or bodies like the Horticultural Society. Wallace and Bates depended on personal enterprise and the market economy alone.

They sailed for Brazil in the spring of 1848. Wallace stayed until 1852, while Bates, who more or less accompanied him for the first two years, persevered until 1859. These years were full of hardship, disease, and distress for both men; their experiences were as physically remote from Darwin's *Beagle* voyage as could be imagined. Wallace's younger brother Herbert came out to assist him, only to die in his arms of a fever; towards the end of his time Wallace met and became friends with Richard Spruce, a collector employed by Kew Gardens to prospect for rubber and cinchona as well as botanically unusual species. Wallace, Bates, and Spruce apparently discussed the transmutation of species, for all three had noticed the impossibility of drawing lines of separation between geographical variants and had come to regard the fecund world of nature around them as a continuum, devoid of boundaries. For all the deprivations and ordeals, Wallace found the lush green beauty of the Amazon completely thrilling. In his old age he still spoke of it with admiration and wonder.

This first tropical exploration ended in shipwreck when the cargo ship on which Wallace was returning to England caught fire and sank. Except for a box of drawings and skins, and a parrot which fled the burning ship, all the specimens that were travelling with him were destroyed—all his hundreds of new species, some £500 worth of saleable material, his private collection, and his journals and notes as well, "all lost with the ship."[45] The survivors huddled miserably in a rowing boat for days, nearly charred by the sun and salt, until they were picked up by a passing brig. The next ship Wallace boarded was an old wreck carrying timber from Cuba that creaked and groaned and shipped water, and nearly went down in mid-Atlantic too. When he arrived at Deal ("Oh, beefsteaks and damson tart"), he vowed never to trust himself on the ocean again.

The vow lasted only eighteen months. None of the advantages that he hoped would accompany his return from Brazil ever materialised. There was no income—his best specimens were on the seabed. There was no applause—his *Travels on the Amazon and Rio Negro* (1853), a unique narrative about the largest river system in the world that he pieced together from memory, failed to excite attention and was remaindered by the publisher.[46] Darwin, who read everything in the natural history line, was unimpressed by Wallace's account, barely bothering to make more than a half-page of notes from a borrowed copy and complaining about the lack of facts.[47] Wallace's next book, on palms, suffered much the same fate in Joseph Hooker's botanical hands. There was scarcely any mingling with new intellectual companions either—Wallace found it difficult to get beyond the closed doors of London's most elite learned scientific societies. When he did, he felt out of place. He was alternately shocked and impressed by the showy theatrics of educated men, among them Thomas Henry Huxley, whose "complete mastery of the subject and his great amount of technical knowledge" amazed him. Huxley was so self-confident that Wallace thought he must be several years older than himself. Reading between the lines, it seems clear that he preferred the perils of the rain forest to the predatory jungle of metropolitan science.

In 1853 he embarked again, hoping to recoup his losses and provide for his future by collecting more specimens. This time he ambitiously set sail for Malaysia and Indonesia, "the very finest field for an exploring and collecting naturalist," combining wanderlust with sensible business projections. He anticipated that material from Malaysia would sell well because British naturalists knew very little about these impenetrable regions, dominated for centuries by the trading empires of the Dutch and Portuguese. Moreover, his imagination was fired by the idea of shimmering islands wreathed in creepers, the ginger-haired orang-utan, the vast, interchangeable surfaces of sea and forest, and the elusive bird of paradise, so strange and beautiful. This time, too, he wanted to search out primitive tribes and pursue his ideas about human origins, stirred by his fleeting encounters with the forest peoples of the Amazon as much as by the concept of transmutation he had found in *Vestiges of the Natural History of Creation*. His inclinations drew him towards the study of mankind in its widest sense. His reading of *Vestiges* told him that mankind emerged from apes, perhaps from the mysterious orang.

The enterprise was dangerous. Only a man who had hacked his way through the South American rain forest could have faced it with any degree of optimism. Even before he left London he felt as if he were negotiating with his own variety of foreign tribe, placating the elders of Victorian science, promising gifts, observing and bargaining with chieftains. At last he managed (with the help of Sir Roderick Murchison at the Royal Geographical Society) to get a free government passage on a ship to Singapore, from where he travelled independently to Borneo. Together with a young English assistant called Charles Allen (according to Wallace more of a burden than anything else) and a Malay servant called Ali, who afterwards took Wallace's name, he set sail to the various islands in turn, pushing as far east as the Moluccas and Papua New Guinea, crossing and retracing his steps according to the seasons and the accessibility of particular sites. Counting his voyages up afterwards he thought he must have made some sixty or seventy separate expeditions, travelling about fourteen thousand miles within the region. These eight years of wandering, as he later put it, "constituted the central and controlling incident" of his life. He abandoned himself to the risks, freedom, and fecundity of tropical island life as few other Victorians ever had.

His first port of call was Sarawak, the independent sliver of territory on the north coast of Borneo, then ruled by the eccentric English rajah Sir James Brooke. Brooke was by far the most colourful European in the archipelago, a man whom Wallace had met in London and from whom he readily accepted an invitation to visit.[48] Wallace remained in this minute tropical kingdom for fourteen months, collecting a wealth of unusual specimens. There were rarely more than one or two visiting Europeans in the country, and for a while he stayed with Brooke and his assistant secretary in their colonial bungalow. "The Rajah was pleased to have so clever a man with him," said the secretary with just a hint of relief. "It excited his mind and brought out his brilliant ideas. . . . if [Wallace] could not convince us that our ugly neighbours, the orang-outangs, were our ancestors, he pleased, delighted and instructed us by his clever and inexhaustible flow of talk—really good talk."

Brooke arranged for Wallace to rent a cottage from which he could explore and collect on a daily basis. The beetles were truly wonderful, Wallace wrote to his agent Stevens. Wallace also lived for some weeks in a longhouse with a family of indigenous Dyaks, the notorious head-hunters of European legend. This meeting of cultures proved an intensely interesting experience, and Wallace emerged from it convinced of the essential unity of all races of mankind. Like Darwin, he regarded himself as the real curiosity in these undertakings, and told his sister how the wild men of the forest could easily beat him at the children's game of cat's cradle. They were just as sophisticated in mental activity as himself. The only difference was the absence of the trappings of civilisation. It was here that the roots of all Wallace's later thoughts about human progress and the universality of mental powers were laid down.

On another occasion he looked after an infant orang-utan orphaned by his own shotgun. Again like Darwin, Wallace's affectionate feelings for animals created a fertile context for more philosophical juxtapositions. There never was such a baby as my baby, he told his sister. "I am sure nobody ever had such a dear little duck of a darling of a brown hairy baby before."[49] By now he was keen to draw connections between various groups of humanity and between humans and animals. He felt he could see evolution everywhere. Grateful simply for the opportunity to be there, he named one of the world's most beautiful butterflies after the rajah, the giant black-and-golden-green birdwing, *Ornithoptera brookeana*.

Before leaving Sarawak, Wallace composed and dispatched to England the short theoretical paper that Lyell—and then Darwin and Edward Blyth—saw in the *Annals and Magazine of Natural History* for 1855.[50] "Every species has come into existence coincident both in time and space with a pre-existing closely allied species," he stated. Geographical proximity was a measure of biological connection, he declared to Bates in a letter. He was disappointed to hear from Stevens that he ought not to be thinking about such esoteric topics. Several naturalists at the Entomological Society apparently expressed regret that he was "theorizing" when what they really wanted was more facts and specimens.

When Darwin initiated an occasional correspondence with Wallace, both of them were pleased. As always, Darwin desired skins, bones, and information, even though he exclaimed that the carriage from Singapore "is costing me a fortune!" Courteously, Darwin relayed the fact that both he and Lyell had been interested by Wallace's article, a compliment that was not hard to give and which Darwin felt sure would gratify its recipient. His letter was a model of friendly encouragement: "I can plainly see that we have thought much alike. . . . In regard to paper in Annals, I agree to the truth of almost every word." However, with Lyell's advice to push on with his book ringing in his ears, Darwin delicately established his seniority in that area by remarking that he was in fact composing a large book on species and varieties.[51]

Either Wallace was too untutored in the ways of the world to notice or he was too caught up in his own thoughts to pay strict attention. Darwin's words revealed their meaning only in retrospect; then as now, men could write at some length to each other without necessarily recognising what was really being said. Wallace had no reason to suspect Darwin was warning him off. In any case, Wallace was glad to hear that Darwin was dealing with the problem of species and varieties. Wallace told Bates that such a book would save him a lot of bother. He was plainly glad to have established contact.

Even making allowances for Wallace's memory afterwards playing tricks, it is clear that ideas about geographical distribution were as important to him in generating the idea of natural selection as they had been to Darwin. In Malaysia, Wallace said, he was particularly interested in the territorial boundaries separating the many different human tribes that he encountered. These boundaries were etched by warfare and village custom as well as by geographical features like ocean straits, rivers, and mountain ranges. The human divisions that he was identifying appeared to be paralleled by similar dividing lines between groups of animals and plants originating from either Asia or the Pacific.[52] Soon Wallace was on the point of discerning an imaginary line marking the contact between each great set of species. He discovered that animals and plants seemingly never crossed that line—there were in the archipelago invisible boundaries between aboriginal floras and faunas, dividing lines that inscribed the site of the creation of animals and plants and marked their ability to push outwards as far as time and circumstances allowed. Geography emerged as a vital key to origins. Wallace became convinced that the spatial arrangement and competitive struggle between living beings went a long way towards explaining the origin of species.

And it is equally clear that Malthus's writings were important. In February 1858, at the end of a long week spent collecting specimens on the island of Gilolo (Halmahera) in the Moluccas, Wallace thought he might have located the dividing line between indigenous and invading humans. The original population of Gilolo, he noted, was exceedingly small, vulnerable to every blow of nature. These indigenous peoples would not survive much longer, at least not while incoming Malays and Papuans continued to prove such assertively dominant immigrants.

His conclusions were reached during a bout of malaria. Fifty years later he recalled that all he could do while waiting for the shivering fits to play out their daily cycle was to lie down and think.

> One day something brought to my recollection Malthus's *Principles of Population,* which I had read about twelve years before. I thought of his clear exposition of "the positive checks to increase"—disease, accidents, war, and famine—which keep down the population of savage races to so much lower an average than that of more civilized peoples. . . . It then suddenly flashed upon me that this self-acting process would necessarily improve the race, because in every generation the inferior would inevitably be killed off and the superior would remain—that is, the fittest would survive.[53]

Perhaps inspired by the fever, Wallace worked out the scheme in his head. Impatiently, he set sail back to Ternate, to his rented house, with its bed, books, food, quinine, and scientific records. In the calm of the evenings, he said, he devoted the next few days to writing his essay. He compared his ideas with his private notes on Lyell, Lamarck, and *Vestiges,* signed it "Ternate, February 1858," and put it in the next post to England.[54] Above all, he wanted Lyell to read it. He chose to send it to Darwin because of Darwin's expressed interest in the subject and his proximity to Lyell.

Wallace naturally remembered this act of creative thought as his greatest intellectual achievement. Modestly, he never pretended to comprehend how it had come about. Many years later, he tentatively suggested that he and Darwin had arrived at the same conclusion because of their mutual fondness for collecting beetles.[55]

But the real machinery below Wallace's thoughts—as Darwin's—had been operating for years. Wallace, like Darwin, had grown up in a Malthusian universe. He was well primed to see the tribes of Gilolo falling prey to the same inexorable rules as the people of Wales or Leicestershire; and sufficiently versed in anti-Malthusian socialist doctrines to know how far he could pursue the idea.[56] The harsh pressure of numbers, agricultural hardship, cycles of economic boom and bust, the role of natural law, and the vagaries of financial success and failure—Wallace had encountered them all. Competition and the differential survival rates dictating a community's future progress came effortlessly to mind.

And if not from Malthus direct, there was always his well-thumbed copy of Lyell's *Principles of Geology* to hand (a book which travelled with him to Malaysia), or the second edition of Darwin's *Journal of Researches,* in which Darwin briefly alluded to the everlasting Malthusian war in nature.[57] Perhaps he recalled conversations with Rajah Brooke about low birth rates among the Dyaks and precarious food supplies on

Aru. He had access to the library of Mr. Duivenboden, the cultivated Dutch merchant who was Wallace's host in Ternate.

Even so, the parallels between Wallace's and Darwin's thoughts are no less remarkable for their cultural symmetry. A common political, intellectual, and national context linked the two inseparably. Their experiences of geographical exploration and travel in the early imperial era, their various connections with competitive, commercial Britain, their mutual appreciation of the marvels of nature and overwhelming desire to understand them, their unfaltering belief in human unity, the reciprocal letters, shared reading material, and concurrent preoccupations came together in a single cathartic burst of identity: two thoughtful men who would rather face the ocean than the crowd, looking deep into nature.

## V

Darwin barely had time to consider what to do next. He glumly sent Wallace's essay and his own covering letter to Lyell. Within twenty-four hours another kind of crisis erupted.

Family illness descended without warning. First, his daughter Henrietta, aged fifteen, came down with a raging temperature and sore throat. Darwin and Emma feared she might have caught diphtheria, the frightening new disease invading Britain in epidemic waves from France during 1858 and 1859, as deadly in its way as the continuing threat of a military invasion under Louis Napoleon's orders. The county of Kent was already under siege. "No actual choking, but immense discharge & much pain & inability to speak or swallow & very weak & rapid pulse, with a fearful tongue," Darwin wrote in consternation to Hooker.[58] Both parents took turns to nurse her. Apprehensively, they asked their Mackintosh relatives, who were guests in the house, to return home; and just as apprehensively, they called for Emma's older sister Elizabeth Wedgwood to come over from Hartfield, near Tunbridge Wells, to help with the nursing. Unintentionally making a bad situation worse, George's headmaster wrote from school to say their second son had caught measles, another dangerous disease in the years before antibiotics and mass vaccination programmes. At any other time they would have brought him home. Now, they asked Mr. Pritchard to keep him at school in isolation.

The following day the baby was taken ill with fever. This baby, Charles Waring Darwin, was their tenth and last child, at that point around nineteen months old. Although Emma must have long before then been ready to call a halt to childbearing (she was forty-eight when the child was born), she and Darwin enjoyed and loved little Charles as devotedly as any of their children. Conception so late in the day was probably something of a weary surprise to her: it had been five years since Horace, the previous child, was born. Emma may have been a little disappointed, too, with the arrival of yet another boy, making four in a row. But in all events, she had been pleased with the new baby, a fat placid soul as Darwin eventually recorded. She and Darwin were greatly worried by the intensity of his terrible fever. They urgently summoned Edward Illot, the medical surgeon from Bromley, to give his advice.

Darwin was in no state of mind to make any kind of balanced assessment about competing claims over his or Wallace's priority. Still, he was correct about Lyell in one sense. Loyally, the geologist rose up in the strongest possible way to defend his friend's interests. Intellectual property rights were an important issue in Lyell's eyes, and their defence demanded quick and decisive action. Lyell was a hard man to contradict when he was involved in priority disputes.[59] By return of post he strongly recommended that Darwin publish a short statement of his own.

Sitting up with the sick baby, Darwin agonised. He said that Lyell was proposing the impossible. "I shd be *extremely* glad *now* to publish a sketch of my general views in about a dozen pages or so. But I cannot persuade myself that I can do so honourably. . . . I would far rather burn my whole book than that he or any man shd think that I had behaved in a paltry spirit." He told Lyell that Wallace had tied his hands by sending the essay to him first. Now that he had read it, he could hardly justify any move to publish his own statement. "I cannot tell whether to publish now would not be base & paltry: this was my first impression, & I shd have certainly acted on it, had it not been for your letter." His mind dwelled on the letter he had already begun to write to Wallace, a letter in which he gave up all his priority.[60]

But the dilemma would not go away. The thought of publishing something was undeniably attractive if only it could be done honourably. Hesitantly, he asked Lyell to discuss it with Hooker so that he could get the judgement of "my two best & kindest friends." As the baby worsened, he thought about it ceaselessly, although to very little effect. In the end he wanted "to banish whole subject. . . . I am worn out with musing."

Back came another letter a few days later. Lyell and Hooker suggested publishing Darwin and Wallace together, a compromise that accommodated everybody's needs as best as possible. With hindsight, it was a gentlemanly solution. Darwin's priority would not be lost, and he could carry on writing his big book; Wallace's views would be published in a way that would greatly enhance their interest and acceptability.

At the time, it certainly looked like the most straightforward and acceptable course of action that could have been devised. And in the years that followed, no two authors thrown together in such a fashion tried harder than Darwin and Wallace to treat each other fairly. Each regarded the other with respect, admiration, and generosity.

Yet for a while the proposal trembled on the edge of audacious skulduggery. No pair of practised fixers could, if they wished, have cooked up a better scheme for promoting Darwin's interests. First and foremost, Wallace did not know anything about the proposal. His private communication to Darwin on a natural history matter, sent out to Lyell for comment, was to be announced without his knowledge and as an accompaniment to writings about which he knew nothing. On the face of it, it looked as if Lyell and Hooker were suggesting that their friend Darwin—a man at the heart of scientific society—should not lose out to an interloper. On the broader scale, they may well have felt compelled to safeguard the values of elite Victorian knowledge—the science of accredited experts, authenticated fact, proper sequences of logical inference, and trustworthy sources—from outsiders like Wallace.

Briskly, Lyell and Hooker proposed reading Wallace's essay and some extracts from Darwin's unpublished manuscripts at a forthcoming meeting of the Linnean Society of London and then sending them forward for publication. Hooker was on the council of the Linnean and was aware that a recent meeting had been cancelled because of the death of Robert Brown, a former president. The rescheduled meeting was coming up in the following week, on 1 July 1858, at the empty end of the scientific season, and Hooker suggested squeezing the Darwin-Wallace material onto the programme. Lyell was due to present an obituary speech about Brown at the same meeting.

They chose the Linnean for entirely opportunistic reasons. Lyell, Hooker, and Darwin were all fellows of the society and council members (Darwin was elected to the council in May 1858). Hooker virtually ran the journal and saw the programme secretary constantly. All three were friends of the current president, Thomas Bell, and other officials and members. With these connections Lyell and Hooker could reasonably expect to have their way, much more so than if they had set their sights on the Royal Society of London, for example, hemmed in with the formal structure of timetables, referees, and the unspoken conventions appropriate to the leading natural philosophical body in the country; or the Zoological Society, where the atmosphere was edgy and the fellows prone to argue. Elsewhere in London, the Botanical Society was almost moribund, the Royal Institution in Albermarle Street preferred lecturers to present their own results, the Geological Society did not usually regard living organisms as suitable topics, and the British Association for the Advancement of Science held its meeting annually, in a different city every year, with the timetable prepared months in advance. Nor were periodicals a pertinent outlet. As Lyell and his circle were coming to accept, the first announcement of serious scientific results belonged not in the monthly or fortnightly magazines, where the aims were broadly cultural, and not in popular natural history weeklies, where the material was often unsubstantiated and (in the eyes of the elite) sometimes notably off-beam, but in the presentation of a theory at a learned society, in front of an educated audience, and then publication at a measured pace in that society's learned journal.

Wallace, on the other hand, was known among the Linnean fellows only as a purveyor of specimens, whose closest relationship was with William Wilson Saunders, the wealthy entomologist who was one of the Linnean's longstanding vice-presidents. Over the years, Saunders purchased a huge quantity of Wallace's tropical insects through Samuel Stevens. Nearly two-thirds of the annual *Journal* were filled with the ensuing lists and catalogues. But for various reasons, Wallace would not be elected a fellow of the Linnean until 1871.[61] Under normal circumstances any contribution of his to a learned society would have to be submitted on his behalf by a fellow. Since Wallace's customary publishing domain was the world of popular magazines like the *Annals and Magazine of Natural History,* Lyell's and Hooker's actions implied that he was lucky to hitch a ride on Darwin's well-cut coat-tails.

No other venue could conceivably offer such obliging attention. Furthermore, Hooker and Lyell seemingly believed no delay could be permitted. For all they knew, Wallace might have sent a copy of the same article or a longer, more detailed version elsewhere for publication. Unless they moved quickly, they must have thought, control of the situation would be lost.

Before Darwin could reply to this startling suggestion, baby Charles became very ill indeed. He had scarlet fever of the most violent kind: "the deadliest of all the fevers," declared the *Lancet.*[62] Time after time, the older Darwin children had wrestled with the same disease and survived. But three infants in Downe died that month, and families all over the country were encountering the tragedy of losing first one, then another child to the epidemic. Darwin and Emma could hardly bear to see their child's suffering. They were almost grateful when, after nearly a week of misery, he died on the evening of 28 June. "It was the most blessed relief," wrote Darwin sorrowfully, "to see his poor little innocent face resume its sweet expression in the sleep of death."[63]

The two were distraught. This was the baby of their middle age, their last child, the one they felt most relaxed about, with a "remarkably sweet, placid & joyful disposition." It seems that the baby may have been slightly retarded, although there is no actual evidence to suggest a disability like Down's syndrome. He was "backward in walking & talking, but intelligent & observant," said his father. There could be nothing prettier, remarked Darwin, than his passion for Joseph Parslow. When the butler came into a room, little Charles would stretch out his arms to be carried. He would lie on Darwin's lap for any length of time, gravely studying his father's face or waiting for a game. When Henrietta later came to write her mother's life story she probably underestimated both the love that Emma and Charles felt for him and his mental capacities. She thought he was born without his "full share of intelligence" and that after the first sorrow Emma was "thankful" at his early death.[64]

This unhappy event brought all Emma and Charles's combined family miseries to the surface again. Death knitted them together in the cruellest of ways, reviving old griefs with the new. Such events evoked the most intense emotions in Victorian family life.[65] "Our poor little darling," Darwin called him. Unwilling to let the memory go, he wrote an account of the baby on a scrap of paper. He tucked it away in the same envelope as his wistful memorial to Anne, his unfading "little angel," who had died in his arms at Malvern of another terrible fever in 1851. If he had any inclination to think about his theory of natural selection at this time, he might easily have reflected on the melancholy fact that his ideas of struggle required the death of the weakest individuals, even of his own babies. His theory was a bleak theory of elimination.

"Thank God he will never suffer more in this world," he told Hooker. "I cannot think now on subject, but soon will."[66]

## VI

He dispatched some manuscript material about evolution to Hooker on 29 June, at the end of the long sad following day. It was an odd, mixed bundle for the purpose, a very hasty culling of paperwork for a major turning point in biological science. He was exhausted.

> I have just read your letter & see you want papers at once. I am quite prostrated & can do nothing, but I send Wallace & my abstract of abstract [*sic*] of letter to Asa Gray, which gives most imperfectly *only the means of change & does not touch* on reasons for believing species do change. I daresay all is too late. I hardly care about it. But you are too generous to sacrifice so much time & kindness.—It is

> most generous, most kind. I send sketch of 1844 *solely* that you may see by your own handwriting that you did read it.—I really cannot bear to look at it.—Do not waste much time. It is miserable in me to care at all about priority.[67]

As he said, his part of the parcel contained a handwritten copy of a letter sent in 1857 to the botanist Asa Gray, and the early sketch on evolution that he completed in 1844.

Gray, the professor of botany at Harvard University, was one of Hooker's closest botanical colleagues and had corresponded with Darwin since 1855. The three had eased agreeably into a three-sided correspondence about botanical geography, classification, and plant anatomy.[68] Gray was devout—a committed Congregationalist—and uncommonly talented in science, by far the most prominent botanist in the United States, and on an intellectual par with Louis Agassiz, his zoological counterpart at Harvard. Towards the middle of 1857, Darwin had been so impressed by Gray's wide-ranging knowledge and his skilful analysis of knotty botanical problems that he confided in him, anxiously joking, "I know that this will make you despise me." When Gray wrote back encouragingly ("can you get at the *law* of variation?"), Darwin responded with a long letter on 5 September 1857 describing his theory of natural selection in detail, including his most recent ideas about the divergence of species, which he thought explained the origin of separate lines of descent.[69]

Spelling it all out for Gray in September 1857 was a difficult task, but one he was keen to attempt if he could get the honest opinion of such an exceptional man (any comparable thoughts of describing the theory to Wallace during the same summer never crossed Darwin's mind). Sentence by sentence, Darwin drew up a statement for Gray, which the young schoolmaster in Downe village neatly transcribed before it was sent to Harvard. It was the retained draft, creased and stained with use, that Darwin now dispatched to Hooker.[70] He scribbled a few hurried corrections and notes on the bottom (getting the original posting date wrong in the process), and sent it away.

It was necessary to include the Gray letter, Darwin believed, because the 1844 sketch, although much longer and more detailed in every way, had in many places been superseded by the sustained attention he had brought to bear on the theory since then. In particular, Darwin's barnacle studies had intervened. These encouraged him to shift his focus away from species originating in geographical isolation, as he imagined the Galapagos finches to have done, towards a more active, competitive arena in which constant variation, struggle, and divergence were the primary agents, and where most speciation would take place in large mixed populations, with numbers high and pressures intense.[71] Such views had been reinforced by his development, in or around 1856, of a "principle of divergence," a fundamental process that he believed acted in conjunction with natural selection, the one major conceptual adjustment he made during his long investigations—"The more diversified the descendants from any one species become in structure, constitution and habits, by so much will they be better enabled to seize on many and widely diversified places in the polity of nature, and so be enabled to increase in numbers."[72] Everything he had studied in the fourteen years since writing the sketch supported this basic vision. Although his overall arguments stayed pretty much the same—these were commitments Darwin would never relinquish—the manner in which he wished to present his mature theory was markedly different.

The sketch, in short, served mainly to establish the important fact that Hooker had read it sometime after it was finished in 1844—there on the manuscript pages were Hooker's pencil marks and queries. There was no point in Darwin's sending extracts from his current manuscript, the big book on species lying half completed on the table in his study, because Hooker's and Lyell's aim was to establish that a written account of Darwin's views had been read by a third party more than ten years ago, in 1844 or 1845, and that an independent letter was similarly read by Asa Gray at least a year before Darwin received Wallace's essay. The issue at the front of their minds was priority. In their eyes priority conferred ownership and secured an author's claim to originality. Such affirmations lay deep at the heart of the scientific process, where natural philosophers strove to identify the laws governing the universe—laws that promised a unique insight into the truth. Although this was to be a double paper (often called a joint paper), the underlying message was to be of two independent workers caught in a single, unexpected thunderclap.

Darwin had no time to contemplate the storm of outrage that once surrounded publication of *Vestiges;* no time to reflect on Emma's religious feelings, or to fret about literary perfection. There was no time to write to Wallace, the absent, unknowing trigger of all this activity. Despite the years he had devoted to analysing species, the only material Darwin could present to the public at such short notice was a letter to an overseas correspondent and an out-of-date sketch. "I always thought it very possible that I might be forestalled, but I fancied that I had grand enough soul not to care," he admitted woefully to Hooker. "But I found myself mistaken & punished."

## VII

Late in the evening of 30 June 1858, Lyell and Hooker forwarded the paperwork to the Linnean Society's secretary, John Joseph Bennett. Either Bennett or George Busk, the other secretary, would read it aloud at the meeting. Wallace's essay needed no further attention. The same could not be said for Darwin's contribution. Mrs. Hooker had spent all the afternoon copying extracts from Darwin's handwritten materials, extracts presumably chosen by Hooker, intending to turn them into a form suitable for reading aloud. Silently, she made a few helpful changes. One result of this methodical afternoon was that Darwin had little idea of what actually went forward until he saw printed proofs several weeks after the event. Lyell and Hooker added a short note of their own to introduce the papers. Afterwards, Lyell recalled that there was such a rush that he could not remember whether he had gone to Hooker's, or Hooker had come to him, or whether they spoke at all until an hour or two before the meeting.

The rush did not signify any lack of attention to detail. In the introductory matter, Hooker and Lyell subtly justified their friend's position and, by implication, their own. Wallace's work was praised and then delicately moved aside, merely the stimulus, they hinted, which encouraged Darwin to make a preliminary announcement.[73]

The material which followed reinforced the impression. The articles were arranged in alphabetical order by author, as was customary at the Linnean Society for double contributions. On this occasion, the alphabet coincided impressively with chronology. First came extracts from Darwin's sketch of 1844, then Darwin's September 1857 letter to Asa Gray, and, at the end, Wallace's February 1858 essay. Darwin's priority reverberated from every page. Even Darwin winced when he saw the layout some weeks later. He had assumed that his remarks would appear as a kind of appendix or as footnotes to Wallace.[74] Privately embarrassed, he was relieved he had not personally supervised this printed reversal of fortunes.

The actual reading of the papers, on Thursday 1 July 1858, was subdued. Hooker and Lyell were present; Darwin, and of course Wallace, absent. Some twenty-five fellows came, not to hear about evolutionary theory, but to listen to Lyell praise Robert Brown's career. Thomas Bell, the president, was chairman for the evening. There were two unnamed guests from overseas, and J. J. Bennett, the secretary: perhaps thirty all told. By chance, Wallace's natural history agent Samuel Stevens was there.[75] By chance, Darwin's friend William Carpenter was present, and William Fitton, his geological acquaintance from Gower Street days.

Daniel Oliver and Arthur Henfry, two men who later became committed evolutionists, attended, and Cuthbert Collingwood, a future opponent. Huxley, shortly to become Darwin's most public defender, was not present, for he was not yet a fellow, nor were Erasmus Darwin or Hensleigh Wedgwood, Darwin's brother and cousin. "No fourth individual had any cognisance of our meeting," said Hooker, forgetting that this blanket of silence also embraced Wallace, the catalyst for the whole affair.[76]

The paper, too, was long and not immediately easy to understand. Darwin and Wallace's articles were first on the programme (Hooker had persuaded George Bentham to step down)[77] and were followed by five other papers on conventional botanical and zoological topics. By the end of the evening, as Bentham noted, the audience appeared fatigued. Darwin's and Wallace's proposals failed to ignite any late-night debate or controversy.

Since Bell usually encouraged discussion—an innovation introduced with his presidency—it seems that their radical ideas did not make much of an immediate impact.[78] "No semblance of discussion," said Hooker, although he thought he remembered that "it was talked over with bated breath" at tea afterwards. Later, as an old man, with half a century of battle metaphors running through his mind, he subsequently declared that "the subject [was] too novel and too ominous for the old school to enter the lists before armouring."

In his declining years Hooker probably exaggerated both the silence and the teatime whispers. Most of the fellows in the audience that day would have known the importance of what was being suggested; and a moment's reflection would have ensured that they recognised at least some of the wider implications. But George Busk, the zoological secretary of the society and Huxley's closest friend, said nothing. Bennett, the general secretary, said nothing. Clever, philosophically minded Bentham, a nephew of Jeremy Bentham, was silent. Even William Carpenter, one of the most advanced physiologists of the age, with strongly naturalistic views of his own and well versed in *Vestiges'* evolutionary thinking, kept his peace as Bell hurried them along to the end of the evening. All of them might have spoken their minds more freely if Hooker and Lyell had not thrown their powerful weight behind the argument. Lyell overawed the fellows, claimed Hooker with a certain amount of satisfaction; and Lyell's, Hooker's, and Darwin's reputations probably inhibited any vulgar show of dissent from the floor. The fellows of the Linnean were polite to a fault. Only the rough and tumble of debates at the Geological Society and the British Association actively invited scientific lions to roar.

Later on, Bell captured the general air of unruffled, clubbable stability.

"The year which has passed," he remarked in his presidential address in May 1859, "has not, indeed, been marked by any of those striking discoveries which at once revolutionize, so to speak, the department of science on which they bear."[79] Accurate enough in the short term, Bell's remark was destined to become known as one of the most unfortunate misjudgements in the history of science.

The Linnean fellows dispersed that night not so much aghast by new ideas as wearied by the length and amount of information presented.[80] Much of the concept of natural selection probably went over their heads. There was insufficent opportunity to concentrate, no discussion to set the blood racing, no authors on the podium to fence and dodge difficult questions or to raise the blood pressure of theologically inclined naturalists. The fellows, moreover, were respectable men, unlikely to raise a rumpus. Hooker and Lyell had chosen exactly the right venue for well-behaved, impassive silence.

For Darwin, the evening marked the end of two of the most dreadful weeks of his life. Unfairly, he began to imagine Wallace's letter had forced him into premature publication; and he began recasting his own role in the proceedings from possible villain to potential victim. "I do not think that Wallace can think my conduct unfair, in allowing you & Hooker to do whatever you thought fair," he told Lyell. Above all, he hoped his own actions would be understood as a genuine wish to behave honorably. "My plans of publication are all changed," he explained to an old Shropshire friend, Thomas Campbell Eyton. Nothing, in short, had happened the way he had wanted.

But he had at last spoken.

*chapter*
2

# "MY ABOMINABLE VOLUME"

ALL THAT REMAINED was to tell Wallace. This was possibly the trickiest part of the whole episode. Without question it would be a difficult letter for Darwin to write. The turbulence of the last two weeks had taken place entirely in Wallace's absence and without Wallace's knowledge, a state of affairs that reeked of *Hamlet* without the prince. His choice of words would set the agenda for Wallace's response.

He raised the problem with Hooker first, hoping that Hooker might write to Wallace as well. "I certainly shd. much like this, as it would quite exonerate me," he observed. "If you would send me your note, sealed up, I would forward it with my own, as I know address &c."[1] Hooker was already putting the Linnean Society papers through the editing process for the society's *Journal,* and quite rightly thought that Wallace should be told.

When Hooker sent over his explanation ("perfect, quite clear & most courteous"), Darwin's relief was palpable. Darwin promptly enclosed Hooker's note with his own and posted the two letters to Singapore.[2] They would eventually get into Wallace's hands, somewhere further east of Singapore, some four months later, via local sailing vessels and steamer. The last leg, Darwin imagined, would be by prau.

Then he worried. The next few days were the worst, full of a nagging suspicion that he had not acted completely fairly and wondering if he had allowed himself to be bounced into slightly disreputable proceedings. He had no idea how Wallace would view the situation. Theories were deeply personal properties in the tightly governed world of nineteenth-century science, inextricably tied to the names of their creators, and not freely available for someone else to place before the public, and certainly not without permission. The republic of letters, despite a growing rhetoric of open information, was individualistic. Darwin knew a number of naturalists who would be outraged if he or Lyell had tried the same priority-sharing treatment on them, plenty of established figures who might have called his motives sharply into question. Restlessly, he told himself that everything had been done with the highest moral intention. It was obvious that six months or more must pass before he could receive a word in reply.

As it happened, Darwin's worst fears were unnecessary. When Wallace read the letters sent from Down House he admitted he was "very much surprised to find that the same idea had occurred to Darwin."[3] Privately, he may have been greatly disappointed. But he replied immediately, displaying all the warm liberality of spirit that Darwin and other future scientific friends came to value dearly. Only the letter to Hooker now survives. "It would have caused me much pain & regret had Mr. Darwin's excess of generosity led him to make public my paper unaccompanied by his own much earlier & I doubt not much more complete views on the same subject," Wallace informed Hooker. "I must again thank you for the course you have adopted, which while quite strictly just to both parties, is so favourable to myself."[4]

Modestly, he accepted the lesser role of co-discoverer that was thrust upon him. Perhaps he realised there was little else he could do. What was done was done. By the time he knew about the dual announcement he was hardly in a position to make a fuss, and his innate good manners probably told him to acquiesce graciously.

Furthermore, he may have recognised that he would be much better off accepting the curious turn of events and enjoying the limelight. It was in his interest to make the best of the new situation. His essay would otherwise scarcely have been made public so rapidly or so advantageously. If he had sent the paper to his friend Henry Bates, still swatting mosquitoes on the Amazon, or to his London agent Samuel Stevens, his ideas would not yet have reached any of the experts who governed the sciences of the day and whose opinion he sought. If he had sent it directly for publication to Edward Newman at the *Zoologist,* or to some other natural history magazine, he would not have had any guarantee that he would be published. Now, his name was coupled with Darwin's, and he was about to appear in a prestigious learned journal, professional advantages that he appreciated intently. Perhaps, too, Wallace found it relatively easy to share the honours. Unlike Darwin he had not yet invested his life and soul in the theory.

Even so, he probably felt a stab of regret to find he was not alone in his ideas. To his lasting credit, he never afterwards displayed the smallest flicker of resentment. As with Darwin, the unexpected collision revealed his finest qualities.

## II

Duty done, Darwin lost no time in getting the children away from infection. "You may imagine how frightened we have been," he wrote to Fox after the baby's funeral. "It has been a most miserable fortnight." Tense with worry, he propelled the whole ailing entourage out of Downe. He aimed for recuperation in some attractive—and distracting—spot. "To the day of my death I shall never forget all the sickening fear about the other children after our poor little baby died."[5] Emma was exhausted, he noted. The rest of the household fared no better, with Henrietta fractious in her convalescence, and Jane the nursery maid (Parslow's daughter) showing signs of what might similarly turn into scarlet fever. The personal strain of the last two weeks left him feeling like "living lumber." He scarcely had sufficient emotion left to absorb the news of his sister Marianne's death that same week, aged sixty. He and Erasmus agreed that Marianne's children should be looked after by Susan and Catherine Darwin, the remaining Shrewsbury sisters, and Susan, who had always shown a particular fondness for these nephews, brought them back to the old family home under her responsibility.[6] Marianne's husband, Henry Parker, had died several years before. Even though people's relationships with death and dying have changed over the centuries, and the experience of grieving in Victorian times was bound by many different systems of class and belief, these deaths, coming so close together, were still a harsh reality for the Darwins. Emma could probably take consolation in Christian assurances about immortality. Her church's doctrines assured her that she would meet her children and other loved ones in heaven. Darwin confronted mortality in solitude and isolation. Old or young, death came knocking.[7] In retrospect, it seems possible that Emma may have suffered twice over from not being able to share religious consolation with her doubting husband.

But recuperate they did. After a cautious journey made in stages to Portsmouth, and then across to the Isle of Wight, with seven children in tow ranging from William, aged nineteen, to Horace, nearly seven, plus a separate carriage full of trunks and maids, a wheeled invalid chair and pet kitten for Henrietta, and one of the indispensable Thorley sisters drafted in for the emergency as a temporary governess, they began to feel easier as soon as they arrived. This tiny offshore island had long been a summer location for royalty, poets, and gentlefolk. Above all, it was quiet, and as Darwin told Fox when they came to rest in the King's Head Hotel,

Sandown, "suits us very fairly." They found a villa to rent in Shanklin further down the coast, "the nicest sea-side place which we have ever seen." The rented villa was nicely situated at the foot of the cliffs on the edge of an immense sandy bay. Darwin hoped the sea air would blow away some of Henrietta's listless malaise. The little boys, he said appreciatively, were very happy in the sand.

Gradually, his private tensions began to unwind. He wandered up through the nearby chine, a narrow cleft in the hillside, to the old houses at the top, making his way along the chalky paths to savour the space and the ocean view. Such open vistas were favourites of his, ideal for soothing a troubled mind. The cliffs here were steeply picturesque and on a sunny day reminded well-travelled visitors of the Riviera or the Bay of Naples, and variously enticed Keats, Tennyson, and Prince Albert. The prince consort had just purchased Osborne House, on the other side of the island, to rebuild as a holiday palace for his growing family. Darwin tramped for miles, glad that his thoughts were taking a happier turn. The loss of his child and of his theory were not to be taken lightly. They both seemed less overwhelming when he was out in the fresh air.

Sometimes William accompanied him over the springy turf, identifying plants with a copy of George Bentham's *British Flora* specially bought for the purpose, and naming them for his father's edification. "Willy charged into the Compositae & Umbelliferae like a hero," Darwin burst out in parental pride in a letter to Hooker one evening. William was about to go to Cambridge University with an entrance scholarship to Christ's, his father's old college—a kind of personal resurrection, Darwin told Fox sentimentally, even down to the academic cramming with a tutor beforehand and taking rooms on the same college staircase. Wildly ambitious for the future, he talked to his son about becoming a lawyer, "my dear future Lord Chancellor." Tentatively, he explored the novelty of treating William as a grown-up equal. Less eager to participate in lord chancellor talk, but with the good humour that came naturally to all the Darwin boys, William bowed to the inevitable. This so-called equality mostly involved helping his father with his researches.

Almost immediately Darwin also began writing again, first at the King's Head in Shanklin, and then more consistently for an hour or so every morning in a room in the villa in Sandown. He did not particularly wish to write: his thoughts were heavy. But it took his mind off his troubles and gave him something to do.

More to the moment, Hooker was pressing him for a proper scientific paper on natural selection, one which would set out his completed system and fulfill his claim to priority. Boldly, Hooker suggested that Darwin provide another article, of thirty pages or so, for the Linnean Society *Journal*, an abstract, as it were, of the long manuscript that Hooker knew was under way at Down. Affectionately, he put his scientific weight behind his friend and gave him a shove in the appropriate direction.

Anguished squeals shot out from the Isle of Wight. "I can hardly see how it can be made scientific for a Journal, without giving facts, which would be impossible." "How on earth I shall make anything of an abstract in 30 pages of Journal I know not."[8] Undaunted, Hooker persisted. The *Journal*, he said stoically, would accommodate whatever Darwin supplied, 100 pages, or even 150. The reluctant author saw the necessity to agree.

Without knowing it, he began what was eventually to be the *Origin of Species*.

## III

At home towards the middle of August 1858, the words poured out of him as if floodgates had broken. Hooker's simple request, coupled with the high emotion of previous weeks, released years of pent-up caution. Darwin was not yet sure what form this writing would take—he called the manuscript an "abstract." But he was bent on producing something for immediate publication. "I have resolved to do it, & shall do nothing till completed," he told Fox. Clearly, the thought of Wallace energized and electrified him. "I am almost glad of Wallace's paper for having led to this."

Always a hard worker, he worked harder than he had ever done before. Day after day, he shut himself in his room surrounded by the clutter of a lifetime of research—notebooks, old sketches and essays, the unfinished manuscript that was interrupted by Wallace's letter, flaps of paper poking out from the backs of books, a welter of pencil jottings, many of them illegible even to him, torn-up correspondence arranged in heaps according to the subject matter, and the silver snuff-box safely recharged in an accessible pocket. He did not use a desk. Instead, he sat with a board across his knee in a big upright armchair, the only chair in the house that accommodated his long legs, raised high off the ground by the addition of an ugly iron frame and castors. Day after day, he filletted, docked, and embellished his twenty-year-old project, bringing the full weight of mature understanding to bear on every word. Although the text was meant to be only an abstract, he wanted to make it as perfect as possible. There was no more room for postponing, no more hedging his bets.

And how he wrote. All the years of thought climaxed in these months of final insight. Alone in his study, secure in his downland ship, pampered by his wife, and insulated from the worries of the world, a sad and reticent man, so nearly preempted in his attempt to rewrite the story of nature, Darwin saw further and more clearly than ever before. Hooker's pressure to be brief helped no end. His mind was lucid, his pen sharp. The steady support he received from friends and family was similarly sustaining.

But the fire within came from Wallace. For so long, Darwin had been hemmed in by anxieties, always circumspect, outwardly conventional, and striving for scientific completeness—the search for flawlessness that marked him out from the start and which, in middle life, was driving him to exhaustion. Now every impediment was pushed aside. Whereas the process of being forestalled might have destroyed a lesser spirit, Darwin emerged resolute. Steel glinted. Wallace's essay gave him the edge he needed. "You cannot imagine what a service you have done me in making me make this abstract," he said to Hooker in October, "for though I thought I had got all clear, it has clarified my brains much, by making me weigh relative importance of the several elements."[9]

Some of this newfound confidence probably lay in the way that his and Wallace's Linnean Society paper was being received. Darwin had imagined the crisis of publication would be far worse than it was. The double paper appeared in the Linnean Society *Journal* (in the zoological section) in August 1858. During the next two or three months it was reprinted either in full or in part in several popular natural history magazines of the day.[10] A number of people made their views known in letters, reviews, and journals. There were more notices than usually assumed.[11]

Richard Owen, for example, referred to the paper in his presidential address to the British Association for the Advancement of Science at Leeds in September 1858, praising Wallace's explanation of the way varieties replace one another, although hastily adding that there was no reason to think that this accounted for the origin of species.[12] Owen's published address had a wide circulation. The British Association deliberately aimed itself at the respectable, middle-class public who formed the larger community for science, and through the work of Owen and others promoted a form of intellectual engagement in which the grand design of the Creator usually played an obvious part. Indeed, the British Association was to become the theatre in which the evolutionary debate would be played out in front of the public for a decade or more. Yet, should he have wished, Owen could have been much harsher, more savage altogether, in immediately crushing the notion of evolution by natural selection. A longstanding acquaintance of Darwin's, he was the leading figure in British natural history sciences, superintendent of the animal and plant collections at the

British Museum, and a noted comparative anatomist. His philosophy of nature would ordinarily have predisposed him to dislike evolutionary proposals, for he advocated an idealist vision of the relationships between animals in which anatomical similarities were understood as expressions of the underlying plan of the Creator—although, here and there, he also put forward developmental connections as reflected in some of the more unusual reproductive cycles of lower animals.[13] In his address, however, Owen seemed almost positively inclined. Another acquaintance of Darwin's, the botanist Hewett Cottrell Watson, added an excitable word or two about the new theory to the next volume of his series on British plants, *Cybele Britannica.*[14] And when extracts from Darwin's and Wallace's papers were reprinted in the popular magazine *Zoologist,* only a few correspondents raised their eyebrows. "Is this wise?" asked one country gentleman. "Is it in accordance with the spirit of modern science?" There was nothing here to worry Darwin unduly.

One reviewer pushed hard enough to chafe. Samuel Haughton, the professor of geology at Trinity College Dublin, sneered unpleasantly.

> This speculation of Messrs. Darwin and Wallace would not be worthy of notice, were it not for the weight of authority of the names [i.e., Lyell's and Hooker's] under whose auspices it has been brought forward. . . . If it means what it says, it is a truism; if it means anything more, it is contrary to fact.[15]

Darwin copied this out to send to Hooker as an omen of things to come. Yet far from destroying his confidence, Haughton's words seem to have tickled the anti-establishment recesses of his mind.

Elsewhere, and completely unknown to Darwin, the fellows of the Linnean Society received and read their number of the *Journal* without any evident alarm. Despite Haughton's outburst, there were increasing numbers of thoughtful men and women in Britain and Ireland unfettered by the constraints of traditional natural theology, religious dogma, or biblical literalism—individuals who would still, nonetheless, regard themselves as responsible members of society and believers in some form of deity. A young naturalist called Alfred Newton, a junior fellow at Magdalene College, Cambridge, sat up late into the night clutching his copy of the *Journal.* "I shall never forget the impression it made on me," he wrote afterwards. "Herein was contained a perfectly simple solution of all the difficulties which had been troubling me for months."[16] Within the week he persuaded his college friend, a trainee ordinand, Henry Tristram, to agree, and Tristram prepared a short paper on the birds of North Africa for the influential ornithological journal *Ibis*. Rather like catching a disease, these two young men claimed they developed in the space of a few days "pure and unmitigated Darwinism."

Gray and Hooker began mentioning the idea of natural selection in print, telling the botanical community that a more complete announcement from Darwin was on its way. Hooker admitted to Gray, "I must own that my faith is shaken to the foundation & that the sum of all the evidence I have encountered since I studied the subject is in favour of the origin of species by variation." Gray replied in a similarly open-minded way. They were relieved to be able to talk about natural selection in public. By drawing them in while his ideas were still secret, Darwin had effectively tied their hands, neither allowing them any practical use of the theory in their work nor the freedom to discuss it with other naturalists. Released from this unvoiced commitment, Hooker published comments on Darwin's and Wallace's evolutionary views in the substantial essay on Tasmanian plants that he was compiling, part of a botanical catalogue that was delayed and not published until the closing months of 1859. There, he announced his support for "the ingenious and original reasonings and theories by Mr. Darwin and Mr. Wallace." Before then he told Gray:

> I am very busy with the Introductory Essay to Flora of Tasmania, a kind of composition I find most hard—I have to make large concessions to Darwin's doctrines of "Natural Selection" and have altogether modified my opinion much on the subject of hybrids,—varieties—returning to parent form—& many other cardinal points. . . . Most thankful I am that I can now use Darwin's doctrines—hitherto they have been secrets I was bound in honor to know, to keep, to discuss with him in private & to combat if I could in private—but never to allude to in public, & I had always in my writings to discuss the subjects of creation, variation &c &c as if I had never heard of Natural Selection—which I have all along known & feel to be not only useful in itself as explaining many facts in variation, but as the most fatal argument against "Special Creation" & for "Derivation" being the rule of all species.[17]

Unaware of Hooker's prudence, Darwin read these Tasmanian proofs with mounting satisfaction. "You cannot imagine how pleased I am that the notion of natural selection has acted as a purgative on your bowels of immutability," he said, more graphically than usual. In this book Hooker proposed that variation and transmutation might well be the answer plant taxonomists had been seeking for the botanical confusion they saw around them.[18]

Theological beliefs did not appear to hinder Gray's evaluation of natural selection either. Gray referred to the Linnean Society papers in complimentary terms in the concluding part of his study of Japanese plants.[19] Sheepishly admitting his pleasure in stirring things up, he also set out to agitate the members of a Harvard University science club in April 1859 by outlining Darwin and Wallace's argument "partly to see how it would strike a dozen people of varied minds and habits of thought, and partly, I confess, maliciously to vex the soul of Agassiz with views so diametrically opposed to all his pet notions."[20] A small but significant part of the Massachusetts intellectual community, including Louis Agassiz, E. S. Dixwell, Joseph Lovering, and Benjamin Pierce, went away that night having been the first in America, after Gray, to hear of Darwin and Wallace's work. Gray was right about annoying Louis Agassiz. Agassiz was thoroughly irritated by the evening's proceedings and impulsively added a jibe at Darwin and Wallace to his renowned *Essay on Classification.* The two Englishmen were foolish, Agassiz roundly declared, if they believed that a longer time would "do what 30,000 years has not done already."

Agassiz did not regard Darwin and Wallace's scheme with any degree of favour—and was not likely to. Ever since 1846, when he had emigrated to Boston from Switzerland, he had been the leading naturalist in America, a man with a world-wide scientific reputation, professor of zoology at Harvard University, charismatic, devout, and highly intelligent. At this point, in 1858 and 1859, he dominated American intellectual life. He was well known as believing that all living beings, including humans, were created by divine fiat. Species were thoughts in the mind of God, he announced in his *Essay on Classification.* Evolution in any form, whether it was Darwin's or any other, was sacrilegious. This *Essay on Classification,* first published in 1857, and republished with anti-evolutionary comments in London in 1859, had been a sophisticated piece of biological thinking, in which he presented the philosophical reasons for regarding species as fixed and stable entities.[21] All classification schemes would be useless, Agassiz argued, if species were always changing. Furthermore, the natural world was so delicately balanced, every organism depending on another in the web of what would come to be called ecological relationships, that all species must, by logical implication, remain constant. They must have been created at the same time, in the shape that they now hold, and all together. Agassiz sincerely believed that the natural world, and all its parts, was a beautiful and divinely inspired orchestral composition.

For these reasons, Agassiz and Darwin were evidently going to be implacably opposed on the question of evolution, not only in their divergent worldviews but in what might constitute the basic principles of biology. Despite their early friendly connections and a genuine regard for each other's competence, Darwin regretted Agassiz's metaphysical approach. "Utterly impracticable rubbish," he had said when he read the *Essay on Classification*. He knew that he would never convince a man like Agassiz. So did Asa Gray. At that Harvard Science Club meeting, Gray deliberately confronted the main source of opposition that Darwin was ever going to encounter in America—an immensely powerful opposition, at that.

Last of all, Lyell nobly grappled with ideas that he found deeply disturbing. "Lyell's thoughts," said Charles Bunbury, his brother-in-law, "are at present very much engaged by Darwin's speculations on the great question of species in natural history."[22] Bunbury said sympathetically that Lyell would not be able to give up the special status of mankind without a tremendous personal struggle.

Bunbury had few such problems himself. "However mortifying it may be to think that our remote ancestors were jelly fishes," he confided to his diary after reading the Linnean Society paper, "it will not make much difference practically."[23]

The same audacious thought crossed Huxley's mind. "Wallace's impetus seems to have set Darwin going in earnest," he scribbled to Hooker in September 1858.

> I am rejoiced to hear we shall learn his views in full at last. I look forward to a great revolution being effected. Depend upon it, in natural history, as in everything else, when the English mind fully determines to work a thing out, it will do it better than any other.[24]

Near and far, known and unknown, these men sensed that Darwin and Wallace had something to say. Even before his abstract was finished, Darwin's personal project was taking on some of the trappings of a collective enterprise and gradually being picked up by a community of friends who operated in a well-organised scientific context that teemed with books and journals, private correspondence, societies, review journals, dinner parties, and speech-making occasions. Friendship was a potent weapon in the process of evaluating new ideas and making decisions. Darwin was far too sophisticated a thinker not to recognise and appreciate it.

At Christmas, he paused. "I never give more than one or two instances," he moaned gently, "& I pass over briefly all difficulties & yet I cannot make my abstract shorter, to be satisfactory, than I am now doing."[25] He had written 330 folio pages and estimated he would require another 150 to 200 to finish. This was hardly an abstract. Gingerly, he began calculating the most efficient size of printer's type for squeezing as many words as possible onto a page of the Linnean Society's *Journal*. Equally gingerly, he tested the temperature of the publishing water at Kew.

He asked Hooker how the Linnean editor would cope with a very long paper in the *Journal;* or what Hooker might feel if Darwin took it elsewhere. Sheet by sheet, Darwin's abstract had turned into a monograph, and the monograph into a book. It looked to him as if a separately published volume was the only answer. "I am thinking of a 12mo volume like Lyell's 4th or 5th edition of Principles," he suggested.

Hooker understood the message only too clearly. He gave the abstract one last try, and asked if Darwin would like him to apply for a government grant so that the Linnean Society could issue a separate supplement.

Darwin rejected the offer. During the New Year holiday he gauged his text's appeal among a throng of visiting relatives ("I think my book will be popular to a certain extent") and decided that an independent volume would probably cover a publisher's costs. If this turned out to be a miscalculation, he was "prepared to subsidise it" himself. He would not publish in the Linnean Society's *Journal.* He would stand alone.

When at long last, late in January 1859, a kindly letter arrived from Wallace, Darwin burst out in a fever of relieved, impassioned activity. Without Wallace, he acknowledged gratefully, he could never have brought his work to such a pitch. Without Wallace's handsome acquiescence, he could never have hoped to publish without being haunted by guilt. "He must be an amiable man," he declared.

"Thank God I am in my last chapter but one," he wrote back to Malaysia.[26] He was convinced that he must make the "everlasting abstract" into an independent volume. After all the years of hesitation, and the upsets of the last six months, he was finally on course for the *Origin of Species.*

## IV

From then until May 1859, when the manuscript was finished, he worked incessantly. He overhauled earlier chapters, completed remaining ones, and wrote a rousing conclusion. Again and again he murmured, "I fear I shall never be able to make it good enough." Again and again, his friends answered questions, read sections of the text, wrote letters, and encouraged him. "It is a mere rag of an hypothesis with as many flaws & holes as sound parts," he wrote to Huxley. "My question is whether the rag is worth anything?"[27] Patiently, they supplied whatever he needed. "A sort of vague feeling comes over me that I have asked you all this before," he said to Gray at one point. "If I have, I beg very many apologies."

Now, with a book clearly in mind, Darwin's overall focus became much sharper. Rigour and discipline pervaded his daily work. In practical terms, he cut down each pre-existing chapter from the big natural selection manuscript and relentlessly omitted all footnotes and citations of sources. Afterwards he regretted losing so much of the solid scientific evidence he had struggled to collect. Then he added and rearranged material to make a more compelling argument, couched in terms that were greatly improved by being compressed. Competition and catastrophe had shaken him out of his usual sense of himself. He felt closer to his material, more in tune with the raw brutality of nature, than ever before. Trenchantly cutting and weaving in this manner, he produced a five-hundred-page volume in thirteen months, an "abstract" of his theories only by way of leaving out documentation and lengthy provisos. The origin of his *Origin* was surgical indeed.

As a writer, too, he discovered unplumbed depths. His voice was in turn dazzling, persuasive, friendly, humble, and dark. Hardly daring to hope he might initiate a transformation in scientific thought, he nevertheless rose magnificently to the occasion. Being stuck in Down House was the best thing that could have happened to him. Pleasingly localised as his book was in manner, it reached out across national and chronological boundaries. His imagination soared beyond the confines of his house and garden, beyond his debilitating illnesses and the fragile health of his children. At his most determined, he questioned everything his contemporaries believed about living nature, calling forth a picture of origins completely shorn of the Garden of Eden. He abandoned the image of a heavenly clockmaker patiently constructing living beings to occupy the earth below. He dismissed what John Herschel devoutly called the "mystery of mysteries." Darwin's book implicitly laid claim to Adam and Eve, as time and again he showed how nature was cruel and full of blunders. The natural world has no moral validity or purpose, he argued. Animals and plants are not the product of special design or special creation. "I am fully convinced that species are not immutable," he stated in the opening pages. No one could afterwards regard organic beings and their natural setting with anything like the same eyes as before. Nor could anyone fail to notice the way that Darwin's biology mirrored the British way of life in all its competitive, entrepreneurial, factory spirit, or that his appeal to natural law unmistakably contributed to the general push towards secularisation and supported the claims of science to understand the world in its own terms. As well as rewriting the story of life, he was telling the tale of the rise of science in Victorian Britain.

Another kind of narrative emerged as well. Darwin wrote as he always wrote, in the same likable, autobiographical style he had developed during the *Beagle* voyage and brought alive in his *Journal of Researches*.[28] Much later on, Francis Darwin said this pleasant style of writing was characteristic of his father in "its simplicity, bordering on naiveté, and in its absence of pretence. . . . His courteous and conciliatory tone towards his reader is remarkable, and it must be partly this quality which revealed his personal sweetness of character to so many who had never seen him."[29]

This artless intimacy was familiar to generations of English readers through the pastoral writings of Gilbert White and gently humorous autobiographical stories like *Tristram Shandy.* Here Darwin spontaneously tapped into well-known and unthreatening literary genres. Although his theories might frighten, his style was thoroughly sympathetic and genial, creating a distinctive magic between author and reader. He appeared in his book just as he was in life—a reputable scientific gentleman, courteous, trustworthy, and friendly, who did not speak lightly of the momentous questions coming under his gaze, a champion of common sense, honest to his data, and scornful of "mere conjecture." This humane style of writing was one of his greatest gifts, immensely appealing to British readers, who saw in it all the best qualities of their ancient literary tradition combined with contemporary gentlemanly values. It served him well during the controversial years to come. In particular, it defused any possible personal animosity. In effect, it made him.

And what a book it was. Few scientific texts have been so tightly woven, so packed with factual information and studded with richly inventive metaphor. Darwin's literary technique has long been noted for echoing *Great Expectations* or *Middlemarch* in the complexity of the interlacing story lines and his ability to handle so many continuous threads at the same time. Darwin was crafting a lasting work of art.[30] More than this, his imaginative powers were to captivate generations of readers. Modestly, he said only that his book could be read as "one long argument."[31]

The structure of that argument was significant. Although Darwin deliberately made the steady procession of chapters look as if it reflected the day-to-day sequence of his research—giving an impression of a relatively uncomplicated progress from facts to ideas—the real story had been quite different. Natural selection was not self-evident in nature, nor was it the kind of theory in which one could say, "Look here and see." Darwin had no crucial experiment that conclusively demonstrated evolution in action.[32] He had no equations to establish his case. Everything in his book was to be words—persuasion, revisualisation, the balance of probabilities, the interactions between large numbers of organisms, the subtle consequences of minute chances and changes. Like Charles Lyell in his *Principles of Geology,* he had to rely on drawing an analogy between what was known and what was not known, in Darwin's instance by making a link between what took place in farmyards and what might be presumed to happen in the wild. He depended on probabilities. He relied on techniques in which the accumulation of factual examples progressively weakened a reader's resistance. Case after case was "quite inexplicable on the theory of independent acts of creation," and he called attention to some fifty or sixty biological phenomena that in his view simply could not be explained by special creation. Of course, underneath, he was aware that his data were inseparable from his theory—that neither could exist in his mind without the other. "How odd it is that anyone should not see that all observation must be for or against some view if it is to be of any service!" he once wrote.[33] And he understood that one of his greatest problems was to define and stabilise the very knowledge that he was attempting to introduce. These kinds of scientific argument were relatively untried in the nineteenth century.[34] In an era when natural philosophers were consciously coming to rely on idioms of prediction, experiment, demonstration, and discovery, when accredited truths of nature were established by seeing and believing, Darwin's approach was doubly unusual.[35] He was inviting people to believe in a world run by irregular, unpredictable contingencies, as well as asking them to accept his solution for the simple reason that it seemed to work.

In this respect, the attention he lavished on facts was a highly effective procedure. Without accredited facts, he told himself, his argument would amount to little more than another *Vestiges,* or another version of Lamarck, a cautionary thought he had kept in mind ever since the day of *Vestiges'* publication. Furthermore, the full weight of facts helped to distinguish his proposal from Wallace's. And his emphasis on facticity, as it were, conformed to the most acceptable contemporary methods of science and eased the starkness of the theory he was proposing.[36] As Darwin saw it, much of his originality and power to persuade thus lay in his mountain of scrupulously considered data. So he made sure that his readers would understand the years of effort that he had invested in establishing the accuracy of the information he presented. He cited by name the experts with whom he corresponded, often quoting from their letters. He described what he had seen with his own eyes during his own experiments. He authenticated the observations of unknown practical men by adding a few words of personal validation from himself or from others, characterising them as a "celebrated raiser of Hereford cattle," or a "skilful pigeon-fancier." He cited "careful observers," and "the good observer."

This sense of personal verification and adjudication was one of Darwin's most visible traits as an author. Politely, gently, resolutely, he ushered his readers around his study, his garden, his circle of correspondents, his greenhouse, his social interactions with the landed gentry and their gamekeepers, and the learned societies he frequented and whose journals he studied. Underneath, he was making a programmatic statement about his authority to speak on the issue in hand and inviting his readers to trust him. The anonymous author of *Vestiges* had little such access to the opinions of experts. Nor had Wallace. Yet the naturalist at Downe had time in abundance, a world-wide series of contacts and introductions, and a sufficiently large income to indulge any amount of research. Through his facts, Darwin conveyed his high place in the structures of hierarchical, imperial England. He sought to surpass his rivals through the quality of the supporting evidence for his work.

## V

In the first four chapters of this book—not yet called *On the Origin of Species*—he took infinite care to set out his wares according to the philosophical rules of the day.[37] The sheer variability of organisms came first. "Breeders habitually speak of an animal's organisation as something quite plastic, which they can model almost as they please," he stated, and quoted Sir John Sebright, who claimed with respect to pigeons that "he would produce any given feather in three years, but it would take him six years to obtain head and beak."[38] To this, Darwin added a matching account of variability in wild animals and plants. All his notes about barnacle innards, wild horses' stripes, primroses, and oxlips took their place. Privately, he characterised this as a "short & dry chapter."[39]

The key point came next. There was an analogy, he claimed, between what a farmer or breeder could do and what might happen among wild animals and plants.[40] His whole concept of natural selection rested on this analogy—an analogy between selective processes taking place under either "artificial" or "natural" conditions. As it turned out, his work was criticised on the validity of exactly this point (among many other future criticisms), not least by Wallace, who thought there could be no direct comparison between wild and domestic precisely because domestic animals had been removed from their natural environments.

In the pages that followed, Darwin explained what he meant. Across the green fields of Britain, all nature was at war with itself. The living world teemed with deadly competition and slaughter, the same elemental energies, red in tooth and claw, that Tennyson characterised in *In Memoriam*. "What war between insect and insect, between insects, snails, and other animals with birds and beasts of prey—all striving to increase, and all feeding on each other or on the trees or their seeds and seedlings, or on the other plants which first clothed the ground and thus checked the growth of the trees," wrote Darwin.[41] God's harmony was an illusion. Unsure whether he would be believed, he produced a plethora of examples of strife in nature. Malthus's principle of population was his justification.

> It is the doctrine of Malthus applied with manifold force to the whole animal and vegetable kingdoms, for in this case there can be no artificial increase of food, and no prudential restraint from marriage.[42]

Limited resources, limited places in nature, and continued natural fecundity gave rise to a battle for survival. Well-adapted variants would be the only ones "selected" to survive. Here, said Darwin, was the origin of new species.

> It may be said that natural selection is daily and hourly scrutinising, throughout the world, every variation, even the slightest; rejecting that which is bad, preserving and adding up all that is good; silently and insensibly working, whenever and wherever opportunity offers, at the improvement of each organic being in relation to its organic and inorganic conditions of life.[43]

In following chapters he worked his way through nearly every aspect of nineteenth-century biological thought, explaining how even the most intractable natural history puzzles could be explained—"descent being on my view the hidden bond of connexion which naturalists have been seeking." In each area of thought he brought a wide range of phenomena under a single explanatory umbrella, unifying and giving fresh meaning to previously disparate data. This unexpected unification had from the start impressed him strongly and was still his main reason for believing in the truth of his theory. "My theory gives great final cause," he had written as early as 1837, "I do not wish to say only cause, but one great final cause."[44]

Embryology became intelligible—"Embryology rises greatly in interest, when we thus look at the embryo as a picture, more or less obscured, of the common parent-form of each great class of animals." Darwin was proud of this part of his argument, which he asked Huxley to read before publication. "The facts seem to me to come out very strong for mutability of species," he had told Hooker.[45]

Palaeontology, comparative anatomy, and taxonomy would also be transformed, he wrote in anticipation. The anatomical resemblances sought by taxonomists were not just abstract notions, nor were they the physical expression of some divine plan drawn up by the Creator, as Agassiz or Owen suggested. Instead, the resemblances were caused by genuine affinity. Furthermore, "descent" explained the existence of vestigial organs like the appendix in human beings—they were anatomical remnants left over by history. If comparative anatomists were to follow Darwin's scheme in full, the reason for finding rudimentary hind legs in snakes, for example, would soon become clear.

Similarly, the web of geographical patterns that plants and animals traced over the globe could be explained on the grounds that species spread and changed. At every point, the notion of ancestry connected previously disparate facts and opened up new perspectives. The practical naturalist in him emerged and spoke plainly—the barnacle scholar, the pigeon-lover, the plant experimenter, and *Beagle* collector, the traveller at last approaching his goal. His theory's value, he was arguing, lay in the way it explained and united so many different features of the natural world.

Ruminatively, he here and there acknowledged the problems that his anthropomorphic language would generate. Often he veered too close to personifying natural selection. While this was perhaps unavoidable in the general sense, he frequently gave the impression that he regarded natural selection as an active agent, an all-seeing farmer in the sky, as it were, who deliberately chose the variants that were to succeed. Only a few months afterwards Darwin admitted to Lyell that this was not his intention and that he ought to have used a more neutral expression like "natural preservation." He and Wallace were to discuss this difficulty at length. The same entanglement occurred with the word "adaptation," which in Darwin's hands hinted at some form of purposeful strategy in animals and plants, the exact opposite of what he meant. Later, he used "contrivance" as a partial solution. Over and over, Darwin struggled with words. The language he knew best was the language of Milton and Shakespeare, steeped in teleology and purpose, not the objective, value-free terminology sought (although rarely found) by science.[46]

He was not even able to speak of "evolution," as such, because at this time the term was mostly used to describe the embryological process of a gradual unfolding of hidden structures; it was the ensuing debate around his published work that gave the word its modern meaning.[47] In the *Origin of Species* Darwin referred to "descent with modification." Equally, he did not use what ultimately became the most famous phrase of all, "survival of the fittest." This was coined a few years afterwards, by Herbert Spencer in 1864, at which point Wallace suggested Darwin should use it.[48] All these verbal ambiguities would lead readers in directions that Darwin did not fully intend.

Unusually for a scientific book, Darwin also provided a frank discussion of the many stumbling blocks that he thought would occur to readers. "Some of them are so grave that to this day I can never reflect on them without being staggered," he admitted. "I have felt the difficulty far too keenly to be surprised at any degree of hesitation in extending the principle of natural selection to such startling lengths."[49]

This confession was his most adroit step so far. He expected a barrage of challenges and intended to provide the answers straight away. In fact, he found the difficulties easy to list, in much the same way as he had once confidently jotted down the inconveniences of getting married. Disadvantages always made themselves obvious to him. These were fresh in his mind. If organisms are constantly changing, where are all the intermediate forms? Have transitional species ever been found in the fossil record? How can complicated organs ever come into existence by stages? "Is it possible to believe that the eye with its admirable correction for spherical & chromatic aberration, & with its power of adapting the focus to the distance, could have been formed from the simplest conceivable eye, by natural selection?"[50] How do the specialised hierarchies of castes emerge in an ants' nest or a beehive? Were instincts created individually for each species by God? It was very difficult to explain how sterility between incipient species might arise when they were originally members of the same interbreeding population. And—as Agassiz pointed out—what might happen to biological classification schemes if species and varieties are forever in flux? One by one, he proposed answers.

With profound deliberation, however, he did not include the two difficulties that would have occurred to everybody. He avoided talking about the origin of human beings and he avoided God. He remembered the bitter furore over *Vestiges*. He remembered the years he had spent worrying about divine intervention. No matter how seriously and cautiously he might treat evolutionary questions himself, he knew that anything he said was bound to ignite furious controversy, and anticipating just such a response, he had long ago drained his manuscripts of any reference to a Creator or human ancestry. He had no intention of reintroducing them now.[51] In this book, he was completely silent on the subject of human origins, although he did refer in several places to mankind as an example of biological details.[52] The only words he allowed himself—and these out of a sense of duty that he must somewhere refer to human beings—were gnomic in their brevity. "Light will be thrown on the origin of man and his history," he declared in the conclusion. When he needed to, he spoke cautiously of the Creator, aware that his book might otherwise be labelled atheistic. But he was careful not to allow the Creator any active role in biological proceedings.

He purposefully avoided the first origin of life, too. For a book that would claim in its title to address the origin of species, Darwin's text refused to propose any theory of absolute origins. He had no systematic history of beginnings to offer, no primeval soup or creative spark, and only at the end of his book did he mention the likelihood of all ancestral organisms originating in one primordial form. Such ancient origins, he privately believed, were lost in the mists of time and were essentially unreclaimable. His story was not about the start of life but about the processes that governed organisms during their life spans.

By the end, he had set out one of the most densely impressive proposals of the century. Although he did not compare his work directly with that of those who had gone before, his theory was nonetheless distinctive. He differed from Lamarck, or even from his evolutionary grandfather Dr. Erasmus Darwin, in that he eschewed any "doctrine of necessary progression" and inner striving towards perfection. While Darwin certainly allowed some place in his scheme for the direct effect of the environment on organisms—the inheritance of acquired characteristics that was popularly assumed to be the main feature of Lamarck's system—he always regarded the chief difference between them to be that he, Darwin, did not allow his organisms any future goal, any teleology pulling them forwards, or any internal force that might drive the adaptive changes in specific directions. On the contrary, Darwin's scheme of evolutionary adaptation was based entirely on contingency. Organisms shifted randomly. Darwin could never understand why Lyell, or any number of commentators over the next few decades, failed to see the contrast as plainly as he did.

He differed from Robert Chambers's *Vestiges* in the solidity of his factual information and tightly organised mechanism for change. Darwin's theory was more strictly limited in scope than *Vestiges'* all-embracing system of development, and perhaps regarded by his friends as more scholarly for that reason. Certainly Darwin considered it a great advantage that he avoided discussing the beginnings of the earth, or the start of life, or the future of humanity—omissions that undoubtedly made his text dull by comparison with *Vestiges* but, in return, gave him a superior rank in conventional scientific circles. They also differed in detail. Where Chambers had seen the links in the fossil record as if each species passed through stages representing life forms below it in the scale, Darwin presented the history of life on earth as if it were a metaphorical tree, growing and branching from some ancestral base, and explained how this might happen according to his principle of divergence.[53] "My views are very different from those of that clever but shallow book, the Vestiges," he assured James Dwight Dana.[54]

And he managed to differentiate himself from Wallace, at least in two respects. Wallace had suggested that domesticated organisms were raised in circumstances that rendered any comparisons between them and wild organisms invalid. Darwin, on the other hand, put this comparison at the foundation of his argument for the origin of species. In addition, Wallace had spoken mostly about the replacement of species by other species, groups by groups, rather than the individual changes that preoccupied Darwin. Neither of these differences was regarded as important by anyone other than Wallace and Darwin themselves.

Taking the distinctions and comparisons together in Victorian context, Darwin's writing was ultimately both unique and part of a larger corpus of pre-existing evolutionary thought. There could be no mistaking the weight of thought that lay behind every word, the judicious strategies, the powerful, transformative metaphors, his notion of a "great tree of life," the interlocking double-punch of detail and breadth of vision. Although he subsequently complained that he had been rushed into the *Origin of Species,* that it was nothing but an abstract, that his evidence was truncated and his footnotes and sources were omitted, it was undeniably his masterpiece.

"When the views entertained in this volume on the origin of species, or when analogous views are generally admitted, we can dimly foresee that there will be a considerable revolution in natural history," he declared fervently in the closing pages. "I look with confidence to the future, to young and rising naturalists, who will be able to view both sides of the question with impartiality."

> When we no longer look at an organic being as a savage looks at a ship, as at something wholly beyond his comprehension; when we regard every production of nature as one which has had a history; when we contemplate every complex structure and instinct as the summing up of many contrivances, each useful to the possessor, nearly in the same way as when we look at any great mechanical invention as the summing up of the labour, the experience, the reason, and even the blunders of numerous workmen; when we thus view each organic being, how far more interesting, I speak from experience, will the study of natural history become![55]

All his hopes came to a crescendo. Simultaneously domestic and universal in tone, Darwin had achieved something extraordinary. One particular country lane that he visited on walks around Downe filled his mind.

> It is interesting to contemplate an entangled bank, clothed with many plants of many kinds, with birds singing on the bushes, with various

> insects flitting about, and with worms crawling through the damp earth, and to reflect that these elaborately constructed forms, so different from each other, and dependent on each other in so complex a manner, have all been produced by laws acting around us. . . . There is a grandeur in this view of life, with its several powers, having been originally breathed into a few forms or into one; and that whilst this planet has gone cycling on according to the fixed law of gravity, from so simple a beginning endless forms most beautiful and most wonderful have been, and are being, evolved.[56]

He hardly anticipated how austere, tragic, and supremely beautiful his work would appear to others.

## VI

Long before the end of May, Darwin needed to stop. He had been buoyed up by mental activity alone, and the perpetual effort ate away at his health. He began vomiting again.

His usual resource was to cut adrift from his work and take a few days' holiday or retire to the water-cure for a while. But with the *Origin of Species* in hand he resented doing this. Every day was needed, he said. Impatiently, he experimented with nostrums for dyspepsia advertised in the pages of newspapers. "I am taking Pepsine," he informed Fox. "I think it does me good & at first was charmed with it."

In the end, however, the best treatment was getting away from it all. He recognised that his visits to the water-cure were becoming more and more essential, having increased the length of his stays from a few days in the spring of 1858, to a week in October, then a full two weeks in February 1859. He no longer travelled all the way to Malvern. In 1857 he had located a new establishment, Moor Park, near Farnham in Surrey, much closer to home and without any of the sad memories associated with his daughter Annie's death. He and Emma had not yet been back to Malvern to visit their daughter's grave. "Old thoughts would revive so vividly that it would not have answered," he confided to Fox, "but I have often wished to see the grave. . . . The thought of that time is yet most painful to me. Poor dear happy little thing."[57] But he retained all his old confidence in water treatment and "no faith whatever in ordinary doctoring."

Moor Park was run by a young couple, Dr. Edward Lane and his wife, accompanied by Mrs. Lane's widowed mother, Lady Drysdale. The spa was a handsome building, the former home of Sir William Temple and full of interesting associations with Jonathan Swift and eighteenth-century literary life.[58] The spacious grounds, lakes, and Dutch parterres laid out by Temple were a major part of the attractions; and Darwin considered the drive from Downe to Farnham very scenic, passing along the foot of the North Downs and skirting the Devil's Punchbowl, with a convenient stopping-off point for lunch at his sister Caroline's house at Leith Hill (Caroline had married Emma's brother Jos Wedgwood). Inside the building, much of the Lanes' success lay in making medical therapy akin to an exclusive house-party. Lady Drysdale maintained polite society habits during treatment, and her lively personality dominated the household arrangements. She was "a great reader, a great whist-player, and the active capable housekeeper of the great establishment."[59] Eventually Lane would concede that he could not afford to carry on in such expensive surroundings and would sell up to run a more economical property in Epsom.

Darwin liked the Lanes and their gracious treatment. Dr. Lane was "too young," he laughed, but that was his only fault, for otherwise he was a "gentleman & very well read."[60] Lane did not believe in all the fringe therapies that Dr. Gully pressed on water-cure patients at Malvern either—no clairvoyance, mesmerism, or homeopathy, and no demand that Darwin must give up snuff, although Lane did point out the dangers of addiction. Instead, Lane attributed most disorders to imperfect digestion, an attitude that seemed entirely sensible to Darwin, and prescribed a wholesome diet, sitz baths, daily showers, and the diversions of pleasant scenery and company.[61] What with the walks, baths, music, agreeable conversation, and a nightly game of billiards, Darwin's time at Moor Park was dedicated to the relaxation he could not achieve at home. "It is really quite astonishing & utterly unaccountable the good this one week has done me," he exclaimed after his first visit in 1857.[62]

As time went by, he became a complete convert to Lane's relaxed therapies. Full of enthusiasm, he sent his daughter Henrietta for treatment in the summer of 1857 to see if it would revive her spirits in the same way that it had helped him, and he furthered his own therapy by getting himself a billiard table for Down House just like the one at Moor Park.

Henrietta felt no better when she returned. Darwin, on the other hand, had given himself a wonderful time poring over billiard advertisements in gentlemen's magazines, consulting clubbable friends, and weighing up various impractical alternatives, in the end opting for a full-sized slate-bedded table from the London firm of Hopkins and Stephens and paying for it by selling his father's gold watch and some bas-reliefs of Greek figures (probably by Flaxman) inherited from the Wedgwood side of the family. The table cost much less than the sum realised on these family heirlooms, and his highhanded disposal of such valuable decorative objects caused some resentful mutterings. "I suppose it was partly his fondness for money that made him do things that we his children thought had better not be done when we grew up and found out about them," observed his son Francis Darwin later on. "For instance he sold a gold watch given to his father by Lord Powis. The beautiful Flaxman things & the Barberini vase [a china reproduction of the original Portland vase, a renowned Wedgwood triumph] were all sold at a nominal sum, part of the money (or all?) being spent on a billiard table." Francis complained that the Flaxman reliefs meant nothing to his father, and were usually covered in dust and placed far too high to see properly. His father's philistinism obviously pained him. "It is certainly curious that so affectionate & sympathetic a man should have had so little love of heirlooms."[63]

When the billiard table arrived in packing cases on a cart from London, its new owner threw himself enthusiastically into the construction process, right down to the last turn on the complicated screw-levelling apparatus that ensured English gentlemen could play on a flat surface despite the irregularities of country house floorboards. One diagram that Darwin sent to his son George at school was so detailed he must have spent the best part of the morning on his hands and knees inspecting the work underneath as it proceeded. Darwin's passion for his new toy knew no bounds. Eagerly, he bought himself an illustrated book ("stunning") full of coloured diagrams of various cue strokes. He learned the different moves, practised diligently, and boasted of well-executed "caroms" to Fox, pleased that his eye had not deteriorated too far from the sporting days of his youth. Soon, he and Parslow were taking a game or two every evening. There was always a ready excuse. "I find it does me a deal of good, & drives the horrid species out of my head," he said cheerfully to Huxley. Thereafter regular rounds at the table drew father, sons, visiting scientists, and butler together in an amiable web of sporting recreation.

At Moor Park he most of all enjoyed being on his own. During one visit early in 1858, a few months before Wallace's letter entered his life, he had loitered for hours in the furthest reaches of the park, which was "very wild and lonely, so just suits me." Inevitably, something tiny would catch his attention: the ants under the Scotch firs were perhaps of a sort not seen at Downe; or the wild clover was visited by different insects; or a woodpecker gave away its hiding-place with telltale chips of bark on the ground. He scribbled notes on any available bit of paper. "It has been stated that woodpeckers remove fragments. In 2 cases I can say this false for such fragments guided me to discovery of nest," he jotted on a crumpled slip that was evidently found at the bottom of a pocket. He was "all eyes," said Edward Lane after one of these visits, always in tune with the tiniest minutiae of life.[64]

Unable to resist a small experiment during one walk, Darwin transferred a few red ants from one nest into another. "I pass my time chiefly in watching the ants," he wrote to his son William, "& I find that though many thousands inhabit each hillock, each seems to know all its comrades, for they pitch unmercifully into a stranger brought from another ant-hill."[65] He thought these red ants were the slave-making species *Formica sanguinae,* and told Emma he was going to send a specimen to the British Museum for confirmation. "I had such a piece of luck at Moor Park," he mentioned to Hooker a few days later. "I found the rare Slave making ant, & saw the little black niggers in their master's nests." Up until then he had thought the custom of making slaves—the brutal practice that so incensed him on H.M.S. *Beagle* when first landing in Brazil—was confined to the human race. Yet he was forced to acknowledge its wider existence after opening up red ants' nests and finding captured black ants in each. "Any one may well be excused for doubting the truth of so extraordinary and odious an instinct as that of making slaves," he would write disapprovingly in the *Origin of Species.*[66] His understanding of these biological relationships was probably influenced by his former colonial experiences. Recalling the horrors he encountered in South America, he regarded the red ants' behaviour as enslavement rather than any comparable social structure such as symbiosis or co-operation. He felt very uncomfortable providing biological parallels for human practices he found abhorrent.

Some of the most intensely private moments of Darwin's middle life came during these solitary walks at Moor Park, moments when he drew fresh inspiration from the beauty of his surroundings and recaptured an uncomplicated joy in nature. He needed these moments more and more. Behind the questioning philosopher lay both an anxious pedant and a sensitive soul, different sides of his character clashing awkwardly as he forced himself to continue writing. He was a driven man. Only an impending physical collapse could force him to slow down. Yet like many driven men he felt uneasy when peace and quiet were actually made available. He could not rest. He could not stop. And evolutionary theory in itself brought him to a knife edge. His mature years were wrapped up in describing the black forces beneath nature's surface. Sometimes he recoiled from seeing nature the way his selection theory demanded. "What a book a Devil's chaplain might write on the clumsy, wasteful, blundering, low & horridly cruel works of nature!" he once exclaimed to Hooker. "My God how I long for my stomach's sake to wash my hands of it—for at least one long spell."[67] Often he worried about the sheer magnitude, the philosophical effrontery, of what he was proposing. He was rewriting the greatest story ever told, offering his contemporaries another Eden, a secular testament for the times. The tension partly expressed itself in his health.

Not surprisingly, his religious position troubled him too. He found his beliefs were increasingly difficult to pin down, sometimes starkly uncompromising, sometimes genuinely responsive to the idea of a deity of sorts. At this time of his life, he said he felt torn both ways. Although he wrote with conviction about a godless universe, he retrospectively thought that while he was writing the *Origin of Species* he probably retained some residual faith: enough that he deserved "to be called a Theist."[68] At least he knew what duty was, he said to Emma. He had no desire to present himself in his book as an outright atheist. He did not wish to slap some of his oldest friends in the face.

He worried about Emma's feelings as well. He sensed that she was concerned about the implications of his views, not so much in the wider cultural sphere, although it seems likely that public strife within the Victorian church perplexed and disappointed her, but much more in the immediate family context. She knew all about his theories. For a long time now she had recognised these ideas as an integral part of his existence and accepted them in much the same way as she accepted his illnesses or the outlandish sequence of his scientific hobby-horses, barnacles, yew-trees, pigeons, and all. They were part of her married life. They amused her at times. She was apprehensive about them at others. Always, she saw how they kept him occupied and fulfilled. Certainly she never prevented his engagment with far-ranging intellectual explorations or hampered his publications because of private disapproval. She never stopped him seeing Huxley, for example, a man who would come to be regarded by some as the devil's disciple. Perhaps she prayed for his soul on Sundays—the letters she wrote him soon after their marriage displayed that likelihood in tender detail. Affection carried the day.

In turn, Darwin knew he could depend on her good sense, her unflappable kindly nature. He was sure that she would support him, no matter what. But it was precisely because of this that he recoiled from exposing her to the full consequences of his own bleak universe. The steady, married love that had grown up between them guaranteed that he would not deliberately hurt her feelings. In a world where the cold stones of graveyards haunted the memory, Darwin realised that he might easily be accused of taking away from her and countless other men and women all hope of heavenly reunion with loved ones, all consolation in the idea of an afterlife.[69] In the kindest possible way, he tried to compromise. He had to say what he needed to say in the *Origin of Species*. But he backed away from any obvious confrontation with the church or with those who were sincere believers. It seems probable that he avoided talking to her about it.

At Moor Park some of the conflict was soothed. "The weather is quite delicious," he told Emma in April 1858.

> Yesterday after writing to you I strolled a little beyond the glade for an hour & half & enjoyed myself—the fresh yet dark green of the grand Scotch firs, the brown of the catkins of the old Birches with their white stems & a fringe of distant green from the larches, made an excessively pretty view.—At last I fell asleep on the grass & awoke with a chorus of birds singing around me, & squirrels running up the trees & some Woodpeckers laughing, & it was as pleasant a rural scene as ever I saw, & did not care one penny how any of the beasts or birds had been formed.[70]

This peaceful country-based cure took him back to a time when his thoughts were altogether less stressful.

## VII

During these interludes at Moor Park, Darwin's old conviviality revived too. He appreciated the diverse company gathered for medical treatment, writing little character sketches of the other patients for Emma and the children's entertainment, and as often as not persuading unsuspecting guests to answer natural history questions for him. A Hungarian attaché to the embassy in Paris promised to send him information about native breeds of horses in Budapest. On another visit a young Irishman taught him "some capital billiard moves." He joked with Lady Drysdale about their mutual compulsion to get to railway stations hours before the train was due to leave ("please to tell Lady Drysdale that I reached the station only 14 minutes before the train started & I should like to know when she will ever have such a triumph as that"); and discussed an immense range of novels with Mrs. Lane, who turned out to be as devoted to the light reading supplied by Mudie's Circulating Library as Darwin unashamedly was. "It often astonished us what trash he would tolerate in the way of novels," lamented his son George later on. "The chief requisites were a pretty girl & a good ending."[71]

Darwin certainly appreciated the undemanding qualities of the fiction stocked by Charles Edward Mudie. Five guineas' annual subscription to this lending library brought him a parcel of up to six recently published books for borrowing every month—fiction and nonfiction books appropriate for middle-class drawing rooms all over the United Kingdom.[72] The choice was not always the customer's. Mudie's staff supplied whatever was new and available, and then filled up the order with miscellaneous works of a solidly respectable nature. Darwin found this no particular hardship. The firm's backlist was crammed with contemporary memoirs, history, belles-lettres, science, religious works, travel, and adventure, as well as fiction in all categories; and much of Darwin's general reading was obtained from Mudie's nonfiction shelves. He was also a member of the London Library, a private library club that similarly dispatched titles by post to country subscribers.

The essence of Mudie's appeal lay in the word "select," for the firm's owner excluded topics and authors that were deemed unsuitable: no longer would the head of a Victorian family have to worry about the blushes of his womenfolk during unsupervised reading. While every Literary and Philosophical Society in every provincial town ran a lending library of sorts, and books themselves were getting cheaper to buy by the year, large centralised concerns like Mudie's, which lent, on subscription, all the most desirable publications, were already sweeping the board. With his vast range of stock and efficient distribution system that expanded dramatically through the century to encompass the furthest reaches of empire, Mudie probably did far more to educate middle-brow tastes than any poet or philosopher. The firm sent boxes of books to nearly every corner of the globe and to nearly every provincial bookshop in Britain, sometimes directly to subscribers like Darwin, or increasingly on a franchise basis, contributing in a businesslike way to the creation of the mass reading audience that typified the Victorian era. Every book that Mudie purchased was one that people wanted to read. Every copy was sent out on loan five, ten, or twenty times until its popularity waned and it was placed on the backlist or sold in the secondhand department. The depot in London was beseiged by readers hoping to be first at the counter for the next installment of *Orley Farm* or *Adam Bede*. In this regard, Mudie was one of the first bookselling magnates to build his business on an audience-led response to literature. As a result, his financial power over publishers was supreme, at least during the middle decades of the century, and even leading figures like Eliot, Kingsley, Thackeray, Bulwer Lytton, Tennyson, and Trollope on occasion wrote specifically in the three-volume format he required for his list.[73] Books were purchased by Mudie in bulk directly from the publisher, on his own terms. Three-volume novels (three-deckers or yellow-backs) became his preferred item, and these constituted nearly half the stock of the giant depot in New Oxford Street. A million volumes on the shelves, declared the advertisements.

Darwin liked these three-decker novels. He found them relaxing, barely registering the titles in their seamless web of deserted sweethearts, secret weddings, wicked cousins, mistaken identities, and the age-old quests for love and passion. His critical faculties were suspended. In these pages he did not have to examine the accuracy of facts or delve far beneath the surface: the books did not require any of the penetrating scrutiny he employed for other kinds of reading matter. If he was given a scientific thought to analyse, his mind was alert, clear, and concise. But given a character in a novel, his responses were entirely predictable. The sillier the story the better, said his children pityingly.[74] At home, Emma would read such novels out loud at regular intervals while he lay on the sofa and idly smoked a cigarette. If he drifted off to sleep there was no difficulty in catching up with the plot. At Moor Park, he and Mrs. Lane swopped their thoughts about the latest offerings. He took a vivid interest in the plots and characters, treating them like real events and real people—traits amusingly derided by his wife and children but nevertheless showing his ability for appreciative immersion in the lives of those men and women who interested him. For this reason he liked long-running serials or family sagas in which he could worry about a heroine's future almost as if she were a daughter. He felt that looking at the end of a novel was a particularly feminine vice; and would, with a laugh, insist that his tastes put him quite beyond the literary pale.[75] "He especially enjoyed a pretty heroine," said Francis. "In this he resembled Uncle Ras who was so much influenced by the heroine that he was a very untrustworthy guide in the matter of novels."[76]

Not everything was easy reading. Darwin paid intelligent attention to authors of the day, receiving from Mudie's titles by George Eliot, Bulwer Lytton, a Trollope or two, and Charles Kingsley. He engaged with the literary works enjoyed by Emma, Erasmus, his sisters, and the Wedgwoods.

Romantic novels occupied him in other ways too. When his niece Julia Wedgwood (Fanny and Hensleigh Wedgwood's daughter, by then around twenty-five years old) wrote a novel in 1858, called *Framleigh Hall*, Darwin enthusiastically threw himself into the dual role of literary critic and supportive uncle. Julia's parents disapproved of the book. They were puzzled by her disagreeable hero and depressing plot, a sort of anti-romantic novel, if truth be told. Her sisters merely dismissed it as boring. Darwin and Erasmus were the only ones in the family to provide Julia with the encouragement for which she yearned, and she brought the manuscript with her to Down House to consult Darwin in preference to her father. Rashly she showed her father the next novel, *An Old Debt*, published in 1859 under the pen-name Florence Dawson. "Pray write something more cheerful the next time," she was told.[77] Thereafter Darwin felt protective about Julia's literary career, successfully combining the indulgence of a close relative with the intellectual attention he usually reserved for male colleagues. Julia, he was beginning to find, interested him greatly.

However, as Darwin confessed to Mrs. Lane and other understanding female friends, most of the acclaimed contemporary classics "cheated" him of his happy ending.

Good humour also usually warmed his blood sufficently to flirt gently from time to time with ladies visiting Moor Park for the waters. Not quite fifty years old, he enjoyed the company of Georgiana Craik, the twenty-seven-year-old daughter of George Craik, professor of English literature at Queen's University, Belfast. She was often taking the cure at the same time as he was. In 1857 Georgiana Craik published a sentimental novel of the kind Darwin favoured, *Riverston,* and she planned others for the future. Darwin gallantly explained natural selection to her over dinner. Miss Craik objected by asking why all the intermediary stages were not found in the fossil record, a response that he jotted down as a good example of criticism he was likely to receive. "I like Miss Craik very much, though we have some battles & differ on every subject."

Georgiana Craik never put Moor Park into any of her novels, or Darwin either. But her future sister-in-law the novelist Dinah Mulock Craik based one of her short stories on experiences clearly related to a visit to the Lanes round about this time. The relaxed, cultivated air at Moor Park, thought Dinah Mulock, encouraged romance, and she wrote a catchy little tale about a water doctor and two male patients vying for the attentions of an elegant widow.[78] It was no surprise to Moor Park readers that the charming doctor won the heroine or that sexual attraction emerged as the best medicine.

Oddly enough, in real life, Dr. Lane innocently contributed something to the same highly charged atmosphere, although he later tried to suppress all records of the incident. In 1858, a few months after Darwin first joined the Moor Park clientele, one of Lane's female patients claimed an adulterous affair with the doctor and published sensational extracts from her diary in a daily newspaper. The patient's outraged husband promptly filed for divorce, a legal action only made possible by the divorce reforms of 1857. The case scorched through the *Times* for weeks in 1858, ending only when the lady's diary of assignations was judged to be completely imaginary—an instance of hysteria.[79] Lane fought hard to save his professional reputation, embarrassed by the knowledge that prospective patients could read in the *Times* that Dr. Lane "paid great attention to all the ladies."[80] Loyally, Lane's regulars continued to support him. No less loyal, Darwin was nonetheless transfixed by the newspaper details. He met at Moor Park another purported lover who was cited in the case ("a very sensible nice young man"), and wrote to Fox in astonishment about the woman's mental state, enclosing a clipping from the *Times*. He was glad to see Lane stoically living down the ensuing scandal.

Possibly Darwin could see where some of the dangers lay. While he was at Lane's water-cure he met a Miss Mary Butler, another patient of Lane's, who charmed him with bright anecdotal chatter, piling her salt beside her plate on the dining table as he did, and delivering a lively series of unbelievable ghost stories. According to his son Francis, Darwin was always susceptible to a pretty woman with plenty to say. When he talked to a woman who pleased and amused him, the combination of raillery and deference in his manner, said Francis, was delightful to see.[81] Darwin explained natural selection to Mary Butler as well, and undertook to supply her with the autographs of famous naturalists, simply tearing off the signatures from letters when he got home. Sitting between her and Miss Craik, he protested happily that he did not know what he might not come to put in his book: "honeysuckles turning into oaks would be a mere trifle & new species springing up on every railway embankment."[82] Mary Butler's company made Darwin's days at the water-cure pass by much more lightly, full of the concentrated intimacy that shared attendance at medical institutions is likely to generate. He welcomed her attention, was flattered at being thought interesting. At least two of his visits were deliberately timed to coincide with her schedule.

In May 1859, when the *Origin of Species* manuscript was finished, Darwin shot off to Moor Park for a week of rest and hydropathy. The last chapter caused him "bad vomiting" and "great prostration of mind & body." He went again in July with the sole intention of driving species out of his head. His list of "Things for a week" indicated that he took the plan seriously—stationery, cigars, snuff, "book to read," shawls, towels, and a hair glove for the invigorating rubs Parslow gave him after the baths. He took Trollope (*The Bertrams*), Kingsley (*Yeast*), Bulwer Lytton (*The Caxtons*), and the first volume of Eliot's *Adam Bede*. He also included Dinah Mulock's romantic melodrama *Agatha's Husband*, ideal for filling the mind with inconsequential thoughts. "Entire rest & the douche & *Adam Bede* have together done me a world of good," he said on his return.

Ants did the trick too. Darwin very much wanted to discuss in the *Origin of Species* the way red ants enslaved the black ones that lived alongside them. What instinct led them to do this? Wandering along in the Moor Park conifer plantations one day, he came across a trail of red ants migrating from one nest to another about 150 yards away. Many of the ants carried cocoons, which greatly impeded their progress, and several lost their track and went off at a tangent. Others carried in their jaws captive black ants in order to restock the new nest with slaves, as he surmised.[83] After a few small experiments in adding and removing cocoons (could there be a "blundering instinct"? he asked himself), Darwin chose one specific ant, identifiable by its cargo, and decided to follow it as far as it went.

Just then, a tramp came by. For a shilling, he agreed to help. George Darwin liked to tell the tale afterwards:

> An ant was indicated & they both squatted down to watch their respective ants, and shuffled on from time to time as the ants proceeded. The place was the side of a country road. A carriage was heard approaching with the horses trotting, as it drew near the horses were slowed to a walk. My father kept telling the man "Now you mustn't look up," & so they both sat there looking intently at the ground & shuffling along alternately. My father's ant came to a bare place just as the carriage was abreast of them, & [he] glanced up for an instant, & saw a whole carriage full of people gazing at the pair intently with their mouths open with astonishment at the apparently insane proceeding.[84]

Amused by his own passions, and late for dinner, Darwin entertained his fellow water-cure patients with his account of the ant and the carriage. The torments of writing the *Origin* were not so all-devouring that he failed to see the funny side of its author.

## VIII

Yet who would publish such a book? Darwin consulted Lyell about possible publishers, making a specific trip to London for the purpose. He hoped to find a good general firm with a reliable niche in scientific affairs. Hesitantly, he asked Lyell if John Murray might be interested, the same John Murray who published all of Lyell's books and who in 1845 had issued the second edition of Darwin's *Journal of Researches*.[85] Although Lyell once referred to Murray behind his back as a "tradesman," he thought this an excellent idea.[86] He hurried round to Albermarle Street the next day, intending to pay Murray one of his most persuasive social calls.

Murray was ideal for several reasons. First, he and Darwin had enjoyed a businesslike relationship over the *Journal of Researches*, which, in Murray's sensible hands, ran to two additional reprints over the intervening years. Small royalty payments still arrived at Down House on a regular basis. Moreover, Murray was a man of parts, interested in science, especially geology and chemistry, and well accustomed to initiating shrewd publishing moves like the Home and Colonial Library, a series of edifying works for the middle classes, and the famous *Handbooks,* the first holiday guidebooks for Victorians, predating Baedekers by a few years. Murray personally supervised successive volumes on his travels through Europe.

More than this, however, Murray was rapidly becoming one of the more important scientific publishers of the Victorian era, astutely expanding the empire of his father, also called John Murray, who had once held sway over the glamorous circle of Byron, Moore, Croker, Lockhart, and Southey, and had established the firm's literary and political magazine, the *Quarterly Review.* Even while Darwin deliberated, the younger Murray was moving into science, picking up a useful line in government publications from the India Office and Kew Gardens, and arranging to publish the annual reports of the British Association for the Advancement of Science. He liked best to publish books of an instructive or improving nature, particularly travel narratives, always reliable sellers during the nineteenth century. Wiry, canny, and indefatigable, Murray was an expatriate Scot, an Anglican of low church persuasion and a staunch Tory: yet he opened his premises in Albermarle Street to authors of all shades of opinion, maintaining his father's reputation for disinterested professionalism. A bit of controversy was never bad for business.

Murray agreed to Lyell's plan. Flustered, Darwin insisted that Murray should read some of his chapters before making any formal agreement to publish. Murray might be so shocked that he would throw the manuscript in the fire at Albermarle Street as the elder John Murray did with Byron's scandalous *Memoirs.* Yet within hours, Darwin's butler, Parslow, was on the train to London bearing a brown paper package.

Even so, Darwin was agreeably forestalled by a contract arriving in the post before this reading could be arranged. Privately, Darwin may have wished that Lyell had been a little less convincing—a book on the natural origins of species was not likely to sell well, he thought ruefully, and Murray might easily suffer a loss. Murmurs of indecision emerged from Downe. "Some parts must be dry & some rather abstruse," he told Murray on 2 April 1859. "Forgive me for adding that if my book does prove a failure you will not find me of an avaricious nature." He promised to keep the length down to an economical four hundred pages. He volunteered to pay for some of the proof corrections. He wondered if the accounts might look more attractive if he purchased one hundred copies for himself at cost.

Diplomatically, the publisher steered Darwin through these conflicting feelings. Nor was Murray as completely swayed by Lyell's advocacy as Darwin feared. Murray sent the manuscript of the *Origin of Species* out to be refereed by two of his most valued friends, George Frederick Pollock, his father's former adviser, and Whitwell Elwin, the editor of the *Quarterly Review.* Although Murray probably never seriously contemplated rejecting it, the process was significant in helping him assess the book's likely commercial impact and the financial considerations that would determine the terms of his contract with Darwin.

Murray evidently read parts of the work for himself as well, for he remarked frankly to Pollock that he thought Darwin's theory was as absurd as contemplating a fruitful union between a poker and a rabbit. Pollock countered by saying that the book would be much discussed. Pollock said he admired the way "Mr. Darwin had so brilliantly surmounted the formidable obstacles which he was honest enough to put in his own path."[87]

Much more critical comments came from Murray's second reader, Whitwell Elwin. Elwin was a cultivated man who lived a dual life as a high-profile literary editor and a vicar of a rural parish in Norfolk. His tastes were eclectic, a useful talent when obliged to write articles to fill sudden gaps in the *Quarterly Review,* and they enabled him to mix easily with scientific and literary authors alike. He was only forty-three when he read the *Origin of Species* manuscript, younger and more a man of the world than Darwin, and, as an ordained priest, well attuned to the theological movements of the day. At that time he was working on a refutation of Baden Powell's book *The Order of Nature,* in which Powell attacked the idea of miracles, while also abridging David Livingstone's *Missionary Travels* for a popular edition. His verdict had always been essential to Murray, his approval vital.

Elwin did not approve. The vicar in him responded, not the cosmopolitan literary Londoner. Darwin's text was a "wild and foolish piece of imagination."[88] Everything about it distressed him. "At every page I was tantalized by the absence of proofs. . . . It is to ask the jury for a verdict without putting the witnesses into the box. . . . For an outline it is too much, & for a thorough discussion of the question it is not near enough."[89]

Elwin recommended that Darwin should confine himself to pigeons. If the book was about pigeons, he said, it would be reviewed in every journal in the kingdom. Such a book would be a "delightful commencement." All the rest should be abandoned.

Darwin snorted in disbelief. "I have done my best," he informed Murray decisively. "Others might, I have no doubt, done the job better [*sic*], if they had my materials; but that is no help." Luckily, Murray agreed. He put Elwin's opinion to one side, and accepted the manuscript. Darwin told Murray to expect the first six chapters during the following week. Both of them had gone too far to be halted by a few words of opprobrium.

## IX

The summer of 1859 passed in a blaze of proof-reading. All Darwin's doubts about his writing style returned with a vengeance. "There seems to be a sort of fatality in my mind leading me to put at first my statement and proposition in a wrong or awkward form," he reflected afterwards.[90] He blackened the galleys with corrections, inserted new information wholesale, and rewrote entire paragraphs. "On my life no nigger with lash over him could have worked harder at clearness than I have."

Emma helped whenever she could. She read the *Origin* in full during the proof stage and loyally tried to help her husband convey his thoughts accurately to readers. There is no evidence that Emma tried to censor his text. On the contrary, the two of them discussed awkward sentences in the evenings, until they found a form that captured what he was really trying to say.

Lyell studied the proofs after travelling round the continent on his summer holidays. And Georgina Tollet, a longstanding friend of Emma's from Staffordshire, came to stay and worked them over for style. Darwin barely mentioned this feminine assistance, no doubt considering it a private affair through which he could ensure that his grammar and spelling were adequate; "this lady being excellent judge of style is going to look out for errors for me," he told Murray. Nevertheless, she helped him tighten his argument. "Georgina Tollet came last Sat.," Emma said to William. "She reads a great deal of Papa's MS & is very useful to him in making him explain things that are not quite clear."[91]

Hidden female assistance like this was commonplace for the period. Though Darwin regularly asked friends like Hooker and Lyell for advice on inaccuracies, tone, and style, he never requested male colleagues to dedicate themselves to routine editorial tasks. At that time, too, manuscripts were hardly ever corrected by staff at a publishing house. Editing was mostly carried out at home by the authors' wives, sisters, daughters, and nieces—by a roomful of readily available household experts. This feminine, home-based input has only recently been recognised as a significant factor in Victorian publishing history.

Certainly Darwin knew he needed assistance. "I find the style incredibly bad, & most difficult to make clear & smooth. . . . How I could have written so badly is quite inconceivable, but I suppose it was owing to my whole attention being fixed on general line of argument, & not on details."[92] Georgina Tollet earned her eventual gift of a silver vase. "One lady who has read all my M.S. has found only 2 or 3 obscure sentences," Darwin proudly announced. He was annoyed to hear from Kew that Mrs. Hooker thought one chapter of the *Origin* very unclear in places.[93]

There were all the customary setbacks and crises of authorship to deal with as well. Hooker inadvertently destroyed part of Darwin's chapter on geographical distribution, absent-mindedly putting the handwritten manuscript in the drawer at home reserved for his children's drawing paper. The pages were irretrievably scribbled over by the time Hooker remembered. "I have the old M.S.," Darwin said after this confession. "Otherwise the loss would have killed me!" Hooker could only groan to Huxley, "I feel brutified, if not brutalised, for poor D. is so bad that he could hardly get steam up to finish what he did. How I wish he could stamp and fume at me—instead of taking it so good-humouredly as he will."[94] Hooker compensated by doing an exceptionally thorough job in commenting on the remaining pages, while Frances Hooker performed valiant feats on the English. But Huxley's unsympathetic cackle came straight from Victorian nursery rhyme. "What do you say to standing on your head in the Gardens for one hour per diem for the next week?"[95] The *Origin of Species* was much more of a collaborative effort than has ever been suspected.

Interruptions there were. There was the summer cricket match at Down House to organise, an annual contest between gentlemen and players, loosely interpreted as the big houses against the villagers (married against single, in later years), and played on the large home meadow owned by the Darwins. Once again Parslow showed his versatility as a scientist's butler by manoeuvring his young gentlemen into an unassailable position. George Darwin, then aged fourteen, and some of the teenaged Lubbocks were useful batsmen. Erasmus came down from London for the social side of the occasion. "Our cricket match went off brilliantly & Mr Parslow's party beat Mr Reeves' all to nothing," Emma told William after the match. "Aunt Susan, Uncle Ras & Uncle Hensleigh came on Sat but unluckily your father had an attack yesterday & was in bed most of the day, he is tolerable now but is going back to Moor Park tomorrow, as he has been failing some little time."[96]

Drama over the children's governesses added to his tension. Ineffective, melancholy Miss Pugh left the household early in 1859. Henrietta Darwin, aged sixteen, recounted how she would sometimes sit at meals with tears running down her face, and even Emma noticed her eccentricities wearing the family down. Emma located another placement for Miss Pugh (with the help of the Hookers) and promised her an annual holiday at Darwin's expense.

Then Madame Grut arrived, fresh from Switzerland, ready to teach French and German conversation. Henrietta considered her unbearable and made no pretence of hiding her feelings. In the end, Darwin sacked her—an event he disliked at the best of times, and on this occasion fraught with other grievances. Henrietta viewed the departure as a personal triumph. "Solemn events have happened," she informed her older brother William.

> This is how it came about. On Monday at breakfast Mama said very civilly that she wanted some alteration in Horace's lessons. Mrs Grut was evidently miffed at that, & then I said I thought *s'eloigner* wasn't to ramble, very mildly, & that miffed her again & she made some rude speech or other "Oh very well if I knew better than the dictionary" . . . Nothing more came of it then & all went smooth till I went up to my German lesson in the evening. When I came in I saw there was the devil in her face, well she scolded the children a bit & then sat down by me, when I showed her my lesson (a bit of very bad French) she said, if I knew better than she did it was no use her teaching me & so on & so on, till it came to a crisis, & she worked herself up into a regular rage. . . . I left the room then, & went downstairs to tell my injuries. When Papa & Mama heard all about it they settled she shd go at once, so Papa wrote her a letter telling her she shd have her £33 & nothing more. . . . then Papa was to go upstairs & deliver the letter. . . . Papa got *such* a torrent, telling him he was no gentleman, & white with passion all the time, wanting to know what she had done, what he had to accuse her of—telling him he was in a passion—she would give him time to think. . . . We had a very flustered tea, & all evening we sat preparing for the worst, what we shd do if she refused to go out of the house etc. However she did turn out much milder & sent us a letter to say she wd go on Wednesday.[97]

After Madame Grut's departure, Miss Pugh returned for a few weeks and then the invaluable Miss Thorleys, Catherine and Emily, came in sequence as temporary replacements. Eventually a Miss Latter was hired, whom Darwin called "a very clever lady."[98] She fitted in well enough to last until 1860, when she moved on to become a mistress in a school in the neighbourhood.

A few paternal pleasures eased in with the late-summer sunshine. Darwin was pleased by the boyish fervour with which Francis, Leonard, and Horace (aged eleven, nine, and eight) took up beetle collecting, and told Fox how his blood boiled with the old ardour when the boys captured a small but unusual ground beetle, *Licinus silphoides,* in the garden—"a prize unknown to me." Francis remembered Darwin's enthusiasm for a long time afterwards. "I have a vivid recollection of the pleasure of turning out my bottle of dead beetles for my father to name, and the excitement, in which he fully shared, when any of them proved to be uncommon ones."[99] Remembering his own thrill at seeing his name in print at an early age, Darwin wrote a short letter about the beetles to the *Entomologist's Weekly Intelligencer,* a cheap news-sheet, as if it were from the boys. "We three very young collectors," it began and was signed Francis, Leonard, and Horace Darwin. It announced the capture of *Licinus silphoides.*[100]

## X

Towards the end, the constant pressure of correcting proofs frayed his nerves. Perhaps the book was coming too close to publication for comfort. Modesty and ambition fought for the upper hand, reticence and impatience rubbed uncomfortably together. He did not court publicity, he told Fox. Fame was not what drove him. "If I know myself, I work from a sort of instinct to try to make out truth." Yet this was only part of the story. Any truth that lay at the bottom of his theories would be a barren kind of truth, he thought, if it remained unknown to the scientific public. For this reason he remained concerned that the forthcoming book would be only an abstract, deprived of the evidence that he regarded as a necessary part of his argument, stripped of all his properly cautious scientific caveats—and he promised himself that he would eventually publish the whole of his long manuscript.

Nor was his customary desire for privacy an easy companion for the self-promotion and exposure the forthcoming book would undoubtedly require. "I am becoming as weak as a child," he groaned to Hooker, "miserably unwell & shattered." Vomiting started again. "I have been so wearied & exhausted of late," he complained in September 1859. "I have for months doubted whether I have not been throwing away time & labour for nothing." He was in an absorbed, slavish, overworked state, he told Fox in another letter. "My abominable volume . . . has cost me so much labour that I almost hate it."

Supportively, Darwin's friends rallied round. Lyell went out of his way to encourage and assist. It was not yet clear to Darwin whether Lyell would accept the concept of natural selection, for Lyell was genuinely worried by the spiritual consequences of what Darwin was proposing. He was finding it hard to go "the whole orang," as he sighed to Huxley. Yet evolutionists "cannot be pooh-poohed & ought not to be so."[101] Darwin and Lyell exchanged many letters during this period. "I am foolishly anxious for your verdict," Darwin admitted. "I regard your verdict as far more important in my own eyes & I believe in eyes of world than of any other dozen men [*sic*]."

Lyell responded in the way he knew best. He praised Darwin's forthcoming book at a public lecture during the British Association meeting in September 1859, not only giving the book advance publicity but also indicating that he would stand by Darwin in the face of the world. This meeting was in Aberdeen, and Lyell had been invited to meet the prince consort, the honorary president of the British Association that year, during it. Mary Lyell breathlessly recorded every detail of their lunch at Balmoral, naturally placing him at the centre of events, although Queen Victoria remembered it rather differently, noting in her diary that the day was burdened with "four weighty omnibuses laden with philosophers & savants."[102]

In his lecture Lyell talked about the striking new evidence about the early history of mankind that was emerging from the silts and gravel beds of northern France. Two English naturalists, Hugh Falconer and Joseph Prestwich, were among the first to uncover Stone Age tools—worked flints—that seemed to push the first appearance of human beings back to a time contemporaneous with extinct fossil mammals. Such an early start for humans created many puzzles, not least in practical geological terms. Yet excavations in Abbeville and other localities in France were revealing examples of these apparent tools embedded in the same deposits as the bones of extinct animals. Lyell summarised the results for his British Association audience, which also provided him with a suitable opening for a few words on Darwin. The forthcoming book, Lyell declared, would throw "a flood of light on many classes of phenomena connected with the affinities, geographical distribution, and geological succession of organic beings, for which no other hypothesis has been able, or has even attempted to account."[103]

For a pre-publication salvo this could hardly be bettered. It looked to Darwin as if Lyell would go with him a good part of the way. "I do thank you for your euloge at Aberdeen," he said appreciatively. "Now I care not what the universal world says. . . . You would laugh if you knew how often I have read your paragraph, & it has acted like a little dram."[104]

Eventually, the page proofs came to an end. Darwin added two longish quotations as epigraphs to the volume, one audaciously taken out of context from William Whewell's treatise on astronomy, suggesting that God worked through general scientific laws rather than through direct intervention, and another from Francis Bacon on the nobility of the search for knowledge. No man could search too far, he transcribed from Bacon feelingly, and dated it "Down, Bromley, Kent, October 1st, 1859."[105]

At the last minute he adjusted the title according to Murray's recommendation. Darwin's first suggestion was rather too complicated: "An Abstract of an Essay on the Origin of Species and Varieties Through Natural Selection." Common sense surely suggested to Murray that the words "abstract," "essay," and "varieties" should go, and that "natural selection," a term with which Murray thought the public would not be familiar, ought to be explained. The agreed-upon title was, however, hardly less cumbersome—*On the Origin of Species by Means of Natural Selection, or the Preservation of Favoured Races in the Struggle for Life.*

To this title Darwin attached his name—an obvious but crucial step. By putting his name to his theory, he categorically distanced his book from the anonymous *Vestiges,* still the most well-known evolutionary tract, and, if necessary, from the flourishing genre of scurrilous, nameless political pamphlets that attacked the Victorian state. Darwin was no radical pamphleteer hiding behind a veil of anonymity. His whole style of scientific endeavour was invested in properly acknowledged authorship. His name was one that could be published and trusted, and his book was intimately bound up with its author's identity. On the title page, a list of his public credentials followed—his M.A. from Cambridge University and membership of learned societies. Wealth, class, education, the respect of his peers, and a recognisable code of gentlemanly conduct were here fused with his views and with himself. In so doing, he symbolically took full responsibility for his work.

On 1 October 1859 he recorded in his diary, "Finished proofs," and calculated that the writing process had taken thirteen months and ten days to complete. On 2 October he was off. Exhausted and sickly, he made his way to a water-cure establishment in Ilkley, at the foot of the Yorkshire moors. Edward Lane and Moor Park were not doing him so much good as usual. This time he decided he needed dramatic intervention. "I am worn out & must have rest. . . . Hydropathy & rest—perhaps that will make a man of me."[106]

"Is there any chance of your being at Ilkley in beginning of October?" he wrote tentatively to Mary Butler. "It would be rather terrible to go into the great place & not know a soul. If you were there I should feel safe & home-like."[107]

*chapter*

3

# PUBLISH AND BE DAMNED

O*N THE ORIGIN OF SPECIES* was published in London on 24 November 1859, while Darwin was taking the water-cure at Ilkley. It was a very ordinary-looking volume bound in sturdy green cloth, 502 pages long, and somewhat expensively priced at fourteen shillings, not nearly as gaily decked out as Murray's red-and-gilt version of Darwin's earlier *Journal of Researches* and nothing like the pocket-sized duodecimo Darwin had at first proposed.[1]

The author's serious intent was obvious. There were no eye-catching natural history illustrations, no pedigree fatstock emblazoned in gilt on the cover, not even a frontispiece of an evocative prehistoric scene as there might be today in a book about evolution. For a volume that described the teeming fecundity of life on earth, the pages were curiously devoid of living beings. But it was a fair specimen of nineteenth-century typography, well printed on decent paper, and serviceably bound. The book's unassuming demeanour suited its author perfectly. "I am *infinitely* pleased & proud at the appearance of my child," Darwin told Murray when his advance copy arrived in Yorkshire. "I am so glad that you were so good as to undertake the publication of my book."[2]

Unassuming or no, this book transformed his life. Of course, he expected controversy, although even in his gloomiest moments he could not have begun to imagine the convulsions of public opinion, praise, and denigration that would follow. From the start, he was prepared to go to any lengths to give his theory the best support that he could provide.

But there was more than this. That November he chose the kind of man he wanted to be—he chose to dedicate himself to his book, to placing his views as fully as he could before audiences that he as yet hardly envisaged, prepared to influence and urge to a degree that would become second nature to him, displaying a deepening of purpose and strength of character that he rarely acknowledged even in his most private correspondence and yet that marked the rest of his days. His active intervention in the post-publication process was hidden but intense. Paradoxically, the intimate process of writing personal letters, one individual speaking to another, became an integral part of his public voice, an activity that could be just as shrewd and tactical—even predatory—as any polemic dreamed up by Huxley.[3] Without moving out of his home, Darwin came to dominate through letters. Promoting the finished book became the directing theme of the life to come as completely as his earlier years had been governed by constructing the theory.

He deliberately set about persuading people to consider his point of view. The reception of his book, as he understood it, would be a social process, depending at first on the force of his arguments and then progressively intertwining with the reactions and support of his friends, the networks of evaluation and accreditation pervading literary London, and his own influence on these. What he did not know then was the way in which it would move out from the elite audience he primarily intended to address at home and abroad to increasingly diverse sections of the reading public. His words would spread through journals, newspapers, public lectures, controversial tracts, and freethinking magazines at the same time as great cultural shifts became manifest—shifts in the status of science, in religious belief, in the impact of publishing, education, and social mobility. The story was not a straightforward triumphant advance, nor was it predictable. Darwin could perhaps only dimly foresee that the rest of his life would be given to sustaining this one demanding publication.

Such circumstances would also change him from a quietly methodical naturalist, content to influence if possible only his own small circle, into a brilliant and subtle strategist. During the months following publication, he developed a style of personal presentation that materially advanced his cause. He showed himself constantly alive to the situations in which his readers and correspondents might find themselves, sensitive to embedded religious and cultural beliefs, quick to applaud any faltering steps in his direction, patient in explaining the same points over and over, resourceful and diplomatic, learning to use his words to further his aims, sometimes with craft and flattery, and often with genuine warmth. He realised that responses to his book would be affected by a whole range of factors and recognised that the dissemination and judgement of any new scientific argument relied in part on the ways in which particular groups of individuals organised the intellectual tasks on which they were engaged. So he refined his own techniques to pinpoint those who should be stirred into action, the most useful journals to approach, the best reviewers to court, the fiercest critics to disarm. He became skilled at marshalling his friends into an effective army. Much of this recruiting process was already familiar to him through his focus on collecting information by correspondence. Nevertheless his aims were different now, given fresh significance precisely because of the book's importance to him. Overnight, Darwin's focus shifted from private effort to public persuasion.

With the advance copy of the *Origin of Species* safely in his hands at Ilkley, he swung into the first of what would become many campaigns of action. He intended sending presentation copies to anyone who might be a significant help to him, and had already ordered eighty books from Murray beyond the twelve free copies due to him as author. He drew up a list of recipients and arranged that these volumes should be distributed direct from Murray's office in London during the closing weeks of November 1859. The high cost of this scheme made him flinch a little. But publication was his symbolic point of no return. Each copy was inscribed by Murray's clerk on the flyleaf "With the author's compliments" or "From the author." For this reason personally signed copies of first-day editions are never seen.[4]

Over the next two weeks, in between unpleasantly therapeutic cold baths, Darwin wrote letter after letter, each one delicately tailored to its recipient and aiming to defuse the worst of the anticipated criticisms. Few Victorian spa towns can have seen such a systematic publicity operation rolling out from its watery doors. Even his oldest friends were enticingly invited to agree with his proposals. "I shall be curious to hear what you think of it," he told Fox, "but I am not so silly as to expect to convert you." Suitably self-deprecating letters went to Wallace ("I hope there will be some little new to you, but I fear not much"), to Huxley ("I know there will be much in it which you will object to"), to Leonard Jenyns ("I know perfectly well that you will not at all agree with the lengths which I go"), to Thomas Eyton in Shropshire ("My book will horrify & disgust you"), to Bunbury ("If you are at all staggered I shall be quite interested"), to his old university teacher John Henslow ("I fear, however, that you will not approve of your pupil in this case"), to Adam Sedgwick ("You might think that I send my volume to you out of a spirit of bravado and with a want of respect, but I assure you that I am actuated by quite opposite feelings"), to Hugh Falconer ("Lord how savage you will be . . . how you will long to crucify me alive!"), and to young John Lubbock at High Elms ("I daresay when thunder and lightning were first found to be due to secondary causes, some regretted to give up the idea that each flash was caused by the direct hand of God").

Tired, and no doubt disarmed by so much apology, he let his guard drop. He sent the most honest of all to Asa Gray.

> Let me add I fully admit that there are very many difficulties not satisfactorily explained by my theory of descent with modification, but I cannot possibly believe that a false theory would explain so many classes of facts as I think it certainly does explain. On these grounds I drop my anchor, and believe that the difficulties will slowly disappear.[5]

Digging conscientiously around in his past for long-lost allies, he sent a copy to John Maurice Herbert, once a jaunty Cambridge University student and fellow member of the self-styled Glutton Club, now a senior judge on the county circuit. And of course there were copies for the family, Wedgwoods and Darwins alike; for friends and neighbours in Downe village like John Brodie Innes and George Norman; for his old *Beagle* captain Robert FitzRoy, moored to a desk in Whitehall; and for the two medical men closest to him, Edward Lane and Sir Henry Holland. Darwin looked forward to hearing what Holland thought. He was a clever man, much respected in the family circle, and widely influential elsewhere. It would be unforgivable if Darwin failed to send him a copy. Pleasant memories demanded the same for Edward Lane. It appears that he also showed his advance copy to Mary Butler while they were both taking the waters.

As for the rest, this was his chance to write directly to the men who would shortly be judging the *Origin of Species* in reviews and articles, the men of the day who would discuss his ideas in London clubs and at dinner parties. He knew that any serious contribution to intellectual thought largely depended on this form of elite, verbal assessment. Indeed, he saw it was a key feature of the Victorian publishing process. For many years now Darwin had participated in exactly the same buzz of knowledgeable chat and correspondence. Through a well-established cycle of discussion and authentication, concentric networks of specialists usually talked things over and came to a verdict.[6] The same members of the intelligentsia that had previously engaged with *Vestiges* and with Buckle's and Spencer's developmental laws would at least consider Darwin's views thoughtfully, although they might not agree with them. He planned to activate "every scrap of influence" he had gathered over the years.

He therefore sent presentation copies to many of the most interesting, modern-minded men in Britain, some of whom he scarcely knew, such as Charles Kingsley and Herbert Spencer, while also judiciously covering the handful of celebrities with whom he was acquainted, such as Sir John Herschel and Lord Stanhope, the historian. To these he added distinguished foreign naturalists and as many leading biologists as he could identify, each of whom he hoped would see his work as a real contribution to the intellectual problems they faced in natural science. Were there "any good & *speculative* foreigners to whom it would be worth while to send copies?" he inquired of Huxley. "If you write to Von Baer," he continued, "for heaven's sake tell him that we should think one nod of approbation on our side, of the greatest value; and if he does write anything, beg him to send us a copy, for I would try and get it translated and published in the Athenaeum and in 'Silliman' [Benjamin Silliman, editor of *American Journal of Science and Arts*] to touch up Agassiz."[7] In the end, his list of eighty or so names embraced most of the major geologists, naturalists, and biologists in the world, as well as individuals based in all the main natural history institutions in Europe, North America, and across the British empire, reflecting a high geographical and international spread, including Henri Milne-Edwards, Louis Agassiz, James Dwight Dana, Joachim Barrande, Johannes Steenstrup, François Pictet, Isidore Geoffroy Saint-Hilaire, Julius Carus, Richard Owen, John Phillips, Carl von Siebold, Jean Louis Quatrefages de Breau, and elderly Heinrich Bronn.[8] Even though these naturalists were unlikely to accept Darwin's proposals, they would be drawn into the resulting debate. A man who is given a copy of a book by its author, accompanied by a charming letter, finds it that much harder to attack or denigrate.

In passing, Darwin's list of names reflected other social factors. Hardly any unknown people were listed to receive presentation copies. Apart from sending the *Origin of Species* to William Tegetmeier, his trusted pigeon friend, and to one or two noted breeders and practical men who had helped him, Darwin signally failed to distribute copies to any of the scores of people drawn from the vigorous subcultures of natural history, horticulture, and agriculture in Britain, and elsewhere, who had supplied relevant information. Everyone who received the book as a gift was someone of influence or a close personal friend. In this, Darwin's immediate concern was to place his views in front of those judges who mattered most. Still, the omission made his ulterior purpose manifest: he did not bother with people who could not make a difference. Tellingly, Darwin did not send a copy to Robert Grant, the man who first introduced him to the idea of evolution in Edinburgh. He did not send one to Harriet Martineau, the committed Malthusian author and intimate friend of his brother's. Darwin evidently weighed each recipient's influence in the balance and discarded those he considered least important to his cause. Several of these omissions were rectified when the second edition came out.

At Murray's end, business steamed ahead. The publisher sent nearly forty copies out to review journals. He further gave a few copies of his own to friends such as Elwin, of the *Quarterly Review,* and perhaps Michael Faraday.[9] Here and there, other pre-publication copies slipped beyond Darwin's control. Erasmus Darwin sent one to Harriet Martineau, who now lived as a reclusive invalid in Ambleside in the Lake District. He may have known that Martineau was not on Darwin's list at Murray's.

Before publication, Darwin was broadly able to shape the initial distribution of his work, and, as he hoped, at least something of the initial response. All this changed as publication day loomed. With four days to go, an adverse blast in the *Athenaeum* rocked his composure. The anonymous reviewer (John Leifchild) scorched through his advance copy, highlighting its dreadful implications: "If a monkey has become a man—what may not a man become?" The reviewer declared that Darwin's book was almost too dangerous to read. It should be put in the safe hands of theologians and left "to the mercies of the Divinity Hall, the College, the Lecture Room and the Museum."[10]

Immobilised in wet sheets at Ilkley, his face covered with eczema, Darwin was furious. He saw that all his cautious circumlocutions were blatantly ignored by the *Athenaeum*'s reviewer. "The manner in which he drags in immortality, & sets the Priests at me & leaves me to their mercies, is base," he raged. "He would on no account burn me; but he will get the wood ready & tell the black beasts how to catch me."[11] It seemed to him that the *Athenaeum* was hell-bent on sensationalising the topic even before other people could read the text. Caught by surprise by the violence of the review, and by the strength of his own reaction, Darwin fired off bitter complaints by letter to his most sympathetic anti-clerical friends. All of a sudden, he hated being at Ilkley. His health worsened, he sprained an ankle, his face was as bad as he ever remembered, and his temper short. It was "odious," he moaned.

That first review rankled and rankled. For years he carried a grudge against the anonymous reviewer, whom he wrongly identified as Samuel Woodward on the basis of something that Hooker let slip. For years, too, the unfortunate Woodward did not understand why Darwin had suddenly turned so frosty. Long afterwards, Darwin still felt raw, and he never could bring himself to acknowledge that perhaps he was over-reacting and that the first unfriendly review invariably hurts the most. From then on, he regarded the *Athenaeum* with suspicion.

Yet as Murray predicted, it made no difference in the end. While Darwin wrote his letters and grumbled about the Yorkshire rain, Murray organised the customary "sale dinner," one of the signposts of the publishing year when agents and bookshops placed bulk orders for items on his forthcoming list. "All the principal booksellers were invited. The new books of the season were introduced to them and offered on specially favourable terms."[12] Murray's ledgers show that he took orders for the *Origin of Species* at this sale, held on 22 November 1859 at the Albion Hotel in central London. Of the thirty books about to come out under Murray's imprint, Darwin's vied for top billing with two other future classics of the era, Samuel Smiles's *Self-Help,* the homely bible for social betterment, and Leopold McClintock's *Narrative of the Discovery of the Fate of Sir John Franklin,* the final denouement in the tragic saga of Franklin's quest for a northwest passage. McClintock's heroic tale promised to be the story of the season: Murray hoped for orders for more than seven thousand copies. Samuel Smiles's *Self-Help* was likely to be equally successful in matching the pulse of the nation. No other nineteenth-century publishing firm ever issued three such books on the same day—Darwin, McClintock, and Smiles—books that so accurately represented the Victorian frame of mind.

The only serious competition that Murray forecast was Dickens's *A Tale of Two Cities,* a surefire hit from Chapman & Hall. Other notable books already in print from rival firms were John Stuart Mill's *On Liberty,* issued earlier in 1859; Tennyson's *Idylls of the King,* clearly a market leader; George Eliot's *Adam Bede;* and *The Virginians* by Thackeray. Confident in his author's selling power, Murray increased the print run of the *Origin of Species* to 1,250.[13]

At the trade sale Murray took orders for 1,500 copies, some 250 more than the actual number printed. This satisfying statistic gave rise to his remark that the book "sold out" on the day of publication although the overall number scarcely matched the many thousands of copies of novels sold by Chapman & Hall or Richard Bentley in a single year, or the 60,000 copies of the solidly religious Bridgewater Treatises that had accumulated on the nation's shelves by 1860.[14]

The most striking thing about Murray's sale was not so much that Darwin sold out—common enough if the market was judged correctly—but that five hundred went to Mudie's Circulating Library. Such a high figure was gratifying for Murray, although also modest enough when compared with Mudie's purchase of 3,520 copies of David Livingstone's *Travels* in 1857 or his order for 3,000 copies of McClintock's *Narrative* on the same day as Darwin was up for bids. Publishers, authors, and readers all recognised the might of the subscription libraries.

Of these Mudie's was the mightiest.[15] His purchase guaranteed Darwin a broad audience, an audience moreover that would repeat and repeat while each copy of the book was lent to a number of subscribers in turn. Although Mudie undoubtedly banked on Darwin's previous attractions as a natural history writer rather than wanting to thrust a subversive tract into God-fearing homes across the country, no one needed a soothsayer to explain that the lending library would propel the *Origin of Species* into a much more popular general category than its subject matter initially suggested. Five hundred copies of the *Origin of Species* might well be read by some two thousand subscribers. Unfortunately the actual circulation figures are unknown for all Mudie's volumes.

Somewhat unexpectedly, too, given the prudish reputation that surrounded Mudie's fiction list, adding the *Origin of Species* to the Oxford Street depot was not too much of a gamble. As a reader himself, Mudie was greatly interested in contemporary American transcendental philosophy and in the nature of the human mind, a private passion reflected in the content of his nonfiction catalogues, which usually offered a selection of evolutionary and early psychological texts.[16] Like Murray, he moreover appreciated the commercial benefits of controversy. Mudie's Library, hitherto much underestimated in the popularisation of science in general, was to become an important force in the dissemination of evolutionary views for the next twenty or thirty years.[17] The firm's quarterly catalogues of books from 1865 or so were to offer most of the more well-known post-Darwinian titles, including all the new editions of the *Origin of Species* and *Vestiges,* and books by Huxley, Lubbock, Tylor, and Lyell, as well as an extensive range of criticism and alternatives. Mudie's admiration for Darwin as the centrepiece of a publishing phenomenon ran sufficiently high for him at a later date to invite the naturalist to dine (Darwin declined).[18] From the start, however, Darwin acknowledged the effect that Mudie had on enlarging his readership. With this purchase, Mudie made it possible for Darwin's book to reach a far wider public than either author or publisher had contemplated.

On the day after the sale, Murray relayed the good news to Ilkley that the print run was exhausted.[19] Another surprise followed. Instead of merely reprinting from the set-up type, Murray said he could admit a few corrections if Darwin wished to send them, and could therefore issue the book as a second edition (technically, it ought to have been called a corrected reprint). Murray wanted to fulfil the orders as soon as possible and asked Darwin to send him any changes immediately.

Darwin was deeply pleased. So was Emma, the patient ghost behind his never-ending struggle for perfection. "It is a wonderful thing the whole edition selling off at once & Mudie taking 500 copies," she remarked to William with understandable satisfaction. "Your father says he shall never think small beer of himself again & that candidly he does think it very well written."

On 1 December 1859, Murray started setting up the slips for the new edition. On 3 December, Mudie's Select Library advertised that the *Origin of Species* was available to be borrowed.

## II

Nearly two hundred letters and a torrent of reviews passed through the author's hands during the next six months. These months were among the most demanding he had ever experienced.

The expected storm of controversy did not crack open immediately. On the contrary, the first effects were unexpectedly personal, for publication brought with it a spontaneous alteration in the way his closest friends regarded him. Darwin's colleagues wrote to him straight away, full of a new kind of admiration. For a fraction of a second there was just the slightest hint of incredulity—incredulity that the "dear old Darwin" of their correspondence possessed such hidden powers of genius. Even though each of them already understood the bare bones of Darwin's theory—several had read parts of the volume before publication, and two of them were actively involved in putting the theory before the public at the Linnean Society—it seems that none quite anticipated the final majesty of his vision. One by one, they expressed delighted surprise at the perspectives he revealed. One by one, they offered intellectual respect and started to align themselves behind him.

Lyell fairly bristled with affectionate pride. Whatever misgivings he may have continued to feel in private, he believed that Darwin's ideas were too important to remain secret. He was truly pleased to see the book published. "Right glad I am that I did my best with Hooker to persuade you to publish it without waiting for a time which probably could never have arrived tho' you lived till the age of 100," he gloated. It was a grand work, he said. "A splendid case of close reasoning & long sustained argument throughout so many pages."[20]

Lyell was much inclined to dissect the finer points of detail, and these conversations were vital in helping Darwin clarify the text of the forthcoming reprint and the many subsequent editions. Kindly and proprietorial, Lyell looked through all the reviews and correspondence, making notes in his scientific journal about the problems that bothered him most and talking them over with Darwin and other colleagues.[21] Darwin's book in fact marked a real turning point in Lyell's mental existence. For so long Lyell had been shackled to revising edition after edition of his geological textbooks, and he badly needed something different to fire his interest. The *Origin of Species* raised his temperature again. Suddenly he found his mind racing with inquiries and tricky problems to resolve. He bombarded Darwin with letters, earnestly seeking answers to a whole range of difficult questions.

Symbolically, too, the occasion was charged with meaning. Like many notable couples in science, like Freud and Jung, or Manson and Ross, master and disciple silently began changing places. Where Lyell's *Principles of Geology* had presented the young Darwin with the gift of scientific insight and showed him how to think about the natural world, Darwin's *Origin of Species* now offered Lyell fresh intellectual purpose. In a way, these books embodied the creative stimulus that each man exerted on the other. Nearly three decades after Lyell first shaped Darwin's intellectual horizons, the *Origin of Species* in turn opened up for Lyell the most stimulating areas of thought that he encountered in his later years. As such, their relative positions shifted. The torch was passed from one to the other, and accepted, a heartfelt act of homage between them. Sensitively—almost tenderly—they recognised each other's central place in their intellectual lives.

Daringly, Lyell decided to rewrite his *Manual of Geology* to accommodate the new evolutionary view and joked about his and Darwin's mutual descent from tadpoles. Dean Milman, Lyell reported cheerily in the first few weeks after publication, said that the writing of the *Origin of Species* was in itself enough to refute the possibility of such a froggy ancestry.[22] In the event, Lyell's proposed pages on natural selection never made it into the *Manual,* although it turned up in different form in his subsequent book on the antiquity of mankind. Darwin's gratitude was rapid and real. "I fully believe that I owe the comfort of the next few years of my life to your generous support & that of a very few others: I do not think I am brave enough to have stood being odious without support. Now I feel as bold as a Lion."[23]

Other friends saw the great originality and power in Darwin's scheme without necessarily conceding the point. The palaeontologist Hugh Falconer read his presentation copy as eagerly as the rest, yet was implacably opposed to its overall argument: "I have been dying like all the world besides to see your book upon species . . . the wicked book which you have been so long a-hatching," he wrote to Darwin. Falconer never accepted evolution. But he never considered giving up his friendship with Darwin either; and he vigorously defended Darwin's right to be heard. He was no more shocked by the *Origin of Species* than Hooker, who labelled it "glorious." Falconer ended up regarding natural selection with genial amusement. When he subsequently offered Darwin a living specimen of an unusual lizard, he could not resist the opportunity to tease. "In your hands it will thrive—and have a fair chance of being developed without delay into some kind of the Columbidae, say a Pouter or a Tumbler."[24]

Hooker willingly accepted the new view of nature. He had lived with it in letters for so long that he was not likely to turn his back on Darwin now. "What a mass of close reasoning on curious facts and fresh phenomena. . . . I see I shall have much to talk over with you," he said. "I expect to think I would rather be author of your book than of any other on natural history science."[25]

"I really think it is the most interesting book I ever read," wrote Erasmus, echoing the chorus of friendly approval. Harriet Martineau was ecstatic. "One might say 'thank you' all one's life without giving any idea of one's sense of obligation. . . . we must all be glad that he has set the world on this great new track," she wrote to Fanny Wedgwood.[26] And shortly afterwards, Darwin heard from Huxley, a wonderful whoop of praise and pleasure. It was a noble book, Huxley cried; no book had made such an impression on him since he had read Karl von Baer's embryology nine years before, a high compliment from the man who translated von Baer into English and whose embryological understanding was firmly based on the German's doctrines. As Huxley remembered it, the beauty of Darwin's theory flashed on him like lightning showing the way home in a storm. "How extremely stupid not to have thought of that!"[27] He was struck by the theory's elegance and economy.

Huxley's praise was evidently sincere. For more than a decade he had longed to find order in nature where apparent chaos reigned. Darwin's naturalistic proposals provided just the kind of intellectual synthesis and challenge he most enjoyed, and his friend's approach and aims chimed impressively with his own. Like Lyell's, his viewpoint materially changed. "I am prepared to go the stake if requisite in support of Chap. IX," he declared, referring to the chapter where Darwin wrote of the imperfection of the fossil record: "Depend upon it you have earned the lasting gratitude of all thoughtful men." Yet he was not prepared to accept all of the book uncritically. There were points on which he disagreed or reserved his judgement. He did not like Darwin's insistence on infinitesimal, gradual change, feeling that Darwin gave himself an unnecessary burden by not allowing even the smallest jump in the fossil record, and promoted the point endlessly in future correspondence. He was not sure what to make of the three central chapters on variation, instinct, and hybridism, either, issues that emerged again and again in talks with Darwin and others. He never accepted Darwin's proposals about the way sterility between previously fertile individuals might emerge in order to make a self-contained breeding population. In these hesitations, Huxley revealed the roots of important criticisms he was to develop much more strongly over the next few years. Even during his first headlong lunge, he emerged as a disciple who did not fully accept the doctrine in all its parts. It was a double-edged sword that he handled with flamboyant panache. He referred to the theory as a "working hypothesis," a phrase that satisfied him rather more than it did Darwin.

He was certain of one thing. Darwin would encounter "considerable abuse & misrepresentation." Some of Darwin's friends, he laughed, possessed just the right amount of combativeness to go to war on his behalf: "I am sharpening up my claws & beak in readiness."[28]

Darwin sighed with relief. He had worried about losing Huxley's approval—or worse, turning him into an enemy.

> Like a good Catholic who has received extreme unction, I can now sing "Nunc dimittis" [Lord, now lettest thy servant depart in peace]. . . . Exactly fifteen months ago, when I put pen to paper for this volume, I had awful misgivings, & thought perhaps I had deluded myself like so many have done; & I then fixed in my mind three judges, on whose decision I determined mentally to abide. The judges were Lyell, Hooker & yourself. It was this which made me so excessively anxious for your verdict. I am now contented, & can sing my nunc dimittis. What a joke it will be if I pat you on the back when you attack some immoveable creationist![29]

Attacks were not long in coming, either. These too were at first shockingly personal. The full blast of "odium theologicum" arrived in a letter from Adam Sedgwick, Darwin's old teacher and professor of geology. Aged seventy-four, still lecturing to undergraduates at Cambridge, Sedgwick was never likely to agree with the *Origin of Species* despite all his avuncular kindnesses to Darwin in the past. His theologically attuned understanding of nature had strengthened with the passage of years; and his dislike of Lamarck, *Vestiges,* and any other transmutationary doctrine was expressed in regular, increasingly irritated supplements to his *Discourse on the Studies of Cambridge.* He opened his most powerful fire in a monster preface attached to the 1850 edition. For Sedgwick, evolution spelled theological mayhem. The purpose of science, he had declared for forty years or more, was to keep mankind faithful to the path of God.

Even so, Darwin was unprepared for the closely written assault which appeared on the table in Yorkshire. "I have read your book with more pain than pleasure," Sedgwick wrote.

> Parts of it I admired greatly; parts I laughed until my sides were sore; other parts I read with absolute sorrow; because I think them utterly false & grievously mischievous— You have *deserted*—after a start in that tram-road of all solid physical truth—the true method of induction—& started up a machinery as wild, I think, as Bishop Wilkin's locomotive that was to sail us to the moon.[30]

The professor then launched into a stern Sunday sermon, reminding Darwin there was a moral part to nature as well as a physical: "a man who denies this is deep in the mire of folly." To accept Darwin's argument would "sink the human race into a lower grade of degradation than any into which it has fallen since its written records tell us of its history." Moreover he thought Darwin indulged in a tone of triumphant confidence, "a tone I condemned in the author of the *Vestiges*." Sedgwick advised Darwin to accept God's revelation. If he did, the two naturalists might eventually meet in heaven. The agitated old cleric felt no need to spell out the alternative.

Thoroughly taken aback, Darwin showed the letter to Emma. Stoically, she supported her husband. But she in turn refused to show it to Henrietta Darwin, probably thinking she was not old enough to know the full extent of her father's heresy.[31] Hellfire was a chastening thought for Victorians.[32] Sedgwick's affable comment at the end about being a "son of a monkey" only made the laceration worse.

Darwin's mettle was up. Even an old friend and teacher like Sedgwick could be swept aside if necessary. "I never could believe that an inquisitor could be a good man," he told Lyell, "but now I know that a man may roast another and yet have as kind & noble a heart as Sedgwick's."[33]

Close behind was Robert FitzRoy with an equally emotional postal onslaught. "My dear old friend," the captain wrote tempestuously, "I, at least, *cannot* find anything 'ennobling' in the thought of being a descendent of even the *most* ancient *Ape*."[34] Another letter from FitzRoy appeared in the correspondence columns of the *Times* a few days later, signed with the pseudonym Senex.[35] Darwin's heart sank. He knew it was from FitzRoy because of the repetition of the argument about primitive mankind. "It is a pity he did not add his theory of the extinction of

Mastodon &c from the door of the Ark being made too small," he grumbled to Lyell. "What a mixture of conceit & folly, & the greatest newspaper in the world inserts it!"[36]

All this was no more than he anticipated. There were several conservatives among his friends who were bound to hate the book. But the Victorian establishment was full of exceptions. A kindly letter arrived from Rev. Charles Kingsley, who wrote from his country parsonage to acknowledge his presentation copy of the *Origin of Species*. "All that I have seen of it *awes* me," Kingsley said. "Both with the heap of facts, & the prestige of your name, & also with the clear intuition, that if you be right, I must give up much that I have believed & written."

Kingsley was a good naturalist, well versed in geology and zoology, the author of a respected seaside book as well as renowned for his reformist theological tracts and state-of-the-nation novels. Huxley greatly admired his intellect, saying that Kingsley's inquiring mind was attractively open to new ideas, quite the opposite to men of Bishop Wilberforce's ilk. Kingsley also knew Darwin's brother Erasmus slightly, and Fanny and Hensleigh Wedgwood through the Christian Socialist movement. He appeared sincerely interested in Darwin's arguments, telling Darwin he could imagine an all-wise, all-powerful deity making organisms that make themselves. Indeed, he was the first clergyman to see in Darwin's schemes an internal beauty that could be shared by science and spiritual revelation alike. To Kingsley, scientists who tried to explain away the existence of the spirit would be as intellectually stunted as clerics who attempted to dismiss the truths of science. "I have gradually learnt to see that it is just as noble a conception of deity to believe that he created primal forms capable of self development . . . as to believe that he required a fresh act of intervention to supply the lacunas which He himself had made."[37]

As it happened, nothing could have been further from Darwin's intention. Natural selection was a phenomenon that could never be governed, or set into motion, by a Creator. Kingsley had misunderstood that the main point of Darwin's book was to remove the Creator from nature.

Nevertheless, Darwin snatched eagerly at these proffered plaudits. Urgently, he asked if he could quote Kingsley's letter in the forthcoming reprint of the *Origin of Species*. He hoped to show that theological shrieks from the *Athenaeum* or from FitzRoy and Sedgwick were utterly unjustified—that at least one prominent (if rebellious) Church of England parson did not condemn him outright.

Genially, Kingsley agreed, soon appearing in print as an unnamed "celebrated cleric" in the new edition of the *Origin of Species* (published 7 January 1860), and in all editions thereafter.[38] It was Kingsley who provided the few consolatory words of devotion in Darwin's book in which it was suggested that it was possible to believe in God as the ultimate author of evolution, conciliatory words that Darwin would otherwise never have allowed and certainly did not deliver in his own voice. And it was Kingsley who became the first theologian publicly to endorse evolution in the closing pages of the *Origin of Species* itself, an early indication of the remarkable flexibility of the Anglican Church when faced with evolutionary issues.

Darwin went on to match Kingsley's pious phrases with two adjustments of his own. Where, in the final lines of the first edition of the *Origin of Species,* he had written of life being breathed into a few primordial forms, he now altered it to read "the breath of the Creator," a concession that he later regretted. He also added an extra epigraph to the two already in the front of the volume, extracting a passage from Joseph Butler's *Analogy of Revealed Religion* in which Butler stated that the word "natural" indicated the existence of an intelligent agent just as much as the words "supernatural" or "miraculous"; in other words, God could work through scientific laws as effectively as through divine omnipotence.[39] Unaccountably stung by the *Athenaeum* review, and annoyed by what he regarded as the religious prejudices of at least two of his oldest acquaintances, Darwin here defiantly stepped out on the long journey of compromise that would in the future come to plague him.

But he could hardly have arranged it better if he had tried. Even before the reviews began, he had his shotguns loaded.

## III

He brought the Ilkley trip to a close on 7 December 1859 and travelled down to London for a day or two at Erasmus's house to make calls on friends and relations. The eczema and stomach troubles had gone. He felt much better now the book was out. In between half hours with Lyell ("a complete convert"), Carpenter ("he reviews me in the National . . . but the last mouthful chokes him"), Huxley ("It will be God's blessing if I do not become the most conceited man in all England"), Dr. Holland ("going an immense way"), and John Edward Gray of the British Museum ("attacked me in fine style"), he visited Murray in the famous drawing room at Albermarle Street, the first time author and publisher had met face to face for a decade. It was a satisfying occasion, the one gratified by the sales record, the other relieved to have vindicated the decision to go forward and publish. They discussed Darwin's corrections for the impending edition. Many of these were handed over there and then. A print run of three thousand copies did not seem nearly so outlandish a proposition as before. Together they laid plans for the future: translations and overseas editions if possible, and another printing of the *Journal of Researches* (in a green binding to match the *Origin*) to capitalise on public interest.

On this visit Darwin negotiated the terms on which he would ever afterwards continue with Murray. Warily, he cut off any opportunity for the publisher to make unfair profit out of his hard work. He remembered how Henry Colburn had fleeced him over the first printing of *Journal of a Naturalist*. If there was to be either profit or risk, he wanted it shared equally. In fact, he was among the first Victorian authors to negotiate what is now known as an advance against royalties. Forty years later, Murray looked back on his arrangements. "The system on which Mr. Darwin preferred to be remunerated for his books, was to have an estimate made of each edition as it was printed, and to have his share of the prospective profits paid him in anticipation."

> This is an unusual method of payment, if for no other reason because no one can tell exactly beforehand what an edition will produce, as there are several unknown factors in the calculation. . . . Mr. Darwin constantly inquired into details—no angry or irritable word ever seems to have passed between him and his publisher. He did not blindly accept facts and figures which came before him; he investigated them all, and questioned when he was in doubt; but his questioning was always that of frankness and courtesy.[40]

He also received some good advice. Murray felt that the large book on natural selection lurking in the back of Darwin's mind was not much of a publishing proposition. It would run to three volumes, Murray protested; it was relentlessly over-detailed. Why destroy future sales of the *Origin of Species* with the dead weight of a fully academic treatise? Murray's sound head for business told him that Darwin ought to stick with his *Origin* in the same way that Lyell had stuck to the *Principles of Geology*. Gamely, he suggested that Darwin think of a series of smaller books based on his accumulated materials. As a result, Darwin agreed to split up his work in the manner Murray recommended, seemingly envisaging something like his earlier trio of books on the *Geology of South America* and forgetting how that particular geological set had quickly become a hated trial to him. But he would not have contemplated abandoning his big book on species any other way. The material must come out in some form or another, he thought. The detailed evidence on which his argument was based must be laid before an audience. He had promised this in the introduction to the *Origin of Species*. With the plan settled, he returned to Downe more content with life than he had felt for a long time.

Only Richard Owen made him uneasy. Darwin could not gauge what

Owen's judgement was at all when he paid a social call during his London stay. Up until then the great naturalist had occasionally remarked that he looked favourably on "continuously operative creating forces." These remarks were more or less consistent with the views he half-expressed about Darwin's and Wallace's Linnean Society papers; and Darwin had never yet believed Huxley's bloodcurdling stories about the black deeds Owen got up to behind other people's backs. Owen wrote to Darwin in a pleasant fashion after receiving his copy of the *Origin of Species,* and Darwin genuinely wanted to know what he thought. Owen, more than anyone, could influence the direction in which the diverse body of European naturalists might swing.

This visit did not augur well. On the surface, everything seemed as usual. Yet Darwin sensed rocks close to shore. First, Owen informed him that he thought natural selection was completely wrong on biological grounds. He believed that organisms revealed in their anatomy a plan, and that a purely mechanical force like natural selection was hopelessly inadequate to explain it. This was not just prejudice. Owen brought a fine knowledge of comparative anatomy to bear on Darwin's theory and found it wanting.

The book also hit Owen at a particularly tense time. All his hopes for the future were currently in doubt, especially his longstanding dream of a splendid new natural history museum built to his own design.[41] This plan was thwarted at every step by Huxley and Hooker, who both strongly objected to institutional centralisation, not least the ill-effects of putting power over all the natural history sciences into one man's hands—especially if they were Owen's. Huxley had already protested to Parliament several times about Owen's proposed natural history museum, sending in petitions signed by Darwin and other naturalists. These petitions delayed the new museum until the 1880s.

That was not all. Huxley hotly criticised Owen's classification scheme for mammals—a scheme based on the comparative anatomy of the brain. From Owen's perspective, it looked as if Huxley and his cronies went out of their way to block his path.[42] Nor did Owen welcome the emergence of Darwin as a major intellectual rival in Britain. While he could praise the philosophical syntheses of Louis Agassiz in America, Lorenz Oken in Germany, and Georges Cuvier in France, these were scholars who dominated or had dominated other national arenas. Owen considered himself the leading figure in British natural history. Taking these factors together, his response to Darwin's *Origin* was bound to be jaundiced.

Most of all, however, Owen's pride was injured. There, in the black-

and-white print of the *Origin of Species,* were cruel words labelling him as an old-fashioned creationist. "All the most eminent palaeontologists, namely Cuvier, Owen, Agassiz, Barrande, Falconer, E. Forbes, &c.," said Darwin, "have unanimously, often vehemently, maintained the immutability of species."[43] Owen was annoyed beyond measure; offended.

And so, as Darwin told Lyell afterwards, the atmosphere was cool when he went to call.[44] Smoothly disagreeable, Owen sprinkled doubt on the accuracy of Darwin's facts and cast a patronising eye over his style of reasoning. "If I must criticise, I shd. say, we do not want to know what Darwin believes & is convinced of, but what he can prove." Miserably, Darwin agreed. He would try to reduce the quantity of "believes" and "convinceds" next time around. "You will then spoil your book," cooed Owen. The charm of the writing, he said perversely, was that it was the "voice of Darwin himself." To cap it all, just before they parted, he said he wanted to mention the *Origin of Species* in some essays he was editing for publication. Owen intended saying that "the best attempt to answer this supreme question in zoology has been made by Charles Darwin."

The man was exasperating; and Darwin afterwards wished he had said what was in his mind instead of shaking hands and leaving on cordial terms. A short discussion about bears and whales particularly irritated him. Taking his information from Samuel Hearne, an old-time Canadian trapper, Darwin stated in the *Origin of Species* that black bears often swim for hours with their mouths open catching insects. Though he rarely hazarded a guess about the actual course of evolution, he risked it with this graphic example. "I can see no difficulty in a race of bears being rendered by natural selection, more and more aquatic in their structure and habits, with larger and larger mouths, till a creature was produced as monstrous as a whale."[45] Just before their meeting, Owen sent word (via Lyell) that Hearne's story was incorrect, and Darwin deleted the offending sentence from the new edition. During the visit to London, the facts were again adjusted. This time, according to Owen, Hearne was correct. "I am to send him the reference, & by Jove I believe he thinks a sort of Bear was the grandpapa of Whales!"

Polite as this meeting was, it signalled the beginning of the end for the two naturalists, not only in private terms but also in regard to larger movements already under way. The *Origin of Species*—and all that it represented—pushed itself forcefully between them. Owen and Darwin were soon engaged in a quarrel that drew in other men and other issues and became integral to the polarisation of debate about evolution. It seems that

Darwin's book was already requiring the people closest to him to take sides, to align themselves with like-minded friends or to drop longstanding acquaintances. This was the last time Darwin and Owen ever spoke.

The year ended with an important coup. On 26 December 1859, the *Times* ran a favourable review, ridiculously favourable for such a conventional paper. Darwin was sure the article must be by Huxley, but was equally sure that Huxley had never before been asked to write for this closely guarded mouthpiece of the British establishment. The *Times* was tradition itself, regularly satirised by Trollope as "Jupiter Olympus"—the very voice of the gods. The paper ran book reviews only once or twice a month. Darwin had never contemplated the possibility that a good review would land on the breakfast tables of the ruling classes on the day after Christmas.

"Who can the author be?" he innocently inquired of Huxley.

> The author is a literary man & German scholar.—He has read my book attentively; but what is very remarkable, it seems that he is a profound naturalist. He knows my Barnacle book, & appreciates it too highly.—Lastly he writes & thinks with uncommon force & clearness; & what is even still rarer his writing is seasoned with most pleasant wit. . . . Who can it be? Certainly I should have said that there was only one man in England who could have written this essay & that *you* were the man. But I suppose I am wrong, & that there is some hidden genius of great calibre. For how could you influence Jupiter Olympus & make him give 3½ columns to pure science. The old Fogies will think the world will come to an end. Well whoever the man is, he has done great service to the cause, far more than by a dozen reviews in common periodicals. . . . If you should happen to be acquainted with the author for Heaven-sake tell me who he is.[46]

Hooker broke the secret a few days later. Huxley had been to a jolly Christmas party where Samuel Lucas, the normal *Times* reviewer, admitted he had no idea what to say. The two men struck a bargain over their glasses of hock and agreed that Huxley should write the review. It would be submitted as if it were by Lucas while Huxley pocketed the fee, just the kind of cloak-and-dagger intrigue that Huxley relished.

The joke was even better when Mrs. Hooker declared she recognised Huxley's work from "the very first sentence." Darwin and Emma laughed conspiratorily, for Huxley told them that the opening paragraph was actually written by Lucas in order to keep up appearances. They could not bring themselves to confess to Frances Hooker. Huxley was such a rogue, said Darwin: such a "good & admirable agent for the promulgation of damnable heresies."

His circle of intimate friends was emerging as an exceptionally gifted and powerful team. Somehow he had produced a theory that inspired passions and commitments these men were prepared to defend to the hilt.

## IV

As the new year took hold Darwin watched his book become a minor sensation. There were plenty of notices in the newspapers. The *Press* (10 December 1859), *John Bull* (24 December 1859), the *Daily News* (26 December 1859), the *News of the World* (8 January 1860), the *Morning Post* (10 January 1860), the *St. James' Chronicle* (10 January 1860), the *Patriot* (19 January 1860), and the *Guardian* (8 February 1860) all carried brief accounts, only one of which (the *Daily News*) was completely hostile. The *Critic* was as critical as its name suggested; the *English Churchman* surprisingly less so. The *Spectator* was restrained, disclaiming any "judicial pretensions" and making a clear distinction between Darwin's scholarly approach and the writings of Lamarck and *Vestiges*. Although the *Saturday Review*, fondly known to Victorians as the "Saturday Slasher," slashed away, Darwin felt that the anonymous reviewer was issuing "some good & well deserved raps" on one or two geological points. There was not much to send him rushing for shelter.

The monthly, quarterly, and fortnightly journals gathered pace soon after. A significant number of these longer reviews were by his friends: Huxley contributed a review to an early number of *Macmillan's Magazine;* Robert Chambers appeared in *Chambers's Edinburgh Journal;* Hooker in the *Gardeners' Chronicle;* and Carpenter in the *National Review.* They all recommended the *Origin of Species* for careful consideration and noted the author's respectability and status as a naturalist. For the most part, however, these initial assessments were muted, a sure reflection of Darwin's high reputation and perhaps also of the physical substance of the book. Nearly all these early reviewers found the going tough. The abstruse nature of the *Origin of Species* was a central feature in the first written accounts of its power and authority. Even the bad reviews were subdued.

Darwin naturally read these notices with great curiosity. Since he mostly knew the identity of the anonymous authors, he was eager to see what they were willing to say in print. One or two amusingly convoluted letters arrived in the post at Down House around now as people who knew him socially tried to smooth over the possible result of their public utterances. John Crawfurd, who delivered an unfavourable judgement in the *Examiner* for December 1859, wrote to Darwin apologetically to explain why he could not make any approving remarks.

It was obvious, however, that the book was shaping up to become a publishing phenomenon. This aspect of Darwin's public trajectory has often been underestimated. Indeed, what has retrospectively come to be known as the Darwinian revolution was, in reality, as much about the transmission of texts across the world and throughout society as it was about science's ultimate agreement about the validity of natural selection. Darwin and his theories—and then the Darwinian movement as a whole—benefitted enormously from the unprecedented surge in publishing activity in the middle decades of the nineteenth century. Over and above the new ideas that the *Origin of Species* presented, Darwin's volume took its place in a far broader cultural movement that heralded the age of mass media and growing audiences for print materials.

Darwin, and Darwin's friends, were quick to recognise the power of the press. That sector of society which read and reviewed books was more influential than ever before. And the advent of relatively inexpensive printing and the spread of literacy in the nineteenth century, in both Europe and America, were giving several previously mute parts of the social body a voice and an arena in which they could be heard.[47] As their band-wagon started moving, evolutionists swiftly commandeered a slice of the Victorian publishing world, ensuring that the *Origin of Species,* as well as the bundle of reformist ideas with which it would come to be associated, took a substantial place in the spectrum of printed materials characterising the period. While Darwin's words carried an undeniable impact in themselves, he and his friends willingly participated in the revolution in communications that would encourage the circulation of scientific ideas through the common intellectual context and across national and social boundaries.

At that time, review journals were the primary medium for a British author seeking approval. In 1859 and 1860 such journals were proliferating like flies, some hoping to rival the comprehensive coverage of the *Athenaeum* or the *Literary Gazette,* others competing with establishment monoliths like the *Quarterly,* the *Edinburgh,* and the *Westminster,* each of which maintained huge readerships and well-defined political allegiances. Along with the review journals came new magazines catering to the burgeoning middle classes that were stimulated by cheaper and more efficient printing machinery, better distribution systems, and ever-widening constituencies as public education roared onwards and polite society discovered it had time on its hands. Some directed attention to useful activity. Many purveyed fiction and poetry. One by one they opened up different parts of the market by undercutting cover prices and providing a much more varied choice of approaches and contents than the older literary reviews. New subjects ranged from high-society gossip and fashion to money-spinning innovations like illustrations and serialised novels. Many publications were small, independently owned concerns like Dickens's *Household Words* and *All the Year Round,* or Thackeray's *Cornhill Magazine,* which was founded in January 1860. Still others were the in-house products of forward-looking publishing firms hoping to capitalise on market trends, like *Macmillan's Magazine,* first published by Alexander Macmillan in November 1859, and the stable of titles published by Harper's in New York, such as *Harper's Bazaar* (from 1867). Similar expansion took place in titles intended for occupational groupings such as engineers or schoolteachers; and cascades of cheaper materials were directed at the middle classes and artisans, pioneered by Charles Knight and his *Penny Magazine,* William and Robert Chambers's *Chambers's Journal,* and Lord Brougham's pamphlets from the Society for the Diffusion of Useful Knowledge.[48] All took advantage of the rapidly diversifying audiences, incomes, and inclinations that were exposed to view when the last remnants of publishing taxes were lifted in the 1850s and excise duty on paper was abolished in 1860. Soon, an entrepreneurial newsagent, William Henry Smith, started up business in the Strand in London's urban centre. Anyone walking into Smith's shop in 1860 could find 150 different newspapers and magazines for sale every week. W. H. Smith's railway bookstalls and circulating library followed shortly after.[49] Mudie's Circulating Library pursued the same successful trajectory.

Much of Darwin's sudden impact—and his continuing impact—was the result of having produced a book of wide general interest just as this wave of periodical reading matter burst into nineteenth-century homes. His volume was subjected to more popular attention than almost any other scientific book. Even *Vestiges'* huge circulation figures pale beside the multitude of public arenas in which the *Origin of Species* ultimately made an appearance.[50] Before the end of Darwin's life, the *Origin* had been discussed in more than one hundred journals and newspapers in Britain alone, ranging from the *Daily Telegraph* to the *Economist, Family Herald, Punch,* and the Roman Catholic *Tablet.*[51]

As was usual at that time, the majority of these reviews were unsigned, a stylistic device ostensibly intended to permit freedom of speech, but more usually exploited as an opportunity for unattributed criticism or unabashed self-promotion, sometimes achieved by authors fulsomely reviewing their own books. Actual anonymity was always relative. Only one reviewer, for example, eluded Darwin's hunt for identities during December 1859. This was the clever writer in the *Saturday Review.* Darwin would have liked to know who he was. He made pertinent remarks about geology and seemed to know a fair bit about natural history. The author has not been traced to this day (though he could well have been Thomas Rymer Jones). These reviewers—mostly men—were almost inevitably drawn from the literary stratum of London, men of letters who were able to earn a living at their trade and also enjoy the respect of intellectual society. It is a measure of Darwin's lengthy social tentacles, and the tightly confined elite reviewing circle to which he belonged, that out of forty-three substantial reviews of the *Origin of Species* written before the end of 1860, only six authors, including the mysterious *Saturday Review* critic, remained unidentified by him.[52]

A tidal wave of books and commentaries was to come later. For the time being, Darwin began collecting the reviews of his *Origin*, systematically tearing articles out of magazines and newspapers, scribbling comments in the margins or on slips of paper pinned to the text, intending to talk them over with Lyell and other friends and keep them for future revisions. Many came to him as offprints bound in paper wrappers or in the original volumes of those journals to which he subscribed. Tidy-minded as always, he arranged them by size on his study bookshelves—quarto or octavo, each numbered in sequence, the newspaper clippings stashed away in a drawer for Parslow or the children to glue into leather-bound albums for him at their leisure. He compiled an index for ease of access.

The outpouring of comment—good, bad, and never indifferent—scarcely diminished over the next two or three years, providing the startled author with a remarkable documentary repository of the widest possible range of opinion drawn from the widest possible spectrum of readers. Darwin's collection eventually amounted to 347 reviews and 1,571 general articles, as well as 336 items kept separately because of their large size. He maintained two hefty volumes of newspaper clippings.[53] Across the top of a handlist prepared for him by his son Francis later in life, he noted bitterly: "Some of the most absurd or unjust articles were by Harvey & Westwood in *G. Chronicle* 1860s, Wollaston in *Annals & Mag of Natural History* & Houghton & Owen, see remarks on latter in *Saturday Review.*—Hopkins in *Fraser* says my view differs in nothing from Lamarck."[54]

This private brooding over his book's reception stood at odds with his external demeanour. "There has been a plethora of reviews, and I am really quite sick of myself."

## V

All the while, Huxley's brio knew no bounds. These six months after publication were as much his as they were Darwin's. Enthusiastically he accepted commissions all over the place for talks and reviews, rapidly creating a public profile for himself as a major proponent of the new ideas. It is not known for sure when he first described himself as Darwin's "bulldog," but he certainly used the expression in the 1870s.[55] After writing the anonymous review for the *Times* he declared to Hooker that "whatever they do, they *shall* respect Darwin and be d—d to them."[56]

Even so, it is not completely clear just how much some of Huxley's earliest activities actually helped Darwin along.[57] In London in February 1860 he delivered a show-stopping Royal Institution lecture on Darwin's theory during which he pulled a handful of flapping pigeons out of a wicker basket like a conjurer for the occasion. Darwin had pinned high hopes on the evening. Royal Institution audiences usually included an influential mix of fashion and brains, and Huxley had done well to get the topic onto the winter programme, which mostly relayed recent advances in physics and chemistry to the intelligentsia. Darwin made a special trip to London to attend and arranged for his poultry friends to supply the magician's basket of pigeons. Afterwards he wished Huxley had refrained from pointing out all the difficulties of natural selection quite so efficiently. Loyally describing it as "very fine & very bold," Darwin privately considered it a hollow triumph. "Huxley made a great failure of the R.I. lecture," Hooker explained *sotto voce* to Asa Gray, "which was a great pity, as he intended to have backed the book but unfortunately managed to damage it."[58]

The situation did not improve much when Huxley published his next review, his third on the *Origin of Species* in as many months, this time for the *Westminster Review.* He had studied the book thoughtfully in the interval. In the *Westminster* he gave his verdict as a professional naturalist by admitting he could not fully accept the principle of natural selection.[59] Despite all the merits that he freely acknowledged, his deep admiration for Darwin's theory, and his expressions of commitment to the overall thrust of the volume, he felt the case must ultimately remain unproven until Darwin showed how varieties turned into reproductively self-contained species. What could possibly make them become infertile with their closest relatives? he asked. To Huxley's mind this breeding problem was almost insuperable, and he was never satisfied by the compromise solutions offered by Darwin. As for himself, Darwin wished Huxley could concentrate rather more on the good points brought forward in the *Origin*. Huxley did not advertise the marvels of embryology, for example, in the way that he was well qualified to do. Applause in that quarter, Darwin thought, would help his cause materially. "Embryology is my pet bit in my book, and, confound my friends, not one has noticed this to me," he grumbled to Hooker.[60]

On the other hand, Huxley welcomed the *Origin of Species* as ammunition for promoting science at the expense of the church, and the principles of naturalism over theologically based concepts. Fat purple-clad bishops were to him an irresistible quarry. The ensuing review in the *Westminster* was the most powerful ever written in Darwin's defence. The *Origin* was a "veritable Whitworth gun in the armory of liberalism," Huxley declared. "What is the history of every science but the elimination of the notion of mystery or creative interferences?" He claimed that Darwin's book swept away old-fashioned theological obfuscation.

> Who shall number the patient and earnest seekers after truth from the days of Galileo until now, whose lives have been embittered and their good name blasted by the mistaken zeal of Bibliolaters? . . . It is true that if philosophers have suffered, their cause has been amply avenged. Extinguished theologians lie about the cradle of every science as the strangled snakes beside that of Hercules, and history records that whenever science and dogmatism have been fairly opposed, the latter has been forced to retire from the lists, bleeding and crushed, if not annihilated; scotched if not slain.[61]

With rhetoric like this blaring from the *Westminster*'s pages, Huxley intentionally began to construct a Darwinian attack on terms that he favoured. He called on the imagery of warfare to make his point, talking of a battle between science and religion, between faith and reason, old against new, stultifying tradition holding back innovation. Recklessly he propelled Darwin and Darwin's book into the centre of the changing times. Warfare presented a specially vivid image to a nation steeped in the dreadful aftermath of the Crimea, still reeling from the Sepoy uprising, and fretting on a daily basis about the possibility of invasion from France. Every young man who joined the Volunteer movement in those years, who armed himself to the teeth with a pitchfork or rabbiting gun, who drilled in uniform and boots and was fully prepared to defend British countryside against French infantrymen or die in fields of waving corn, could identify with Huxley's oratory. To liken Darwin to a Whitworth gun—the most advanced weapon then developed in Britain—was a compliment with immediate relevance. Individually, too, Huxley's whole existence seemed to revolve around battle: battles against unfavourable family circumstances, battles against snobbery, battles against reactionary scientists, or those he considered oily or dishonest. Huxley was quick to take offence, always argumentative and ready to fight.[62] Sometimes taken aback by Huxley's keenness in this direction, Darwin would joke that his friend's spirits picked up wonderfully if a row was in the offing.[63] The image of warfare reflected the man's chosen path through life.

At home, Darwin felt a twinge of embarrassment to be the focus of so much eloquence. "A *brilliant* review by Huxley with capital hits," he remarked to Lyell. "But I do not know that he much advances subject."

Every grain in the balance, he said, was beginning to count. Serious criticisms were emerging. Francis Bowen in the *North American Review* was "clever & dead against me"; François Pictet in the Swiss *Bibliotheque Universelle* opposed him, although "most candid & fair"; John Duns in the *North British Review* declared "its publication is a mistake"; and Samuel Haughton, in the Dublin *Natural History Review*, accused Darwin of doing no more than reviving Lamarck's theory.[64] Word got round that the great John Herschel called natural selection the law of higgledy-piggledy. "What exactly this means I do not know," wailed Darwin, "but it is evidently very contemptuous."[65] A note from William Whewell confirmed the cautious Cambridge response. Anecdote had it that Whewell refused to allow a copy of the *Origin of Species* to be placed in the library of Trinity College.[66]

All these comments came from men who were highly regarded in their own fields. Not surprisingly, Darwin found the situation disheartening. He especially caught his breath over a review by Thomas Wollaston in the *Annals and Magazine of Natural History*. Before this, he had hoped Wollaston would go some of the way with him, for the entomologist was flexible in his attitude to species, flexible enough to have bantered with Darwin about transmutation at a weekend party at Down in 1856 when the guests "made light of all species & grew more & more unorthodox."[67] Instead Wollaston was furious. He ignored the book's best points, said that Darwin did not understand the definition of species, complained about the personification of selection, and dwelled, like Huxley, on the difficulties: "Would not one step more plunge us headlong into the Nebular Hypothesis and the whole theory of Spontaneous Generation?"[68] Wollaston could see no reason to abandon the idea of divine creation and plenty of dangers in any alternative. Darwin said it was perfectly clear to him that the review was by Wollaston. The text was "so full of parentheses" it could not be by anyone else. "The stones are beginning to fly," he told Hooker.

Similarly dismissive opinions surfaced in a range of other periodicals. The *Christian Observer* insisted that domestic animals varied only because God made them capable of doing so. The *Rambler*, a liberal Catholic journal, rebuked Darwin for philosophising in areas where science ought not to trespass. The *British Quarterly* uneasily indulged in a monstrous fantasy in which a monkey proposed marriage to a dainty young lady in a crinoline—a notion guaranteed to strike horror into the moral heart of Victorians. Aghast at the theological implications of a natural origin for species, the physician Charles Robert Bree promptly issued a pamphlet, *Species not transmutable, nor the result of secondary causes; being a critical examination of Mr. Darwin's work entitled "Origin and Variation of Species."* In this pamphlet Bree hinted that Darwin must have invented the "celebrated cleric," cited in the second edition of the *Origin of Species:* Bree thought no priest worth his cloth could possibly claim that evolution was compatible with Christianity. Aghast, in his turn, at the accusation of lying, Darwin retorted that Bree "has not the soul of a gentleman in him."[69]

Even in those quarters where he most hoped he would be read sympathetically, Darwin came up against sharp resistance. Geologists were quick to point out the errors in a particular example he had given to indicate the vast age of the earth. In the *Origin* Darwin suggested that the original chalk beds of the Cretaceous period that had once covered the Weald, in Kent, could have been eroded at the rate of one inch per century, indicating that the process took some 300 million years.[70] "I have been rash and unguarded in the calculation," he admitted afterwards, anxiously consulting Andrew Ramsay and Joseph Beete Jukes, both of the Geological Survey. John Phillips, the professor of geology at Oxford University and president of the Geological Society of London, slammed Darwin's figures in his annual address to the fellows of the society, leading Darwin to halve his estimate in the second edition of the *Origin,* and omit the case entirely in the third.[71] This was public ignominy. He said to Lyell that he had "burnt his fingers so consumedly."[72]

Not to be outdone, Adam Sedgwick discharged an unsigned tirade in the *Spectator,* the second article about Darwin's book in the *Spectator* in the space of six months, reiterating the position he had declared by letter to Darwin at Ilkley. "You cannot make a good rope out of a string of air bubbles," Sedgwick proclaimed.[73] He detested the theory's "unflinching materialism."[74] Marching out of his college to the lecture room, Sedgwick proceeded to let off steam on his university students. That spring he set an examination paper for his Cambridge finalists, in which to pass every candidate needed to demolish the *Origin of Species.*

> Explain what has been understood by the theory of development and transmutation. Give a short synopsis of Darwin's published views on this theory pointing out how far they are to be regarded as inductive, and how far as hypothetical.[75]

The continued attention also sparked a different kind of reaction—one that hinged on Darwin's sudden renown. During those early months, several people claimed to have thought of natural selection first. Patrick Matthew, an obscure but fiery political writer, wrote to the London magazines to draw attention to his book *Naval Timber and Arboriculture,* published in 1831, in which he had indeed described the mechanism of natural selection. "Erasmus always said that surely this would be shown to be the case someday," confessed Darwin.[76] Yet Darwin had never heard of Matthew, telling friends that a treatise on naval timber was way off his usual reading track. He took steps to deal with this source of possible controversy quickly and cleanly. He wrote a brief response for publication and made his excuses politely. The last thing he wanted was another priority dispute. Undaunted, Matthew capitalised on the connection for several years afterwards, much to Darwin's private irritation. Perversely, Darwin cheered up when it became apparent that Matthew was not the only precursor. An eighteenth-century doctor called William Wells was also shown to have described the principle of natural selection in an essay attached to a larger volume, an *Essay on Dew* (1818). Darwin had missed this one too. With some satisfaction he reported to Hooker: "So poor old Patrick Matthew is not the first, and he cannot, or ought not, any longer to put on his title-pages 'Discoverer of the Principle of Natural Selection.' "[77]

A month or so after Matthew made himself known, an Irish doctor, Henry Freke, also claimed priority over Darwin, saying he had published in 1851 an account of animals and plants evolving from a single filament of organised matter. He sent a copy of his article to Down House, but it was so densely metaphysical that anyone might be forgiven for not spotting that it was about evolution: the *London Review* despairingly noted Freke's "verbose, elliptical, repetitive diction." Darwin had overlooked this article as well. "Dr Freke has sent me his paper—which is far beyond my scope."[78] Still, he was not particularly bothered by these precursors who were of such a different mould from Wallace. It seemed to him—probably rightly—that they intended drawing attention to themselves through the publicity surrounding the *Origin of Species.*

When the satires started, it was a sure sign of controversy brewing. "I have received in a Manchester newspaper a rather good squib that I have proved 'might is right' & therefore that Napoleon is right & every cheating tradesman is also right," he reported to Lyell.[79] The squib rang true enough. To apply natural selection to the human race was an obvious step from Darwin's argument for animals and plants, a step that nearly everyone was willing to make in print except for the author.

Manchester gave way to Dublin, the traditional home of satirical pamphlets. Hooker's friend the Quaker botanist William Henry Harvey published a squib for his students at Trinity College Dublin that tore into natural selection and compared humanised frogs to slavering toadies. Harvey's humour was so curt in places that Hooker did not pass a copy of the squib on to Darwin. Circumspectly, he revealed the pamphlet's existence only when Harvey later admitted that there might be some value in natural selection. Before then Harvey wrote long letters to Darwin relaying esoteric theological points and delivered a crushing review in the *Gardeners' Chronicle.* Since the origin of variation remained unknown, Harvey argued, variability at least must be a miraculous event.[80] Darwin felt the blow hard. Harvey's and Wollaston's comments irked him much more than they should have. "Theology has more to do with these two attacks than science," he explained to Hooker.

Then Dublin gave way to London, as the humorous magazine *Punch* opened its long and abundantly fertile association with Darwinism. The animal lurking inside human beings was obvious, screeched *Punch* with delight. Slimy reptiles, old ducks, serpents in the grass: all these were alive and well in fashionable Belgravia. The hazardous process of choosing a spouse appeared soon after. A *Punch* satire entitled "Unnatural selection and improvement of Species" matched fat with thin, tall with short, sweet young lassies with sour old men.[81]

At Down House, eight pages of anonymous botanical verse arrived, neatly copied out so that Darwin could not recognise the handwriting, quizzing and praising the theory alternately. Did eccentric old Dr. Boott ever write poems? he asked Hooker. This book of his, he caught himself thinking, was certainly producing an extraordinary range of responses.

## VI

The *Edinburgh* was harshest of all. In the April 1860 number, Richard Owen exploded in a long, malevolent anonymous review, an article he must have been composing when he talked to Darwin in London. Owen dealt with nine other books and papers in the same review, one of which was Hooker's 1859 essay on Tasmanian plants, in which Hooker attempted to use Darwin's and Wallace's ideas to explain the relationships of antipodean plant species.[82] All Owen's complex grievances poured out, the thwarted natural history museum among them. He was in turn scornful, condescending, rude, and astute.

Of course he was critical. Darwin predicted as much from their conversation in London. Nevertheless every word was a shock. Owen did not mention Genesis—he was no fundamentalist Bible-thumper. Instead, he insisted that natural selection was incapable of doing what Darwin suggested. He dwelled on alternatives, such as anatomical archetypes and underlying plans of creation. He discussed the first, possibly spontaneous, origins of life. Under the guise of an impartial, anonymous third party, he intimated that Professor Owen had already pondered these issues and had come to wiser, altogether more philosophical conclusions.

On and on it went, slicing up Darwin like any carcass on an anatomist's dissecting table. Owen reduced Darwin's facts to nonsense. He doubted Darwin's competence. He took issue with definitions. He ridiculed Darwin's literary style. He sneered at Darwin's friends, swiping at Huxley's Royal Institution lecture and taking unpleasant aim at Hooker—"one of the disciples"—and called Hooker as short-sighted as his master. Bears swimming around in search of fish could not metamorphose into whales: "we look in vain for any instance of hypothetical transmutation in Lamarck so gross as the one above cited."

By chance, Hooker and Huxley were staying with Darwin for the weekend when this review was published. The three of them were astonished as they turned the pages. Further and further they read, Huxley and Hooker incredulously repeating passages out loud, Darwin anxiously wincing. It was good that they were together. Caught off guard in the countryside, each conscious that they had been publicly humiliated in turn, they united in the face of a common enemy.

"It is extremely malignant, clever & I fear will be very damaging," Darwin protested in a letter to Lyell that evening. "He is atrociously severe on Huxley's lecture, & very bitter against Hooker. . . . Makes me say that the dorsal vertebrae of pigeons vary and refers to page where the word dorsal does not appear. Sneers at my saying a certain organ is the branchiae of Balanidae; whilst in his own *Invertebrata* published before I published on cirripedes, he calls them organs without doubt branchiae."[83]

Hooker and Huxley urged Darwin to write an answer, even if only to point out the factual distortions. And Darwin aggressively rattled his sword for a while. He was taken aback by the naked unpleasantness of a supposed friend—that and what he imagined was duplicity. "What a base dog he is. Some of my relations say it cannot *possibly* be Owen's article, because the reviewer speaks so very highly of Prof. Owen. Poor dear simple folk!"

With the visitors gone, and left on his own to reread the review, his response was far more predictable. Admittedly he gave himself the passing gratification of scrawling tetchy remarks in the margins ("False," "false"). But he never seriously contemplated a response. That night he vowed never to answer Owen. In truth he found the decision difficult: he believed that he was unjustly treated. He believed his work deserved a fair hearing from those who were most qualified to comment, not a tirade such as this. Yet he came to the conclusion that dignified silence was far the best course. He was too accustomed to withdrawing from unpleasant confrontations to begin open combat now.

Underneath, however, he was aggrieved. What went for reticence and cautious modesty, in this regard, was perhaps merely the polite face of sullen dislike and scarcely acknowledged feelings of aggression. Years of friendship evidently stood for naught. Years of scientific intimacy were meaningless. "It is painful to be hated in the intense degree with which Owen hates me."

In retrospect there can be little doubt that Darwin exaggerated the situation. He let himself get carried away on a tide of hostility and squeezed it all down into a resentful simmer that continued more or less until the day he died.[84] What he could not forgive was the jibing at his competence—precisely what Owen could not forgive either. Stung as he was, Darwin came to realise that his theory really would divide the world of natural science, really would turn friends into deadly foes. From then on, he knew there would always somewhere be an enemy. "I am thrashed in every possible way," he remarked to Murray a few days later.

Thereafter, Darwin looked at Huxley's writings with new respect, recognising the bitter antagonism against Owen that fuelled him. "Extinguished theologians lie about the cradle of every science as the strangled snakes beside that of Hercules," he read again in the *Westminster.* Absolutely splendid, he declared with a rush of emotion.

## VII

Another review was destined to have a rather different impact. Whitwell Elwin, editor of the *Quarterly Review,* did not care for Darwin's argument in the *Origin of Species* any more now than when he had first read it in manuscript for John Murray. Ever since the *Origin*'s publication he had been searching for someone who would deliver it a crushing blow. He found his reviewer in Samuel Wilberforce, Bishop of Oxford.

Furthermore, Elwin was simultaneously looking for a reviewer for the unconventional volume *Essays and Reviews,* a book that had been published in March 1860.[85] This was written by seven eminent divines who actively deconstructed the Bible. "Seven against Christ," screamed the first reviews; "a work of destruction," pronounced the *Athenaeum.* Elwin, like many of his contemporaries, was shocked at seeing reputable churchmen apparently cutting away at the foundations of Anglican belief. One essayist denied salvation; another claimed miracles could not occur; yet another that faith should be strong enough to stand alone without the Bible or church doctrine.[86] Although the authors put into words only what some factions of the established church had long said in private about the metaphorical nature of the New Testament and compression of time in Genesis—issues that had been discussed at least since Strauss's pioneering *Life of Jesus*—the commotion at seeing such subtle and apparently heterodox thought broadcast in print was immense. Victorian science was intimately wrapped up in the ensuing turmoil. The church authorities clearly felt that such unfettered liberal ideas would lead towards heresy, perhaps even atheism, with all that this threatened for the traditional status quo.[87] Rationalism was a dangerous pursuit, churchmen like Elwin thundered from their village pulpits. Not willing to be overwhelmed by these blasts from the altar, many ordinary people welcomed the idea of free inquiry into religious matters and cast a cynical eye on the church's stuffy conventions. Nevertheless, Elwin viewed *Essays and Reviews* and the *Origin of Species* with horrified dismay. He thought each volume concealed beneath its innocent cloth covers dangerously subversive tendencies.

Somewhat unexpectedly, Murray at first thought Huxley might review the *Origin of Species* for the *Quarterly* and recommended him to Elwin.[88] Then Elwin toyed with the thought of doing it himself, always the best way for an editor to get the review he really wants. Yet he seized gratefully on Murray's next suggestion—Samuel Wilberforce, Bishop of Oxford and son of the anti-slavery campaigner William Wilberforce, well known for his intelligent interest in natural science and already an occasional writer for the *Quarterly*.

If anyone could have stepped out of a Trollope novel at this point it was Wilberforce. A bulky, vigorous, self-assured man, he was widely tipped to become a future Archbishop of Canterbury. Like his fictional counterpart Archdeacon Grantly of Barchester, Wilberforce's worldly ambitions were all too obvious. He relished his acquaintance with William Gladstone, the member of Parliament for Oxford University, and with Prince Albert. He enjoyed sitting on the bishops' bench in the House of Lords. He participated in clerical in-fighting with a satisfaction that unashamedly coloured his diocesan activities.[89] By popular account Trollope had taken him as the prototype for the energetically domineering archdeacon in the Barchester series. Real life was hardly less droll. Known to the public as Soapy Sam, Wilberforce would explain that the nickname came from his often getting into hot water but always coming out clean. Less charitably his opponents claimed he was too slippery to catch any hold. To Elwin's gratification, he agreed to review both books for the *Quarterly*.

Many delays followed. The bishop was preoccupied with the first ecclesiastical stirrings against *Essays and Reviews*.[90] Not least of his problems was the fact that five of the seven reverend authors were Oxford men ostensibly under Wilberforce's theological jurisdiction, including such leading figures in the university as Benjamin Jowett, Mark Pattison, and Baden Powell. Wilberforce felt he needed to take a strong line against these errant members of his flock. His fiercest contempt was reserved for the "scarcely veiled atheism of Mr. Baden Powell," the Oxford professor of geometry, an ordained priest who in his contribution to the volume calmly refuted all Paley's traditional evidence for the existence of God. Furthermore, Baden Powell mentioned Darwin's *Origin of Species* favourably. This was "a masterly volume," said Powell, "a work which must soon bring about an entire revolution of opinion in favour of the grand principle of the self-evolving powers of nature."[91] Powell died very soon after *Essays and Reviews* was published. Had he lived, Wilberforce thought, he could well have got him prosecuted for ecclesiastical heresy. Luckily for Powell, death conveniently removed him from the bishop's theological firing line.

When Wilberforce turned to the *Origin of Species* he naturally viewed it in the same incendiary light. He submitted a review to the *Quarterly* that was decidedly against Darwin, written with a witty punch that made it the unmistakable product of his pen. He wrote candidly about evolution's dangers. "Now we must say at once and openly . . . that such a notion is absolutely incompatible not only with single expressions in the word of God but with the whole representation of that moral and spiritual condition of man which is its proper subject matter."[92] He reminded readers of Darwin's "ingenious grandsire," the first Erasmus Darwin, and quoted a satirical passage from the *Anti-Jacobin* directed against Erasmus Darwin's evolutionising, where lazy monkey-men might accidentally rub off their tails by sitting round the cave fire all day long. He poked fun at the *Origin*'s facts, reducing them to absurdities. "Now all this is very pleasant writing, especially for pigeon-fanciers," he sniffed comically.[93]

However, he included some specialised anatomical arguments which must have originated from discussions with his friend Richard Owen. Here and there, Wilberforce slipped in Owenite barbs—Darwin's disinclination to mention precursors, for example, and his tendency to depend on individual avowals, "I believe," and "I trust." These came directly from Owen's unpleasant review in the *Edinburgh*.

Glad enough to have his pages filled, Elwin arranged to run it in the forthcoming July 1860 issue of the *Quarterly*. "I think our next number will be sufficiently modern," he reported to Murray.

## VIII

Matters quickly came to a head. Six months of reviews and increasing private tension within the intellectual community culminated in an explosive debate about the *Origin of Species* at the British Association for the Advancement of Science meeting held in Oxford in June 1860. The confrontation was not so much in itself. Bishop Wilberforce argued with Huxley. Huxley argued with Bishop Wilberforce. Underneath, however, the fight was seen as centring on who had the right to explain the origin of living beings—should it be theologians or scientists? At a stroke, Darwin's book, and Darwin's theory, became public property, and the event was to etch itself into the collective memory as a defining moment in Victorian history, one that exposed the painful questions about faith and the position of mankind with which the nation was struggling. While the *Origin* did not generate these questions, this Oxford meeting brought them into the open. And the debate between Huxley and Wilberforce famously came to symbolise the perceived conflict between science and religion, such a powerful contemporary image that it, in turn, contributed materially to the doubts that many Victorians felt about their faith, while reinforcing the convictions of as many others.[94] Although Darwin had gone to such trouble to banish apes and angels from his text, they were present at Oxford, moving their feet restlessly, itching to take over.

As with all such turning points, a host of smaller conflicts and anticipatory differences lay behind it. One matter of significance was the nature of the British Association meeting itself. These annual, week-long events were highly organised, respectable affairs to which members of the interested public, amateur devotees, and scientists of all persuasions came with their families for edification and "rational entertainment." Visitors would take rooms in an inn or stay with friends in nearby houses, and would participate in all the outings, exhibitions, and entertainments laid on by local committees. The programme of scientific talks tended to present little more than summaries of the course of recent research in various disciplines, spiced up with appearances by celebrated speakers and an occasional aristocratic guest such as the prince consort (who had opened the Aberdeen meeting the year before). Well attended, well financed, well settled in the academic and public circuit, these meetings represented a relatively new phenomenon in nineteenth-century Britain—the annual congress. The organisers' object was to consolidate the public prestige of science.[95] Intellectual novelty was neither required nor expected.

In this regard, the city of Oxford made the meeting grander and more important than usual. The British Association president that year was Lord Wrottesley, a good astronomer, former president of the Royal Society, and the recipient of an honorary degree from the university a few months in advance of the association's visit. Other well-known Oxford men played an active part in the proceedings. Wilberforce joined the local organising committee chaired by the professor of botany, Charles Daubeny.[96] Gladstone was due to attend. The exuberant new Museum in Parks Road, designed by Benjamin Woodward, ornamented according to John Ruskin's naturalistic principles, and brought into being at vast expense by the efforts of Henry Acland, the university professor of medicine, was completed just in time to be used as a prestigious venue for association talks and lectures. With its fern and fossil-decked Gothic architecture, this edifice amply conveyed the university's desire to nurture the sciences much more positively than before. At a time when almost two-thirds of the graduates of Oxford took holy orders, the university's science teaching then lagged far behind Cambridge, and the plan to establish a museum had been by far the most ambitious step taken thus far by scientific dons.[97] To complete the effect, the oldest and grandest university buildings were opened for sightseeing, the college halls and libraries made themselves available, and a whirl of fêtes, dinners, teas, supper-clubs, marquees in college gardens, river walks, tours of the quadrangles, and excursions to local beauty spots were arranged. Because it was Oxford, a larger number of prominent people planned to attend. Because it was Oxford, the intellectual fare was sharper. The whole business promised an agreeable week of mingling with friends accompanied by interesting diversions.

Wilberforce already placed private emphasis on the forthcoming meeting. He hoped to dine with senior politicians and clergymen and display his diocese in a favourable light. It was an opportune moment for him to do so. Ever since his arrival in Oxford in 1845, he had failed to manage this doctrinal hotspot quite as effectively as he wished. John Henry Newman had converted to Roman Catholicism in a blaze of publicity, and the "Oxford movement" had proved tricky to contain. Even now, in 1860, Edward Pusey was still inclined to mutiny. Moreover, the trend had created a personal difficulty for Wilberforce when all three of his brothers converted to Rome, followed by their wives, and then his brother-in-law, the future Cardinal Manning.[98] He was increasingly perceived as a peculiar kind of Anglican bishop who could surround himself with Catholic colleagues and relatives. With rationalism squeezing him on one side and Catholicism on the other, Wilberforce was dismayed to discover that Frederick Temple, one of the seven contributors to *Essays and Reviews,* was going to preach the British Association's Sunday sermon to the assembled notables. What might Temple, an outspoken, muscular Christian, be audacious enough to say in Wilberforce's own pulpit?

A final contributory factor was Darwin himself—or rather Darwin's absence. Darwin did not attend the meeting. This made all the difference. Without him events could take their course much more effectively. If he had been there, he would have cramped everyone's style—speakers would have politely deferred to him, he would have tried to soothe or to minimise inflammatory situations. In his absence, controversy could run wild, causes could be snatched out of thin air, scientific flags waved, tables thumped, exaggerations tossed around with abandon. His absence, in other words, was the most pertinent feature of all.

At first he thought he might attend. Despite lingering concern over Henrietta, who was not yet recovered from a fever earlier in 1860, and continuing stomach problems of his own, he inquired about a set of rooms in Magdalen College and discussed the possibility of sharing with Hooker. The last time Darwin had gone to a British Association meeting at Oxford had been in the 1840s, when he enjoyed a family picnic with Henslow, Hooker, and Emma and took sightseeing tours in the surrounding countryside, secretly pleased by Hooker's engagement to Frances Henslow. It had been a happy time. If he did decide to attend, he told Hooker cautiously, Emma was willing to come with him.

Early in June, Henslow unwittingly gave Darwin a taste of what he could look forward to. In a long letter written from Cambridge, Henslow described a recent meeting of the Cambridge Philosophical Society at which the scientific professors debated Darwin's views.[99] Sedgwick was apparently "temperate enough" about Darwin. But Sedgwick was furious over Baden Powell's lofty dismissal of miracles in *Essays and Reviews* and raged about Powell's praise for the *Origin of Species*. He plainly considered Powell's utterances a devious Oxford plot against Cambridge common sense. William Clark, the Cambridge anatomy professor, spoke severely against Darwin, and together they whipped up a noisy antirationalist storm.[100] Most of these dons were ordained in the Church of England, as was customary at both Oxford and Cambridge Universities, and as conservative men of science they ultimately grounded their work in doctrines of divine order and the created plan. They deplored the way that Darwin's book—and *Essays and Reviews*—pushed into territories that they considered inappropriate locations for rational inquiry.

At that meeting of the Cambridge Philosophical Society, Henslow vigorously defended Darwin's right to investigate the question of living origins, although he, like the others, balked at jettisoning divine creation. In this, Henslow showed the mettle that his friends still admired. Elderly he might be, but he retained his inner fire. Yet his affection for Darwin evidently pushed him further than his heart would otherwise have taken him. He explained to Darwin that old-fashioned doctrinal rivalry between Oxford and Cambridge was as much to blame for the hullabaloo as anything, and he went on to smooth Sedgwick's ruffled feathers by marking up for him a copy of Owen's *Edinburgh Review* malediction to show that Darwin had been unjustly treated. At the same time, Henslow was doing even more. In his botany classes at Cambridge he had introduced the students of 1860 (the last class he taught before his death) to Darwin's principles.[101] While telling them of his own unshakeable religious faith, he nevertheless encouraged them to respect intellectual endeavour wherever it might lead. Faced with a painful split of loyalties, both in class and in the meeting hall, Henslow bravely defended his protégé's work.

Sedgwick grumbled and sent a terse statement of his criticisms of Baden Powell to the *Cambridge Herald*—a local newspaper eagerly read by college fellows when their own doings were reported.[102] Feelings ran high. Then, when John Phillips, Oxford's professor of geology, visited Cambridge shortly afterwards to deliver the annual Rede lecture, his audience of Cambridge dons was only too pleased to applaud Phillips's attack on Darwin's natural selection. Phillips was fortified with theological points on miracles derived from Wilberforce and opposed to Baden Powell.[103] In that lecture Phillips moved well beyond the disapproval he adopted in February when first discussing the *Origin of Species* as president of the Geological Society. He was to become one of the most formidable opponents of evolution.

With Henslow's letter in hand, Darwin quailed at the thought of dealing with these combative old boys in person. The prospect of greeting Sedgwick or Phillips at Oxford dismayed him. If he went, he could hardly avoid Owen. He would certainly be called upon to defend himself in public, to stand up and speak in front of a mixed, and generally uncomprehending, audience. He would be obliged to make inconsequential chit-chat about his book at soirées. As Hooker scornfully reminded him, British Association meetings were full of "toadying & tuft hunting & buttering."[104] Darwin did not seek that kind of celebrity.

His stomach saved him. Two days before the meeting his health broke down completely. Never had a bout of sickness been more welcome. He rushed to the safety of the water-cure, not risking the possibility of a sudden recovery. There, at Edward Lane's new hydropathic establishment at Sudbrooke Park, he could hide away. Whatever might happen would happen without him.

## IX

At Oxford evolution was an issue from the first evening. Darwin's book and the stream of comments and reviews published over the previous six months ensured everyone had something to say. Lord Wrottesley's opening address, as bland and all-encompassing as these functions require, was charged with emotion in his call not to forget the wonders of God's creation.

The initial papers the next day did not disappoint. The usual crowd-pullers of geology and geography lost some of their customary attraction as attention lurched decisively towards the life sciences. All the rising young men of science, and those members of the public who hoped for controversy, followed Huxley, Hooker, and Owen into Section D, held in the new scientific Museum, a relatively arcane division usually devoted to dull talks on botany, zoology, and physiology, over which the aged Henslow presided. Any fireworks would be here, the youngsters thought, with Huxley and Owen on the platform together.

Audiences were an essential part of British Association meetings, particularly at Oxford. Shared events like these thrived on spectacle, gossip, and controversy. In a very real sense, participants demanded something different every time, and local organisers would strive to produce for them increasingly striking occasions year by year, offering outings to technological marvels like Clifton Suspension Bridge in Bristol, candlelit caves in Derbyshire, artillery displays at Portsmouth, or yachting excursions to the Isle of Wight. The press regularly attended and reported these events. In return, visitors required scientific celebrities to do what was expected of them—to declaim, smoulder, preach, or shock. As a result, British Association meetings were very much led by the audience, an unusual phenomenon that rarely occurred in the nation's other scientific societies.

So the speakers in Section D were quick to establish where they stood on the *Origin of Species*. Charles Daubeny introduced the topic directly. Since Daubeny was principal local organiser of the meeting, a vice-president of the association, and Henslow's respected counterpart as professor of botany at Oxford, his viewpoint was sure to be taken seriously. Cautiously, he came down on the side of the angels. Some light-hearted talk about monkeys ensued. Then Owen jumped to his feet to assert there was no anatomical evidence for evolution, and that the brain of a gorilla was different from the brain of humankind, a point he had made constantly over the previous years. Human brains, he stated, possessed one tiny structure, the hippocampus minor (a fold in the layers at the base of the brain), that was never found in apes. For Owen, mankind's moral and theological distinctiveness was embedded in this small feature of brain anatomy. He staked his classification scheme for mammals on this difference, and one or two others, in cerebral anatomy; and it was on these points that Huxley had attacked him in print several times before. Huxley replied sharply that morning, letting his tongue run away with him. The anatomical differences were not that large, he insisted. Facetiously, Huxley remarked that churchmen would have little to fear "even if it should be shown that apes were their ancestors."[105]

The exchange set the tone for the rest of the meeting. As Hooker told Darwin, "You & your book forthwith became the topics of the day."

The following day's big draw was John William Draper, the chemist and historian, head of the medical school of the City University of New York although originally from Liverpool. Draper held interesting views about natural changes, sometimes applying these to chemistry, sometimes to the progress of nations, sometimes to human physiology and growth and development. The title of his paper, "On the Intellectual Development of Europe, Considered with Reference to the Views of Mr. Darwin," suggested he would favour progress and transmutation in human beings, and it did eventually supply much of the groundwork for his big book on the advancement of human societies published in New York in 1863. Since he was also well known for his denunciation of organised religion, making a particular target of Roman Catholicism, Draper looked ripe for a rousing presentation.[106] A murmur ran through the crowd that Wilberforce or Owen would use Draper's talk as an excuse to criticise the *Origin of Species*. It seemed as if a clash might be coming.

Huxley nearly did not attend. He suddenly wearied of the extended goings-on, and he may well have been irritated by the assumption he would necessarily be involved if there was a fight to be had. Long afterwards he confessed to Sir James Crichton-Browne that he suffered a great deal from stage-fright during those early years.[107] If he went to Draper's talk, he would have to participate. Far better to get out of the city for a day in the country with his wife.

By chance he met Robert Chambers in the street, who "broke out into vehement remonstrances, and talked about my deserting them." Chambers was in buoyant mood, fresh from secretly revising the eleventh edition of *Vestiges* for publication later in the year and keen to see the idea of transmutation defended. Any ground captured by Darwin's clique would be ground for his book too, and he had already added several, mostly appreciative, comments about the *Origin of Species* to the new edition.[108] He thought Owen's *Edinburgh* article showed that *Vestiges'* "law of development" was worth renewed attention; and at the very least, republishing *Vestiges* in 1860 would capitalise on renewed evolutionary excitement. Most of all, Chambers longed for someone to denounce Wilberforce and all his theological kind. Sedgwick's attack on *Vestiges* more than fifteen years before still grated. He turned the force of his persuasive ardour on Huxley, and on the spur of the moment Huxley changed his mind. So on Saturday, 30 June 1860, Huxley took his place on the platform with the speakers and other organisers of Section D, in front of a large audience in the lecture hall of the university Museum. He was glad to see that Owen was not there.

Irresistibly drawn by the prospect of heated exchanges about monkeys and the heavenly host, an unusual number of people arrived at the museum along with Huxley, including a noisy posse of undergraduates staying on through the holidays, so many that the organisers felt obliged to move the session to a much larger room on the west side of the building (which later became the library). The *Athenaeum* said the audience was immense, and other reports put the figure somewhere between four hundred and seven hundred. Hooker thought there were nearly a thousand people present.

Draper was on first. He revealed himself a keen cultural evolutionist, describing the advance of human society as it was released from what he called a thoroughly benighted Catholic past, an embryology of nations, so to speak. Draper said that human progress depended on science vanquishing theology. If anyone was likely to push a bishop into a rash response it was he.

But after an hour and a half on the intellectual march of the ancient Greeks, the audience squirmed with relief as Draper returned to his seat. "For of all the flatulent stuff and all the self sufficient studies—these were the greatest. It was all a pie of Herbert Spencer & Buckle without the seasoning of either," objected Hooker.

Henslow, as chairman, let one or two members of the public ask questions before raising an inquiring eyebrow at Huxley. Would he like to comment? A few unruly spirits at the back were already chanting "Mawnkey, mawnkey!"[109] Huxley shook his head. The eyebrow moved to Wilberforce.

This was the moment Wilberforce was waiting for. Theatrically rising to his feet, he took control of the meeting. He discoursed eloquently for thirty minutes or more, powerful, argumentative, amusing, his voice filling the hall with confident assertions and making good use of the witticisms he felt constrained to omit when in the pulpit. Draper was forgotten as Wilberforce transformed his still-unpublished review of the *Origin of Species* into his text for the day. It was one of his best secular sermons, carried along by peals of laughter from the students and punctuated by attentive silences. Darwin's facts, Wilberforce proclaimed, did not warrant his theory. Evolution by natural selection was unphilosophical. The line between humanity and animals was obvious and distinct. There was no tendency on the part of lower organisms to become self-conscious intelligent beings. "Is it credible that a turnip strives to become a man?"[110]

Perhaps it was the laughter that loosened Wilberforce's grip. At some point he turned with a flourish to Huxley and alluded to the remarks made the day before. Was Huxley related on his grandfather's or grandmother's side to an ape?

In truth no one could afterwards remember exactly what Wilberforce did say. One witness, possibly no more reliable than the rest, recorded that Wilberforce expressed the "disquietude" he should feel if a "veritable ape" were shown to him as his ancestress in the zoo.[111] These words have the ring of a Victorian bishop about them. Still, the literal words were not important. Wilberforce used the age-old formula of delivering an insult disguised as a friendly jest. The gibe was understood by every member of the audience. They smelled blood. So did Huxley. "The Lord hath delivered him into mine hands," he apparently whispered to Benjamin Brodie on the bench beside him. As the audience fell silent, Huxley stood up on the platform.

He kept them waiting until the end. First, he repudiated the arguments Wilberforce employed. It was clear to him that most of the biological points were derived from Owen. So he stated that brains do not differ very much across the animal-human divide. He commended the way Darwin's theory organised the chaotic data of natural history. When the riposte came, it was so quick that only a few people towards the front can have possibly heard. Like Wilberforce's sally, the form of words was almost immaterial.

> If I would rather have a miserable ape for a grandfather or a man highly endowed by nature and possessed of great means and influence, and yet who employs those faculties for the mere purpose of introducing ridicule into a grave scientific discussion—I unhesitatingly affirm my preference for the ape.[112]

With this, the accumulated force of the theological, scientific, and social upheavals of the previous twenty or thirty years came to a head. Ever afterwards, Huxley was credited with having said that he would rather be a monkey than a bishop.

One by one, Hooker, Henslow, and John Lubbock rose to speak over the clamour of the crowd. Their arguments in defence of Darwin were just as powerful as Huxley's, perhaps more so; and each man felt afterwards that he had personally made the best points of the day. Undergraduates waved their programmes and cheered, people pushed in and out of the hall, the heat rose, bonnets bobbed and weaved, and one over-stretched scientist's wife in the early months of pregnancy gave up completely and fainted. Many people felt that the bishop had been ill treated—that Huxley was much too vulgar in his reply. Somewhere in the audience, young Henry Tristram, who had already published one of the first articles to use natural selection as an explanation for biological problems, changed his mind about Darwin and declared to Alfred Newton, who was sitting next to him, that he was from now on an anti-Darwinian. He said he objected to seeing a guardian of the nation's soul shouted down by a mob hailing "the God Darwin and his prophet Huxley."[113]

Also in the audience, Robert FitzRoy tried to make himself heard. It may have seemed to him that he was personally responsible for this confrontation between faith and science, between God and his former shipmate. Hepworth Dixon, the editor of the *Athenaeum,* was one of the few who caught what he said. FitzRoy shouted that he "regretted the publication of Mr. Darwin's book, and denied Professor Huxley's statement that it was a logical arrangement of facts."[114] He waved a copy of the Bible aloft—a futile exercise noticed by the distinguished mathematical physicist George Johnstone Stoney, and recorded by him in an account written many years after the event. According to Stoney, FitzRoy declared he had "often expostulated with his old comrade of the *Beagle* for entertaining views which were contradictory to the first chapter of Genesis," and implored the British Association audience to believe God's holy word rather than that of a mere human on the question of creation. The scene was corroborated by Julius Carus, the biologist who afterwards translated many of Darwin's books into German. "I shall never forget that meeting of the combined sections of the British Association when at Oxford 1860, where Admiral FitzRoy expressed his sorrows for having given you the opportunities of collecting facts for such a shocking theory as yours."[115] The room fell silent. The pathos of FitzRoy's intervention cannot have escaped at least some members of the audience. Since disembarking from the *Beagle,* FitzRoy had become enmeshed in the financial and mental problems that would eventually drive him to suicide. Seared with emotion, he slumped back in his chair almost unheard.

Afterwards, young naturalists like George Rolleston, Philip Sclater, William Henry Flowers, Michael Foster, and Ray Lankester talked melodramatically of a great victory, enthusiastically hero-worshipping Huxley and embracing the principles he endorsed. Some of the less scientific members of the audience were also impressed. W. F. Fremantle, son of the politician, declared it was "one of the most memorable events of my life." Lubbock said he was proud to have played a part, however small.[116] Hooker glowed with triumph about smashing the "Amalekite Sam" with botanical facts—an exaggerated impression, no doubt, but Hooker's blood was up. Henslow, too, made some spirited remarks before dismissing the assembly with "an impartial benediction."

The bishop's friends were equally convinced that Wilberforce had the best of it. His supporters "cheered lustily" and appeared bullishly satisfied afterwards. This well-distributed sense of success was important. Every speaker believed that he had won—that he had conquered. As the day wore on into evening, and various members of the audience elaborated on the scene, the conversations probably became more sensationalised, more easily turned to enhance particular points of view, more polarised and emphatic: Wilberforce versus Huxley, the church versus science, old versus new, rationality against obscurantism, "the triumph of reason over rhetoric," said Robinson Ellis, future professor of Latin at Oxford, the burnished sword of faith confronting a hotbed of disbelief. The clash was understood as a struggle between titans.

The significant thing was that a contest had taken place. This occasion presented a clearly demarcated display of the respective powers of conflicting authorities as represented in two opposing figures. Wilberforce and Huxley were perceived as fighting over the right to explain origins—a dispute over the proper boundary between science and the church that seemed as physically real to the participants and to the audience as any territorial or geographical warfare. Each side was convinced that its claims about the natural world were credible and trustworthy, that its procedures were the only valid account of reality. As it happened, these opposing forces were unequally balanced in Victorian England. Science at that time held little innate authority in itself, and its status was sustained mainly through the rhetorical exertions of its practitioners, among whom Huxley would come to shine, whereas the church was the strongest body in the nation, attracting and retaining the very best intellects of the age. Afterwards, it was rumoured that Huxley's victory for science was falsely embellished by science's supporters.[117] In this dispute, the challenge was clear. Any success for the Darwinian scheme would require renegotiating—often with bitter controversy—the lines to be drawn between cultural domains. Science was not yet vested with the authority that would come with the modern era. Its practitioners were exerting themselves to create professional communities, struggling to receive due acknowledgement of their expertise and the right to choose and investigate issues in their own manner.[118] As Wilberforce demonstrated, that authority currently lay for the most part with theology. The gossip running through the crowd afterwards quickly crafted an epic narrative, a collective fiction with an inbuilt meaning much more tangible and important than reality. All felt they were witnessing history in the making.

A public polarisation of opinion had emerged. The issue became excitingly simple. Were humans descended from monkeys or made by God?

The point was not lost on the absent author. "Was Owen very blackguard?" he asked Huxley.

> How durst you attack a live Bishop in that fashion? I am quite ashamed of you! Have you no reverence for fine lawn sleeves? By Jove you seem to have done it well. . . . I would as soon have died as tried to answer the Bishop in such an assembly.[119]

*chapter*

4

# FOUR MUSKETEERS

THE BISHOP STORY ran and ran. When Darwin's son William went back to Cambridge in the holidays for some extra mathematics coaching, his tutor told him an anecdote going the rounds. Two Cambridge dons had happened to be standing near Wilberforce just after the exchange with Huxley. One of them was the blind economist Henry Fawcett. Fawcett was asked whether he thought the bishop had actually read the *Origin of Species* and said in a loud voice, "Oh no, I would swear he has never read a word of it." Wilberforce bounced round with an awful scowl ready to lash into him, but noticed at the last moment that Fawcett was blind. For the sake of politeness, the bishop had to bite his tongue. Darwin enjoyed the story, repeating it to Huxley and Hooker vebatim.[1]

Another story came direct from Huxley, who said he felt no personal animosity towards Wilberforce at all. A guest at the Oxford house where Huxley had been staying thought he must be the bishop's son—with their strong, square, belligerent faces there was a striking physical similarity later brought out by matching *Vanity Fair* cartoons.[2] "Was the argument a family spat?" innocently inquired the guest.

Other entertaining reports trickled in. Darwin's friend John Brodie Innes, the vicar of Downe, met Wilberforce at an ecclesiastical house party. Innes showed the bishop a letter from Darwin about the clash. "I am very glad he takes it in that way, he is such a capital fellow," Wilberforce said.[3] In London Erasmus snickered that Wilberforce probably "had no idea he would catch such a tartar."[4] And a little later on, Prince Albert amused himself by appointing Huxley and Wilberforce as joint vice-presidents of the Zoological Society for a concurrent period of office.

Altogether, something of an air of boyish contest prevailed among these respectable men of parts, almost as if they were standing on the perimeter of a schoolboy scrap, cheering on rival sides before they all went off to lunch. Given the small social circle in which they operated and the pervasive cultural norms derived from common educational and political institutional structures, the singularity is not perhaps unexpected. The two main protagonists were never estranged—though never intimate friends either—and both came to treat their Oxford collision with nonchalant satisfaction.

So when the bishop's review in the *Quarterly* reached Down House, Darwin was prepared to be amused. As he read, he scribbled casually in the margin, "What a quibble," "Rubbish," "All a blunder." Only the last part gave him pause for serious thought, and that not for his own sake but for Lyell's. Darwin was well aware how hard it was for Lyell to go even as far as he was going, and he handled Lyell's religious hesitations about evolution sympathetically. Wilberforce's closing words deliberately confronted Lyell with the full enormity of what he was doing. "No man has been more distinct and more logical in the denial of the transmutation of species than Sir C. Lyell," Wilberforce thundered. "We trust that he still abides by these truly philosophical principles; and that with his help and with that of his brethren this flimsy speculation may be as completely put down as was what we may venture to call its twin though less instructed brother, the *Vestiges of Creation*."[5] The bishop knew exactly what he was doing. Lyell's public support gave Darwin's theory a great deal of credibility. Darwin told Hooker, "The concluding pages will make Lyell shake in his shoes. . . . By Jove if he sticks to us he will be a real Hero."[6]

Generally speaking, the whole affair continued to entertain. Richard Owen was inevitably drawn in, not only because of the anatomical information he fed Wilberforce for his review and the speech at Oxford, but also because Darwin and Huxley now considered him a villain of the highest order. Hugh Falconer sent Darwin a lively description of Huxley snarling around Owen's heels and basting the "Saponaceous Bishop." Darwin rubbed his hands. "I must say I do heartily enjoy Owen having had a good setting down—his arrogance and malignity are too bad," he responded.[7]

Shortly afterwards, Lyell made a humorous slip, misreading a word in one of Darwin's letters. Where Darwin wrote "natural preservation," Lyell thought it read "natural persecution."[8] Persecution caught the current mood to a hair, said Darwin jovially. Yet Owen's review still throbbed painfully in his mind, associated with what he now considered all the black arts of religious prejudice as represented by the *Athenaeum* and

Bishop Wilberforce. "Owen will not prove right when he said that the whole subject would be forgotten in ten years," he remarked defiantly to Asa Gray.

> My book has stirred up the mud with a vengeance; & it will be a blessing to me if all my friends do not get to hate me. But I look at it as certain, if I had not stirred up the mud some one else would very soon; so that the sooner the battle is fought the sooner it will be settled,—not that the subject will be settled in our lives' times. It will be an immense gain, if the question becomes a fairly open one; so that each man may try his new facts on it pro & contra.[9]

He was grateful for everything his friends were doing. "From all that I hear from several quarters, it seems that Oxford did the subject great good.—It is of enormous importance, the showing the world that a few first-rate men are not afraid of expressing their opinion. I see daily more & more plainly that my unaided book would have done *absolutely* nothing," he told Huxley.[10] "I shd have been utterly smashed had it not been for you & three others."[11]

## II

Going public in this fashion helped Darwin and his book immeasurably. John Murray was right to believe that controversy was good for business. The general attention ensured that Darwin's views—as well as those of his supporters and critics—were far more widely broadcast than many other scientific concepts of the era, circulating first among members of the literate reading public and then progressively reaching most sections of society before the end of the 1870s. "Darwin's book is in everybody's hands," said George Henry Lewes in the *Cornhill* in 1860.

Of course, Lewes did not literally mean everybody. He meant the educated reading classes.[12] The first visible responses to Darwin, for the most part, were emerging from the upper reaches of British society, the community to which Darwin belonged and to whose members he implicitly addressed his words.[13] Lewes understood perfectly the manner in which a handful of leading figures shaped cultural points of view. He, like others, assumed that the majority of important new ideas would originate in the university-educated sections of society and diffuse outwards and downwards along the social, geographic, and economic scale. And Darwin's effect can indeed be tracked through Victorian audiences in such a fashion, making waves like a pebble thrown into a pool. On the other hand, Lewes merely reiterated the views of his time and place. There was no single British culture. To see the nineteenth century in terms of high learning alone was to give priority to a set of values that obscured other moral codes, other political commitments, as well as disguising the backbreaking labour of the masses. People from all walks of life were reading and thinking seriously about Darwin. Moreover, Darwin's work did not generally lose its theoretical content as it percolated through the nation's consciousness.[14] Quite the reverse. Readers outside the elite community confronted many of the issues ignored by its author and integrated them into systems of thought that, on occasion, included resistance to dominant authority. An anonymous writer in the *Saturday Review* put his finger on the people's pulse more accurately than Lewes when he remarked that the controversy "passed beyond the bounds of the study and lecture room into the drawing-room and the public street."[15]

Nevertheless, it was to Darwin's friends that the first wave of positive responses must be attributed. For it was obvious that Darwin's theories were as useful to them as they were to his theories. Over the following decades, Darwin's defenders came to occupy influential niches in British and American intellectual life. Together, these men would also control the scientific media of the day, especially the important journals, and channel their other writings through a series of carefully chosen publishers—Murray, Macmillan, Youmans, and Appleton. Towards the end they were everywhere, in the Houses of Parliament, the Anglican Church, the universities, government offices, colonial service, the aristocracy, the navy, the law, and medical practice; in Britain and overseas. As a group that worked as a group, they were impressive. Their ascendancy proved decisive, both for themselves and for Darwin.

Darwin's opponents failed to achieve anything like the same command of the media or penetration of significant institutions. Opponents did not unite with the same *esprit de corps*. In fact the community that grew up around Darwin in the wake of the *Origin of Species* was a notable feature of the period. Within a year after publication, it was nearly impossible to break into Darwin's tightly integrated group without some expressed homage to evolution.

This circle turned the sociable aspects of nineteenth-century life to good use. They became intimate. Huxley, Hooker, Busk, Tyndall, Lubbock, and the rest, even including Falconer, despite his never taking to evolutionary theory, were warm-hearted, garrulous beings who talked and dined together, exchanged letters, swopped natural history specimens, asked for photographs to display on each other's walls, stood godfather to rounds of children, established supper clubs so that they could keep in touch, exerted patronage, read proofs, discussed each other's work, and commiserated, supported, and congratulated one another in turn. Their wives paid each other morning calls; the men toured the Alps or rented summer houses in the Lake District near enough to walk over for a late breakfast. As the years went by, deaths or an occasional marriage knitted them more closely. Not quite an intellectual aristocracy as seen in other kinship networks of the United Kingdom, these intimacies bound a very small group of scientific Victorians as securely as any tribe in the rain forest—a tribe that constantly adjusted its complex web of relationships both inside and outside, and with adjacent and overlapping groups.[16]

Likewise, the maturing personal relationships between four figures in particular were crucial to the dissemination and acceptance of natural selection. Lyell, Hooker, Huxley, and Gray instinctively moved together. Their combined effect was formidable. To be sure, the adrenaline of pursuing a mutual goal and attacking a common enemy occasionally masked their differences. Despite their shared commitment to Darwin, they were sometimes poles apart in their individual points of view. Where Lyell allowed himself to be flattered by the attention of society, Hooker preferred to exert influence behind the scenes. Asa Gray believed in God, America, and design, while Huxley was alternately witty and bullying.

Almost immediately, these four musketeers divided up the intellectual world between them. Lyell took on the geological history of mankind and energetically visited the gravel beds of the Somme and Abbeville to examine flints and animal bones and to question palaeontologists such as Falconer, William Pengelley, and Joseph Prestwich, who were more familiar with the sites than he was. He was to make this subject his special contribution to the evolutionary story. The human beings who had used those flint tools had presumably lived at the same time as cave bears and mammoths. That startling possibility had first been floated by Boucher de Perthes some fifteen years before, but was not then accepted because of the uncertain nature of the evidence and the way it contradicted conventional views about the arrival of mankind on earth. While few geologists at that time believed seriously in a literal biblical flood that separated the realm of Genesis from that of modern mankind, they nevertheless found it convenient to consider the watery remains of the glacial period as a dividing line between past and present. In his *Principles of Geology,* Lyell had described just such a cold watery period separating the ancient world from the modern—the period later known as the Ice Age. On that point of view, humankind appeared on the earth only after the cold and ice had gone, when conditions were assumed to be more suitable for humans. Moreover, most people (even geologists) found it acceptable to believe that the first humans had appeared perfectly formed, as the biblical story declared. Whereas Lyell and others intellectualised the issue and equivocated as to how exactly humans might first have risen from the ground, the traditions of religious art fixed the point in Western culture more literally, moving from the earliest depictions of Adam and Eve in the Garden, through images of wildmen and so-called primitives, to the remarkable scenes portrayed in Franz Unger's geological treatise *Die Urwelt in ihren verschiedenen Bildungsperioden* of 1851 ("The Primitive World in Its Different Periods of Formation"), in which Unger's artist depicted a perfectly formed first family. Lyell's attention to flint tools and the likely barbarism of early humanity caused a perceptible stir. Such a major reassessment of human antiquity, with all its implications for overturning conventional thought, opened the door to fundamental religious difficulties.

These conundrums haunted Lyell. They were the start of his longlasting engagement with the subject of human antiquity that finally turned him into a reluctant evolutionist.[17] Still wavering privately over the implications of Darwin's scheme, Lyell dedicated himself to a wholly engrossing new area of research. He put aside his aging *Principles of Geology* and began working on a major new tome, the *Antiquity of Man*, published in 1863, one of the notable milestones in his own evolutionary journey and an important landmark in Darwinian affairs.

While Lyell grappled with antiquity, Hooker aimed at the empire of botany. At Kew Gardens in London, Hooker occupied a position in institutional science that allowed him, in his own sphere, to make as much of an agitation as Huxley. In his day botany was one of the strongest imperial sciences, rivalling only astronomy in its perceived importance to the British economy, and operating through an increasingly integrated system of colonial gardens and overseas university departments, all held together by the growing authority of the Royal Botanic Gardens at Kew, itself a formal wing of the British government. The extent to which government botanists underpinned the economic growth of the developing empire is often forgotten today. Tea, coffee, rubber, sisal, sugar, teak, mahogany, cinchona, cotton, and flax—all these were brought into commercial operation through the colonial botanic garden system. At the autocratic, metropolitan hub sat the Hookers of Kew, father and son, the two most powerful botanists in the world in the nineteenth century.[18]

Hooker made good use of this institutional base to distribute and defend Darwin's views. He persuaded the editors of botanical magazines in Britain and Europe to run favourable reviews of the *Origin of Species* (writing them himself if necessary), encouraged the staff of colonial gardens to discuss the *Origin* in their local scientific societies (again offering them a piece for their journals should they need it), ensured that Darwin was studied as closely in Calcutta, Sydney, and Cape Town as he was in

Britain, added remarks about Darwin's work to numerous official letters spinning out across the globe, and took it upon himself to convince his most distinguished botanical friends, including George Bentham, William Henry Harvey, Alphonse de Candolle, and Charles Naudin, each of whom hesitated to go the whole way with Darwin. His correspondence with these figures probed many of the difficulties that naturalists then felt about the practical mechanics of evolutionary theory, and Hooker showed several of these letters to Darwin for his information. Hooker further turned his position in London's administrative circles to good purpose by mingling regularly with museum trustees, members of Parliament, and colonial governors. At Kew, he began directing a small programme of botanical investigations that would ultimately document many of the adaptive strategies of plants that substantiated Darwin's theories. Although his role was not as likely to attract public attention as either Lyell's or Huxley's, he provided the strength of purpose, patronage, solid government contact, bureaucratic status, and geographical breadth essential for consolidating a lasting transformation in science. He was Darwin's rock; and Darwin depended on him with an intensity he hardly showed for any other man. Much later on, he said warmly, "There never was such a good man for telling me things which I like to hear." In a letter to Jean Louis Quatrefages written on 5 December 1859, he called Hooker "our best & most philosophical botanist."

Asa Gray became gatekeeper for North America. He ensured that everything from Darwin's pen that was destined for the Americas passed through his own capable hands, a privilege he guarded zealously. Like Hooker, Gray was captivated by the insights offered by the *Origin of Species* and promoted the book at every opportunity. There was perhaps little else that Gray could do: he could hardly back out now that Darwin's 1857 letter to him had been published in full evolutionary context in the Linnean Society *Journal.* Like Huxley, he had previously disliked transmutation and had vehemently rejected *Vestiges* when it was published in 1845 in the United States. Now, come what may, he found himself thoroughly caught up in the evolutionary camp. Yet Gray had never been comfortable with the dry intellectual tools of his trade and was independently coming to see plant species as disconcertingly fluid units, not easy to define at the boundaries. He was a confirmed empiricist, one of the few hardline empiricists in the transcendental mist of Emerson's, Thoreau's, Agassiz's, and Lowell's America, rejecting idealist ideas about "abstract types," scorning Romanticism in the sciences while appreciating the transcendentalists' adherence to the divine in mankind.[19] To him, the *Origin* represented the first serious alternative to Agassiz's metaphysical biology.

So he volunteered to arrange an American edition of the *Origin of Species* to be published as soon as possible, opening negotiations on Darwin's behalf with Ticknor & Fields, the Boston house with which Gray had good relations. A number of pirate copies were already circulating in New York, rushed out in bulk by the firm of Appleton's in the first few months of 1860 without Darwin's knowledge. In actual fact, the house of Appleton was doing nothing illegal. The firm, founded by Daniel Appleton and then run by his sons, published hundreds of books of an educational nature and many local versions of overseas titles. In those days of cut-throat commercial markets, publishers were unhampered by overseas copyright agreements and authors were not protected by legislation enforcing foreign contracts or royalty earnings. The firm that got the book on the shelves was the one to make the profit. These unauthorised reprints of Darwin's *Origin* (evidently printed from a single copy purchased in London and rushed across the ocean) had their own bibliographic idiosyncrasies. The first had two quotations facing the title-page, the second had three. A third issue, whose status is uncertain, bears the words "Revised edition."[20] All were published by Appleton's before June 1860 and bound in a greyish-brown pimpled cloth.

Darwin clearly needed a friend like Gray there on the ground to protect his interests. Gray abruptly prevented any further entrepreneurial reprints of the *Origin* by negotiating with William Henry Appleton in person, promising that in exchange for a proper publishing contract and token fee he would supply a fully authorised text, complete with Darwin's endorsement. William Appleton agreed, and the first edition that Darwin approved was published later in 1860, carrying on the title the words "New edition, revised and augmented by the Author."[21] For this Darwin shared with Gray a cheque for £50 ("pin-money for Mrs. Darwin") and received a solitary copy of the volume. He cared far more about producing a supervised version than for the lucre. "Most sincerely do I thank you from my heart for all your generous kindness & interest about my book," he wrote to Gray. After this shaky start his relationship with Appleton's prospered.[22] The firm became the primary agent for bringing Darwin's writings and Darwinism in general to American shores.

Gray smoothed the path for all subsequent editions of the *Origin of Species* in America, several of which put American readers well in advance of details not yet published in Europe. With Gray's encouragement, Darwin added an important historical introduction ("Preface") as well as a supplement indicating his additions and alterations. In this preface Darwin answered his earliest critics and attempted to provide some of the absent acknowledgements that British reviewers had cynically noted. He assessed the evolutionary work of his immediate precursors, cautiously referring to Lamarck as "this justly celebrated naturalist" and mentioning "how completely my grandfather Dr. Erasmus Darwin anticipated these erroneous views in his *Zoonomia* . . . published in 1794." He made sure to praise Spencer and a number of other authors, including Isidore Geoffroy Saint-Hilaire, and finished with a quotation on the persistent types of animal life from Huxley's Royal Institution lecture. This historical sketch was not published in England until 1861.[23]

Moreover, Gray stoutly defended Darwin against American attacks and wrote three important reviews in as many months for leading North American journals. He pushed himself forward as a major intellectual rival to Louis Agassiz, tussling with Agassiz over the definition of species in a series of well-attended public meetings in Boston during 1860 and 1861, questioning whether species were metaphysical constructs, created by God according to a transcendent plan, as Agassiz declared, or whether they arose by natural means from the processes of variation and adaptation, as Darwin propounded. Gray shamelessly enjoyed these fights, a continuing contest inextricably bound up with his power struggle with Agassiz; and he found an authoritative ally in William Barton Rogers, the geologist later to become first president of the Massachusetts Institute of Technology. These two readily understood that Agassiz was the only man in America to possess the stature and influence to crush theories that did not meet his approval; and the resulting controversy in Massachusetts rivalled anything that Huxley and Wilberforce could provide in Britain. Rogers argued violently with Agassiz in a series of four evening meetings at the Boston Society of Natural History, showing that Darwin's views would not collapse like a pack of cards under Agassiz's wrath as had other transmutationary theories like *Vestiges*. Gray harassed Francis Bowen, who opposed Darwin on philosophical grounds, at the American Academy of Arts and Sciences.

Gray launched his reviews in the *American Journal of Science and Arts* in March 1860. This journal was run by James Dwight Dana and Benjamin Silliman, Jr., two clever brothers-in-law at Yale, whom Gray, Darwin, and Agassiz knew well. Agassiz retaliated by writing a bitter commentary on species, also published in the *American Journal*, and reiterated his definition of divine creation in various natural history periodicals. Gray responded with another, this time in the form of a dialogue, which included a measured response to Agassiz. Every one of Gray's words "tells like a 32-pound shot," said Darwin appreciatively.

Gray plagued Agassiz privately as well, hounding him at Harvard University seminars, and pursued him through the scientific journals of the

East Coast. His aggressive mood was strengthened by the admiration for the *Origin of Species* expressed by Jeffries Wyman, the Harvard professor of anatomy whom Gray thought "the best of judges." Dawdling around the log fire in Wyman's college rooms at Christmastime in 1859, a group of friends had grown "warm discussing the new book of Mr. Darwin's." James Russell Lowell, Henry Torrey, and Charles Eliot Norton were there with Wyman and Gray.[24] They knew that their Harvard colleague Louis Agassiz would be up in arms. The *Origin of Species*, Agassiz had said dismissively during those first weeks, was "poor—very poor."[25] Eagerly Gray set about using Darwin's work to attack Agassiz's absolute monarchy. Month by month, he sent Darwin copies of local reviews, clippings from Boston newspapers, and verbal reports about his progress with natural selection in the New World.

The regard was mutual. Darwin admired Gray's tactical successes and valued his philosophical acumen. "I declare that you know my book as well as I do myself; & bring to the question new lines of illustration & argument in a manner which excites my astonishment & almost my *envy!*" he said as Gray's reviews of the *Origin* came out.[26] Gray's talents were wasted on plants, he joked. "You ought to have been a lawyer, & you would have rolled in wealth by perverting the truth, instead of studying the living truths of this world."[27] Later, Darwin made the point again with feeling. "I said in a former letter that you were a lawyer; but I made a gross mistake, I am sure that you are a poet. No by Jove I will tell you what you are, a hybrid, a complex cross of Lawyer, Poet, Naturalist, & Theologian!—Was there ever such a monster seen before?"[28] Soon he was convinced that "no other person understands me so thoroughly as Asa Gray. If I ever doubt what I mean myself, I think I shall ask him!"[29]

## III

And after the Oxford debate, the fourth musketeer, d'Artagnan, found his focus. Huxley opted for apes. Apes and evolution let him channel his gifts into a single, high-blasting campaign, one that simultaneously allowed him to further his work in the biological sciences, criticise theological authority, advance the cause of young professionals like himself, tackle enemies like Owen on their home territory, promote a naturalistic approach to the living kingdom, and feed his lust for life and combat. By becoming the front man for human evolution from apes he could fight hard for all the things he believed in. The others were happy enough to let him blaze away. He reigned supreme over what can only be called the marketing of evolutionary theory—a heady publicity campaign for a reformed, fully scientific, rational England, where power should be wrestled out of the restrictive hands of the church and aristocracy and reestablished on what Huxley regarded as suitably clear-headed principles. Biology played a crucial role in this vision. Huxley considered evolution by natural selection to be the best argument yet for cutting ecclesiastical claptrap away from science. For him, it opened the door to a properly naturalistic consideration of the origins of living beings and mankind. More than this, despite his reservations, he regarded it as a good hypothesis—one that worked and provided explanations.

Huxley intuitively recognised that an open battle over the *Origin of Species* would be advantageous for all concerned. While it would be going too far to claim that he did not care whom he fought—he was not a complete bully, and it is clear that he mostly opposed those who, in his eyes, were endangering the integrity and rationalism of scientific thought—it must be said that he appreciated a good rival. Darwin knew he was lucky that such a volatile man felt prepared to assist him. If circumstances had been different, Huxley might have been as ready to attack as to defend. "For heaven's sake don't write an anti-Darwinian article," Darwin said at an early stage in their relationship. "You would do it so confoundedly well." Years later, with scores of Huxley's destructive essays and articles behind them, he could still say the same. "There is no one who writes like you. . . . If I were in your shoes, I should tremble for my life."[30]

Huxley was lucky in finding Darwin too, for through him he established himself as a leading publicist for science. The Oxford meeting had marked a turning point in his career as surely as it did for evolutionary theory. At Oxford Huxley saw the power of the crowd. He saw the effect of wit and daring. Before that, his scientific work was flashy but diffuse, a series of anatomical studies whose undeniable value was weakened by the lack of any overall direction and whose impact was limited to the domain of scientific experts. The *Origin of Species* allowed him to reveal his flair.

His fight with Owen was central in this, and the course of evolutionary theory would have taken a far more circuitous path if Huxley had not found such an elevated target to topple. Furthermore, he deliberately chose to use the British Association for the Advancement of Science as a major platform. From year to year much of the Darwinian debate took place in the shape of snarled abuse delivered by either Owen or Huxley in front of delighted British Association audiences.

At first, it did seem as if Huxley was out of control. In September 1860, just after the Oxford clash, his son Noel died of scarlet fever. Huxley raged against the death with such ferocity that Darwin and Hooker exchanged anxious remarks. "I know well how intolerable is the bitterness of such grief," Darwin wrote consolingly.[31] The loss of his daughter

Anne was still distressing. "To this day, though so many years have passed away, I cannot think of one child without tears rising in my eyes; but the grief is become tenderer & I can even call up the smile of our lost darling with something like pleasure."

During this highly charged time, Charles Kingsley emerged as Huxley's saviour, sending him letters that in the end brought him back from the edge of the abyss. But Kingsley could not rekindle in Huxley any glimmer of traditional religious belief. That was gone forever, evaporated with the last breath of his boy. All through this anguish, however, Huxley wrote like a madman, spewing out venom against Owen and the injustices of the world in the most violent scientific paper he ever composed. Single-handedly (that was the way he liked it), he smashed into the old-fashioned museum regime that Owen represented. He claimed that the laboratory-based, investigative study of living beings that he championed was the only appropriate vehicle for the changing times.

In this paper Huxley fulfilled the threat he made at Oxford that he could prove Owen's statements about brain anatomy wrong. Not just wrong, Huxley asserted—dishonest. The man was a liar, he said, and produced a list of previously published anatomical observations by other scholars that contradicted every one of Owen's assertions about the structure of the human brain. Huxley aimed particularly at Owen's description of the hippocampus minor as a feature unique to mankind. On the contrary, declared Huxley, the hippocampus minor was present in all the higher quadrumana: "anyone who chooses to take the trouble to dissect a monkey's brain, or even to examine a vertically bisected skull of the true Simiae, may convince himself." Professional expertise and reputations were at stake here, and both Huxley's and Owen's passions ran deep. Although looking like little more than a trivial argument over factual observations, it actually represented a clash of fundamentally opposing systems of thought. Empirical data could not—and would not—resolve the issue about the hippocampus because the disputants did not agree about their relevance.[32]

Huxley published his article in January 1861 in the first number of his own journal, the relaunched *Natural History Review,* which he and a group of progressive scientific thinkers had agreed to purchase as a commercial proposition, ostensibly with a view to making money but mostly to provide an uninterrupted outlet for their radical, mainly evolutionary ideas. The co-owners included Lubbock, the botanist Daniel Oliver, and anatomists George Rolleston, George Busk, and William Carpenter, a conspicuously pro-Darwin team. Huxley made a few extravagant, token gestures as editor at the outset, claiming that the *Review* would be completely impartial, and then immediately turned it into a powerful mouthpiece for his form of biology; and even though the journal died after little more than four years, it was, in its short lifetime, a pungent instrument for naturalism in science. "The tone of the Review will be mildly episcopophagous," Huxley told Hooker, "and you and Darwin and Lyell will have a fine opportunity if you wish of slaying your adversaries."[33] In its first years, the *Natural History Review* published all the most important brain and ape articles of the period, as well as essays and reviews promoting Darwinism in general. Darwin contributed a review or two himself and followed the journal's fortunes.[34] "What a complete & awful smasher (& done like a 'buttered angel')," he said encouragingly after reading Huxley's opening shot against Owen. "By Jove how Owen is shown up. . . . What a canting humbug he is."[35]

Egged on by Darwin, Huxley made brains and toes the subject of his lectures to various audiences during the spring of 1861. Cheekily, he told his patrons at the Royal Institution that they were indisputably related to apes, for their big toes were nothing other than poor copies of prehensile apish thumbs. He continued the line in lectures to working men delivered not far away on the other side of Piccadilly. This second batch of lectures was part of a regular series delivered by several old hands at the School of Mines in Jermyn Street, where Huxley taught natural history and purveyed cheap public instruction at sixpence a time. The lectures were notable in content and as a phenomenon. It was striking to see a large number of London working men voluntarily giving up their scant leisure time, after a working day of twelve hours or more, to attend an academic class in the School of Mines, and paying a sixpenny entrance fee out of a weekly wage of some thirty shillings. Huxley's series ran from February to May 1861 under the title "The Relation of Man to the Rest of the Animal Kingdom." The lectures were lively and informative, attracting a large clientele. Lyell, who visited in March, was astonished at the magnitude and attentiveness of the crowd.

"My working men stick by me wonderfully, the house being fuller than ever last night," Huxley said to his wife. "By next Friday evening they will all be convinced they are monkeys."[36]

## IV

But the man at the hub of this enthusiastic support felt oddly restless. Darwin could not settle at all during the post-publication period. He tried starting the next book, the long disquisition on variation he had promised Murray, but his heart was not in it. He felt chained to the *Origin of*

*Species* even though it was physically gone from his desk. His days rose and fell with the afterswell, while he dealt with a huge correspondence and the increasing number of reviews that his book evoked. At the same time, he kept up his letters to Lyell, Hooker, Huxley, and all. Were they converts or "perverts" as Lyell engagingly put it? Privately he was beginning to feel what every author experiences after a long project is completed. He was drained.

Four substantial reviews of the *Origin of Species* appeared in March 1860, eight in April, five in May, and three in June, each with important reservations, criticisms, or misunderstandings to deal with. These writings uncovered minor gaps or flaws which he wished he had spotted earlier. Shocked disapproval began making itself felt; and a raft of scientific objections was floated, sometimes only a disguise for more fundamental religious or metaphysical opposition, at other times exposing genuinely puzzling phenomena that natural selection did not appear to answer. Darwin, after all, was asking a great deal of his audience. He was inviting them to believe in what was then thought to be unbelievable.

Military metaphors peppered his thoughts. He talked of "buckling on my armour" and the long, uphill fight. Sedgwick, he complained, "has been firing broadsides"; Gray was "fighting splendidly"; Lyell "keeps as firm as a tower." Attacks were "heavy & incessant of late." "I am getting wearied at the storm of hostile reviews; & hardly any useful."[37]

There was good news from Wallace, however. Darwin received a letter from him in May 1860 that eased his mind. The letter itself has been lost, but Darwin, in his reply, expressed his pleasure at Wallace's "too high approbation of my book."

> Your letter has pleased me very much, & I most completely agree with you on the parts which are strongest & which are weakest. . . . I think geologists are more converted than simple naturalists because more accustomed to reasoning. Before telling you about progress of opinion on subject, you must let me say how I admire the generous manner in which you speak of my book: most persons would in your position have felt some envy or jealousy. How nobly free you seem to be of this common failing of mankind.—But you speak far too modestly of yourself;—you would, if you had had my leisure, [have] done the work just as well, perhaps better, than I have done it.[38]

Wallace's feelings about the *Origin* can be gleaned more directly from other letters written to Bates and his old school friend George Silk. He told Silk, "I have read it through five or six times, each time with increasing admiration. It will live as long as the 'Principia' of Newton. . . . Mr. Darwin has given the world a new science, and his name should in my opinion, stand above that of every philosopher of ancient and modern times. The force of admiration can no further go!!!"

To Bates he explained, "I know not how, or to whom, to express fully my admiration of Darwin's book. To him it would seem flattery, to others self-praise; but I do honestly believe that with however much patience I had worked and experimented on the subject, I could never have approached the completeness of his book, its vast accumulation of evidence, its overwhelming argument, and its admirable tone and spirit. I really feel thankful that it has not been left to me to give the theory to the world."[39] Politely, generously, and with an undeniable whiff of relief, the two co-authors took up the positions that they were to hold relative to each other for the rest of their lives.

Still, perhaps Darwin was relieved to have got Wallace off his conscience. Almost immediately he turned to organising translations of the *Origin of Species*. He very much wanted European naturalists to consider his arguments properly. France and Germany were first in his mind. Six months beforehand, he had distributed presentation copies in both countries with the hope that some eager young naturalist in Berlin or Paris might request permission to translate his volume.

At that time, the initiative for foreign publication usually rested with the translator, who was expected to get permission directly from the author and negotiate a contract with a local publisher. So Darwin welcomed the news that Heinrich Bronn, a distinguished philosophical naturalist, would undertake a German edition. Bronn was a geologist of note who knew Darwin's geological work of old. He also held a good reputation in the scientific world for his inquiries into the elemental laws of matter, in which he drew parallels between organic and inorganic phenomena and thought about branching trees of fossil development. Though elderly, he was sprightly, and he reviewed the *Origin of Species* relatively favourably when it came out. "I have had this morning a letter from old Bronn (who to my astonishment seems *slightly* staggered by Nat. Selection) & he says a publisher in Stuttgart is willing to publish a translation & that he Bronn will to a certain extent superintend," Darwin informed Huxley.

Nevertheless Bronn had intellectual preoccupations of his own that he hoped to explore through his translation of Darwin's book. He was fascinated by the controversy emerging in Paris between Louis Pasteur and Felix Pouchet over the possible creation of life in a laboratory test-tube. Could living globules emerge out of disconnected organic materials, as Pouchet's *Heterogenie* of 1858 claimed? Or did every living being—even the smallest germ—need to be produced by another living being, as Pasteur tried to demonstrate? The controversy hinged on what kind of experimental evidence would satisfy inquiry on the question. Pouchet's and Pasteur's rival experiments pointed in several directions at the same time, and the interwoven religious and political situations in Catholic France were no less complex.[40] Pouchet's close association with philosophical materialism and his disregard for traditional forms of religious belief made his claim for the chemical origin of living beings highly suspect in the eyes of at least some of the general public.

Bronn vividly saw the point at issue. For him, evolution must go hand in hand with spontaneous generation, although he was not inclined to believe in either. But a word-for-word translation of the *Origin* was not what he had in mind. Instead, he diligently put back into the book the controversial themes that Darwin deliberately left out. Bronn's translation included many philosophical asides and disquisitions on the first origin of life. Furthermore, he added a final chapter of his own, in which he drew attention to the religious difficulties in fully accepting Darwin's views. Until Darwin could take purely inorganic matter and make a living creature, Bronn said, readers must consider descent with modification an unproven suggestion.

When it was published in 1860, by the firm of Schweizerbart in Stuttgart, this free-ranging translation consequently alerted German-speaking readers to the most provocative aspects of the book—either to the satisfaction of philosophical radicals or the deep misgivings of more conservatively minded thinkers. The German scientific public, already relatively familiar with notions of metamorphosis and transmutationary ideas, from Goethe's work through to *Vestiges*, encountered Darwin's ideas in a form that diverged considerably from the author's intention.

Darwin had scarcely expected a translator, however eminent, to adjust the *Origin*'s argument to suit himself. Armed with some heavy German dictionaries, he struggled through Bronn's pages to see what had been done. Even the title indicated some of the difficulties inherent in moving ideas and metaphors from one cultural context to another. Darwin's "favoured races" was translated by Bronn as "perfected races"; his "struggle for existence" was "struggle for survival."[41] Bronn's final chapter was particularly densely written, and in desperation Darwin finally asked Camilla Ludwig, the new German governess at Down, to turn it into English ("very difficult" Darwin said to Lyell and offered to lend him Miss Ludwig's version). Gradually, Darwin became aware that Bronn simply left out those sentences of which he did not approve—for example, "Light will be thrown on the origin of man and his history." He fretted about Bronn's literal translation of the term "natural selection." Earlier he had written to Bronn, "I cannot help doubting whether 'Wahl der Lebensweise' expresses my notion—it leaves the impression on my mind of the Lamarckian doctrine (which I reject) of habits of life being all important." As Darwin understood it, *Wahl der Lebensweise* more or less meant "choice of lifestyle." Bronn took the hint. Darwin was glad to see that he settled on *naturliche Zuchtung,* meaning natural breeding or cultivation, which caught his intention well enough.

Uneasy about this turn of events, he cast about for another German translator. After several years he located Julius Victor Carus, a younger, altogether more compliant naturalist who said he believed in natural selection. In 1867, after Bronn's death, Carus produced an amended version of Bronn's translation, working closely with Darwin by correspondence.

France also looked promising at the start with Madame Belloc (grandmother of Hilaire Belloc) offering to translate the *Origin of Species* soon after publication. Madame Belloc probably contacted Darwin through Mary Butler, his water-cure friend.[42] But she retreated when she noticed the weight of the subject matter: "on reading it, she finds it too scientific," reported Darwin. He then thought he might have found a substitute in Pierre Talandier, an explosive Frenchman whiling away his time as a political exile from the Second Empire by teaching languages at the Royal Military College, Sandhurst. Yet Talandier could not get a publisher in France to touch him, a situation probably caused at least as much by his political stance as any perceived dangers in Darwin's book. "I have had endless bother about French translation, between two stools, which makes me gladder to close with any one for German translation," Darwin murmured to Huxley.[43]

In the end, Darwin's book was put into French by Clemence Royer, a Frenchwoman living in Geneva, and published in 1862.[44] It is not entirely clear when Darwin realised that she intended to translate his book, since translators were every bit as opportunist and unregulated as publishers and it seems that she notified him of her forthcoming volume only shortly before publication. His pleasure was short-lived. Royer went much further than Bronn in changing the substance of what Darwin said. When the book came out in 1862, Darwin complained that she turned the *Origin of Species* into a travesty of his views. Royer, naturally enough, felt she had enhanced what was already there and knew her intellectual ground. She was well acquainted with the work of European political economists, possessed good English, and mixed with many of the naturalists and anthropologists in Geneva who corresponded with Darwin and reviewed the *Origin,* including Jules Pictet, Édouard Claparède, and Carl Vogt, himself the translator into German of Chambers's *Vestiges.* Several members of this circle had left France after the upheavals of 1848. The new conservatism of nineteenth-century Paris made it an uncomfortable place for liberal, politically active, left-wing thinkers like these to live, and Vogt and Royer were among those who moved to Geneva and created a high-minded intellectual coterie in exile. At this time in her life Royer advocated social progression, women's rights, and advanced views on scientific philosophy, very daring and confrontational in all areas. This included her private life, which involved living openly with a married man, a mirror image of George Eliot's circumstances in England. Royer was impressed with the *Origin*'s implications for human society. "One could say that this is the universal synthesis of economic laws, the social science par excellence, the code of living beings for all races and all times."[45]

Her translation was arresting. First of all, she added a long anti-clerical preface attacking Catholic and Protestant alike. If she offended Swiss sensibilities, she wrote, an "Oxford bishop has provided me with the example." She explored eugenics and the perils of nineteenth-century marriage, emphasising the need for choice and good breeding, and making her point by using the emotionally loaded phrase *election naturelle* for "natural selection." She added footnotes that over-ruled Darwin's cautious apologies. She considered Darwin was wrong to speak of a universal war in nature, and referred throughout to *concurrance de vie* instead of the struggle for existence. She added a quantity of Lamarckian ideas about inbuilt progress and organisms striving to adapt to circumstances. Her title included the non-Darwinian phrase "laws of progress."

It was the most unusual reconstruction of his work Darwin ever faced. She must be "one of the cleverest and oddest women in Europe," he exclaimed crossly. "Almost everywhere in the *Origin* when I express great doubts she appends a note explaining the difficulty or saying there is none whatever!" He could not laugh off these distortions. In 1865 he was still struggling to come to terms with her adjustments. Emma told their daughter Henrietta that he "is at work today on the verdammte Mlle Royer whose blunders are endless."

Perhaps his predicament was made easier by the fact that Royer was a woman. Darwin's men friends rallied round, eager to dismiss her as a mere eccentric whose views were too absurd to heed. Édouard Claparède wrote to say that he had tried to prevent Royer from "disfiguring your work more completely."[46] "Mlle. Royer is a singular individual whose attractions are not those of her sex," he mysteriously explained. The same air of baffled incredulity underpinned Ernest Renan's aphorism that Royer was "almost a man of genius." It seems clear that Royer was defying convention. Women in those academic circles were expected mostly to facilitate the unimpeded flow of their menfolk's scientific ideas by translating, editing, proof-reading, and suchlike. The anticipated norm was a demure willingness to let the male author speak for himself, as Sarah Austin exemplified in her best-selling translations into English of Humboldt's writings, or as Emma Darwin and Frances Hooker displayed. To rewrite and to politicise was unacceptable—unacceptable whatever the sex of the translator. Nevertheless, any book might quickly turn into a spectacle if a female translator stepped out of place. To be sure, Royer was inaccurate, misinformed, and following a cause. Yet Darwin and his friends may have found it a relatively simple matter to link these faults with her gender and dismiss her evolutionary outbursts as a feminine curiosity.

Even so, a French *Origin of Species* was appreciated by anthropologists, zoologists, geologists, botanists, and anatomists living in France, Belgium, Russia, and Switzerland, including Charles Brown-Sèquard, Édouard Lartet, and Jean Louis Quatrefages, and was discussed from time to time at meetings of the Société d'Anthropologie in Paris and elsewhere. The experimental botanist Charles Naudin was particularly grateful to have the *Origin* in his native tongue, for he was not easily able to comprehend Darwin's letters to him. Letters from Down House presented him with a fatal combination of bad handwriting, the English language, and complicated botanical concepts.

Yet the *Origin of Species* was not destined to have anything like the impact on French science that it did in Britain or Germany during those first years. French naturalists never took easily to Darwin's ideas, and if they wished to endorse evolution they usually opted for a generalised form of Lamarckism. In partly Lamarckising Darwin's work, Royer may ultimately have done Darwin a good turn.[47] Her version of the *Origin* was also distributed among French social scientists.[48] Later on Darwin corresponded with Royer in guarded fashion about a second edition, which was published in 1866, and a third in 1870. Each of these later editions troubled him, and the third offended him. He insisted that she change or remove several of her notes—and she grudgingly complied while complaining that the changes weakened the argument.

Taken together, these overseas publications necessarily fell into other cultural contexts and became associated with different issues. In retrospect it seems likely that Darwin was unprepared for the way his writings would be reinterpreted. German, French, and American readers were coming to grips with an *Origin of Species* different from the one he thought he had written. Bit by bit, these foreign editions and translations may have forced him to acknowledge the independent life of his child.

Last but not least, attention came from family quarters as well. Darwin's niece Julia Wedgwood was by now an aspiring literary critic, one of a new breed of women in Britain who took up the suffrage cause a little later in the century. Where a female commentator like Royer was perceived by Darwin as unusual, even a nuisance, Julia had the advantage of being a relative. In 1860 Julia wrote a thoughtful analysis of the *Origin of Species* for *Macmillan's Magazine,* following on from Huxley's review in that journal. The article was too obscure in places for Darwin—Julia was strongly influenced by the charismatic preacher Thomas Erskine, with whose followers she spent part of 1859, and by Frederick Denison Maurice, the founder of Christian Socialism, a friend of her parents, who was striking out on an independent theological path. Julia's article took the form of a conversation about the proper boundaries of science and religion: "Can you receive as truth something that on the other side of the boundary becomes utter falsehood?" asked one protagonist. In the guise of another speaker, Julia defended her uncle from the charge of atheism. Darwin was touched.

> I think that you understand my book perfectly, and that I find a very rare event with my critics. . . . Owing to several correspondents I have been led lately to think, or rather to try to think over some of the chief points discussed by you. But the result has been with me a maze—something like thinking on the origin of evil, to which you allude. The mind refuses to look at this universe, being what it is, without having been designed; yet where one would most expect design, viz. in the structure of a sentient being, the more I think on the subject, the less I can see proof of design.[49]

Slowly, it began to dawn on him that no matter what he did, he was doomed to be a public figure.

## V

From the time of the Oxford meeting his home life had not been entirely easy. His oldest daughter, Henrietta, was perpetually unwell, not fully recovered from what had been diagnosed as a typhoid-like fever. These lingering bouts of low-level fever worried Darwin greatly. Such illnesses could easily flare up into killers, as was demonstrated by Prince Albert's death from typhoid at Windsor Castle in 1861. Less severe forms like Henrietta's tended to drag on and on, leading to physical weakness, occasional feverish relapses, and poor digestion—a sickly, chronic state that usually drove sufferers into long-term invalidism. He therefore took Emma and the children for an extended summer holiday in 1860 to Emma's sisters, Elizabeth Wedgwood and Charlotte Langton, whose houses were next to each other in Hartfield, Sussex. Any visit there was comfortably familiar, but this did little to help Henrietta. For a month or more, while he wrote letters and dealt with translators, the family sank into dealing with the ups and downs of convalescence.

Without really realising it, Darwin yearned for something fresh and interesting to do. He needed a distraction. At Elizabeth's house he embraced the chance to get away from his book and his letters. He wanted to be alone in the country.

Escaping the family one day, he went looking for native orchids in the boggy Sussex hollows, and he stumbled on another kind of plant he had never much noticed before, the tiny insect-eating sundews. Fascinated, he dug one up to take back to the house for a few small experiments, and watched, intrigued, as the sticky leaves curled around a meal of flies. The phenomenon was well known to botanists but not to Darwin. By the end of the afternoon he was caught as surely as any fly. Working on these living plants over the next few months gave him endless, uncomplicated pleasure—the pleasure of working with his hands again, of observation and speculation unhampered by the burden of Henrietta's ill health or the species problem, of seeing nature in action again after the tedious months of writing.

The first batch of plants (the common *Drosera rotundifolia*) that he collected from Hartfield amply demonstrated the sundew's sensitivity. The small, round, reddish leaves were covered in sticky hairs or tentacles, rather like a miniature sea anemone, which flexed and bent to snare any unwary insect that landed. Darwin wondered about *Drosera*'s ability to sense the presence of an insect and how a flying animal could be trapped by such slow-moving leaves. He wondered how far the hairs were truly sensitive and if they also performed the digestive function. He puzzled over the fact that the plants were evidently adapted to eat meat instead of making their own nutrients out of sunshine and water in the usual manner. Were the sticky hairs acting like misplaced roots, perhaps, adapted to suck nutrition from flies rather than the soil? Or were they really like sea anemones, the reef-building coral polyps that he had loved to think about on the *Beagle* voyage? All in all, the plants displayed amazingly animal-like functions and responses.

Niggling natural history questions like these were always good for him. His spirits lifted as he commandeered a kitchen shelf for a temporary laboratory and prowled Elizabeth's house looking for flies; "a little botanical work as amusement," he told Lyell, making fun of his own enthusiasms. Willingly, he allowed his natural inclination towards anthropomorphism to take over. "At present he is treating Drosera just like a living creature,"

Emma wrote to Mary Lyell, "and I suppose he hopes to end in proving it to be an animal."[50] This became evident at mealtimes when Darwin regulated the plants' feeding regime as if he were a Victorian zoo-keeper. Just before lunch on 17 July 1860, he placed four dead flies on the sundew's leaves. By suppertime the outer rim of hairs was curving over. At breakfast the next morning he reappeared beside the plant to give it a spider. At 10:00 a.m. he marked individual leaves with different-coloured threads from Emma's sewing basket, feeding one leaf with a piece of paper, another with dry shavings of wood, another with a bit of feather, the last with shreds of moss. He was back an hour later with a piece of raw meat.

As the days passed, he experimented with other materials, here a bit of bathroom sponge, there a gnat's wing. Touching the leaves with a sewing needle failed to raise any effect. It looked as if *Drosera* could identify its proper food and reject inappropriate stimuli. At teatime one afternoon a week later, the leaves with flies were "splendidly curled in." The insects were "*well* embraced."

These plants returned with the family to Downe. They subsequently provided Darwin with hours of indoor amusement while Henrietta continued poorly and the weather was rainy. Appropriately, if rather bizarrely, many of Darwin's silent fears and preoccupations of recent months seemingly began interlocking. The question of stomachs persistently seeped into his thoughts—Henrietta's stomach, his own stomach, leaves as stomachs. These *Drosera* leaves appeared to carry out some of the functions of an animal digestive system, something like human stomachs turned inside out and exposed to view. To investigate his leaves was to search for the essence of the Darwin family weakness, the bane of his life. It seems as if he may have contemplated more than a simple problem in plant physiology.

*Drosera* did not give up its mysteries lightly. Try as he might, Darwin could not understand how the plants distinguished food from indigestible alternatives. At home at Down House he tested them with everything that came to hand. In August he tried feeding leaves with drops of milk, olive oil, white of egg, and gelatin, then moved on to syrup, white sugar, laundry starch, and gum. In September he plied them with strong tea and sherry. Soon he used himself as a private source of supply, trying human mucus, saliva, and urine.

With the domestic armoury exhausted, he turned to drugs and chemicals. Most of the substances he used were common enough for the period, and several were already in the family medicine chest. At one point he dosed the plants with chloroform, using the drops he kept handy for toothache. Chloroform vapour, he told Hooker excitedly, "paralyses them completely." Every day brought a fresh experiment, another substance, another way of tempting the sundew's appetite. Upstairs, Emma laboured at the same task with Henrietta. She ate hardly anything, Emma reported to William.

The story was much the same under the lens of his microscope. He could not work out what was happening inside the sundew's hairs at the cellular level. Was he perhaps witnessing the absorption of nutritious materials? he asked Henslow.

> When the viscid hairs contract or become inflected they pour out much fluid & the contents of the cells in the footstalks, instead of being a thin pink homogenous fluid, becomes a broken mass of dark red, thick fluid. . . . What has surprised me is that the globules & cylinders of the thick dark red fluid or substance keeps on an incessant *slow* contracting & expanding movement: they often coalesce & then separate again; they often send out buds, which rapidly increase at the expense of the larger parent mass; in short endless slow changes in form.[51]

He hoped he was seeing the digestion process. "Is any such phenomenon known?" he inquired. "It may be quite common, as I am so ignorant of vegetable physiology."

As it happened Darwin was watching a phenomenon not fully understood until the twentieth century in which the wrinkling of the inner cell membrane, special to *Drosera,* provides a large surface area for the transfer of fluids. Darwin noted that what he called "clumping" happened only when plants were exposed to food substances: "A very weak solution of C. of Ammonia instantly sets the process at work." Henslow was baffled by Darwin's terminology of clumps, coalescing globules, and colour changes. Peering down his microscope, Darwin fretted anxiously over the problem, haunted by the notion of digestive processes gone wrong.

The plants accompanied him, too, when the family travelled to Eastbourne in the early autumn for another recuperative holiday for Henrietta. Yet at Eastbourne Henrietta relapsed, so quickly and so completely that Emma and Darwin were frightened she might die. Her stomach pains were "dreadfully fierce." It was "a fearful attack," Darwin told Henslow in panic. "What the end will be, we know not."

Thoroughly alarmed, they summoned Henry Holland from London, who prepared them to expect the worst. For a week they lived on the edge. Then she rallied slightly. "We have had such a week of misery as I did not know man could suffer," Darwin managed to write to Asa Gray afterwards. "As much misery as man can endure."

> My daughter grew worse and worse, with pitiable suffering, so that all the Doctors thought we should lose her. But the stoppage is over, & she has rallied surprisingly; but whether there is much organic mischief & what the final result will be cannot be known, till the miserable issue is decided. But she is quite easy now, & one comes at last to care only for that; & we have managed to conceal from her, her extreme danger.—You are so kind & sympathetic that I have not resisted telling you our unhappiness.—We shall not be able to remove her home for several weeks even if the case is not worse than the Doctors now hope & believe.[52]

The experience unsettled him much more than he admitted. Death breathed so close behind. With three children already gone, he said he dreaded the thought of losing Henrietta as a fourth. He could not see what possible prospect lay ahead for her now. Even a long period of convalescence might still leave her as a semi-invalid.

From then onwards, he and Emma viewed Henrietta's health with anxiety, an understandable response at a time when the nation was ravaged by unmanageable disorders and infectious disease. The fear they experienced at Eastbourne never really left them. Henrietta did make a full recovery in the end and ultimately lived to a great old age. But her parents worried over her smallest infirmity for years to come, solicitously hiring bed-carriages for journeys, keeping her indoors most of the time, consulting different doctors, sending health bulletins around the family, and devising cautious routines that permitted only half an hour downstairs in the evenings, all of which tipped relatives and friends into regarding her as a permanent invalid. Henrietta emerged at the end of this illness not as a healthy young women of seventeen yearning to pick up the threads of the social life she was missing, but as a delicate daughter, prone to unspecified infirmities and relapses. At least she recognised the self-absorbed nature of these preoccupations when, as a mature woman, she said her mother had been "a perfect nurse in illness." Being ill under Emma's supervision was an opportunity for a daughter to be cherished, perhaps in contrast to Emma's normally undemonstrative manner. "We are cool fish, we Darwins," said Leonard many years later. The same opportunity to be physically close probably also applied to her father. When she was ill, Henrietta received Darwin's undivided attention. She remembered how Darwin would stop working in his study and come into her room with such an expression of tender sympathy and emotion that she said she could scarcely bear to see it. "Both parents were unwearied in their efforts to soothe and amuse whichever of us was ill; my father played backgammon with me regularly every day, and she [Emma] would read out to me."[53]

Although it does seem likely that these early illnesses indicated some lingering internal problem, Henrietta probably saw little incentive to abandon the semi-invalid state. Later on, her own marriage was partly built on giving and receiving the same kind of all-consuming medical attention.

Darwin's unease about Henrietta merged into unease about sundews. Back at home in Down House, Darwin could no more diagnose the feeding requirements of his insectivorous plants than get to the bottom of his daughter's illness. While Emma rubbed castor oil onto Henrietta's abdomen (the latest medical recommendation, Hooker assured him), Darwin calculated and measured and dripped different solutions of chemicals onto his *Drosera* leaves. Then he moved on to poisons, ordering arsenic, henbane, and strychnine from William Baxter, the Bromley chemist. He made, in his study, weaker and weaker solutions of poison in order to test the detective powers of the leaves: how small a dose would kill them? He began corresponding with a forensic scientist in London who specialised in detecting small doses of arsenic in murder cases and was interested to discover that his plants were much more sensitive to poisons than people. His own stomach deteriorated as if in sympathy. "I sometimes suspect I shall soon entirely fail; my stomach now keeps bad nearly all day & night." Unaware of these parallels with Victorian detective fiction, he read *The Woman in White,* Wilkie Collins's mystery story. "The plot is wonderfully interesting."[54]

It might almost be said that focusing on leaves served to displace Darwin's wretchedness about Henrietta. Teasing and torturing the plants filled his mind. He could not easily contemplate the thought of Henrietta dying, or the thought that through inheritance he might be the cause of her weak stomach. Perhaps these emotions were funnelled into the escape route of natural history, just as he had once escaped the realities of another daughter's death by immersing himself in the anatomy of barnacles. Perhaps the plants themselves reflected his own fears. Not long after this, he discovered he could kill them.

All his work, however, was called into doubt by a single stray human hair that drifted onto an unprotected leaf. To his surprise, the sundew thought that this too was food and began curling over the hair, sparking a chain of worrying doubts in Darwin's mind. Were the tentacles stimulated by weight rather than by chemicals? All his results so far indicated that something chemical was going on. Disconcerted, he tried feeding them with pieces of cotton thread, less than 1/50 of an inch long, and then with Emma's hair, finer than his own. Obligingly, the tentacles curled over. He was nonplussed. "I cannot persuade myself that it is the weight of 1/78,000 of a grain of solid substances which causes such plain movement; nor that it is in most cases the chemical nature; & what it is, stumps me quite," he complained to Hooker.[55] He mentioned the same doubts to Lubbock and Lyell.

> I will & must finish my Drosera M.S. which will take me a week, for at this present moment I care more about Drosera than the origin of all the species in the world. But I will not publish on Drosera till next year, for I am frightened & astounded at my results. . . . All this dreadful illness for last six months (& that wicked dear little Drosera) has made any progress in my larger book almost nothing.[56]

He brought the project to an end in February 1861 when he went to deliver some remarks on the subject at the Royal Society's Philosophical Club, a dining club of scientific gentlemen that ate together an hour or two before Royal Society meetings.[57] Faced with an intellectual cul-de-sac his attention was temporarily exhausted.

For most men this would perhaps have finished the matter. Yet Darwin's hobby-horses never really came to a halt in so definite a manner. Every so often, for years afterwards, he would ask botanical friends about insectivorous plants and would beg unusual specimens from Kew to investigate. Eventually, in the 1870s, he was able to make a full study of meat-eating plants. Published in 1875, this was to be one of his most engaging books. "By Jove I sometimes think Drosera is a disguised animal!" he told Hooker.

## VI

Even if he had not felt obliged to put sundews aside, other concerns began to crowd in. When he was in London for the Philosophical Club, he heard that Huxley's wife was still very depressed over the death of their son despite the arrival of a new baby. Huxley asked Hooker and Darwin to stand godfather to the new baby, assuring them he would not care a fig if the words of the church service stuck in their throats. Huxley chose the name Leonard for the baby because, so the story goes, it contained within it the dead child's name, Noel. Both Hooker and Darwin agreed to be godfathers, telling Huxley they would simply ignore the religious strictures. Leonard Huxley, who grew up an unconcerned atheist in the centre of this scientific circle, later became Hooker's, Darwin's, and Huxley's biographer and edited several collections of letters.

Darwin was concerned about Henrietta Huxley. When he saw Huxley he suggested that she might like to visit Down House with the children for a rest in the country. "Do you not think a little change would be the best thing for Mrs. Huxley, if she could be induced to try it?" Down House was like a sanatorium already, he said, with his own Henrietta sickening upstairs. Mrs. Huxley (Nettie) would feel absolutely at home.

> I have been talking with my wife & she joins heartily in asking whether Mrs. Huxley would not come here for a fortnight & bring all the children & nurse. But I must make it clear that this house is dreadfully dull & melancholy. My wife lives upstairs with my girl & she would see little of Mrs. Huxley, except at meal times, & my stomach is so habitually bad that I never spend the whole evening even with our nearest relations. If Mrs. Huxley could be induced to come, she must look at this house, just as if it were a country inn, to which she went for a change of air.[58]

Nettie Huxley had probably never received such a gloomy invitation before, but to everyone's surprise she accepted. She wrote to Emma to warn her that she was rarely downstairs much before one o'clock, to which Emma replied that this was the usual state of affairs at Down House. She would only be following suit.

With their timetables so pragmatically agreed, the two wives spent the best part of March 1861 in each other's company at Downe. Before then they hardly knew each other individually (Emma confessed that "it will be rather serious her coming without Mr H."). The visit established the basis for a long-lasting family friendship. Emma and Nettie talked comfortably about their children and the usual childhood diseases, read books, and played music together. It may also have been a relief to realise that they were not alone in sharing their married lives with an intrusive third party like science. Best of all, they discovered a shared admiration for Tennyson. Emma had long despaired of Darwin's dismissive attitude to Tennyson, and smiled to hear Mrs. Huxley's frank declaration that Darwin had "no poet's corner in his heart." Henrietta Darwin enjoyed Tennyson too. So the women spent their evenings round the springtime fire, snugly wrapped in invalid shawls, reading aloud. They embraced the latest instalment of *The Idylls of the King*, Tennyson's risqué account of Vivien's seduction of Merlin. "Only fancy Mr. Darwin does not like poetry," Nettie wrote in mock horror to her husband. "I fear he has not so good an opinion of you since I mentioned your taste for it. Let us pray for his conversion & glorify ourselves like true believers."[59]

## VII

In May, Henslow died. For Darwin it was like the death of a parent.

Henslow was aged only sixty-four (a year older than Lyell), and his death was unexpected, at least to his scientific acquaintances. A month beforehand, he had published an animated note in *Macmillan's Magazine* correcting a misapprehension about his views on species and defending Darwin. He said he respected Darwin's right to his opinion, but believed that evolution could never be proved. "God does not set the creation going like a clock, wound up to go by itself," he told his brother-in-law Leonard Jenyns privately.[60] For him to fail so suddenly was a great surprise. But there was time enough for Hooker and his wife, Frances (Henslow's daughter), to travel from Kew to Ipswich to be at Henslow's bedside; time enough for Jenyns to get over from Bath to say a fond farewell (Henslow was married to Jenyns's sister); and time enough for frail and elderly Sedgwick, Henslow's devoted old university colleague, to arrive by railway train from Cambridge, a long journey across country, to make his adieus. Sedgwick was moved to tears by the occasion and whispered words of faith into Henslow's ear not knowing if they were heard.[61] Hooker asked Darwin if he would come as well.

After an embarrassing misunderstanding in which Darwin thought Henslow was already dead, Darwin made his excuses and declined.

> If Henslow . . . would really like to see me I would of course start at once. The thought had [at] once occurred to me to offer, & the sole reason why I did not was that the journey, with the agitation, would cause me probably to arrive utterly prostrated. I shd be certain to have severe vomiting afterwards, but that would not much signify, but I doubt whether I could stand the agitation at the time. I never felt my weakness a greater evil. I have just had a specimen, for I spoke a few minutes at Linnean Society on Thursday & though extra well, it brought on 23 hours vomiting. I suppose there is some Inn at which I could stay, for I shd not like to be in house (even if you could hold me) as my retching is apt to be extremely loud.[62]

He felt guilty for months afterwards. He had never before used illness in such an obvious way to avoid personal obligations. By rights, he should have made every conceivable effort to attend Henslow's deathbed. Henslow had made him what he was, not only by giving him the chance of a lifetime with the invitation for the *Beagle* voyage, but also by his kindly attentions and support thereafter. The refusal was seemingly grounded in the same inner tightening Darwin had felt on his father's death in 1849, not as intense as the despair occasioned by the children's deaths, yet involving other disturbing feelings of emotional debts unpaid. Henslow's demise ended another chapter in his life. His absence was a poor return for Henslow's friendship.

The excuse, moreover, put Darwin in an awkward position. Hooker knew he had been well enough to go to a Philosophical Club dinner in London; well enough for a speech at the Linnean Society; and well enough to entertain Mary Butler, his Ilkley water-cure friend, at home in April, with an evening of spirit-rapping and table-turning in the Down House drawing room. "She tried mesmerizing Franky," Emma reported, "in which I think she would have succeeded but as he is such a nervous subject I did not much encourage it."

But he was too sick for Henslow. Afterwards, full of remorse and wanting to compensate for his selfish behaviour, he tried to make amends in a brief account of Henslow's influence on him that he wrote at Jenyns's request for a memoir that Jenyns was preparing. Darwin's recollections glowed with affection and respect. He dwelled on the qualities that he most admired in Henslow—modesty, sympathy, and self-effacement. Many years later, it struck George Romanes that Darwin was actually describing himself, an "uncanny description," said Romanes, of Darwin's own virtues. The identification between pupil and teacher was indeed close. "Poor dear & honoured Henslow," Darwin subsequently wrote to Hooker. "He truly is a model to keep always before one's eyes."

Guilt, death, ill health, and suffering haunted him. He was taking life much more badly than Emma had seen for some time, and she wondered if she was neglecting him because of her concern about Henrietta. She wished he could find some form of solace in religious belief. Around now, as she had done once or twice before, she wrote him a letter, choosing to write to him rather than to speak, in order to express herself carefully. She felt a letter was the best way to ensure his attention.

> I cannot tell you the compassion I have felt for all your sufferings for these weeks past that you have had so many drawbacks. Nor the gratitude I have felt for the cheerful & affectionate looks you have given me when I know you have been miserably uncomfortable.
>
> My heart has often been too full to speak or take any notice. I am sure you know I love you well enough to believe I mind your sufferings nearly as much as I should my own & I find the only relief to my own mind is to take it as from God's hand, & to try to believe that all suffering & illness is meant to help us exalt our minds & to look forward with hope to a future state. When I see your patience, deep compassion for others, self command & above all gratitude for the smallest thing done to help you I cannot help longing that these precious feelings should be offered to heaven for the sake of your daily happiness. But I find it difficult enough in my own case. I often think of the words 'Thou shalt keep him in perfect peace whose mind is stayed on thee.' It is feeling & not reasoning that drives one to prayer. I feel presumptuous in writing thus to you.
>
> I feel in my inmost heart your admirable qualities & feelings &

Charles Darwin in 1863, five years after announcing his theory jointly with Alfred Russel Wallace at the Linnean Society of London. The photograph was taken by his son William.

(Gray Herbarium, Harvard University)

A rare page from the original manuscript of the *Origin of Species*. Darwin destroyed most of the rest.

(Karl Pearson papers, University College London)

The first copy of the *Origin of Species* was sent to Darwin at the water-cure in Ilkley, Yorkshire.

(Darwin collection, courtesy of the Syndics of Cambridge University Library)

"But with regard to the material world, we can at least go so far as this—we can perceive that events are brought about not by insulated interpositions of Divine power, exerted in each particular case, but by the establishment of general laws."

W. WHEWELL: *Bridgewater Treatise.*

"To conclude, therefore, let no man out of a weak conceit of sobriety, or an ill-applied moderation, think or maintain, that a man can search too far or be too well studied in the book of God's word, or in the book of God's works; divinity or philosophy; but rather let men endeavour an endless progress or proficience in both."

BACON: *Advancement of Learning.*

ON

THE ORIGIN OF SPECIES

BY MEANS OF NATURAL SELECTION,

OR THE

PRESERVATION OF FAVOURED RACES IN THE STRUGGLE FOR LIFE.

BY CHARLES DARWIN, M.A.,

LONDON:

JOHN MURRAY, ALBEMARLE STREET.

1859.

John Murray, Darwin's publisher, at his desk at Albermarle Street.

(G. Paston, *At John Murray's* 1932)

Science books were popular with Victorian readers. A scene in the British Library.

(Wellcome Library, London)

Keep

9 St. Mark's Crescent N.W.
March 2nd.

Dear Darwin,

I am very glad you like my notion about the catterpillars. It is a kind of "forlorn hope", but fortunately it can be easily tested.

I dare say you are right about sexual selection in butterflies, but I still think that protective adaptation has kept down the colours of the females, because the Heliconidæ and Danaidæ are almost the only groups in which the females are generally equally brilliant with the males.

I can tell you several
in the East who would
observe "expression" for you

Darwin's correspondence was a key research tool. This letter shows his pencil markings and the annotation "Keep."

(Darwin collection, courtesy of the Syndics of Cambridge University Library)

The efficiency of the Victorian postal system made it possible for Darwin to develop an extensive correspondence network.

(Mary Evans Picture Library)

Many relatively unknown naturalists helped Darwin. William Tegetmeier (left) experimented on poultry and pigeons.

(E. W. Richardson, *A Veteran Naturalist* 1916)

At Down House Darwin bred pigeons and poultry himself.

(*Illustrated London News*)

Alfred Russel Wallace, who put forward a near-identical theory of natural selection, photographed in Singapore in 1862.

(J. Marchant, *Alfred Russel Wallace: Letters and Reminiscences* 1916)

Wallace's hut on Waigiou Island, Indonesia. "Here I lived pretty comfortably for six weeks."

(A. R. Wallace, *The Malay Archipelago* 1869)

Wallace's copy of the *Origin of Species* in which he substituted the words "survival of the fittest" for "natural selection."

(Keynes collection, courtesy of the Syndics of Cambridge University Library)

84 NATURAL SELECTION. CHAP. IV.

struggle for life, and so be preserved. How fleeting are the wishes and efforts of man! how short his time! and consequently how poor will his products be, compared with those accumulated by nature during whole geological periods. Can we wonder, then, that nature's productions should be far "truer" in character than man's productions; that they should be infinitely better adapted to the most complex conditions of life, and should plainly bear the stamp of far higher workmanship?

by survival of the fittest, nature

It may be said that ~~natural selection~~ is daily and hourly scrutinising, throughout the world, every variation, even the slightest; rejecting that which is bad, preserving and adding up all that is good; silently and insensibly working, whenever and wherever opportunity offers, at the improvement of each organic being in relation to its organic and inorganic conditions of life. We see nothing of these slow changes in progress, until the hand of time has marked the long lapse of ages, and then so imperfect is our view into long past geological ages, that we only see that the forms of life are now different from what they formerly were.

"survival of the fittest"

Although ~~natural selection~~ can act only through and for the good of each being, yet characters and structures, which we are apt to consider as of very trifling importance, may thus be acted on. When we see leaf-eating insects green, and bark-feeders mottled-grey; the alpine ptarmigan white in winter, the red-grouse the colour of heather, and the black-grouse that of peaty earth, we must believe that these tints are of service to these birds and insects in preserving them from danger. Grouse, if not destroyed at some period of their lives, would increase in countless numbers; they are known to suffer largely from birds of prey; and hawks are guided by eyesight to their prey,—so much so, that on

CHAP. IV. NATURAL SELECTION. 85

parts of the Continent persons are warned not to keep white pigeons, as being the most liable to destruction. Hence I can see no reason to doubt that ~~natural selection~~ might be most effective in giving the proper colour to each kind of grouse, and in keeping that colour, when once acquired, true and constant. Nor ought we to think that the occasional destruction of an animal of any particular colour would produce little effect: we should remember how essential it is in a flock of white sheep to destroy every lamb with the faintest trace of black. In plants the down on the fruit and the colour of the flesh are considered by botanists as characters of the most trifling importance: yet we hear from an excellent horticulturist, Downing, that in the United States smooth-skinned fruits suffer far more from a beetle, a curculio, than those with down; that purple plums suffer far more from a certain disease than yellow plums; whereas another disease attacks yellow-fleshed peaches far more than those with other coloured flesh. If, with all the aids of art, these slight differences make a great difference in cultivating the several varieties, assuredly, in a state of nature, where the trees would have to struggle with other trees and with a host of enemies, such differences would effectually settle which variety, whether a smooth or downy, a yellow or purple fleshed fruit, should succeed.

survival of the fittest,

In looking at many small points of difference between species, which, as far as our ignorance permits us to judge, seem to be quite unimportant, we must not forget that climate, food, &c., probably produce some slight and direct effect. It is, however, far more necessary to bear in mind that there are many unknown laws of correlation of growth, which, when one part of the organisation is modified through variation, and the modifications are accumulated by ~~natural selection~~ for

survival of the fittest,

Mudie's Select Circulating Library distributed Darwin's *Origin of Species* along with fashionable literature of the day. The opening of Mudie's premises in Oxford Street, London, was an important social event.

(*Illustrated London News*)

The Natural History Museum, Oxford, scene of the 1860 British Association meeting where T. H. Huxley clashed with Bishop Wilberforce over ape ancestry.

(*Illustrated London News*)

In 1861 the humorous magazine *Punch* began a longlasting relationship with evolutionary theory.

(Wellcome Library, London)

MONKEYANA.
AM I A MAN AND A BROTHER?

Thomas Henry Huxley (right) and Samuel Wilberforce photographed by C. L. Dodgson (Lewis Carroll) at the 1860 British Association meeting.

(M. L. Parrish Collection, Princeton University Library; and W. Tuckwell, *Reminiscences of Oxford* 1900)

Benjamin Disraeli caricatured in 1864 after he asked the question "Is man an ape or an angel?"

(*Punch*. Wellcome Library, London)

Paul Du Chaillu brought the skins of gorillas to England just as the Darwinian controversy ignited. Victorians were shocked by thoughts of their possible ancestors.

(P. Du Chaillu, *Explorations and Adventures* 1861. Wellcome Library, London)

Popular opinion immediately associated Darwin's theories with ape-men and missing links. Julia Pastrana toured Europe as an exhibit in the 1860s.

(Royal College of Surgeons, London)

T. H. Huxley's polemical picture showing the anatomical relationship between primates.

(T. H. Huxley, *Man's Place in Nature* 1863. Wellcome Library, London)

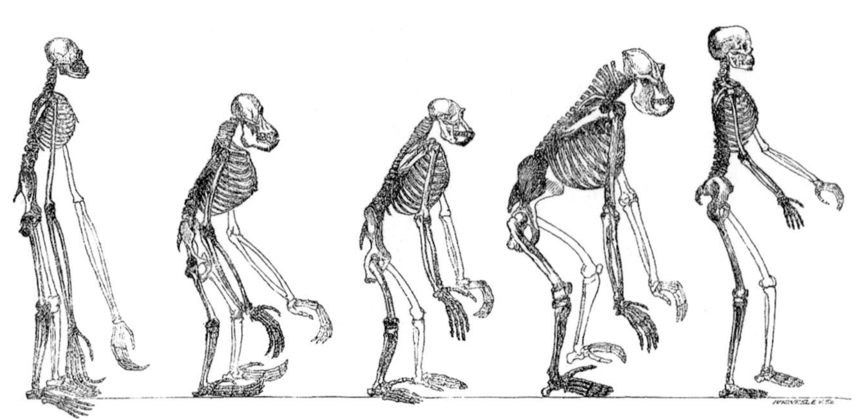

> all I would hope is that you might direct them upwards, as well as to one who values them above every thing in the world. I shall keep this by me till I feel cheerful & comfortable again about you, but it has passed through my mind often lately so I thought I would write it partly to relieve my own mind.

"God bless you," wrote Darwin in the margin. She was asking him to do the impossible. Come what may he was the author of the *Origin of Species.*

## VIII

Early in 1861, Murray asked Darwin for a third edition of his book, plainly confident that it would sell. This was published in April in a print run of two thousand copies, a fully revised edition. The title page indicated that this edition brought the total number of copies printed up to seven thousand. The most significant adjustments Darwin made were to include the historical sketch that he had attached to Bronn's German translation and the authorised American edition of 1860. He added a notice of the forthcoming publication of Asa Gray's pamphlet *Natural Selection Not Inconsistent with Natural Theology.*

Arranging publication of Gray's pamphlet was one of Darwin's more subtle moves. He had been impressed by Gray's positive, thoughtful reviews of the *Origin of Species* in the *Atlantic Monthly.* Keen to spread the good news, Darwin arranged to have these reviews reprinted and distributed in Britain at his own expense, under the title *Natural Selection Not Inconsistent with Natural Theology. A Free Examination of Darwin's Treatise on the Origin of Species, and of Its American Reviewers* (1861). He was a trifle peeved that Murray refused to accept this commission, saying it was hardly a commercial proposition; and instead Darwin got it printed by Trubner's and distributed it personally from Down House, each copy of the pamphlet accompanied by one of his most beguiling letters. He asked Huxley and Hooker to get it mentioned in *Natural History Review* and the *Gardeners' Chronicle* respectively. By reprinting and distributing a favourable review, Darwin was taking the process of persuasion a step further than he had attempted before, and he was pleased when it appeared to be successful. In years to come he was to make the same move several times again, a relatively unnoticed aspect of his propaganda campaign.

In this pamphlet Gray argued that evolution should not be dismissed as an atheistic horror story. He said that Darwin's book ought to be given a fair hearing, free from the blasts of religious prejudice. He discussed the other reviews that had appeared in the United States and showed where their criticisms were answered by Darwin. All this was balm to Darwin's ears. Although Darwin disagreed with Gray's theological compromises—Gray suggested that God might create favourable variations and thereby still oversee the evolutionary process from a distance—he also believed that Gray showed that faith and natural selection were not mutually exclusive: "*far* the best theistic essay," he said, he had ever read.[63] Gray's science was impeccable. His religious integrity was obvious. Darwin persuaded Gray to drop the customary anonymity and to allow his name to go on the pamphlet's title page ("indispensable"). He went to a great deal of trouble to distribute the pamphlet to the people and journals he felt mattered most in Europe. And he felt particular satisfaction in sending presentation copies to Bishop Wilberforce and the *Athenaeum*—satisfaction and smug revenge.

As for the *Origin of Species* itself, Darwin's financial account with Murray was healthy. The first edition brought him a payment of £180; the second, two instalments of £318 6s. 8d and £275 8s., which probably reflected his sudden notoriety and the increased print runs as much as anything. Murray's payment for the third edition of 1861 was £372.[64] Darwin felt he had done his best for Murray and did not produce another full-scale revision until 1866.

## IX

All the while, apes were pushing noisily to the fore. Huxley and Owen's increasingly accessible arguments made these creatures the topic of the day, a striking instance of the way that a particular species of animal can occasionally make manifest all the concerns of a human community unsettled by new ideas.[65] Apes and monkeys (the general public were none too precise about the distinction) soon represented an exotic combination of all the issues involved in the evolutionary debate: fear, disgust, anatomy, theology, mankind's created status, humour, morality, vulgarity, and sheer sensationalism.

Huxley did not have the debate all his own way. As chance would have it, the African explorer Paul Du Chaillu erupted onto the English lecture circuit early in 1861, full of stories about gorillas, a beast almost unknown to specialists and the public alike. Du Chaillu brought with him skins, skulls, and pickled specimens in barrels—the stuff of nightmares for some, horrified attention for others. He was a French-American who had grown up on the west coast of Africa, only twenty years old, and attractively unconventional, appearing on the London stage as a longhaired showman in buckskins; he was a charlatan whose reputation oscillated as wildly as the truth of his stories.[66] On arriving in England he contacted

Richard Owen, saying he wished to place his collection of gorilla specimens—the "Man Monkey" as the popular press had it—at the British Museum's service.

Swiftly, Owen coopted Du Chaillu for his own purposes. These included the hope of silencing Huxley and ramming home the need for a national museum of natural history with himself at its head. Owen arranged for Du Chaillu to speak about gorillas at the Royal Geographical Society on 25 February 1861, one of the most spectacular meetings of the year, attended by cabinet ministers and their wives, as well as scientists and other interested parties.[67] Taken aback by the pelts and skulls that Du Chaillu displayed on the platform, this elite audience turned the speaker into an overnight phenomenon. With Owen's support, Du Chaillu went on to talk at the Royal Institution and the Ethnological Society, and his *Explorations and Adventures* was published later in the spring by John Murray. No doubt Du Chaillu found his relationship with Owen useful for publicity purposes, and he kept it warm with hints of more specimens to come. He hooked Owen for good by giving him a full-size male gorilla skin for the British Museum, complete with a scattering of bullet holes.

Du Chaillu's colourful accounts were perfectly tailored to the stereotypes of his day. He claimed that gorillas were ferocious beasts, that they attacked without provocation, that their roars shook the woods. This news was apparently confirmed by Du Chaillu's sketch of a gunman (himself) about to shoot a rampaging male animal. The scene was full of explicit Victorian imagery. Mankind versus nature; colonial conquest; civilisation versus the wild; humanity confronting the beast within: all these could be read into the picture, and were read. No amount of learned talk about the anatomy of brains could have framed the essence of the ape ancestry question so vividly. It was popularly said that Darwin proposed that Victorian men and women, the presumed flowers of civilised society, were direct descendants of vile beasts like these. Huxley, Owen, Du Chaillu, evolution, monkeys, and gorillas tumbled together in people's minds.

The satirical journals picked up the conceit immediately. "Am I a Man and a Brother?" asked a gorilla in the May 1861 number of *Punch*.

*Am I satyr or man?*
*Pray tell me who can,*
*And settle my place in the scale.*
*A man in ape's shape,*
*An anthropoid ape,*
*Or monkey deprived of his tail?*

*Then Huxley and Owen,*
*With rivalry glowing,*
*With pen and ink rush to the scratch;*
*Tis brain versus brain,*
*Till one of them's slain;*
*By Jove! it will be a good match!*

On and on the *Punch* verses went. The anonymous author was Sir Phillip Egerton, a Tory M.P. with a fine collection of fossil fish but not much of a reputation for wit or poetry. Egerton was a trustee of the Royal College of Surgeons and the British Museum, and, in an institutional sense, he acted as Owen's patron. When Huxley discovered who the author was, he squealed in triumph. For Egerton to defect from the Owenite camp and write such an amusing squib "speaks volumes for Owen's perfect success in damning himself."[68]

*Blackwood's Magazine* continued the apish theme with a rhyme set to a drinking tune.

*Pouters, tumblers, and fantails are from the same source;*
*The racer and hack may be traced to one horse:*
*So Men were developed from Monkeys, of course,*
*Which nobody can deny.*
*Which nobody can deny. . . .*[69]

Irresistibly, Victorian humorists declared that apes were more intelligent than men because they at least knew when to keep silent. Typological satire flooded the pages of *Punch,* with guest appearances from Mr. O'Rilla and Professor Porpus. Soon a Mr. G-G-G-O-O-O-rilla, beautifully dressed in evening clothes, was pictured arriving as a guest at a high-society party. This was quintessential *Punch* display, linking obscure trivia with scientific parody.[70] Appreciative of the public taste for apes, *Punch* dedicated its 1861 Christmas Annual to the gorilla and pictured the magazine's imaginary Mr. Punch playing leapfrog with his alter ego for the year. These popular writings and cartoons not only conveyed the central point under debate and expressed the fears of traditionally minded British readers but also generated a nonspecialist, vernacular idiom for discussing and understanding the question.

Gorillas did not go away. Owen spoke about Du Chaillu's specimens in a lecture in 1861 at the Royal Institution titled "The Gorilla and the Negro" and crossly responded to Huxley's cruel post-Oxford article. The brain question, he said, was not one of fact but of interpretation. Owen claimed his "hippocampus minor" applied to humans alone and was not a term that could be used indiscriminately when speaking about animal brains. Though the gorilla was evidently much closer in structure to mankind than any other ape, this skilled anatomist asserted that its behaviour and its brain anatomy indicated an unbridgeable divide.

Owen's paper made a splash in the pages of the *Athenaeum,* accompanied by drawings of gorilla skulls, eliciting a sharp reply from Huxley criticising points of pictorial detail. Owen responded in chilly tones in the letter columns on 6 April, to which Huxley retorted in the same columns, "Life is too short to occupy oneself with the slaying of the slain more than once."[71]

Darwin and Hooker held their breath. This time Huxley really might have gone too far. He had descended into invective. But when Hooker told Darwin to calm down, the reply came back as smartly as Huxley's. Darwin said he would never forget Owen's behaviour over the *Edinburgh Review.*

> In simple truth I am become quite demoniacal about Owen, worse than Huxley, & I told Huxley that I shd. put myself under his care to be rendered milder. But I mean to try to get more angelic in my feelings; yet I shall never forget his cordial shake of the hand when he was writing as spitefully as he possibly could against me. . . . Oh dear this does not look like becoming more angelic in my temper.[72]

The apish unpleasantness continued in the next number of Huxley's *Natural History Review,* in which George Rolleston, the Oxford professor of comparative anatomy, published a slightly more temperate paper on brains, comparing an orang-utan with a human, but still defending Huxley's position. In a letter to the *Annals and Magazine of Natural History* Owen again pointed out that the issue was not a matter of fact but one of definition. Huxley publicly dissected a spider monkey that had recently died in the Zoological Gardens to substantiate his alternative case.

When Du Chaillu's book was issued early in May, Owen's camp started to waver. John Edward Gray of the British Museum declared Du Chaillu's gorilla story must be fantasy. With an incongruous twist of museum expertise, he pointed out that the donated skin displayed no gunshot holes in the chest as Du Chaillu's picture suggested. The only holes were in the back of the skull, as demonstrated in a diagram he sent to the *Athenaeum.*[73] The implication was that Du Chaillu was a liar, and perhaps a coward as well, for he had evidently shot his specimen in the back. The *Westminster Review* condemned Du Chaillu and ran through the hippocampus controversy again, this time favoring Huxley's view—the author was John Chapman, a friend of Herbert Spencer and George Lewes. Huxley's own journal, the *Natural History Review,* guardedly mentioned Du Chaillu's "rather vivid imagination." In the end, the *Athenaeum* published letters from all the protagonists, each defending himself to the hilt, even from Du Chaillu's brother-in-law, who accused his relative of embroidering the facts. The British Museum rushed out a commercial picture postcard showing the skeletons of a man and a gorilla, amicably standing side by side.

X

Charles Kingsley had the last word. By now Kingsley felt he knew every detail of the controversy, every nuance of the elaborate performances played out on the British Association stage, and every crevice of the personalities involved. The British Association meeting at Cambridge in 1862 provided him with his moment. Coming so soon after the *Athenaeum* articles and letters, this was an excitable meeting, the summit of Owen and Huxley's long-running personal encounter. Huxley was president of the section in which Owen presented two anti-Darwinian papers. One was on the aye-aye, a tree-climbing lemur from Madagascar; Owen claimed that the adaptations that suited this animal for an arboreal, insect-eating life disproved evolution. In the second, Owen dwelled defiantly on the brain again and introduced the age-old question of whether apes have toes or thumbs. Owen appeared to be "lying & shuffling," said Huxley. He made mincemeat out of Owen's thumbs and toes. One by one, Huxley's anatomical friends rose in his slipstream to the attack.

Kingsley produced a privately printed skit during the meeting composed in the style of Lord Dundreary, a comic theatrical character of the season, called *Speech of Lord Dundreary in section D, on Friday last on the great Hippocampus question.* A hippocampus in the human brain was a rum thing, said Dundreary. "I never felt one in mine; but perhaps it's dead and so didn't stir."

> The other gentleman who got up last, Mr. Flower, you know, he said that it was all over the ape everywhere—all over hippocampuses, from head to foot, poor beast, like a dog all over ticks! . . . And Prof. O. said it wasn't in apes at all; but only in the order Bimana, that's you and me. . . . So one must be right, and all the rest wrong, or else one of them wrong, and all the rest right—you see that?[74]

Not long after, Kingsley dreamed up the storyline of his children's book *The Water Babies.* "*Such* a story" said Alexander Macmillan, the publisher, gratefully. *The Water Babies* was published in instalments in *Macmillan's Magazine* beginning in August 1862 and then as a book by Macmillan in 1863. Alongside the religious cleansing and evolutionary transformations experienced by Kingsley's young hero, the soot-blackened chimney-sweep Tom, lay Kingsley's characterisation of contemporary science, in which he brought the thinly disguised figures of Huxley, Owen, and Du Chaillu and the hippocampus debate to a wider Victorian readership. If a water baby existed, wrote Kingsley, it would have to be cut in half, one half for Owen, the other for Huxley. Professor Ptthmllnsprts (Put-them-all-in-spirits) was Huxley. Deftly, Kingsley had Ptthmllnsprts claim nothing was true except for what he could see, hear, taste, or handle.

> He had even got up once at the British Association, and declared that apes had hippopotamus majors in their brains just as men have. Which was a shocking thing to say; for, if it were so, what would become of the faith, hope and charity of immortal millions? You may think there are other more important differences between you and an ape, such as being able to speak, and make machines, and know right from wrong, and say your prayers, and other little matters of that kind; but that is a child's fancy, my dear. Nothing is to be depended on but the great hippopotamus test.[75]

The Owen-Huxley clash lent itself readily to Kingsley's mockery. "If a hippopotamus major is ever discovered in one single ape's brain, nothing will save your great-great-great-great-great-great-great-great-great-great-great-greater-greatest-grandmother from having been an ape too." Kingsley also poked fun at the entrenched positions the two men had adopted. Asked to explain why there were no water babies, the Huxley character rudely retorted, "Because there ain't." No other commentator so succinctly conveyed Huxley's bulldog spirit.

When Linley Sambourne came to illustrate the book in 1886, he included caricatures of Huxley and Owen examining a bottled baby.[76] Some six years after that, Huxley's own grandson, the future biologist Julian Huxley, was sufficiently confused by the literary fame of the episode to ask his grandfather whether he could look at this water baby in its bottle.[77]

In truth Darwin's proposals needed a well-publicised affray like this. Following hard in the footsteps of *Vestiges of the Natural History of Creation* and the Wilberforce-Huxley Oxford debate, the attention generated by apes and the arguments about apes propelled Darwin's ideas about evolution out of the arcane realms of learned journals and books into the ordinary world of humour, newspapers, and demotic literature. Mr. Punch's monkeys and gorillas, Du Chaillu's tall stories, and Huxley and Owen's battle of wits forced the full implications of Darwin's densely packed theory to sink in much more quickly and thoroughly than he could ever have expected.

This time George Henry Lewes got it absolutely right for *Blackwood's Magazine* in 1861. "The Darwinian hypothesis . . . is clamorously rejected by the conservative minds, because it is thought to be revolutionary, and not less eagerly accepted by insurgent minds, because it is thought destructive of old doctrines."[78]

*part two*

# EXPERIMENTER

*chapter*

5

# EYES AMONG THE LEAVES

SOON THE WARMER weather brought out the wild flowers around Downe—a mass of bluebells in the Sandwalk followed by ox-eye daisies in the home meadow and other species a little further away in the woods and fields towards the village of Cudham, where Darwin and Emma liked to walk in the afternoons to a place they called Orchis Bank. Resting on that grassy bank, idly listening to bees humming round the flowerheads, was far nicer than worrying about reviews. "Observing is much better sport than writing," Darwin admitted.[1] The location carried a special place in Darwin's and Emma's affections, for these walks were mentioned time after time in their respective memories of each other, and it may well have been the same sweetly tangled bank, "clothed with many plants of many kinds," that Darwin had in mind when bringing the *Origin of Species* to its close. "Larks abound here & their songs sound most agreeably on all sides; nightingales are common," he recorded appreciatively soon after their move to Down.[2]

He had always found the study of plants to be a pleasant combination of relaxation and interest. Whenever he felt over-stretched or ill, a few botanical investigations usually soothed his troubled mind. After the *Origin of Species* was published they helped him forget how haggard he was becoming: how much he was turning into a letter-writing machine, forever defending his pitch, incessantly chipping away at the walls of resistance, always courting approval, nudging, or explaining. Rambling about the garden or along the Kentish lanes took him away from the ferment whipped up by Huxley and others in London.

He would usually laugh at these placid interests of his and claim he was no botanist, saying he was "a man who hardly knows a daisy from a dandelion." Certainly, he rarely ventured into what he thought of as real botany, the herbarium-based sciences of taxonomy and morphology at which Hooker and Gray excelled. His friends were much better at these than he was. Yet he loved to puzzle over the quietly complicated lives of plants. Ever since coming to Downe, with its fields and lawns set in the chalky countryside, he took daily interest in the green activity going on around him.

But even Darwin was surprised by the ardour for orchids that coursed over him in the middle of 1861, something like an unexpected love affair late in life. Attracted by their beauty and diversity, he pushed his book on variation aside and launched himself into a complete reevaluation of the anatomy and reproduction of these complicated plants, in the end publishing a small monograph on the subject in 1862, his first theoretical work after the *Origin of Species*. In the process, he came to admire orchids with the single-minded devotion that he had once given to barnacles; and he sometimes wondered if his addiction was turning into "another barnacle job." Even he, the most attentive of strategists, could see the incongruity of following the magisterial pace of the *Origin* with a little book on flowers.

The endeavour, as it happened, did turn into another barnacle project, for many reasons and at many levels. First and foremost was the feeling that he was being introduced to something new and beautiful. Everywhere he spoke of his curiosity about orchids, his appreciation, and how "very lucky" he was in his "beautiful facts."

Furthermore, he probably longed to get away from his dull writing work on variation for Murray. He said he felt tired, sometimes bored, with the evolutionary controversy, irked by wrenching answers to critics out of his churning brain. He had written or dealt with three English editions, two American editions, and two translations of his *Origin* in three years. He sensed that his friends would probably carry on promoting evolution without his constant personal intervention, for a while at least, more effectively than he in most cases. He lacked the energy for confronting difficult public situations. This, coupled with an apparent disinclination to expose himself to intense emotion, encouraged a temporary retreat. His greenhouse or his study was a good place to be when the intellectual wind howled around his ankles. "I am got intensely interested," he confided. "I cannot fancy anything more perfect than the many curious contrivances."

Practical investigations therefore seemed particularly alluring. He wanted to pit his wits once again against the native ingenuity of animals and plants, to be able to follow a lucky hunch, and use all his guile and skill to arrive at a satisfactory conclusion. He liked to win—no doubt about that. More than that, he loved the spirit of the game itself. Inside every practical scientist the same pleasure in competing against nature lurks below the surface, the same enthusiasm for experiment, the same satisfaction in dreaming up new gambits and twists that can trick or tease the natural world into revealing its secrets. The sense of personal enjoyment that Darwin derived from research was strong. He relished the creative opportunities that lay in identifying significant and subtle problems that were not amenable to traditional approaches, and felt his theory gave him a way of looking at the world that helped unlock doors. "I am like a gambler, & love a wild experiment," he declared.[3] To the end of his days he said, "I cannot bear to be beaten." Perhaps without really understanding why, he needed to confront the raw material of nature again. His barnacle work had made him feel focussed and purposeful during a difficult time in his life. Orchids looked set to supply a similar mental release. Writing the *Origin,* the book by which he was becoming known to posterity, could almost be thought of as a personal ordeal sandwiched in between.

## II

Not everything came to a peak together. When other plants stirred in the garden, Darwin began a multitude of small experiments that he had been saving up over the dull winter months. He looked briefly at sundews again and investigated a Venus's fly-trap which Hooker sent to him from the Kew glasshouses. Darwin greeted the fly-trap with cries of delight, because this species closed its hinged and spiky leaves over the flies like a mantrap. "How curious it is to see a fly caught & how beautiful are the adaptations compared with Drosera." He speculated about possible connections between sundew hairs and the spines lining the fly-trap's outer edges.

As soon as he could get outside in the weak spring sunshine, he also repeated some observations on garden plants that had been rained off the previous summer. Assiduously, he reactivated his correspondence with Daniel Oliver, the senior curator at Kew Gardens. Oliver was an ideal contact for him, interested in the same kind of living plant functions as Darwin, willing to follow up research topics, and soon to become professor of botany at University College London, where he took many of Darwin's ideas into the world of academic botany. Hooker had introduced them a few months earlier, which proved to be an inspired move. "He must be astonished at not having a string of questions," Darwin cheerily remarked to Hooker. "I fear he will get out of practice!"

Primroses and cowslips bloomed early that year and Darwin went to work cross-pollinating the different kinds. On his instructions, Brooks and Lettington, the Down House gardeners, transplanted a number of wild primroses into an experimental bed in the kitchen garden. Darwin intended to discover whether the two separate forms of flowers—the long-styled (pin-headed) and short-styled (thrum)—were specially adapted to fertilise each other. Although each form of flower was self-fertile, he thought pin-to-thrum matings were more likely, on his theories, to succeed better than pin-to-pin or thrum-to-thrum. A discrepancy in the size of pollen grains suggested that this might be the case. "I do not know whether I shall suceed in making out the meaning of the dimorphism," he told Oliver in April 1861, "but I have not been idle, for I have made much above 100 crosses with the pollen of the different sizes."

He crouched in the flowerbeds transferring pollen grains with a fine paintbrush from primrose to primrose. Then he waited for the fertilised plants to set seed. In May he counted the number of seedpods on each plant. A week or two later, he collected the pods, unzipped them, and weighed the seeds from different batches, using his old apothecary's balance from the Shrewsbury days, having decided that the weight of the seeds would be the best way to establish each plant's relative fertility. By June he had confirmed his expectations. If the two forms were crossed, they were at their most productive. If they were allowed to fertilise their own kind, much less so. Darwin concluded that the flowers were physically adapted to facilitate outbreeding, almost as if the original hermaphroditic condition was gradually differentiating into two sexes. "I think that you will think that I have made out the meaning of dimorphism in Primula satisfactorily, & a very odd case it is & has caused me much labour in artificial crossing," he reported to Hooker. He sent a long, original paper on this subject to the Linnean Society in November 1861. He had created something meaningful out of apparently insignificant researches.

In practical terms, Darwin's experiment in this area was perhaps neither more nor less than any other natural historian or amateur botanist might have pursued for amusement during the early summer months. Yet he brought to it an over-riding preoccupation with the consequences of the results. This feature of his practical work remained constant throughout a long life. Every experiment he was to devise in future years was performed with its ultimate relevance for evolution by natural selection in mind. He rarely tried things out merely to see what might happen—or at least, even his most improbable inquiries were executed with some hypothesis or "wild speculation" to explore.[4] "He often said that no one could be a good observer unless he was an active theoriser," remarked his son Francis much later on.

> It was as if he were charged with theorising power ready to flow into any channel on the slightest disturbance, so that no fact, however small, could avoid releasing a stream of theory, and thus the fact became magnified into importance. In this way it naturally happened that many untenable theories occurred to him; but fortunately his richness of imagination was equalled by his power of judging and condemning the thoughts that occurred to him. He was just to his theories, and did not condemn them unheard; and so it happened that he was willing to test what would seem to most people not at all worth testing. . . . The love of experiment was very strong in him, and I can remember the way he would say, "I shan't be easy till I have tried it," as if an outside force were driving him.[5]

That "outside force" was the notion of natural selection. Furthermore, Darwin must have had a firm belief—or a naïve belief—in the possibility of getting answers from his work. As often as not, experimental science goes nowhere. It either runs into a dead end or provides results that cannot be used in the way that was intended. Experiments frequently fail and a new route into the problem has to be devised. Small-scale investigations like Darwin's carried the additional hazard of being so small as to be insignificant to the larger scheme of things, barely a drop in the intellectual ocean. Nonetheless, Darwin seems to have convinced himself that the hours of work that he dedicated to counting seeds or measuring seedlings would be pertinent. "Another quality which was shown in his experimental work, was his power of sticking to a subject; he used almost to apologise for his patience, saying that he could not bear to be beaten, as if this were a sign of weakness on his part," continued Francis. "Perseverance seems hardly to express his almost fierce desire to force the truth to reveal itself."

This faith in the power of practical work showed Darwin at his most pragmatic. He evidently believed that the best kind of biological science, the science that was taking shape as a discipline during the middle years of the nineteenth century, ought to be based on a culture of tangible experience. He felt confident that the solutions he wanted resided somewhere in his raw materials. His self-appointed task was to find them.

## III

After this encouraging start, Darwin eased into a happy summer of orchids. For it was the existence of even more complicated reproductive adaptations in orchids that so fascinated him. Orchid flowers, which normally contain both male and female reproductive organs, do not usually fertilise themselves, he insisted, despite the proximity of the sexual parts. Instead, he thought the internal anatomy was arranged to ensure that the male pollen must be carried by an insect to another plant or flower, and that the female organs always received pollen from another source. In British and European orchids the process almost always involved bees, first described by Conrad Sprengel in 1793. Darwin thought Sprengel's book was "wonderful," and liked to praise the neglected merits of its author.[6] Long ago Darwin too had noticed the internal structure of wild orchids and surmised that bees and moths carried pollen from one flower to the next.

But this was only the half of it. In his garden or thereabouts he could find several native kinds. The unobtrusive green-winged orchid (*Orchis morio*) came up with the cowslips in the field, and the common orchid, *Orchis mascula,* usually known as Shakespeare's long purples, arrived later, in the wooded area by the Sandwalk, the copse at the end of the Down House estate. There were other sorts on banks and ditches within a short walk; "no British county excels Kent in the number of its orchids," he said appreciatively. The lower lip of the flowers, on which the insects landed, was either lobed or frilled, ornamented with coloured spots and patches, and sported any number of hairs, ridges, keels, wings, or spurs. The thought that these were lures for insects was unavoidable. Further inside, the male pollen masses were attached to thin, flexible stalks, like a pair of fragile, movable antennae, and, just below, what appeared to be the top of the female organs expanded out into a flat sticky plate or rostellum. The size, shape, and positioning of these male and female parts were different in every species. Darwin wondered about their origins, and how the flowers could possibly function. His intention was to show that even the most complex structures and life cycles, even those that depended on completely different organisms such as insects for their fulfillment, could be explained by natural selection. Where most people tended to regard plants like orchids as the handiwork of God, Darwin saw the flowers as a marvellous collection of *ad hoc* evolutionary adaptations.

Armed with an array of tin cans and biscuit boxes to serve as containers, Darwin tramped the countryside in search of orchids, getting Brooks and Lettington to pot up the specimens he brought back. These employees had to stretch their traditional vegetable-garden repertoire to cope with a variety of demanding species that eventually included *Epipactis latifolia,* a woodland orchid which changes shape according to soil conditions, and *Goodyera repens,* a native of Scotland, which requires special bedding-out in peat all on its own. Darwin discovered it was best for his experiments if he dug up the whole plant, roots and all. He needed a ready supply of flowering spikes and access for observations in situ. The difficulties of germination (he did not know that many required the presence of a special fungus) and the long wait for the flowers to appear on mature plants made them an additional challenge.

Inevitably, too, he began corresponding with field naturalists, horticulturists, botanists, and country gentlefolk who might be able to help. These contacts—especially Alexander More, who lived on the Isle of Wight, a site of special botanical interest—sent him native orchids from various parts of the British Isles. Alexander More often put fresh plants on the overnight train to Farnborough, and Parslow, the butler, would ride down to the station to collect whichever carefully sealed package had arrived that day.

Darwin also began manoeuvring among a new group of correspondents, this time botanical women. As a subject, botany then enjoyed enormous popularity with nonprofessionals and was often associated in the public mind with respectable middle-class activities such as gardening and flower-painting. It was also particularly popular with women.[7] On John Lindley's advice, Darwin approached Lady Dorothy Nevill, the political hostess, to ask if she might be able to send him a few hothouse orchids from her collection. Lady Dorothy was an avid reader, writer, and gardener who became very much interested in the *Origin of Species*. Her hothouses—as Darwin hoped—contained unusual plants sent to her by friends from several parts of the world. In his first letter to her, he was beguilingly candid about his exotic desires, saying he would very much like "Limodoridae, Vanillidae &c, especially Mormodes & Cycnoches, Bonatea, Masdevillia, and any Bolbopyllum with its lower lip or labellum irritable."

Lady Dorothy collected famous botanists rather as she collected plants. She evidently considered Darwin a good catch. "I am so pleased to help in any way the labours of such a man—it is quite an excitement for me in my quiet life, my intercourse with him. He promises to pay me a visit when in London. I am sure he will find I am the missing link between man and apes," she told Lady Airlie.[8] In return, Darwin was willing to be caught if it meant that he got his specimens. Over the years Lady Dorothy sent him several rare plants. She also asked for his photograph to hang, as she said, in her private sitting room opposite Sir William Hooker, a trophy gallery to which Darwin assigned himself with a sigh. His other lady acquaintances were usually less demanding.

Through Hooker, Darwin also acquired introductions to a series of highly competent garden directors in the colonies, and to the botanical brothers Roland and Henry Trimen, working in South Africa and Ceylon respectively, each of them a mine of useful information.

Family outings contributed to the cause. During one long and energetic morning while staying at Elizabeth Wedgwood's house, Darwin transplanted *Malaxis paludosa,* the marsh orchid, from one boggy patch to another to see if insects were prepared to visit an intruder. Covered in mud, he returned to say that the insects did. At another time, his son George remembered being sent out to Cudham on a balmy summer night to catch and count the moths visiting plants on Orchis Bank, followed by an evening spent sitting at his father's table surrounded by printed catalogues identifying the species. Once or twice Parslow put on leather gaiters to gather flower spikes from a ditch. When his master's supply ran low, he took the London train to collect choice government specimens from Hooker.

Darwin's neighbours were similarly obliging—his was a passion that could be understood and indulged. George Turnbull, of the Rookery on the other side of the village, let Darwin use his heated greenhouse for the more delicate specimens that came into his hands, and tolerantly watched as his own gardener, John Horwood, was swept into Darwin's voracious information system.[9] Horwood was more knowledgeable than Darwin's own gardeners and had experience with hothouse cultivation techniques. On one occasion, he was the only person able to identify plants that arrived from Kew without their labels. Probably gratified at being able to display his professional expertise, Horwood went on to fertilise *Vinca rosea* for Darwin, a demonstration of technical virtuosity that he undertook only when several contributors to the *Gardeners' Chronicle* recorded constant failure.

Moreover, Darwin was fortunate in being able to tap the horticultural vigour of the times. He benefitted greatly from the surge in botanical expertise that characterised the Victorian temper of life, the new clubs, societies, training schemes, glasshouses, heating systems, illustrated magazines, and libraries. He obtained easy access to an abundance of specialist nurserymen and seedsmen, plant breeders, gardening authors, and skilled practical men, all of whom contributed in one way or another to the great fervour for plants that marked the middle years of the nineteenth century. Rapid developments in glasshouse technology were benefitting amateurs and professionals alike, at the same time as new trade routes and overseas connections made the importation of fancy species more feasible, and the restless urges of the middle classes were spilling into natural history fashions and crazes as never before. Botany, with all its ramifications into colonial enterprise, the plantation system, horticulture, herbarium research, agriculture, recreation, and fashion, was the "big science" of its day as well as a rising popular phenomenon. Plant breeding became an absorbing hobby for some, the backbone of a prime business concern for others. Horticulture diversified into profitable commercial propositions as new forms of stoves, plate glass, iron girders, mechanical ventilation systems, pipework, and the park-keeper's urge for gaily coloured bedding-out schemes in leisure gardens all came into being in the middle years of the century. Although Britons have historically always considered themselves a nation of gardeners, there was something unique about this particular combination of plants, business acumen, public attention, and technology that generated unprecedented attention.

There could hardly have been a better time to take up the study of orchids either. The Royal Botanic Garden at Kew naturally involved itself with the tropical species, making full use of Britain's ascendency overseas to collect a wide variety of species. These were essentially plants for research and colonial display. At the same time, the Horticultural Society in London (soon to add "Royal" to its name) encouraged its members to grow show specimens, building an orchid house in the society's public gardens at Chiswick and stimulating the growth of commercial firms such as James Veitch & Sons, which built up a successful business in breeding new blooms. In 1859 the Horticultural Society awarded its first Certificate of Outstanding Merit at a show to a cattleya bred by James and John Veitch. This previously rather specialised taste for orchids developed into a vogue during the 1860s and 1870s, rather as tulipomania had gripped Europeans in the seventeenth century and pinks and auriculas excited pre-war gardeners in the early twentieth.[10] Reputations soared with the development of a single exceptional bloom. Incomes improved; botanical skill, determination, and personal endeavour were rewarded.

Few enthusiasts succeeded in keeping their orchids alive for very long. Early misunderstandings about potting techniques and temperatures defeated nearly everyone, and fresh specimens were constantly imported to replace the old. Looking around the stalls at one horticultural show, Hooker sadly remarked that England was the "grave of tropical orchids." He referred to the depredations around Rio de Janeiro, an area where Darwin had collected only thirty years before, which was subsequently stripped of all its native orchids, never to reappear.

Still, Darwin had a ready-made resource at his fingertips.[11]

## IV

He had noticed one of the more unusual adaptations before. Standing in his study one day he was tinkering with an orchid flower when the pollen masses suddenly shot out and hit the window: "about a yard's distance," he said in surprise. He tried to detonate other orchids by inserting thin paintbrushes and pencils into the mouth of the flowers. Few of them possessed similar high-velocity devices. But if he looked carefully, the pollen masses all displayed an ability to move on their stalks—a feature which had previously been mentioned in print by William Herbert, Henslow's botanical friend, and some other authors. Even so, Darwin seems to have been one of the first naturalists to ask what this movement might mean in terms of insect visitors. He decided that the pollen masses must bend or shoot themselves into the right position to stick onto an insect's back or head (usually the proboscis) for transfer to another flower. He had inquired about this possibility in the letter column of the *Gardeners' Chronicle* in 1860, and received a number of replies. One answer arrived at Down House only as an empty envelope. Just before he threw it away, Darwin investigated further. Tucked away at the bottom he found a number of insect mouthparts laden with tiny parcels of pollen.

He also noticed a secondary reflex movement of the pollen masses (pollinia) after they were detached from the flower. He timed these movements with his pocket watch, finding they usually took around thirty seconds, just long enough for the bee to which they would usually be attached to push into another flower. His fancy ran away with him.

> A poet might imagine that whilst the pollinia were borne through the air from flower to flower, adhering to an insect's body, they voluntarily and eagerly placed themselves in that exact position in which alone they could hope to gain their wish and perpetuate their race.[12]

More to the point, adaptive evolution, proceeding blindly and without any foresight or plan, could explain an exquisite feature of natural life that would otherwise be regarded as convincing evidence for God-given design.

His work on orchids thereafter deliberately addressed the issue of design. To account for the apparent design of the living world, after all, had been the central plank of his argument in the *Origin of Species* and these plants gave him an opportunity to demonstrate the actual adaptive ingenuity of nature, his answer to William Paley's natural theology, the secular alternative to divine craftsmanship. When he came to publish this work he told Asa Gray decisively that his orchid studies represented "a flank movement on the enemy."[13]

The issue rode particularly high in his mind at this time because of Gray. In a series of letters to Darwin stretching from 1860 to 1862, and emerging out of his reading of the *Origin of Species,* Gray interrogated

Darwin about the origin of design in nature. Gray believed that the apparently perfect adaptions that the natural world exhibited—as in the mutual co-adaptations between insect and plant—reflected the thought and intent of a Creator, however such a deity might be understood. But he also recognised the efficacy of Darwin's notion of natural selection. In letters and reviews he volunteered a compromise solution in which favourable variations were perhaps produced by the hand of God. Natural selection would then pick or select these in the competition for life. Gray, in effect, wanted to have it both ways. God would carry on supervising nature, for it was the Almighty who introduced favourable variants. And natural selection would act just as Darwin had proposed, as an objective, mechanical, winnowing force which tailored organisms to their surroundings. In making this suggestion, Gray was to differ markedly from other creative evolutionists or those who advocated providential evolution. He gave Darwin his due and did not make God the selecting agent.

Darwin understood the dilemma. Gray was not advocating traditional forms of divine creation. The botanist William Harvey, Hooker's friend, had written to Darwin in an altogether stricter manner, saying that "the Divine Creator" could call up "without seed, from the dust of the ground a new organism, by the power of his omnipotent word." Gray was suggesting something much subtler, a form of divinely directed evolution. The threat was not lost on Darwin. If established, it would have destroyed the meaning of natural selection and any hope of a positivistic biology.

But he was frank. "I grieve to say that I cannot honestly go as far as you do about design." His *Origin* had meant to demolish just such a view. By showing how blind and gradual adaptation could produce the same results as the apparently purposeful design that William Paley glorified, and the perfect adaptations in the natural world that the Bridgewater Treatises displayed in all their ramifications, Darwin had intended to challenge the argument that design necessarily indicated the presence of a designer.[14] This inference, in which divine purpose was identified with human intent, had been used for centuries as one of the strongest proofs of the existence of an infinite deity. In natural selection Darwin substituted an alternative hypothesis that was both logically adequate, he thought, to produce the results seen and philosophically more secure.

Nevertheless, his respect for Gray—and, it must be said, his dependence on him—persuaded him to delve further into his own beliefs than he was accustomed to go, one of the few times that he turned his cool analytic gaze onto himself. He was perplexed, he told Gray. "I cannot think that the world, as we see it, is the result of chance; & yet I cannot look at each separate thing as the result of design."[15] In letter after letter they exchanged opinions, each trying to be as honest about himself as he believed the other deserved.

Neither was a trained theologian, and their views changed as they grew older. Certainly Darwin ended up as a nonbeliever. It seemed to Darwin that Gray merely reiterated the age-old claim that the creator can preordain events. What of "necessity & Free-will," and the "Origin of evil"? he inquired, subjects "quite beyond the scope of the human intellect."[16]

> With respect to the theological view of the question; this is always painful to me.—I am bewildered.—I had no intention to write atheistically. But I own I cannot see, as plainly as others do, & as I shd. wish to do, evidence of design & beneficence on all sides of us. There seems too much misery in the world. I cannot persuade myself that a beneficent & omnipotent God would have designedly created the Ichneumonidae with the express intention of their feeding within the living bodies of caterpillars, or that a cat should play with mice. Not believing this, I see no necessity in the belief that the eye was expressly designed. On the other hand I cannot anyhow be contented to view this wonderful universe & especially the nature of man, & to conclude that everything is the result of brute force. I am inclined to look at everything as resulting from designed laws, with the details, whether good or bad, left to the working out of what we may call chance. Not that this notion *at all* satisfies me. I feel most deeply that the whole subject is too profound for the human intellect. A dog might as well speculate on the mind of Newton.[17]

Gray gave him a sympathetic ear. Darwin wrote to Lyell too, an indication of the seriousness with which he regarded the question. Stones fall without God's intervention, he told Lyell. Sparrows die. Every scientist, he said, can accept the idea of gravity as a mathematical law operating according to natural rules. "I cannot believe that there is a bit more interference by the Creator in the construction of each species, than in the course of the planets."[18]

The actual existence, or not, of an Almighty was in this regard irrelevant, Darwin remarked, although he was prepared to let Gray, and doubtless Lyell, carry on believing that a divine authority could at one point have made the laws.

> One word more on "designed laws" & "undesigned results." I see a bird which I want for food, take my gun & kill it. I do this *designedly*.—An innocent & good man stands under a tree & is killed by flash of lightning. Do you believe (& I really shd. like to hear) that God *designedly* killed this man? Many or most persons do believe this; I can't & don't.—If you believe so, do you believe that when a swallow snaps up a gnat that God designed that that particu-

lar swallow shd. snap up that particular gnat at that particular instant? I believe that the man & the gnat are in same predicament.—If the death of neither man or gnat are designed, I see no good reason to believe that their *first* birth or production shd. necessarily be designed. Yet, as I said before, I cannot persuade myself that electricity acts, that the tree grows, that man aspires to loftiest conceptions all from blind, brute force.[19]

Decades afterwards, he said that around this time he probably deserved to be called a theist.[20] In truth, Darwin was profoundly conditioned to become the author of a doctrine inimical to religion. But he only gradually rose to understand the depths of his own implications long after the *Origin of Species* was published. Orchids made him think as hard about his own beliefs as he hoped his future readers would think about theirs.

## V

Over the next few months, Darwin channelled his efforts into understanding the basic anatomy and embryology of each of the major groups of orchids, taking his cue from the evolutionary idea that when an organ arises it is invariably a modification of something else. The task was not especially easy. Orchids were a deceptive group, even with the added benefit of evolutionary theory. He valued Robert Brown's observations on fertilisation and particularly studied Hooker's paper on the British orchid, *Listera ovata,* written in 1854. After several weeks of hard work with his microscope, however, he wondered if even a judicious botanist like Hooker might be wrong about the origin of the spiral vessels inside the stem, and he sent inky diagrams to Hooker for his opinion. Hooker could not work out what his friend meant; he had to dissect a few specimens at Kew to reexamine the parts in a way that an anatomical taxonomist would understand. Not for the first time, Hooker must have wondered what he would be sucked into next. Seeds, barnacles, sundews, species, orchids—Darwin could be a very demanding correspondent. Every postbag brought something different. Hooker's increasing commitments as assistant director at Kew meant these inquiries were taking more time than he could readily spare.

Darwin sent a slew of similar questions to Asa Gray, asking about *Spiranthes,* an orchid fairly common in North America but rare in Britain, called lady's tresses because of the plaited effect of the flowers, and *Goodyera,* a close relative. He questioned whether the spiral arrangement of flowers helped insects steer as they crawled up and around the spike, fertilising flowers as they went. He had never seen an insect visit *Spiran-*

*thes.* "It is no use watching this," he wrote to Alexander More on the Isle of Wight. "I watched it last autumn at Eastbourne till I was sick."[21] His friends and aquaintances began to feel amused by the intensity of Darwin's interest in apparently obscure topics. Yet these precise anatomical investigations helped him reframe what would come to be known as the ecological relationships between species on an explicitly evolutionary basis.

Furthermore, in Darwin's view, the orchid nectary was a modification of one of the petals and ought to secrete some sweetish juice to attract insects. "Unless the flowers were by some means rendered attractive, most of the species would be cursed with perpetual sterility," he remarked. The longer the nectary, the longer the insect's proboscis needed to be—a good example of co-adaptation. Neat as this argument was in principle, Darwin could not establish the existence of nectar. Unwilling to give up his theory, he ate a nectary himself to see if it tasted sugary. A little while later he acquired from the nation's greatest orchid specialist, James Bateman, a specimen of *Angraecum sesquipedale,* which possessed an exceptionally long nectary measuring eleven or twelve inches. "What a proboscis the moth that sucks it must have!" he exclaimed.[22] No such moth was then known to science. Years after Darwin's death, a moth with a twelve-inch proboscis was discovered in Madagascar.

He was so gripped by his passion that he did not stop during the holiday season. He took the family to Torquay in 1861 for the sake of Henrietta's health, bringing with him a selection of potted orchids, and his microscope, all wedged into a wooden crate that had to travel upright. He fussed all the way there. But he said he was glad to get away. "I have been a poor wretch for many months," he told his cousin Fox, and a large glass of port wine on arrival was "a very necessary restorative."[23] At Torquay Henrietta improved in the seaside air. The family trudged over the sands for picnics in nearby coves, leading Henrietta on a donkey, and rented a sea-bathing machine for her to follow fashionable example. "She gets up twice every day now & can walk one or two hundred yards," Darwin reported.

Orchids dominated the rest of his time. He sent the younger boys out over the Devon hills searching for specimens, knowing they would be happy enough scrambling around and getting dirty. Darwin directed affairs from the cliff paths with his walking stick. When William arrived from Cambridge University for a few days, hoping to discuss his future prospects, he too was sent off to participate. Darwin got him dissecting and drawing, glad to have his assistance. William had become adept at botany during the last few years and was accustomed to helping his father with his scientific chores. Docile as always, William probably found it hard to refuse.

Father and son evidently talked to each other as well. During this visit it was agreed that William should not read for the law after graduating from Cambridge but should instead cut his university days short and accept an opening in a banking firm in Southampton that John Lubbock had found for him. This would require leaving university without a degree. The two decided that William would merely postpone his B.A. examination and take it a year late, which he did do, returning to Cambridge for a few days in 1862 to sit his papers and complete the university's residential requirements.

Darwin was grateful for Lubbock's intervention in this affair. Previously he had not known what to recommend to his oldest son, especially as banking appeared to be closed to outsiders. A potential banker had to be invited to join a firm, invariably through the old boys' network, as in this instance, and had to pay handsomely for the privilege, almost like buying a commission in the army. Once in, the new partner could expect a relatively secure and respectable occupation. In friendly appreciation, Darwin sent Lubbock a collection of small insects from Devon collected by his sons from damp and smelly places under logs, confident that one naturalist would know best how to show gratitude to the other.

The bank in Southampton proved rather more difficult to please. Those kinds of privately owned banks depended on their partners' guarantees for the capital reserve, and Darwin was asked to guarantee the sum of £10,000 on his son's behalf in case there should be a run on deposits. Such guarantees remained the custom until the Joint Stock Banks and Companies Act of 1863 allowed limited liability and some redistribution of risk, a change in legislation brought in after a couple of big financial crashes in 1857 and 1858. Wary of the possible risk involved, and taken aback by the size of the sum expected, Darwin instructed his solicitor to look into the matter so that he would know exactly where he—and William—would stand in case of a crash or fraud (Darwin's own London bank suffered from a notorious fraud case in 1858). Privately, he wondered at Lubbock's casual, well-buttressed assumption that such an enormous guarantee was perfectly routine. By writing some hard-headed letters, he bludgeoned the bank's partners down to £5,000.

Otherwise, he was irrepressible. Hooker received increasingly urgent requests for specimens:

> I was really ashamed to bother you about Catasetum; I have now written to Parker & Williams for chance. I shd. be very glad of a Cat-

> tleya or Epidendron & specially (from what A. Gray writes) for an Arethusa or one of that section.—Have you Mormodes with some buds that would complete my desiderata for comparison. If the racemes were cut off & packed in tin-cannister with a little slightly damp moss, they might be sent by post, & I would pay postage; possibly Arethusa might come in pot, as that seems most important for me.[24]

Parcels came and went, tin cans were filled. The pace slackened when Erasmus called by, bringing Hope Wedgwood with him (Fanny and Hensleigh's youngest, roughly the same age as Henrietta). Darwin sent the ladies off on a scenic tour round Devon, Emma, Henrietta, and Hope together, the only time Emma went anywhere without small boys attached to her skirts and a trip remembered with pleasure for that reason by Henrietta. Left alone, the men discussed science at the dinner table, sticking pieces of ice together in order to understand John Tyndall's new theory of glacier motion.[25]

When they returned to Downe, Henrietta was better and Darwin worse. The travelling made him feel older than he was. "I cannot stand such fatigue & am in fact a man of seventy years old." He was actually fifty-two.

## VI

Autumn and winter passed in a haze of family concerns. Emma's sister Charlotte Langton became ill and died in January 1862, aged sixty-five. Charlotte's house was to be sold, breaking up the enclave of Wedgwood sisters at Hartfield, and Emma could not imagine how Elizabeth would cope on her own. Sure enough, after a spell living in a London house close to Erasmus and Hensleigh, in 1868 Elizabeth purchased a property in Downe village, called Tromer Lodge, to be near her remaining sister.

At Down House, influenza prostrated the household. Since Parslow was the only one left on his feet, Darwin sent him to Bromley with the horse and cart to get a doctor to minister to them all. As soon as Darwin recovered, he was struck down by eczema. It was a "miserable time." Even so, decisions were taken. He and Emma decided at last to send Leonard to boarding school, feeling that ill health should not delay his education any longer. Darwin feared that Leonard was "rather slow & backward (in part owing to loss of time from ill-health)." Aged twelve, he followed his brothers to Clapham School at the beginning of 1862. Henrietta felt sufficiently recovered to stay with her London cousins for a few weeks early in the season.

With Leonard and Henrietta gone, Horace emerged as a new source of anxiety. He was the youngest child, then aged ten, and began having what Darwin and Emma described as "fits," or the shakes. They thought at first he might have developed a neurological disorder caused by a bad knock on the head. Soon Darwin got it into his mind that Horace—like the others—was sinking into a hereditary disorder. However, with hindsight it seems equally probable that Horace may have been consumed with a boyish passion for Camilla Ludwig, the young German governess, who went everywhere with them and was a family favourite. Miss Ludwig arrived from Hamburg in 1860, after Miss Latter left, probably applying to become a governess to an English family only because of unexpectedly reduced circumstances. The little information that can be gathered about her indicates that she was apprehensive about the position. To be a governess was a delicate operation in Victorian England, neither a servant nor an equal, yet more intimate with the employer's family than with any other people beyond her own relatives or future spouse.[26] She supervised the children who remained at home, teaching German to Henrietta and Bessy and helping Darwin with his German language researches. It was Miss Ludwig who soothed Horace after every attack of the shakes.

Darwin recounted his worries to William:

> [He] has oddest attacks, many times a day, of shuddering & gasping & hysterical sobbing, semi-convulsive movements, with much distress of feeling. These semi-convulsive movements have been less during these last few days, & are never accompanied by loss of consciousness. Do you remember his being pitched out of the Truck: Mr Headland thinks his brain probably suffered a little concussion; but I cannot help thinking that it is all due to some extreme irritation of stomach. Miss Ludwig is unspeakably kind to him, & he will remain with her all day & night. We shall have no peace in life till the poor dear sweet little man gets better.[27]

After a while it looked as if the mysterious attacks mostly came on after meals, and Darwin took Horace to London to see the general surgeon Mr. Headland, who prescribed "pepsine" for stomach trouble and an occasional blister to draw out any toxins. By now Darwin blamed himself, convinced that he had transmitted by inheritance his "wretched constitution" to the little boy. "We have Horace failing badly with intermittent weak pulse, like four of our children previously," he groaned to John Innes. "It is a curious form of inheritance from my poor constitution, though I never failed in exactly that way."[28] He consulted several other doctors, including Henry Holland, who inquired whether the fidgets were caused by intestinal worms ("Does he pick his nose?"). Dr. Engleheart, the village surgeon, was perhaps the most perceptive in recommending that Miss Ludwig should take a short holiday. Emma saw something in the idea. "Horace's devotion to Miss L. is got to such a pitch that I don't know what he will come to. He can't bear to sit on different side of the table at meals so that he often gives up the fire side for the sake of sitting by her."[29]

> Mr Engleheart is very anxious to get him away from Miss L & he thinks a change or excitement would do him good. I doubt whether it will make much difference about Miss L but I think sometimes his fondness for her agitates him & makes him worse. I think she is very judicious & quick with him.[30]

Shortly afterwards, Emma decided that Miss Ludwig should pay a visit to her mother in Germany and that she and Horace would go to Southampton, ostensibly to see William in his new situation as a banker but really to give Horace something to occupy his mind. It all had to be tactfully planned. "We think it will break his heart much less to leave her here & come to you, than for her to leave first," she notified William. The enforced separation apparently did the trick, and Horace was never so bad again. It was a bonus for the adults to discover William placidly settled in his Southampton banking career. In the interim, Frau Ludwig sent another daughter to fill in as governess for Henrietta and Bessy; and after a suitable interval Miss Ludwig herself was allowed to return. She was missed by Darwin, who liked her a great deal and needed her to translate difficult patches of scientific German. He sent a friendly note to her in Hamburg, packed with news about the family. He called her, unlike all the other governesses who came and went, by her given name, Camilla.

Ironically, what kept Darwin going through all this worry was his interest in the sex life of orchids. Round about now, with a jolt of recognition, he saw that the sexual arrangements of orchids were similar to those of barnacles. Like barnacles, orchids were basically hermaphroditic organisms. Like barnacles, they went to enormous lengths to prevent any form of self-fertilisation except as a last resort. They piled up adaptation after adaptation to ensure that no solitary sexual activity took place—either the male and the female parts ripened at different times, or each barricaded itself behind a complex series of structural modifications that required different kinds of triggers. The flowers were anatomically hermaphroditic but functionally male and female—exactly the same phenomenon he had discovered in barnacles. Sex, in Darwin's theories, always required two.

Furthermore, orchids displayed the same kind of evolutionary sequences he had mapped in barnacles. He did not suspect anything of the kind until he got hold of various species of *Catasetum* from tropical Central America, "the most remarkable of all orchids." Then he realised that the plant that botanists called *Catasetum* was functionally male. The female of the species was mistakenly catalogued as a different plant. A third kind contained both sets of sexual organs in full working order and was probably a direct descendant of the original hermaphroditic organism from which the other two had diverged. What botanists had for years described as three separate genera were reunited by Darwin as the male, female, and hermaphroditic forms of a single species.

He demonstrated the point with a renowned oddity kept in spirits in the museum collection at the Linnean Society, a pickled stem of orchid flowers collected in British Guiana by Sir Robert Schomburgk. On this stem grew three flowers, each apparently from a different genus. "Such cases shake to the very foundations all our ideas of the stability of genera and species," John Lindley, the great authority on orchids, had remarked. Darwin recognised that by some curious chance the single stem replicated in miniature the real separation of the sexes in the wild. In April 1862, he went to London to deliver a short paper on *Catasetum* at a meeting of the Linnean Society. "I by no means thought that I produced a tremendous effect on Linn. Soc.," he told Hooker afterwards, "but by Jove the Linn. Soc. produced a tremendous effect on me for I vomitted all night & could not get out of bed till late next evening, so that I just crawled home.—I fear I must give up trying to read any paper or speak. It is a horrid bore I can do nothing like other people."[31]

Sick as he was, this was an achievement for a man who refused to call himself a botanist. "I am sometimes half tempted to give up species & stick to experiments," he wrote to Hooker after this virtuoso display. "They are much better fun."[32]

## VII

Meanwhile his *Origin of Species* was becoming an object with a life of its own.

First of all, it became clear that Darwin was not the only evolutionist. While his book certainly captured much of the general imagination, and Huxley's publicity machine pumped up a full head of steam, other developmental proposals pushed back into view. The *Origin* stimulated a market for all kinds of evolutionary books and theories. Old arguments reappeared to mix with the new; subsidiary controversies sprang up; publishers raced to turn a quick profit with similar products; and readers developed an obvious thirst for more. A revised edition of *Vestiges* published in 1860 sold extremely well, making it difficult to separate the response to *Vestiges* from that to Darwin.[33] Hard on *Vestiges'* reprinted heels came a matching theological counterblast, a reissue of Hugh Miller's *Footprints of the Creator* brought out by his widow as an "antidote to some of the ill-grounded notions brought forward in other quarters."[34] Mrs. Miller judged her moment well. Flourishing sales brought her the money that had somehow never materialised when her husband was alive.

At the same time, Robert Grant, Darwin's old acquaintance from Edinburgh, published a synopsis of his University College zoology lectures with commendatory words addressed to Darwin at the front. In this dull little volume Grant mentioned his friendship with Darwin, and his—Grant's—longstanding commitment to evolution.[35] Under the circumstances, it seems fair to say that Grant must have hoped to increase his book's sale by capitalising on their former relationship. The two men's positions were pitifully reversed. These lectures were Grant's first publication in twenty years, his income was meagre, his zest for research was dead. He had reached rock bottom. To harness himself to Darwin's ideas must have seemed a last chance for personal renaissance.[36]

Most notably, Herbert Spencer revealed his extreme originality in his *First Principles,* produced in parts from 1860 to 1862. In this work he redefined his own version of evolution, first outlined in articles published in the 1850s. To this he added views sharpened by his reading of Darwin. Like Chambers, the author of *Vestiges,* he seems to have believed that his views were vindicated by Darwin's writings, although he differed from him on several grounds and experienced mixed emotions over the publicity that the *Origin of Species* generated. Chambers and Spencer would hardly have been human if they had not sometimes thought that Darwin had an advantage over them in being so highly visible. It was not only what one wrote, they surely thought, but also who one was and where one came from. In his autobiography Spencer said dismissively that he could not remember his first impressions of Darwin's *Origin* except for annoyance that parts of his own scheme were "wrong." He recorded "gratification in seeing the theory of organic evolution justified."[37]

Cryptic though Spencer's title might have seemed, his *First Principles* addressed the underlying laws of the physical universe and human existence. Every aspect of the world is continuously changing, he stated, and the direction of this change is from simple to complex. Matter always moves from a state of chaos to a state of order, from "indefinite incoherent homogeneity" to "definite coherent heterogeneity." In other words, simplicity becomes complexity, and uniformity becomes variety. He thought this tendency for change occurred everywhere, in physics and astronomy just as much as in biology and human society. Evolution, so to speak, was the opposite of dissolution. Unerringly, Spencer captured the nation's sense of progressive advance and diversification. Unerringly, he argued that the world of religious thought should be separated from that of science, postulating the existence of an "Unknowable power" behind all knowable phenomena. Both science and religion, he said, should recognise the impossibility of defining that power. He was himself a declared atheist.

In essence, Spencer presented a general view of the world that was to have a pervasive effect on late-nineteenth-century thinking. Much of what was ultimately attributed to Darwin was the result of philosophical shifts expressed in one form or another by Spencer. At the very least, most of what Spencer proposed about directional development and progress, although basically Lamarckian or environmentalist in thrust, was conflated in people's minds with the Darwinian impetus. His writings, like Darwin's, turned people's thoughts towards the human condition and the great issues of the day.[38]

Yet Spencer was not blessed with the gift of clarity. He was the "most immeasurable ass in Christendom," objected Carlyle. Huxley and Hooker more or less understood what Spencer meant, and Wallace, too, who sympathetically acknowledged that Spencer sought to get "to the root of everything."[39] Huxley respected Spencer's powers of thought. Because of this respect, Darwin tried to approach his writings with an open mind. But however much he tried, he found Spencer's definitions meaningless: "his style is too hard work for me." Without ever saying it outright, he may have thought Spencer's philosophy too extravagant.

> Herbert Spencer's conversation seemed to me very interesting, but I did not like him particularly and did not feel that I could easily have become intimate with him. . . . After reading any of his books, I generally feel enthusiastic admiration for his transcendent talents, and have often wondered whether in the distant future he would rank with such great men as Descartes, Leibnitz, etc. about whom, however, I know very little. Nevertheless I am not conscious of having profited in my own work by Spencer's writings. His deductive manner of every subject is wholly opposed to my frame of mind. His conclusions never convince me: and over and over again I have said to myself, after reading one of his discussions—"Here would be a fine subject for half-a-dozen years' work."[40]

So he struggled through *First Principles* and most of Spencer's subsequent works as they were published, muttering to himself about "unfounded speculation," unable to suppress a smirk when Huxley quipped that "Spencer's idea of a tragedy was a deduction killed by a fact."[41] Darwin never made any effort to get to know Spencer. On the contrary, apart from borrowing the expression "survival of the fittest" in later years, he went to some trouble to distance himself from Spencer's writings.

Second, by now some of the most notable nineteenth-century thinkers were contemplating the inner recesses of Darwin's theory and pulling out of it some of the threads that would lead them towards the modern world. Whereas theologians naturally continued to take issue with the *Origin of Species'* signification for the human soul, and connected the book's purportedly irreligious position with the growing furore over *Essays and Reviews* and the perils of atheism, several key intellectuals commended Darwin's method of scientific reasoning, a style depending more on the accumulation of probabilities, and on analogy, than on the classic form of proof by demonstration. Although John Herschel might have complained about the law of "higgledy-piggledy," younger men such as Henry Fawcett, the blind economist at Cambridge, and John Stuart Mill compared the new form of reasoning favourably against the old. Mill, who read the *Origin of Species* at Fawcett's prompting, sanctioned Darwin's work in the 1862 edition of his *System of Logic,* and told Fawcett that "though he cannot be said to have proved the truth of his doctrine, he does seem to have proved that it *may* be true, which I take to be as great a triumph as knowledge & ingenuity could possibly achieve on such a question."[42]

That Mill should take an encouraging line so early in the *Origin*'s history meant a great deal in the republic of letters. Above all, Mill's approval showed that the natural history sciences (mostly descriptive and nonpredictive) could be brought into an acceptably rigorous philosophical framework.

> Mr. Darwin's remarkable speculation on the origin of species is another unimpeachable example of a legitimate hypothesis. . . . It is unreasonable to accuse Mr. Darwin (as has been done) of violating the rules of induction. The rules of induction are concerned with the condition of proof. Mr. Darwin has never pretended that his doctrine was proved. He was not bound by the rules of induction but by those of hypothesis. And these last have seldom been more completely fulfilled. He has opened a path of inquiry full of promise, the results of which none can foresee.[43]

Fawcett too was to prove an invaluable support in the University of Cambridge. He had reviewed Darwin favourably in *Macmillan's Magazine* in 1860 and now began teaching and discussing the wider relevance of the *Origin* in the mathematical and economic community. "He belonged to that shrewd, hard-headed, North-country type which was so conspicuous at Cambridge," reminisced Leslie Stephen, who knew him well, "and which, it must be confessed, was apt to be as narrow as it was vigorous intellectually. Fawcett knew Mill's political economy as a Puritan knew the Bible."[44] Taking Fawcett's lead, one or two mathematicians, economists, and statisticians also began considering Darwin's book. In its way, the *Origin of Species* contributed indirectly to some of the major shifts in ideas about probability and the mathematical rules of chance in the nineteenth century.[45] It added weight to an emerging consensus about the value of statistics for tracking random events and indicating hidden trends, shown most noticeably in the growing life assurance market but also supported by the government's interest in maintaining bills of mortality and stricter financial policies; and generally endorsed the important notion in science that the physical universe is at root fluid, subject to change and contingency. Darwin's theories contributed to the scientific movements that ultimately led to the taming of chance, replacing the idea of permanence with relativity.

More obviously, social economists seized on parallels between the organic kingdom and political economy. Competition, struggle, adaptation, success, and extinction—all these concepts moved freely between both domains. They were the Malthusian parallels on which Darwin had first drawn when composing his theory.[46] While many commentators of the period remained divided on Malthus's meaning for human society—to those with working-class sympathies Malthus's principle merely blamed the poor for being poor, a marked contrast to those who applauded it for encouraging responsibility and self-improvement—there could be no denying the concept's status. In one sense it could be said that Malthus's images were turning full circle, for Darwin applied political economy to biology, and now these biological ideas were being reintegrated back into political economy, seemingly providing a "natural" account of the way human populations and social economies were thought to work.[47] Malthus's principles were biologised and then reabsorbed into economic thought. In another sense, the social and the biological were scarcely separable. Malthus's remarks did not so much travel back and forth as exist already embedded in the same cultural context. Either way, Malthus's doctrines looked like incontrovertible laws of nature to a nation steeped in competitive economic activity, buoyed up with Samuel Smiles's anthems of self-help, adaptation, struggle, and survival, and as a political body fully engaged in territorial and commercial expansion.

"It is remarkable how Darwin rediscovers among beasts and plants the society of England, with its division of labour, competition, opening up of new markets, inventions, and the Malthusian struggle for existence," remarked Karl Marx in a letter to Engels in 1862.[48] Marx read the *Origin of Species* soon after publication, noting "the clumsy English style." He understood the *Origin*'s threat to traditional Victorian standards more clearly than most. "Although developed in the crude English fashion, this is the book which in the field of natural history, provides the basis for our views," he continued to Engels. He repeated much the same comment to Ferdinand Lassalle. "Darwin's work is most important and suits my purpose in that it provides a basis in natural science for the historical class struggle."[49] Marx laughed at the British fear of apes. "Since Darwin demonstrated that we are all descended from the apes there is scarcely any shock whatever that could shake our ancestral pride."[50]

The poets were not far behind. Alfred Tennyson apparently ordered a copy of the *Origin of Species* in advance so that he might read it as soon as it appeared.[51] He had long been interested in transmutation in a general manner, using ideas drawn from *Vestiges* for some of the most melancholy parts of *In Memoriam* in which he railed against the heedless forces of nature. And he was well known to have voiced many of the perplexities of the Victorian mind:

*There lives more faith in honest doubt*
*Believe me, than in half the creeds.*

Tennyson could not find the God he knew in the *Origin,* the loving being who directed the world towards beneficent ends. For him, the bleakest aspect of Darwin's work was the widespread cruelty in nature that he described, the "wasteless fecundity" that must end in death.

> An omnipotent creator who could make such a painful world is to me sometimes as hard to believe in as to believe in blind matter behind everything. The lavish profusion too in the natural world appals me, from the growths of the tropical forest to the capacity of man to multiply, the torrent of babies.[52]

Soon afterwards, Tennyson changed one of the biblical allusions in *In Memoriam.* The line "Since Adam left his garden yet," was adjusted to a more scientifically accurate "Since our first sun arose and set."[53] Never willing to accept the full panoply of Darwinism, and always vaguely dissatisfied with the new view of nature, Tennyson contented himself with the conclusion that evolution, if it was true, indicated that better things would come in the afterlife. In 1863 he remarked to William Allingham, "Darwinism, man from ape, would that really make any difference? Time is nothing, are we not all part of deity?" Thereafter, in an effort to gain some real insight, Tennyson read pertinent books and questioned the great thinkers of the day. He was to meet Darwin in 1868. And in 1869 he participated in forming the Metaphysical Society (it included Huxley) for the discussion of these and similar topics. He ended up thinking that the Darwinian theory was for the most part true but that mankind probably stood on one of the lowest rungs of the ladder.[54]

Darwin's book pushed Robert Browning almost as far. The *Origin of Species* appeared midway between Browning's *Men and Women* and *Dramatis Personae,* broadly coinciding with the serious spiritual uncertainties he experienced on his wife's death in 1861. Browning returned to London after Elizabeth Barrett Browning's death and by chance formed a friendship with Julia Wedgwood, Darwin's niece. At one point Julia hoped to marry him, a hope never fulfilled.[55] Many of Browning's realignments of faith during this period came out in his "Caliban upon Setebos," published in *Dramatis Personae,* in which God constructed the world merely as a plaything, devoid of any meaning. Browning made little distinction between Darwin, Lamarck, Spencer, and Chambers. Late in life he issued a formal explanation of his view. "In reality all that seems *proved* in Darwin's scheme was a conception familiar to me from the beginning: see in *Paracelsus* the progressive development from senseless matter to organised until man's appearance. . . . But I do not consider his case as to the changes in organisation, brought about by desire and will in the creature, proved."[56]

Others read the *Origin of Species* attentively. Ernest Renan, author of *Vie de Jésus* (1863)—the book that with George Eliot's translation of Strauss raised troubling doubts in orthodox Victorian minds—had something positive to say. His study of Jesus had deliberately left out the divine, a point noted with interest by Darwin when he and Emma read it. He was as naturalistic in his way as Huxley. "It may be that Darwin's hypotheses on the subject can be judged to be insufficient or inexact; but undoubtedly they are on the road to the great explication of the world and of true philosophy."[57]

George Henry Lewes came to much the same conclusion when discussing the theory of natural selection in his *Animal Life,* published in 1862.[58] "It *may* be true but we cannot say that it *is* true," he reported. Lewes and Eliot were familiar with many of the people involved in the evolutionary debate, especially with Herbert Spencer, a close friend, and they both to some degree had adopted Spencer's doctrines before the *Origin* was published. Lewes, a good naturalist in his own right, was at this point composing an ambitious account of the principles of animal physiology drawing on several of Spencer's concepts; and Eliot in her novels dwelled affectingly on the tangled course of human society. Lewes and Eliot had sat up reading the *Origin* together shortly after it was published. "Though full of interesting matter, it is not impressive, from want of luminous and orderly presentation," Eliot commented in a letter.[59] Yet she drew on Darwin as subtly as any other. She made creative use of her wider understanding of evolutionary development in *The Mill on the Floss* (1860) and then wove stories around heredity for *Middlemarch* and *Daniel Deronda*.[60] Lewes assented to the main thrust of Darwin's arguments, although he explicitly linked them with Spencer's themes.

It was, however, Friedrich Max Müller, the philologist, who pushed Darwin's theories beyond the world of the creative imagination into grand questions of human identity. In the winter season of 1861–62 he galvanised high society with lectures at the Royal Institution on the origin of languages. Müller forced his audience to think carefully about what it is to be human. Had the language of primitive mankind developed from imitations of natural sounds? He thought not. Instead, Müller championed the idea that words were inseparably related to mental concepts. Words served as symbols for things, and there could be no thought without the language to express it. He said it was therefore impossible for language to arise by natural development out of the vocalisations of animals precisely because animals did not possess human concepts. "Language is something more palpable than a fold of the brain, or an angle of the skull," he announced.[61]

Even so, Müller praised other aspects of Darwin's theory, applying the idea of natural selection to the genealogy of ancient languages, as the leading German philologist August Schleicher was to appreciate.[62] Darwin took Müller seriously. He read the printed version of the lectures and discussed them with Hensleigh Wedgwood, the family philologist. "I quite agree that it is extremely interesting," he said of Müller's thesis.[63] Hensleigh thought that human speech could only have emerged bit by bit from animal sounds, a point of view that matched Darwin's ideas. "H. says [Max Müller] is all wrong & has partly converted Papa to that opinion," said Emma in a letter to William.[64]

Quietly, the *Origin of Species* crept into the thoughts of Elizabeth Gaskell. She brooded on what she knew of Darwin's life story, attracted by the tale of the *Beagle* voyage and the possibilities of setting a novel around the pleasant trope of a natural history collector unknowingly capturing a woman's heart. Her meditations carried a personal element. The Gaskell family were distantly related to the Darwins (via the Hollands), and although Mrs. Gaskell did not know Darwin or Emma at all well, probably meeting them only once or twice in as many decades, she was intimate with Hensleigh and Fanny Wedgwood, often sending her daughters Meta and Marianne over to visit the Wedgwood girls. She also knew others in their social circle, like Harriet Martineau and Erasmus Darwin. Meta Gaskell sometimes accompanied the Wedgwood daughters to Down House to stay with their cousins Henrietta and Bessy Darwin. One of those occasions was remembered by Darwin simply because he managed to conquer Meta Gaskell at chess. Julia Wedgwood worked with Elizabeth Gaskell on her biography of Charlotte Brontë.[65]

It was about now that Mrs. Gaskell created Roger Hamley, the hero of *Wives and Daughters,* basing him loosely on Darwin. The story was unfinished at her death, although most of it was published as a serial in the *Cornhill Magazine* from 1864. Darwin's fictional counterpart was a shyly agreeable character, a naturalist and traveller returned, a man whose heart was given to the loving investigation of nature, an experimenter who respected his subject matter. Science, even of the most unsettling kind, could still present a humane and generous face.

## VIII

Near and far, people actively engaged with his text. In Boston, Theodore Parker, the Unitarian divine, amusingly presented "A Bumblebee's Thoughts on the Plan and Purpose of the Universe." In nearby Concord, Henry David Thoreau experienced the *Origin of Species* as a revelatory text. Asa Gray had arranged this transatlantic meeting of minds by sending a copy up to Concord with his brother-in-law Charles Brace. "Never had Thoreau been so captivated by a project," noted William Howarth.[66] Thoreau's early death in 1862 left his views on Darwin's work tantalisingly unformed.

Yet the community of transcendentalists was well disposed to at least some of the *Origin of Species'* proposals. Thoreau apparently discussed theories of development and evolution with Emerson, who was himself inclined to appreciate some form of progression in nature. Before the *Origin of Species* was published, Emerson had talked with Moncure Daniel Conway, the Protestant theologian, about progressive development in nature, where evolution might represent the gradual freeing of human beings from their animal roots, ultimately leading "to a godly state." Conway wondered whether the existence of evil might perhaps reside in those animal roots. He preached a sermon in Ohio about Darwin in December 1859. "This formidable man . . . did not mean to give dogmatic Christianity its deathblow; he meant to utter a simple theory of nature."[67]

A trio of Australian museum men, Ferdinand Mueller, Frederick McCoy, and William MacLeay, were less easy to satisfy. Each condemned evolution outright. MacLeay politely acknowledged his acquaintance with Darwin from their London days. Nevertheless, "I am utterly opposed to Darwin's or rather Lamarck's theory." McCoy refused to let Darwin's books—even his innocuous *Journal of a Naturalist*—into the National Museum in Melbourne. If any Melbourne resident wished to consult a public copy of the *Origin of Species,* he or she had to travel to Sydney, where a solitary volume stood on the shelves of the Mechanics' School of Arts library.[68] And in a little-known burst of museum madness McCoy spent several years fruitlessly attempting to get hold of one of Paul Du Chaillu's gorilla skins in order to stage an exhibit in the Melbourne Museum. One look at the beast, he confidently predicted, would convince anyone that evolution was nonsense. On the other side of the globe, John William Dawson, of McGill University in Montreal, unleashed such a torrent of palaeontological invective that Hooker remarked that "he seems to hate Darwinism."[69]

Even royal minds addressed the question of natural selection in the privacy of family apartments. Queen Victoria spoke for thousands of her subjects when she congratulated her eldest daughter on her fortitude in grappling with the *Origin of Species.* "How many interesting, difficult books you read," she told Vicky, the crown princess of Prussia, in 1862. "It would and will please beloved Papa."[70]

All this for a man who would not—or could not—make a public appearance. In February 1861, John Lubbock stood up in front of Darwin's friends and neighbours at a meeting of the Bromley Literary Institute to explain the absent author's theories of evolution by natural selection.[71]

## IX

The orchid book was published in May 1862 under the title *On the Various Contrivances by Which British and Foreign Orchids are Fertilised by Insects and on the Good Effects of Intercrossing.* Darwin chose the word "contrivances" specially to indicate that there was no purposeful design in the natural world, although in retrospect it was a word no less imbued with intent than "adaptation."

> When this or that part has been spoken of as adapted for some special purpose, it must not be supposed that it was originally always formed for this sole purpose. The regular course of events seems to be, that a part which originally served for one purpose, becomes adapted by slow changes for widely different purposes. . . . On the same principle, if a man were to make a machine for some special purpose, but were to use old wheels, springs, and pulleys, only slightly altered, the whole machine, with all its parts, might be said to

be specially contrived for its present purpose. Thus throughout nature almost every part of each living being has probably served, in a slightly modified condition, for diverse purposes, and has acted in the living machinery of many ancient and distinct specific forms.[72]

Murray gave the book a decorative plum binding (the only one of the Murray Darwins not to appear in green) and embellished the cover with a gilt *Cycnoches,* chosen for its bold outline. Before then Darwin had issued a welter of instructions about the printing and type size, supervising the woodcuts with a sharp eye for detail, hurrying the compositors along, even specifying that "fertilisation" should be spelled with an *s,* not a *z.* He told Murray that he hoped the book would appear sufficently modest. He did not wish to deceive readers into thinking they were buying a lavishly illustrated tome on the orchid fancy. This time he thanked his helpers generously in print.

But he felt oddly nervous about publishing, wondering if he was making a fool of himself.

The subject is, I fear, too complex for the public & I fear I have made a great mistake in not keeping to my first intention of sending it to Linnean Soc.; but it is now too late, & I must make the best of a bad job.[73]

However he ventured to tell Murray that "I think this little volume will do good to the *Origin,* as it will show that I have worked hard at details."[74]

Despite the worries, the results impressed his botanical friends. They noted Darwin's talent for observation. "It is a very extraordinary book!" declared Oliver. "What a new field for observing the wonderful provisions of nature you have opened up," said Bentham. "What a skill & genius you have for these researches. I have tonight learned more than I ever knew before," echoed Gray. From the breezy flatlands of Cambridgeshire, Charles Babington (Darwin's old beetle-collecting rival and now Cambridge professor of botany after Henslow), who was no friend of natural selection, stiffly relayed the news that he thought the book "exceedingly interesting & valuable." Hooker practically burst. "You are out of sight the best physiological observer and experimenter that Botany ever saw." What they recognised, perhaps for the first time, was Darwin's ability to identify what might be valuable in a wide range of differing phenomena and concentrate intently enough on it to see what could constitute the whole solution to the problem in his head.

This orchid book tipped some of the more traditionally minded botanists onto Darwin's side. Admittedly, there was something about nineteenth-century science that enabled British botanists to contemplate evolution in plants far more calmly than zoologists regarded evolution in animals. The rooted nature of plants, their lack of feeling, their dependence on the environment, their obvious fluctuations in numbers, perhaps made the possibility of their evolution a little easier to accept. "Bentham and Oliver are quite struck up in a heap with your book & delighted beyond expression," reported Hooker, glad to see two leading figures capitulate to natural selection.

Bentham made his views known in his 1862 presidential address to the Linnean Society, and again the following year in a valedictory address to the same society. He alluded to John Stuart Mill's verdict. "Mr Darwin has shown how specific changes *may* take place," he remarked. "His is not therefore a theory capable of proof, but 'an unimpeachable example of a legitimate hypothesis' requiring verification, as defined by J. S. Mill in his excellent chapter on Hypothesis."[75] This endorsement from the presidential chair made its mark on the Linnean fellows. In turn, Miles Berkeley, Charles Naudin, Alphonse de Candolle, Jean Louis Quatrefages, and Charles Daubeny began to think there might be something in what Darwin proposed after all. Lyell said that next to the *Origin of Species*, he considered *Orchids* the most valuable of all Darwin's works.[76]

Elsewhere, silence reigned. Darwin's choice of subject matter baffled Victorian readers panting for gorillas and cavemen. His book looked quaint. While the *Times* roared against Huxley's support for "Mr. Darwin's mischievous theory," the source of the controversy appeared to have strolled into a greenhouse. Except for a few reviews in gardening magazines, scarcely any learned appraisals appeared. Few zoologists or natural philosophers noticed the volume. It sold rather too slowly for Murray's comfort.

And in an irritating reversal of Darwin's intentions, one or two commentators misunderstood his theme and treated the book as a testament to God's marvellous ingenuity. The *Literary Churchman* obstinately closed a review with the words "O Lord, how manifold are Thy works."[77] Charles Kingsley said it presented "a most valuable addition to natural theology," telling Huxley that the wisest God was the one who could make all things make themselves.[78]

Worst of all, in Darwin's eyes, the writer and statesman George Douglas Campbell, the eighth Duke of Argyll and lord privy seal, decided to speak out. Argyll was a cultivated man, well versed in the sciences. "I have read Darwin with great interest," he told his friend Richard Owen.[79] Nonetheless he considered Darwin's explanations for orchids "the vaguest and most unsatisfactory conjectures." The metaphysical points about adaptive design and perfection that he then made in the *Edinburgh Review* were too high-flown for Darwin fully to grasp. "The Duke of Argyll has written an article on Supernaturalism in the Ed. rev," said Emma in November 1862.

> He is quite opposed to your father's views but he praises the *Orchids* in such an enthusiastic way that he will do it a good turn. The article is so obscure I cd. not understand it. Hensleigh & Snow have written an article in Macmillan on Max Muller & that I suspect is equally obscure.[80]

Argyll proposed that God made orchid flowers in such a way that humans could explain them in human terms. He went on to develop his thoughts on this kind of creative evolutionism in an influential manner over the next few years, ultimately becoming an important commentator on the ancestry of mankind. Argyll particularly ridiculed Darwin's description of the long nectary of *Angraecum sesquipedale* and the missing moth needed to fertilise it. "Contemptuous," muttered Darwin defensively. "Very clever, but not convincing to me."[81] Darwin recollected that he had once met Argyll at a smart dinner party in London, although he had barely spoken to him.[82]

So he allowed himself a shiver of satisfaction when an anonymous writer launched a retaliatory attack on the duke in the *Saturday Review*.[83] Emma and he enjoyed the brio of the article and sent it to Hooker, who declared it "perfect." The shiver swelled into family pride when it turned out that the unknown author was Darwin's own nephew on the Shrewsbury side, Henry Parker, Marianne's eldest boy, who had left Oxford University and was trying to establish himself in London as a literary man. What an "odd chance," Darwin exclaimed. His life was visibly changing when the younger generation of the family started rising in his support.

All in all, he suspected he might have wasted his time on orchids. "It has not been worth, I fear, the 10 months it has cost me: it was a hobby horse & so beguiled me."[84]

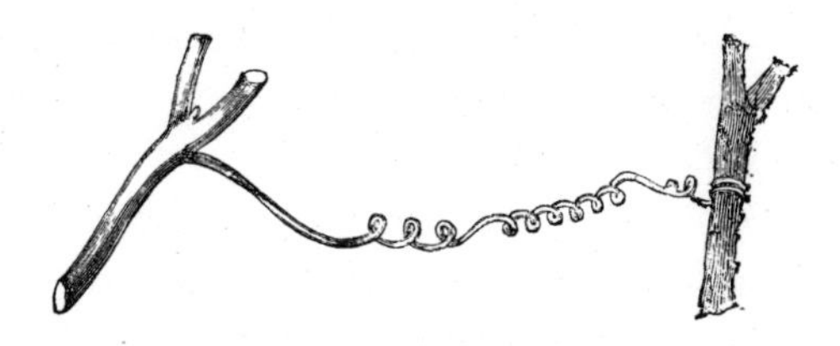

*chapter*
6

# BATTLE OF THE BOOKS

ONE MORNING early in 1862, Alfred Russel Wallace stepped off the train from Dover. His cheeks had sunk, his skin was sallow. His beard could not disguise the fact that his travels in Malaysia had turned him into a wraith. He said afterwards that the effort of bringing all his crates together in London was immense. He specially regretted saying goodbye to the two living birds of paradise that he had brought home. During the voyage he had fed them cockroaches brushed from the ship's biscuits, worried about frosty nights in Egypt, and stayed an extra week or two in Malta so that they could adjust to lowered temperatures. His first task on reaching London was to transfer these "wonderful birds" into the hands of Mr. Bartlett, the senior keeper at the Zoological Gardens. "Attracting much notice," they were the first birds of paradise to be exhibited in Britain, said the *Saturday Review,* and were placed in an indoor cage until a new aviary was ready in the gardens. "Thus ended my Malayan travels," Wallace wistfully recorded.[1]

To return to Europe was for him something of a mixed blessing. He had a lot of catching up to do, not just with the progress of natural selection and the rise of Darwin's *Origin of Species* but also with his professional and private life. All those years of hardship had taken their toll. London came as a physical shock to his system, so drab after the high colours of the East. He had forgotten how grimy the buildings were, how frantic the pace of nineteenth-century life, and he felt oddly disoriented by the horses and carriages rattling past. Damp air seeped into his bones. "I went to live with my brother-in-law Mr. Thomas Sims, and my sister Mrs. Sims, who had a photographic business in Westbourne Grove. Here, in a large empty room at the top of the house, I brought together all the collections which I had reserved for myself and which my agent Mr. Stevens had taken care of for me. I found myself surrounded by a quantity of packing-cases and store-boxes, the contents of many of which I had not seen for five or six years, and to the examination and study of which I looked forward with intense excitement."[2]

For four months he did little except sit in his sister's house and acclimatise. At the same time, he may have mourned a little over ending his travels. But at the first opportunity he got in touch with Darwin, sending along a wild honeycomb from Timor, hoping it "will be interesting to you." It was—and a little later on Wallace forwarded specimens of the bees that made it.

Both men seemed keen to delay a personal meeting. Perhaps too much had happened since their unexpected literary union. Instead, they corresponded in a courteous fashion about the reviews and criticisms of their theory, relaxed enough with each other by mail. It could have been worse. Starved of intellectual contact while he had been away, Wallace longed to discuss the leading elements of his theory with the only other man who truly understood it. He was unfamiliar with the third edition of the *Origin of Species* (the copy Darwin sent must be in Singapore, he said apologetically), but wasted little time in telling Darwin that he felt it was a bad idea to list all the difficulties quite so prominently. "You have assisted those who want to criticise you by *overstating* the difficulties & objections—several of them quote your *own words* as the strongest arguments against you," he said.[3] He unpacked his copy of the first edition from his luggage and started to pick through the points where they agreed or differed.[4]

Wallace visited Darwin at Down House in June or July 1862, although no record of the meeting exists other than Wallace's statement that it took place.[5] Further contact was interrupted by Leonard Darwin's catching scarlet fever, followed by a long convalescent stay by the Darwin family in Southampton and Bournemouth, during which Emma also developed scarlet fever. "We are a wretched family," Darwin declared to Fox, "and ought to be exterminated. . . . There is no end of trouble in this weary world." Then Wallace fell ill with pleurisy. If nothing else, these illnesses provided a convenient talking-point, and the gloomy exchange of news about ailments helped consolidate their developing relationship. The following winter Wallace and Darwin met at Erasmus's house in London for a brief, well-mannered event which confirmed the cordiality each felt towards the other.

Meanwhile, Wallace made himself known to Lyell and Huxley in London, both of whom were eager to meet him. Wallace's originality and fresh perspective appealed enormously to them. As a co-founder of natural selection, he would also, they anticipated, become an invaluable addition to the crusading ranks of biology. These expectations deepened into esteem as they became familiar with Wallace's collecting exploits, his intellectual curiosity, and his courteous character. Lyell valued Wallace's opinion on evolutionary matters almost as if it were Darwin's—sometimes precisely because it was not Darwin's. For Lyell remembered how much Wallace's early articles had stimulated his imagination and inspired him to start a private scientific journal about the problem of species, and he welcomed Wallace's willingness to talk about the human condition and other absorbing topics that Darwin sometimes seemed reluctant to probe. Wallace spoke freely with Lyell about archaeology, primitive culture, language, biogeography, geological astronomy, and the purpose of the universe. The appreciation was mutual. Wallace said of Lyell that "I saw more of him than of any other man. . . . my correspondence with him was more varied in the subjects touched upon, and in some respects of more general interest, than my more extensive correspondence with Darwin."[6]

Huxley brought Wallace up to date with the gorilla wars. Wallace soon considered Huxley's specialist knowledge of the anatomical side of natural history to be unrivalled. He was gratified when Huxley coined the expression "Wallace's line" to describe the biogeographical boundaries between the Pacific and Asiatic faunas.[7] Swept along in Huxley's current, Wallace became a member of the Zoological Society and began enjoying some of the most forward-looking scientific company in nineteenth-century Britain.

His circle also included Henry Walter Bates, back in England from his travels in the Amazon. Hesitantly the two travellers wondered what, if anything, they still had in common, and their relationship, while remaining close, was never again as sympathetic as it had been in Leicester or in the South American jungle. Together they made an appointment to meet Herbert Spencer, whose *First Principles* they read in 1862 and admired. Wallace's and Bates's thoughts were full of the great unresolved problem of the origin of life, "a problem that Darwin's *Origin of Species* left in as much obscurity as ever," announced Wallace, as they went to see Spencer.

Underneath the welcome that Wallace received from the scientific establishment, however, there lurked a hint of rapacity. These men-about-town fell on him with alarming speed, seemingly impatient to suck him dry. They wanted to know what he could offer, which sword he could rattle.

In actual fact Wallace found it hard to identify any suitable role for himself in Britain in the post-*Origin* years. At times it may have appeared to him that he was hardly needed. Much of the evolutionary flare-up had already passed him by. Darwin seemed fully in command of spreading the word at home and overseas, and the public business of defence lay in the hands of individuals like Huxley, who dominated nearly every corner of the controversy and appeared unwilling to share the limelight. Furthermore, Darwin's name was becoming an acknowledged synonym for the theory of natural selection. Any chance of an evolutionary movement called Wallacism—even if Wallace wished for such a thing—had probably disappeared. Several years afterwards when reflecting on his return to England, Wallace said quietly of the theory that Darwin "had already made it his own."

He may have been content to let it go. For a while he concentrated on distributing and classifying his Malaysian collections—three thousand bird skins, twenty thousand beetles and butterflies—and spent the better part of the next five years writing highly informative articles based on these specimens. He contributed important papers to the Zoological, Entomological, and Linnean Societies. Nor did he lose sight of natural selection. In private, he began to probe some of the most enigmatic aspects of his and Darwin's evolutionary scheme, thinking hard about the protective colours of animals, the mental capacities of mankind, and the emergence of early human societies.

But he was unsettled by London's superficial existence. With a start, he remembered exactly why he had abandoned the metropolis for the impenetrable green of Malaysia. "Talking without having anything to say," he later wrote in his memoirs, "and merely for politeness or to pass the time, was most difficult and disagreeable." Lady Lyell's impression when she met him in 1863 for the first time, at a London lunch party hosted by Lyell in their home in Harley Street, was of a very retiring figure, "shy, awkward and quite unused to good society." If truth be told, Wallace was overawed by the Lyells' style of living, with its daunting array of silver forks and high-table conversation. He preferred Lyell to walk over to visit him in his lodgings. He felt more at ease in Huxley's home, a modest villa in St. John's Wood, at that time a cheap suburb of London, where Henrietta Huxley and the children unpretentiously bustled round. Huxley's son Leonard recalled Wallace's unassuming manner during these visits. All in all, Wallace found it increasingly more convenient to shut himself up with his bird skins than to accept dinner invitations.

Darwin was relieved to find him so agreeable—perhaps, as well, so unlikely to be troublesome. He did what he could to smooth Wallace's path over the next few months, helping him submit articles to those scientific societies of which Wallace was not a member and writing supportive letters. They became strong friends, a friendship rooted in goodwill towards each other, their common travelling experiences, and the intellectual rigours of what was now a mutual enterprise far larger than either man individually. Darwin genuinely admired Wallace's mind and always treated him with deference and respect. He hated to differ from Wallace on natural history points. If he did so, he would become anxious about his own powers of reasoning. From time to time, he protested that the theory of natural selection was "as much yours as mine."

Yet it was an odd relationship, as tricky to negotiate as any arranged marriage, and both men felt inclined to tread slowly. Darwin did not let their promising new connection, for example, interrupt his continuing research and writing projects. They never appeared on a scientific podium together, never composed an article together, never rose united to defend their combined theory against detractors or opponents, never sat for a double portrait, although one shrewd photographer with an eye to posterity asked if they would be prepared to do so. United well enough on that single point, they both spontaneously refused the request.[8] Although theirs was a noteworthy partnership, they never performed as a duo. This surely had much to do with a deep-seated wish to acknowledge the other's independent creativity. They affirmed the ties that bound them and, for the most part, went their individual ways.

## II

With his orchid book published, Darwin turned to what he considered his proper task for the future—presenting the full evidence for natural selection. He had promised to supply this evidence in the preface to the *Origin of Species,* and his sense of fair play told him he ought to fulfil the obligation, however much he found it a chore. Yet it was to turn into a massive, unfinished, perhaps unfinishable project. From 1862 until 1868, when the first in this planned trio of books was published, he pushed himself relentlessly. He began with variation in domestic organisms, an amorphous and difficult subject. "Oh Lord what will become of my book on variation," he had exclaimed to Hooker in the middle of his happy experiences with orchids. Now his voice became doleful, as if the fun had stopped and real work was to begin. "It is so much more interesting to observe than to write."

The longstanding theme of variation drew him back into the world of animal breeders, dog handlers, farmers, and horticulturists and allowed him to consolidate all the disparate facts he had been collecting for decades. The result was a two-volume study called *The Variation of Animals and Plants Under Domestication,* published in 1868. He intended following it with a matching volume discussing variation under natural conditions, a book that was never written.

Darwin's friends were admittedly a little bewildered by his self-imposed labours. As with the orchid researches, they wondered whether peas and chickens were worth all that effort. Even Darwin himself, who was dogged by bad health during these years, came to feel that his life might "hardly be long enough" for everything he hoped to achieve. And yet these volumes, and the home-based experiments on which they rested, set him on a new path.[9] He was led to fresh insights that influenced the way he would subsequently think about animals and plants—and humans too. To a large extent he reworked existing material. Details from his earliest notebooks reappeared almost unchanged in the book, and relevant parts of the old manuscript of "Natural Selection" were plundered into oblivion.[10] But Darwin also reformulated his understanding of the processes of reproduction, fertilisation, heredity, and variability, setting out problems that would occupy his mind for years to come.[11]

Most of all, he intended using the new book to provide an extended response to critics of the *Origin of Species.* As countless reviewers told him, he had not explained the cause of variability, how the individual differences in organisms came about. In the *Origin,* Darwin said openly, "Our ignorance of the laws of variation is profound." So in *Variation* he intended explaining what he knew about the source, stability, and transmission of variations, in essence making an attempt to understand the phenomena of heredity some thirty or forty years before the modern science of genetics emerged. Although Darwin had not yet devised the theory of "pangenesis" that would ultimately take pride of place in the book, he felt sure that the origin of variability lay in the unsettling of the reproductive system brought about by factors such as domestication, the processes of out-crossing and interbreeding, changing conditions of life, and sexual procreation (as distinct from parthenogenesis and various forms of vegetative reproduction). He reminded Hooker that "variation is the base of all."[12] He regarded the book on variation as an essential sequel to the *Origin of Species.*

It was a project for him, moreover, that could only be carried out in his home and garden. Much underestimated by Victorian readers, the practical work that lay behind this book on variation was extraordinary. Darwin's book was crammed with experimental results that were derived either from his own investigations in his greenhouse and aviary or gathered by correspondence from associates scattered across the globe. For it, he depended on his circle of correspondents, his library, his friends, his household staff, and his family. Everything was verified by correspondence, reading, and experiment. "You need not believe one word of what I said about gestation of dogs," he told Lyell at one point. "Since writing to you I have had more correspondence with the master of hounds, & I see his record is worth nothing—it may of course be correct, but cannot be trusted." Perhaps even more than when he had composed the *Origin of Species,* he transformed his daily activities into scientific knowledge.

Despite Darwin's calling them "trifling observations," these researches were more sophisticated and to the point than usually supposed. Few authors of the period, for example, troubled to say in what respect offspring resembled their parents, so Darwin undertook cross-breeding experiments in which he tried to break down each plant or animal's defining characteristics into those features that were acquired relatively recently and those that were more ancient, like the black bars on pigeon wings. At other times, he sought information on reversion to the ancestral type by allowing free mating between different breeds of chicken. He looked for atavism (a term first defined in relation to plants and only later applied to animals and humans), in which a well-marked variety produces a throwback, revealing the hereditary past, using his garden laburnums as an example. When considering farm animals, he asked gentlemen farmers of his acquaintance about secondary sexual characteristics and features that were transmitted by one sex alone or were apparent only at particular stages in the life cycle. He thought deeply about potency, believing that older, well-established breeds were more likely than recent variants to transmit their characteristics unchanged, a fact of life that apparently governed the stud books of pedigree cattle.

As for himself, most of his practical investigations now lay in plants. He experimented on plant sterility and fertility, investigated hybrids and hybridization, the effects of inbreeding and outbreeding, the impact of good or poor conditions of life, and phenomena connected with the grafting of plants, monstrosities, and the way highly developed varieties of garden plants rarely reproduce true to kind. Throughout, he read and reread authoritative sources, especially those by Thomas Andrew Knight and Carl Gärtner on inheritance and hybridity. He corresponded with experimental hybridizers of the period, including Charles Naudin, renowned for his interbreeding studies on plants, and the Swiss botanist Carl Wilhelm von Nägeli, who worked on plant hybridity in Munich. Soon he came in touch with August Weismann, in Freiberg, who was interested in similar topics, and began writing to him. Like Naudin, Gärtner, and Gregor Mendel, each of whom investigated the same broad spectrum of hereditary phenomena during the 1860s, Darwin hoped that information derived from careful experimentation with plants—ideal experimental organisms for the period—would open up at least some of the mysteries of inheritance.

Interestingly, he included mankind in his research, regarding human beings as self-evidently variable domesticated organisms. Over these years he consulted anthropologists, medical men, statisticians, anatomists, missionaries, and a variety of printed sources, ranging from travel literature to the *Lancet,* on the question of human variability. When his notes on mankind became too unwieldy, even for a big book like this, he put them aside for a separate future volume, although he did incorporate some of the most striking human examples in relevant places in *Variation*—anatomical variations like polydactyly (multiple fingers or toes), abnormalities like harelip and deaf-mutism, and inherited medical conditions like gout. Darwin's correspondence with medical men such as James Paget was crucial here. It also pleased him to mention some of the cases that his father had discussed with him before his death in 1848.[13]

Notwithstanding the solid intellectual aims underpinning the work, an indisputable air of *Gulliver's Travels* crept in. His letters inquired about geese or goldfish, cross-bred pheasants, cucumbers, or abortive roses. Nothing seemed too absurd to contemplate, however briefly. "What do you say to wheat being grown from oats in the second year?" asked John Innes in 1862. Darwin was specially exercised over honey bees. The presence of neuter bees in every colony made it troublesome to understand how variation and selection (which depend on reproduction) could possibly take place. The drones could not pass their characteristics on to progeny because they left none.[14] "I am half mad on subject."

When he inquired in the *Journal of Horticulture* whether bees could vary, a flurry of negative replies forced him to think hard. Beekeepers from all over the country told him that there were no obvious variations in native stock. As Darwin's web of inquiries spread outwards, even the eminent German apiarian Johannes Dzierzon informed him that variation was negligible. Only one man, a Mr. J. Lowe, an amateur beekeeper from Edinburgh, thought he had once seen a light-coloured variety visiting his hives. In desperation Darwin turned the question on its head: if there were no physical differences, perhaps there might be variations in behaviour? He urgently asked Lubbock to look out for fields of clover.

> I write now in great Haste to beg you to look (though I know how busy you are, but I cannot think of any other naturalist who wd. be careful) at any field of common red clover (if such a field is near you) & watch the Hive Bees: probably (if not too late) you will see some sucking at the mouth of the little flowers & some few sucking at the base of the flowers, at holes bitten through the corollas— All that

> you will see is that the Bees put their Heads deep into the head & rout about.—Now if you see this, do for Heavens sake catch me some of each & put in spirits & *keep them separate.*—I am almost certain that they belong to two castes, with long & short probosces. This is so curious a point that it seems worth making out.—I cannot hear of a clover field near here.—Pray forgive my asking this favour, which I do not for one moment expect you to grant, unless you have clover field near you & can spare ½ hour.[15]

Domestic rabbits gave him trouble too. Digging around in his old notes, he unearthed details about rabbit breeds collected for him by Abraham Bartlett, the senior keeper at the London Zoological Gardens. Long ago, Darwin and Bartlett had together pondered the inheritance patterns of the so-called Himalayan rabbit, a good-looking show breed, with white coat and black ears, nose, tail, and feet. At Darwin's request, Bartlett now began to carry out selected crosses between the Himalayan and other breeds. These crosses sometimes produced coloured animals called blues, a mystifying result that would not be fully explicable until the advent of Mendelian genetics. Darwin was at a loss. "The effects of crossing are sometimes marvellous in bringing out old & lost characters or in producing new characters."[16] But he visited the British Museum's backroom collections to examine pelts. He came to believe the black markings (points) were ubiquitous, as seen in cats, dogs, Chillingham cattle, deer, and horses, and that the silver, bluish-grey coats were reversions to an aboriginal, wild colouring.

Then he concentrated on bones, boiling up rabbit carcasses in an outhouse and measuring the skeletons, convinced that domestic rabbits ought to be bulkier in physique than the wild form. At long last, he found a use for the rabbits he had trapped on the island of Porto Santo, next to Madeira, and brought back alive from the *Beagle* voyage. In 1837 he had sent these to the London Zoo for display. Darwin got in touch with Bartlett again, asking him to kill two of them ("plenty more," said Bartlett) and forward him the bodies. These, he imagined, were feral descendants of imported domestic breeds and ought to have reverted back to the wild type. Bone after bone, he built up his case. Generally speaking, these comparative measurements of rabbit bones required hours of patient work with graduated calipers and a microscope. In exasperation, he laid them out in rows on the Down House billiard table—his favourite table—to make sure that he really could see a chain of variations.

And he returned to pigeons, finally producing a written account of the years of loving attention he had bestowed on them. The actual birds had been dispersed long ago. Even the old pigeon house had been moved away from the stable-yard and abandoned to ivy and the occasional child who might wish to play inside—although Darwin and Emma apparently had at first intended to climb the wooden ladder themselves and watch sunsets from its new vantage point on top of the wall by the kitchen garden.[17] The most tangible relic of Darwin's time as a pigeon-fancier was his set of dismembered skeletons and the dried skins of representative specimens. His pets had become data ready to be turned into ammunition.

Every domestic pigeon, he claimed in *Variation,* was derived from a single ancestral species, the wild rock dove, or *Columba livia.* Many pigeon-fanciers, on the contrary, thought that some eight or nine wild species may be involved. "The amount of variation has been extraordinarily great," Darwin agreed. "Formerly, when I went into my aviaries and watched such birds as pouters, carriers, barbs, fantails, and short-faced tumblers, etc., I could not persuade myself that all had descended from the same wild stock, and that man had consequently in one sense created these remarkable modifications."[18] But his breeding experiments convinced him it was so. Something of this ancestry could be seen in the way the characteristic black bars reappeared on the wings. Each highly developed breed carried within it the residue of its heritage. He said to Hooker it was rather like invisible handwriting that was revealed only at a later stage.

With no birds left at Down House, Darwin necessarily experimented at one remove. He was assisted in this by William Tegetmeier, the naturalist who had previously helped him with similar work on domestic birds during the 1850s. Tegetmeier owned a diverse collection of prize-winning poultry and willingly supplied Darwin with information from his back yard in Wood Green, a suburb of London near Tottenham, carrying out matings between different breeds of chickens or pigeons, confirming results, inquiring among his poultry-owning friends about the validity of results, reading parts of Darwin's manuscript, arranging artists to produce illustrations for *Variation,* and sometimes writing short articles about the Darwinian questions passing through his hands for publication under his own name elsewhere. At every possible point, Darwin consulted him about the implications these breeding experiments held for his theories.

While it remains mysterious quite what Tegetmeier received from the relationship, apart from the claim to be working with an eminent figure like Darwin, Darwin certainly appreciated Tegetmeier's help—he recognised that his statements would be an empty shell without Tegetmeier's practical contribution. In letter after letter he oozed persuasive charm. "I am delighted to hear that you have the Fowls," he wrote in February 1863.

> As soon as you have chickens you could kill off the old Birds. I shd. think the 3 ample.—It would be better to cross some cocks & Hens of the half-breds from the two nests; so as not to cross *full* brother & sister. I have not much hope that they will be partly or wholly sterile, yet after what happened to me, I shd. never have been easy without a trial.—I suggested Turbits [pigeons], because statements have been published that they are sometimes sterile with other breeds, & I mentioned Carriers, merely as a very distinct breed. I thought Barbs & Fantails bad solely because I had made several crosses & found the ½ breds perfectly fertile,—even brother & sister together. Did I send you (*I cannot remember*) a M.S. list of crosses; if so for Heaven sake return it.—I get *slowly* on with my work; but am never idle.—I much wish I could have seen you at Linn. Soc; but I was that day very unwell.—Pray do not forget to ask Poultry & Pigeon men (especially latter) whether they have ever matched two birds (for instance two almonds, Tumblers) & could not get them to breed, but afterwards found that both birds would breed when otherwise matched.—
>
> I hope the world goes pretty well with you.[19]

Occasionally, charm gave way to impatience. "Have you quite thrown me overboard as too troublesome?" Darwin inquired a couple of years later. "I have not heard from you for an age.—I wrote some two months ago asking you to send as soon as you could any extracts & facts, which you told me you had collected, about number of sexes—Also any account of even one or two breeds of the fowl.—as colour of plumage of hen & chickens of Pile Game or Golden Hamburgh.—Will you not aid me so far?"

In his own small way, Darwin was creating a private Garden of Eden, bringing back together the birds and beasts of the domestic world, either on paper or in the flesh, in order to discover their origins. "I am a complete millionaire in odd and curious little facts," he told Hooker, "and I have been really astounded at my own industry."[20]

## III

Signs of Darwin's activity were everywhere. Trays of seedlings stood about in sheds and outhouses. He dotted the kitchen garden with upturned flower pots, each protecting some hidden investigation. He covered Emma's azaleas with unsightly netting just when they came into flower. It was important to use "net" for these, he said, the same material as on ladies' hats, because a bell jar or cloche would create too much moisture. He raided Emma's embroidery basket for coloured wools and silks to use as plant markers, and he shut Hooker's exotic flowering orchids away in the study, saying they were for research rather than ornamenting the drawing room. From time to time he invaded Lettington's and Brooks's rows of runner beans to drape the plants with gauze, the purpose of which was unknown to everyone except the master. Darwin must have tested his gardeners' patience during these experimenting years, for he was liable to step out from behind a bush when they least expected it, curtailing opportunities for leaning on a fork and taking stock of the surroundings. Daily, he issued instructions to his family and staff. "Cover that little Ervum in Sand-walk, on which I have never seen Bee visit," he reminded himself, forgetting that the boys usually ran wild down there and could trample on a delicate experimental project.

The family learned to live with his enthusiasms. Some of the children's earliest memories were of helping Darwin with botanical experiments, and it was probably through these, and an increasing number of distinguished scientists visiting the house, that the youngest ones came to realise their father was famous. Leonard came home from his first term at Clapham School intent on reading the *Origin of Species*—the other schoolboys had presumably talked about it. "I remember my father entering the drawing room at Down, apparently seeking for someone, when I, then a schoolboy, was sitting on the sofa with the *Origin of Species* in my hands. He looked over my shoulder and said: 'I bet you half a crown that you do not get to the end of that book.' " Leonard did indeed give up soon after. Darwin "won his bet but never got his money."[21]

At much the same time, George Darwin asked the school's headmaster, Charles Pritchard, to "read with him" parts of Darwin's *Orchids* and repeat some of the experiments at school. Pritchard was an enthusiastic botanist and noted astronomer, and he had a large conservatory at Clapham Grammar School built to his specifications that included a fernery. When Emma went to call at school ("on the buttering principle," she said, with maternal foresight), she was given a guided tour of the hothouses, "which are very superior & full of curious & beautiful things especially the fernery."[22] Pritchard was probably pleased to encourage the sons of such an eminent author. "We succeeded in fertilising and ripening the seeds of *Oncidium papillo* [one of the hothouse species], but we did not succeed in our attempts to induce the seeds to germinate as his father challenged us to do."[23] To have Darwin's sons at the school, and some direct scientific contact with Darwin himself, however slight, was—if discreetly handled—an excellent advertisement for Pritchard's commitment to scientific education.

Darwin included Henrietta and Bessy in some of his research. He and Henrietta studied the advertisements in the *Gardeners' Chronicle* and *Cottage Gardener* and ordered a Wardian case (glasshouse) to stand in the bay window in the dining room, a complicated piece of Victorian engineering that was kept warm by placing dishes of boiling water under the frame.[24] They intended keeping sensitive plants like the mimosa in it, for amusement on rainy days when they could not easily walk outside. "There is a mysterious box come for you, marked Glass but with a kind of gridiron lid as if it had something alive inside," Erasmus wrote warily from London. The giant glass case was an expensive failure, however. Neither Darwin, nor the girls, nor Parslow could keep it hot enough. The begonias and "very curious Oxalis" with moving leaves that Hooker supplied soon died.

All the time, absorbing new questions kept sidling in. He discussed the composition of manure with neighbours like Innes or Norman as if nothing could be of greater interest to him. He inquired into the parentage of cabbages as if they were wayward sons. He thought cordially about potatoes, remembering the wild *Solanum* he once gathered in the rain on Chiloè Island, and got his gardeners to plant eighteen eating varieties at Down so that the family could compare the shape and taste of the tubers on the table. He worried about infertility in the weeping beech. The effects of the environment on plants gave him much to think about. "I hardly know why I am a little sorry, but my present work is leading me to believe rather more in the direct action of physical conditions," he remarked to Hooker. "Perhaps I shall change again when I get all my facts under one point of view, and a pretty hard job this will be."

In another aside, he lavishly praised the Down House gooseberries, poking fun at Hooker's liking for these old-fashioned fruits. His children shared the repartee, saying they must curb their own appetites to ensure a good supply for their father's friend. Francis Darwin said his earliest memories of Hooker involved this strange taste for gooseberries. "I clearly remember him eating gooseberries with us as children, in the kitchen garden at Down. The love of gooseberries was a bond between us which had no existence in the case of our uncles, who either ate no gooseberries or preferred to do so in solitude."[25] Darwin cultivated fifty-four varieties of gooseberry at Down House, a mere fraction of the hundreds available to European nurserymen, but still sufficient to occupy a substantial area in his fruit garden. Nearly all the varieties listed in the *Gooseberry Register* of 1862 that he discussed in *Variation* have long since disappeared.

When Hooker came to Downe bearing gifts of bananas from the Royal Botanic Gardens, they were christened Kew gooseberries. "Do you not think you ought to be sent with Mr. Gower to the Police Court?" queried Darwin, eyeing one large bunch from the government's greenhouses. "If

Etty lets the Gooseberry season go bye without inviting me I will kill her," Hooker quipped in return.

Darwin also interested himself in the social relations of plants. He became inquisitive about weeds—always high on a gardener's agenda. Considered in an abstract way, in the light of his theory, weeds were highly successful organisms. They were vigorous, invasive, competitive species, the very picture of health and adaptability. One sorry incident in the Sandwalk at Down House illuminated this much more than any theoretical reasoning could have done. The area had been planted by Darwin and Emma in 1843 or so, soon after their arrival in Downe, with a variety of native trees such as hazel, alder, lime, hornbeam, birch, and dogwood, to add to the well-established oaks and beeches that were already there, with a line of hollies down the exposed side next to the valley. Emma liked the natural look under the trees, with wood anemones and bluebells in their turn, set among clumps of wild ivy. She employed a young boy every now and then to pull up the dog's mercury, an invasive, poisonous weed capable of advancing up to three feet a season. One year a new boy misunderstood the orders.

> As my father and mother reached the Sandwalk they found bare earth, a great heap of wild ivy torn up by its roots and the abhorred Dog's mercury flourishing alone. My father could not help laughing at her dismay . . . and he used to say that it was the only time she was ever cross with him.[26]

Darwin asked Asa Gray about the invasive properties of British weeds in the United States of America. Cocksfoot, a meadow grass, was at that time spreading fast through North American orchards, the epitome of an entrepreneurial, colonising species, and regarded by Darwin as representative of a more general, highly vigorous national spirit. British weeds were like British colonisers, he thought. "Does it not hurt your Yankee pride that we thrash you so confoundedly? I am sure Mrs. Gray will stick up for your own weeds. Ask her whether they are not more honest downright good sort of weeds."[27] He advised Julius von Haast to attend specially to invasive European species in New Zealand.

Ceaselessly he returned to the intriguing problem of the sex lives of plants. Particularly, as he worked, he focused on what he and Huxley called "sterility," the mechanisms that incipient species develop to prevent themselves blending back into the populations from which they emerge. This bore on the practical difficulties of keeping species apart. Unfortunately, most varieties could breed perfectly well with another variety of the same species, and it was hard to imagine how a population could ever keep itself sufficiently separate to turn into a new species without invoking some form of geographical or functional isolation. Huxley always mentioned the problem when he talked publicly about the *Origin of Species*. "To get the degree of sterility you expect in recently formed varieties seems to me hopeless," Darwin expostulated to Huxley in December 1862. Nevertheless, that was precisely what Darwin was after.

In particular, he brooded over the reproduction of *Linum grandiflorum*, the old-fashioned garden flax, a staple of Victorian flower beds, that has either a long or a short style inside its pink or scarlet flowers, a dimorphism which mirrored the pin- and thrum-headed primulas he had studied a year or two before. Darwin wondered if these anatomical differences would hold the clue to his problems. It did seem likely that long and short were mutually adapted to ensure cross-fertilisation. According to his theories, long-styled flowers would find it difficult to mate with other long-styled flowers, or short with short. Offspring might be reduced in vigour—a kind of incipient sterility.

Much of this experimental work was tricky to perform. Although the flax species is considered a hardy annual in Britain, the plants required a heated bed. As he followed the fertilisation processes under his microscope, however, he was persuaded that he was witnessing a functional differentiation between otherwise identical flowers. The "reduced fertility" and "diminished vigour of the progeny" threw light on the nature of crossing and hybrids in general. He sent a paper on the subject to the Linnean Society early in 1863 and discussed it in *Variation*.

He then set out to understand the three-way fertilisation patterns of the purple loosestrife, *Lythrum salicaria*. These complex plants made flax look like child's play. "I am almost stark raving mad over Lythrum," he cried to Gray.

> I cannot publish this year on Lythrum salicaria: I must make 126 additional crosses!! All that I expected is true, but I have plain indications of much higher complexity. There are 3 pistils of different structure & functional power & 3 kinds of pollen of different structure & functional power, & I strongly suspect altogether five kinds of pollen, all different in this one species![28]

Baffled, he asked his oldest son William to do the mathematical calculations. William dutifully regarded this as no more tiresome than adding up columns of banking figures. Eagerly Darwin asked for more. Over the summer months of 1862, William recorded in a notebook dozens of investigations on his father's behalf. He spent several weeks studying *Lysi-*

*machia vulgaris* (creeping jenny), comparing it with *Lythrum* for the relative length of stamens and pistil. He drew differently shaped pollen grains from *Lythrum* with the aid of a camera lucida and sent the measurements to his father. "Tuesday," he said in early August, "I examined 102 plants this morning." Many of these notes were intelligible only to Darwin and himself. "25 flowers examined from 5 different plants, of these 25, 113 had 5th stamen longest, 2 had 4 ¥ 5, almost 5 had the pistil in contact or nearly so with 5th stamen, 10 had no points in them," he reported to his father.[29] In 1863 he worked his way through a long list sent by Darwin, headed "List of plants apparently adapted to prevent self-fertilisation."

Unsuspecting Wedgwood nieces were roped in during family parties. After one visit to Downe, Sophy, Lucy, and Margaret Wedgwood (Caroline and Jos's daughters), wrote to tell him that they had collected 256 "specimens of *Lythrum*" in meadows around Llandudno as instructed and sorted them into the three kinds.[30] "We find it rather difficult in gathering to know what are distinct plants and what only offsets." Darwin had the grace to reply, "My dear angels! . . . I never dreamed of your taking so much trouble." In the end, fuelled with information from family and friends, Darwin sent another long paper to the Linnean Society in 1864.

Hooker, Gray, and Oliver were essential soul-mates in this. Late in 1862, Darwin was, however, delighted to locate another botanist cut from the same agreeable experimental cloth. John Scott was a curator at the Botanic Gardens in Edinburgh and much interested in hybridisation and experimental horticulture. Darwin reeled him in quickly. "They give or lend me all plants at Kew; but they are very weak in primulas," he sighed persuasively. "I am sick of ordering plants at London nurseries; I so often get the wrong thing."[31] Flattered, and then intrigued, Scott gave freely of his knowledge. Darwin became fond of him over the years and eventually did much to advance his career.

## IV

Best of all, at the end of 1862, he decided to build a hothouse. He was finding it a longish walk over to George Turnbull's gardens, where the invaluable Horwood patrolled the glasshouses, and he probably sensed that his welcome might be wearing thin.

"Now I am going to tell you a *most* important piece of news!!" he exclaimed to Hooker in December.

> I have almost resolved to build a small hot-house: my neighbours really first-rate gardener has suggested it & offered to make me plans & see that it is well done, & he is a really a clever fellow, who wins

> lots of prizes & is very observant. He believes that we shd succeed with a little patience; it will be grand amusement for me to experiment with plants.[32]

True to form, he balked at the expected cost. He consulted neighbours like Turnbull and Norman, wrote to Hooker about industrial-capacity stoves and to Fox about complicated mechanical ventilation devices with handles that screwed open and shut, generally busying himself with the question. At Down House he already possessed an unheated glasshouse, a brick-based, south-facing lean-to against the orchard wall. What he did that winter was to build the first of several heated extensions. He commissioned Horwood to draw up plans, and during January and February 1863 he watched proprietorially as the building emerged under the hands of the village carpenter. Later, in the 1870s, he added another two bays (perhaps three) and ran hot-water pipework through these and the original unheated house, so that he had a range of houses at different temperatures.

The hothouse was completed in February 1863. Darwin asked Hooker if he might have some plants from Kew—"I long to stock it, just like a school-boy." He was itching to get on. "Would it do to send my box-cart early in the morning, on a day that was not frosty, lining the cart with mats and arriving here before night. . . . there would be about five hours (with wait) on the journey home." Hooker hardly had time to nod in agreement before Parslow was at Kew's front gate with an open-topped cart. When 160 different plants arrived at Down, Darwin felt a trifle ashamed of himself. He made an apologetic remark about depleting the national collections.

Unpacking Hooker's government-issue flowerpots, "gloating" with Henrietta, and contemplating future experiments as he disentangled each plant from its coconut matting was exactly what suited him best. "I have begun dull steady work on 'Variation under Domestication,' " he said to Gray. "But alas & alas pottering over plants is much better sport."[33]

At a stroke Darwin's routine was transformed. Every morning and afternoon he now spent an hour or so in the new hothouse before walking round the Sandwalk. There was always something to see to, something to fire his imagination, something to plan. There he found warmth, plants, and peace. These three gods provided solace as his health deteriorated and his attention wandered from the lengthy *Variation* manuscript. In this regard, it was a great advantage to be a gentleman-amateur. Darwin was not bound by any intellectual constraints. He had no teaching or lecturing commitments like Huxley, no professional occupation to fulfill like

Hooker, no administrative tasks like Gray. He could afford to follow his passion as it directed him.

## V

Here and there, a flash of temper showed. Darwin responded rudely to a gardener writing in the *Journal of Horticulture* early in 1863. The gardener, Donald Beaton, remarked that no plant-breeder of any reputation in England would wish to have his name associated with Karl Gärtner, because, Beaton claimed, most of Gärtner's statements about plant breeding were unsupported by the evidence. Startled, Darwin took the remark as a direct hit at himself. He knew Beaton a little bit, and had corresponded with him about Gärtner's findings: if Gärtner was wrong, so was Darwin. Darwin printed a sharp rejoinder in the same journal, using all his authority as an elite scientist to stun Beaton into silence. He may have done more than he intended, for Beaton died of apoplectic seizure in the autumn.[34]

He scarcely cooled off before the *Athenaeum* published an article on spontaneous generation. Inexplicably, Darwin exploded. This time, an anonymous reviewer of William Carpenter's latest book contemptuously dismissed Darwin and Carpenter as master and disciple. The reviewer made caustic remarks about *An Introduction to the Study of Foraminifera* (Carpenter's book on an order of marine organism) and then suggested that the forces of life must lie in primeval ooze, not in Darwin's "breath of a Creator." The author made it clear that he favoured Pouchet's chemical origin of living organisms, "heterogeny" in the terminology of the day.

Carpenter replied to the *Athenaeum* denying any intellectual debt to Darwin and going so far the other way as almost to claim that his studies of Foraminifera justified an anti-Darwinian inference. He disassociated himself completely from godless ooze. Darwin was surprised, for he counted Carpenter as a friend and a partial convert to evolutionary theory. Losing all sense of proportion, he hit back in April in an article also published in the *Athenaeum*. "A mass of mud with matter decaying and undergoing complex chemical changes is a fine hiding-place for obscurity of ideas," he fulminated against the reviewer.

> Is there a fact, or a shadow of a fact, supporting the belief that these elements, without the presence of any organic compounds, and acted on only by known forces, could produce a living creature?[35]

He went on to defend the way his theory connected "by an intelligible thread of reasoning a multitude of facts." Then he "exhaled" crossly in a letter to Hooker.

> It will be some time before we see "slime, snot or protoplasm" (what an elegant writer) generating a new animal. But I have long regretted that I truckled to public opinion & used Pentateuchal term of creation, by which I really meant "appeared" by some wholly unknown process.—It is mere rubbish thinking, at present, of origin of life; one might as well think of origin of matter.[36]

Usually the soul of patience, Hooker scolded Darwin for interfering in matters that should be left alone.

> I cannot abide this lugging of Science before the public in Times & Athenaeum, & implore you my dear fellow not do so again. . . . The only party that gains by these discussions is the proprietor of the paper, the only one that loses every way, is the maintainer of truth. Science will be much more perfected if it keeps its discussions within its own niche.

Darwin hardly needed to be told this, but he sent a second letter to the *Athenaeum* anyway. His temper was not improved to hear that Richard Owen was the anonymous reviewer. Eventually he calmed down and apologised to Hooker. "You give good advice about not writing in newspapers; I have been gnashing my teeth at my own folly. . . . if I am ever such a fool again have no mercy on me."[37] He told Lyell the same: "I was an ass to write to *Athenaeum.*"

Nevertheless, soon afterwards, he nearly argued with Asa Gray over politics. While the Civil War tore America apart, Darwin's feelings ran high, far higher than commonly thought. North against South, Unionist against Confederate, the shock waves reverberated across the globe. From the first talk of Confederate secession until the surrender of the southern states at Appomattox Courthouse, he was intently engaged: "I never knew the newspapers so profoundly interesting." Every day he would turn first to the reports in the *Times* by William Howard Russell, the special war correspondent who for Britons was the voice of America in the public prints. Darwin passionately wanted the southern states to abandon slavery. He told Gray that he despaired at President Lincoln's initial sidestepping of the slavery issue in order to defend the Union. On the other side of the ocean, Gray ardently supported his president. "The first gun raised my spirits, and they have never flagged since," he declared, letting his Yankee sentiments spill into the open. For him, the Union, not slavery, was the primary issue.

All Darwin's abolitionist fury burst out. "I have not seen or heard of a soul who is not with the North," he wrote emotionally.

> Some few, & I am one, even wish to God, though at the loss of millions of lives, that the North would proclaim a crusade against Slavery. In the long run, a million horrid deaths would be amply repaid in the cause of humanity. . . . Great God how I shd like to see that greatest curse on Earth Slavery abolished.[38]

To be sure, Darwin simplified the issues. Although he supported the anti-slavery cause more completely than any other social principle in his life, it was nevertheless relatively easy for him, quietly situated in an English village and buttressed by a private income, to advocate a moral crusade in America. Conveniently, he forgot the colonial and industrial sources of British economic wealth, forgot that his own stocks and shares rested on the manual labour of railway navvies, miners, indentured millhands, and plantation coolies. Full of humanitarian fervour, he ignored the complex political and personal turmoil through which men like Gray were living. Of course, Gray abhorred the system of slavery. But he found it annoying that his English friend took the virtuous high ground without an apparent thought for the terrible, pragmatic realities of maintaining the Union.

On the other hand, the *Trent* affair of 1861 brought out Darwin's patriotic instincts. Like most Britons, he found it outrageous that a Union ship should intercept and board an English mail boat, even if it carried two Confederate envoys, and said so to Gray. Gray threw Darwin's scientific theories back at him: "We must be strong to be secure and respected—natural selection quickly crushes out weak nations." Gray accused English politicians of being swayed by the mob.

"Hitherto I have been able to write with some sympathy," muttered Darwin to Hooker. "Now I must be silent; for I look at the people as a nation of unmitigated blackguards." The two Englishmen agreed that Gray was "blind to everything & what is worse brags like the greatest bullies amongst them."[39] For a while they exchanged Gray's letters, "as political and nearly as mad as ever," remarked Darwin as he sent one along to Hooker in November 1862. Unmoved, Gray lectured them both from across the Atlantic about the North's ability to persevere alone, and issued testy reminders that articles in the *Times* did not carry as much significance in the United States as Darwin evidently thought. "Homely, honest, ungainly Lincoln is the *representative man* of the country," Gray insisted.[40] Meanwhile, the textile manufacturing regions of Britain encountered economic distress as supplies of raw cotton from the American South dried up. Despite his dislike of the Confederacy, Darwin fired off a donation of ten guineas to the Lancashire Cotton Spinners' relief fund, far more than he usually gave to charity.[41]

The near collapse of Union forces defending Washington in the battle of Bull Run stirred him more than expected.

> I have managed to skim the newspaper, but had not heart to read all the bloody details. Good God what will the end be; perhaps we are too despondent here; but I must think you are too hopeful on your side of the water. I never believed the "canard" of the army of the Potomac having capitulated. My good dear wife & self are come to wish for peace at any price.[42]

It was a tense time, although the two never broke off relations. "We must keep to science, I fear, for we both seem to be getting to think each other's country's conduct worse & worse," Darwin said in April 1863.[43] As the *Times* switched sides and Russell was expelled from American soil, Darwin commiserated with Gray, calling the newspaper the "Bloody old *Times*," as William Cobbett used to do.[44] When the war ended, he reflected sorrowfully on their mutual troubles.

> I congratulate you, & I can do this honestly, as my reason has always urged & ordered me to be a hearty good wisher for the north, though I could not do so enthusiastically, as I felt we were so hated by you. . . . I declare I can hardly yet realise the grand, magnificent fact that Slavery is at an end in your country.[45]

Not far below the surface of this correspondence lay anxious thoughts about Louis Agassiz. Both before and during the Civil War period, Agassiz's definition of species gave scientific credence to the idea of marked human racial differences. His followers, and on several occasions Agassiz himself, claimed Negroes were a separate species from Caucasians, a point of view that had immediate implications for the anti-slavery movement. For a long time now, Agassiz had been convinced that human beings were divided by God into a number of species from the beginning—a logical corollary to his belief in the separate created identity of every living form. Negroes, he proposed, were physiologically and anatomically distinct from whites, created by God to be fixed in their separateness.

So saying, Agassiz joined the controversy over the unity or plurality of the human race, a controversy that had begun decades before Darwin's writings and had spread widely across contemporary culture during the 1830s and 1840s, especially in America when given scientific and anthropological significance by Samuel George Morton's study of skulls and Josiah Clark Nott and George Robbins Gliddon's notorious *Types of Mankind* (1854), which provided a series of apparent biological justifications for black inferiority. Racial biology, politics, and the origin of species combined in ugly fashion during the war crisis, principally because Agassiz's writings lent scientific authority to southerners determined to defend the slave system. Agassiz became dogmatic in his opposition to transmutation, increasingly vehement in his racial, creationist biology. Here was a celebrated man at the head of his adopted nation's intellectual life, a fine naturalist and lover of nature, a staunch supporter of the Union—Agassiz was altogether a puzzle to his colleagues. His students felt the tension. From 1863 or so, some of them began drifting away from Agassiz's museum at Harvard, either attracted by Darwin's alternative universe or developing their own form of non-Darwinian evolution.[46]

Opponents like Gray or Lyell were frankly repelled by his public pronouncements. Gray, whose dislike for Agassiz was now intense, vented his feelings to Hooker.

> This man, who might have been so useful to science and promised so much here has been for years a delusion, a snare, and a humbug, and is doing us far more harm than he can ever do us good.[47]

## VI

In 1863, three significant books written by Darwin's closest supporters sprang from the English presses, each author defending and extending evolution in his own way. These books presented indisputable evidence that natural selection was not a mere flash in the pan. The scope and logic of natural selection was made apparent to contemporaries even if not acceptable to some of them. And as a mode of formal communication the books had a material impact. While newspapers, journals, and reviews were important to Darwin in their day, books were better suited to discuss the multiple issues that were now emerging. The friendly support that Darwin first received from his circles of acquaintance in reviews was, in effect, transformed into a genuinely public space in which other authors and other audiences began to participate in debate and negotiation. These books moved a long way beyond Darwin's original thesis. Contentious and resilient, "Darwinism" began developing as a body of thought.[48]

Lyell headed the charge. By now he was more than ready to make public the accumulated results of his investigations into the early history of mankind. He still believed there was a spiritual side to humanity that defied rational explanation. But he had reached such an advanced stage in his researches that he had to publish, come what may. His letters to Darwin were as warm as ever. A sociable man, he would have liked to see more of his old friend on the scientific circuit. He reminisced sentimentally about their long friendship and promised Darwin that his new book would support evolution. In return, Darwin admired Lyell's courage—he recognised that what had been relatively easy for him to accept was a wrench for the older man. "Considering his age, his former views, and position in society, I think his conduct has been heroic on this subject."[49]

Lyell's title was blunt: *The Geological Evidences of the Antiquity of Man with Remarks on Theories of the Origin of Species by Variation.* It was published by Murray in the first week of February 1863, sold out immediately, was reprinted three times in 1863, and was very widely reviewed. The firm of G. W. Childs in Philadelphia brought out an American edition, and by 1864 the book was available in German, French, and Dutch. "I am reading your book on the Antiquity of Man," wrote William Whewell to Lyell in February, "as all the world has done or is doing." Mudie's Circulating Library bought several thousand to distribute to an audience impatient for all things evolutionary.

In this book Lyell pulled back the curtain on civilisation to reveal the world of human geological history. Until then, the paucity of human fossil remains had suggested that mankind was fairly recent in geological terms, a view that accorded well with the idea that humanity appeared only when the earth reached its modern state after the glacial period, or—for those who believed in the biblical flood—at the point when the waters receded. Lyell pushed the origin of human beings much further back than this watery dividing line, into the geological deep past. His writing, as always, was vivid. His description of Philippe-Charles Schmerling's descent by rope into Belgian underground galleries and cave chambers in search of ancestral bones made readers sit up and take notice. Furthermore, his use of the expression "missing links" in the fossil record lodged permanently in the public mind.[50]

And he fired the imagination of contemporaries as varied as Jules Verne, who used Lyell's enlarged vision of time in *Journey to the Centre of the Earth* (1864), and the geological author Louis Figuier. In the second edition of his illustrated volume *La Terre avant le Deluge* (1867), Figuier jettisoned the Garden of Eden in order to show a savage world inhabited by men and women clothed in skins and wielding stone axes.[51] In short, Lyell's book shattered the tacit agreement that mankind should be the sole preserve of theologians and historians. In a way, he gave the people their geological past. It was the first significant book after Darwin's *Origin* to shake humanity's view of itself.

Aggrieved by the subject matter and confident that humans appeared only recently on the earth, Richard Owen wrote peevishly to Murray saying he "sometimes thought of an essay on the Novity of Man."[52]

As Darwin leafed through Lyell's pages, however, he could not find any clarion call for evolution. Towards the close, Lyell certainly confirmed that he had changed his mind on transmutation since the first edition of *Principles of Geology* and now stood with Darwin. Yet after dwelling on Lamarck, and admitting he had done Lamarck an injustice in the *Principles,* Lyell described the technical objections to the *Origin of Species* very thoroughly. There was a huge gulf between man and beast, Lyell stated. How this gulf was bridged remained a "profound mystery." "Oh," exclaimed Darwin in the margin of his copy.

Darwin had pinned his hopes on Lyell's making an explicit avowal. To that end, he had pushed and nudged and harried and flattered Lyell over the years, all to no avail, it seemed. "I am fearfully disappointed at Lyell's excessive caution," he announced to Huxley. "Deeply disappointed," he told Hooker.[53] The book was a mere "digest," he said forlornly. Embarrassingly, he had invited the Lyells to Down House for a celebratory post-publication weekend.

> The Lyells are coming here on Sunday Evening to stay till Wednesday. I dread it, but I must say how much disappointed I am that he has not spoken out on Species, still less on Man. And the best of the joke is that he thinks he has acted with the courage of a Martyr of old.—I hope I may have taken an exaggerated view of his timidity, & shall *particularly* be glad of your opinion on this head.—When I got his book, I turned over pages & saw he had discussed subject of Species, & said that I thought he could do more to convert the Public than all of us; & now, (which makes the case worse for me), I must in common honesty retract. I wish to Heaven he had said not a word on subject.[54]

His stomach saved him. Emma declared he was too unwell to receive visitors. Shortly afterwards, Darwin wrote an unecessarily candid letter to Lyell, complaining that Lyell was not doing enough to help him. Then he spelled out the differences between himself and Lamarck. Lamarck's volume was "a wretched book; & one from which (I well remember my surprise) I gained nothing. . . . I must add that Henrietta, who is a first rate critic & to whom I have *not said a word* about Lamarck, last night said, 'Is it fair that Sir C. Lyell always calls your theory a modification of Lamarck's?' "[55]

Lyell was upset. He replied that he had spoken out to the "utmost extent" of his tether, "farther than my imagination and sentiment can follow, which I suppose has caused occasional incongruities."[56] From Lyell's perspective, it probably seemed that Darwin wanted to wring the very soul out of him. His ambivalence about evolution was real. As Hooker remarked, Lyell was "half-hearted and whole-headed." Despite all their friendly connections, Darwin never really forgave him for betraying his hopes.

Lyell's book was destined to raise annoyance in other quarters too—annoyance that made the divisions among these Victorian naturalists obvious. One after another, Lyell's friends complained about the *Antiquity of Man* for other reasons. Falconer said that Lyell had stolen his results. Prestwich claimed the same. So did John Lubbock, who had actual evidence that Lyell had appropriated his published work on the lake habitations of Denmark and consequently reviewed the book in icy terms in the *Natural History Review.*[57] Darwin had to intervene between Lubbock and Lyell in order to soothe both sets of ruffled feathers.

Then Richard Owen launched a scattergun fusillade in the *Athenaeum*, hoping to hit any passing Darwinian by declaring that the facts in Lyell's skull chapter (which had been written by Huxley) were false. Casting their woes temporarily aside, the Darwinians rallied round. Falconer attacked Owen in the *Natural History Review*, Hooker defended Lyell ("poor dear L."), and Huxley cheekily supplied Lyell with sufficent material for an anatomical rebuttal. Rolleston backed up Huxley in the *Athenaeum*. Darwin merely declared Owen was the very devil of an adversary. "I am burning with indignation." So when Jacques Boucher de Perthes found a fossilised human jaw in the gravels of Moulin-Quignon—the first human jawbone—the discovery was shot down before it ever flew. Falconer denounced the jaw as a fraud, while Carpenter, no expert, perversely authenticated it as genuine. Owen retaliated by criticising Falconer's account of the newly discovered *Archaeopteryx*, an intermediary bird-reptile, "a strange being à la Darwin." To Owen the fossil bones in no way represented a link between birds and reptiles.

All in all, these talented men stirred up a bitter row over human antiquity and evolution in which each struggled for primacy. Personality and intellectual commitment were the leading forces in appraising these significant fossil discoveries and the right to speak about ancient mankind. "It is wretched to see men fighting so for a little fame," said Darwin, deliberately backing down from the controversy.

## VII

Huxley's book was called *Evidence as to Man's Place in Nature*, published a few weeks after Lyell's. "Lyell's object is to make man old," protested the *Athenaeum*, "Huxley's to degrade him."

First, though, he published his lectures for working men. Huxley sent a copy of this pamphlet to Darwin, who declared it was "simply perfect. . . .

What is the good of writing a thundering big book when everything is in this little green book, so despicable for its size? In the name of all that is good and bad, I may as well shut up shop altogether."[58]

Pleased, Darwin showed the lectures to his daughter Henrietta, who picked up a couple of typographical errors and (with the pompous earnestness of youth) told her father that she wished Huxley would write "something that people can read; he does write so well." Huxley called her Miss Minor Radamanthus after that, an allusion to the eagle-eyed judge in Greek mythology.[59]

His *Man's Place in Nature* was much fiercer. The text showed Huxley at his snarling extreme. He regarded the piece as a definitive rebuttal of Owen's views on monkeys, brains, and humans. Where Lyell hesitated, he zipped along, cynical and acerbic in turn.

But it was a line drawing that actually said it all. On first opening Huxley's book, readers saw exactly what his argument would be. His frontispiece showed five skeletons standing in line, each bony figure leaning slightly forward ready to evolve into the next. From gibbon to orang, chimpanzee, gorilla, and man, the implication could not be plainer. Humans were the result of a series of physical changes from the apish state.

Like all pictures with a political message, this had required careful preparation. Huxley engaged Benjamin Waterhouse Hawkins, the zoological illustrator, to draw the skeletons to scale from specimens at the Royal College of Surgeons and then enlarge or reduce them as necessary to make the point. The pose was the same in each instance, and the sequence (from left to right) an intuitive understanding of the Western way of reading lines of development. It was inspired visual propaganda.

In the text that followed, Huxley showed how mankind must, on all logical biological grounds, be placed with the apes. As usual, Richard Owen was the enemy he openly addressed.

> It is not I who seek to base Man's dignity on his great toe, or insinuate that we are lost if an Ape has a hippocampus minor. On the contrary, I have done my best to sweep away this vanity. I have endeavoured to show that no absolute structural line of demarcation, wider than that between the animals which immediately succeed us in the scale, can be drawn between the animal world and ourselves; and I may add the expression of my belief that the attempt to draw a psychical distinction is equally futile, and that even the highest faculties of feeling and of intellect begin to germinate in lower forms of life. . . . Is mother-love vile because a hen shows it, or fidelity base because dogs possess it?[60]

He knew his audience well enough to include a word or two about primitive human societies and their supposed animalistic propensities, including an unnecessary description of cannibalism, along with an engraving of a so-called butcher's shop in the Congo complete with severed arms and legs. These sensationalist sops to the audience found little favour with Huxley's friends. Lyell said, "I hope you send none of these *dangerous* sheets to press without Mrs. Huxley's imprimatur." Falconer protested that he "would let no young lady look at it."[61] One of Huxley's young acolytes, the comparative anatomist St. George Mivart, disapprovingly remarked he had seen the book for sale on a railway bookstall, available to anyone. Reviewers noted the same flaw. One observed dryly, "We are not yet obliged to be quite on all-fours with Professor Huxley."[62] Another accused him of diving into "the African forests in search of his grandfather."

The book was little more than a crowd-pleaser, although it had the advantage of expressing, in a small compass, the anatomical issues at stake. Yet it was to spread around the world almost as rapidly as Darwin's *Origin of Species*. William Appleton, in New York, produced an American edition. Charles Mudie bought several hundred copies for his lending library, recognising a good title when he saw one. Darwin welcomed it with delight. "Hurrah the Monkey Book has come," he shouted in triumph. "I long to read it, but am determined to refrain till I have finished Lyell, & I have got only half through it. The pictures are splendid."[63]

Quick as a flash, Huxley replied, "Why did not Miss Etty send any critical remarks on that subject by the same post? I should be most immensely obliged for them."[64]

Victorian commentators were equally quick to note that belligerence was the name of Huxley's game. In May a well-observed skit was printed in the magazine *Public Opinion* that played on Huxley's obvious relish for argument, "A Report of a Sad Case recently tried before the Lord Mayor, Owen versus Huxley, in which will be found fully given the merits of the Great Recent Bone Case."[65]

In that skit, Owen and Huxley were depicted as Victorian street traders dragged into a courtroom trial for brawling. All through the fanciful trial they shouted learned insults at each other: "posterior cornu," "hippocampus," and so on. Such vulgarity was "scarcely human," declared the judge, at which the Huxley character laughed rudely. Rowdy barrowboys in the gallery, including Hooker, "in the green and vegetable line," and "Charlie Darwin, the pigeon fancier," cheered Huxley on. The Owen character refused to let Huxley swear the courtroom oath—how could the court take an affidavit from a man who did not believe in anything? he inquired stonily. Huxley's character retorted that Owen was just as bad, for he "does not know a hand from a foot." In the end the judge claimed "no punishment could reform offenders so incorrigible." This barrack-room farce intelligently seized the main points at issue, the political and theological as readily as the anatomical. The inventive author was rumoured to be George Pycroft, a surgeon and fellow of the Geological Society.[66] "It is capital," said Darwin in high good humour. "The more I think of the 'Sad Case' the cleverer it seems."

When the laughter died down, he acknowledged that "a scientific man had better be trampled in dirt than squabble."

## VIII

Close behind came Henry Walter Bates offering butterflies to the cause. Reticent and unassuming, this relatively unknown naturalist supplied the first real evidence for natural selection at work, slipping it into an evocative travel narrative of the Amazon that captivated Victorian readers in 1863 during the same spring publishing season.

Bates had arrived back in England in the summer of 1859 after eleven years of collecting in Brazil, mostly living and working around Ega (Tefé), where Wallace had left him, some two hundred miles upstream from Manaus. Like Wallace, he brought home a lifetime's work in natural history, based in a collection of more than fourteen thousand different species, many of them insects, of which some eight thousand were new to Western science. He was committed to the idea of evolution, which like Wallace he had first learned from Chambers's *Vestiges,* and this commitment was afterwards supplemented by his own investigations. Soon after his arrival in England he read Darwin's *Origin of Species.* Just as was true of Wallace, his years in the rain forest left him unprepared for making his way in the metropolis. After paying off his debts, his profit for those eleven years of collecting barely amounted to £800. He had no visible means of supporting himself. He possessed nothing beyond his skill in empirical natural history.[67]

It was Darwin who threw him a lifeline. The two men had met early in 1861, and Darwin liked him immediately. In a way, Bates was everything that Wallace was not. Bates was never a rival to Darwin—however agreeable that rival might prove. Bates put his ideas completely at Darwin's service. He was grateful for Darwin's favours. He was, Darwin told Asa Gray, "a man of lowly origin, of great force of character, & wonderfully self-educated." To Hooker he said, "What a pity that this man shd. have to work for his daily bread." In return, Darwin was perhaps able to praise Bates more fluently than he managed with Wallace. If Darwin had learned anything from the Wallace incident, it was not to underestimate the talent of his natural history correspondents. At the very least, Bates was eager for Darwin's practical help. Darwin helped him apply for a curator's job at the British Museum and commiserated with him when it went to Owen's candidate, Albert Gunther. He carried on looking for suitable work for Bates until he found him a position as the assistant secretary of the Royal Geographical Society in 1864 (Wallace applied unsuccessfully for the same job). "I am sure he is no common man," said Darwin about Bates to his influential friends.[68]

More especially, Bates arrived in England with two gifts for Darwin. One was the concept of insect mimicry. The other was a collection of butterflies that displayed evolutionary diversification.

Mimicry was an unusual feature of the living world that not even Darwin could have anticipated. While working together in the Amazonian river basin, both Bates and Wallace had perceived that occasional unrelated species of butterfly could superficially resemble each other. Wallace had taken this insight with him to Malaysia, while Bates continued working on it alone in Brazil. Afterwards, in fact, the two felt gently competitive about the notion, each believing he had independently worked out the idea directly from the observations, and each slightly annoyed that his individual work on the subject tended to be confused with the other's. In this regard, Wallace's reputation for benevolence did not apply. Beyond his relationship with Darwin (in which he was notably consistent), he was generally very reluctant to give up his originality to other men. Indeed it may have been the shock of colliding with Darwin over natural selection that sensitised him to future priority issues. He did not argue with Bates as such—he merely voiced his own claim to the idea of mimicry. But he argued fiercely with George Romanes, years afterwards, when Romanes accused him of stealing the notion of "physiological selection." And there was a moment of coolness with Huxley over another minor incident.[69] One small circumstance at the London Zoo suggested the nub of the problem between Wallace and Bates. Charles Lyell one day saw Bates in the Zoological Gardens and hailed him as "Mr. Wallace." When corrected, he blundered on, "Oh, I beg pardon, I always confound you two."[70]

Bates proposed that mimicry functioned as a life-saving disguise. One insect might be unpalatable to birds. If another species came to imitate it, it too would benefit from not being eaten. Once Bates was alerted to the possibility of insects imitating each other for protection he found the phenomenon everywhere, even noticing insects that masqueraded as sticks and leaves. His daily expeditions became treasure hunts, or detective stories, as he turned over leaves and logs, peered under lianas, constantly adjusting his perspective as he searched for the giveaway clue that revealed an insect leg or beetle carapace blending into the tropical background. Back in England, Bates explained to Darwin how this protective mimicry must emerge through adaptation and selection. By eating some insects and not others, birds were naturally selecting the forms and colourings that would survive. The better the mimic, the better the survival rate. He provided a full account in a paper delivered at the Linnean Society in London in 1862. Brimming with enthusiasm, Darwin took the unusual step of asking Huxley if he—Darwin—could write an article describing Bates's work for the *Natural History Review.* In that article Darwin made the evolutionary implications of "deceptive dress" explicit. Published in 1863, it was an unstinting appreciation of another man's work—work that intimately supported his own.[71]

After this there could be no stopping him. Oblivious to the younger man's reluctance, Darwin pushed Bates towards a longer publication on the natural history of the Amazonian basin.

Within a few weeks, Darwin had fixed him up with John Murray, an arrangement that could not have been made without Darwin's vigorous representations to Murray. Darwin read and commented on Bates's manuscript page by page as it was produced and supported him when he wilted under the unaccustomed pressure of authorship. He offered tips on writing, provided money for illustrations, and invited him home for the weekend. He got Asa Gray to arrange for extracts from the book to be released in the United States. There can be no doubting the effort that Darwin poured into Bates's personal and intellectual welfare in order to get this book published. Bates's *The Naturalist on the River Amazons* [*sic*] came out in April 1863, a lasting classic of the genre. No doubt exhausted by the force of Darwin's attention, Bates said feelingly that he would rather spend another eleven years on the Amazon than write a second volume. Darwin declared that Bates was nearly as good as Humboldt and busied himself getting the book reviewed in all the journals.

As it stood, Bates's volume represented one of the subtlest strikes on behalf of Darwinism. Unlike Huxley or Lyell in either tone or content, Bates addressed a different facet of the Victorian mind, one that thirsted for adventure stories and heroic accounts of geographical exploration far beyond Britain's horizons. Bates's account interlaced jungle tales and exotic natural history with descriptions of indigenous Amazonian culture and its Portuguese overlays. One haunting image was of its author, miles from European contact, despondently reading and rereading the advertisements in an old copy of the *Athenaeum* for company. His empathy and interest in the areas and peoples where he lived and worked were evident.

As far as Darwin was concerned, the important material came towards the end. There Bates gave an eyewitness account of the origin of species in nature. Two Amazonian butterflies, the black-and-crimson-spotted *Heliconius melpomene* and the *Heliconius thelixope,* if taken separately, were perfectly distinct species. Bates discovered four or five transitional forms living in particular geographical areas in between the two, each connected by a chain of gradations. The intermediate forms were not hybrids. They were geographical races, each on their way to becoming separate species. "We seem to obtain here a glimpse of the manufacture of new species in nature."[72]

"It is a grand book and whether or not it sells quickly it will last," Darwin congratulated him. "You have spoken out boldly on species; and boldness on the subject seems to get rarer and rarer."[73] His nemesis, the *Athenaeum,* coolly reported that Bates twisted the facts to support Darwin.

Yet butterflies were destined to become evolution's most elegant practical support. During the next few years, Wallace produced his own analysis of insect mimicry as a form of protective colouration, revealing his belief in the efficacy of natural selection and adding the idea that a single species may mimic several different models. To unpick the various configurations of Malaysian Papilionidae as Wallace did required astonishing mastery of detail. Nevertheless, this form of imitation came to be known as Batesian mimicry, and its study was carried forward by Roland Trimen, based in South Africa, another of Darwin's correspondents. Later, the naturalist Fritz Müller, who had the advantage of living in Brazil, began his own long-term investigation into another kind of mimicry, in which the insects converge in looks. Usually known as Müllerian mimicry, this additional demonstration of natural selection was published in 1878. Many of the strongest believers in Darwinism in the 1860s and 1870s turned out to be those who studied species in their natural habitat, the field naturalists like Bates and Trimen, and men with extensive field experience such as Hooker and Wallace.

## IX

What with butterflies, fossil mankind, and apes, Darwin had more than enough to occupy his mind. But his attention continued to linger lovingly on plants. "Are there any Lythraceae in Ceylon?" he asked George Thwaites, the superintendent of Peradeniya Gardens in Sri Lanka. "What I am anxious to know is, whether you find it necessary to grow the different varieties [of Hollyhock] far apart from each other?" he inquired of Charles Turner, a nurseryman at Slough. "The case of the yellow plum is a treasure," he told Thomas Rivers, another nurseryman in Hertfordshire. "I do want very much to know, whether you have sown any seed of any Moss-Roses, & whether the seedlings were Moss-Roses.—Has a common rose produced by seed a moss-rose?" All through the spring and early summer of 1863 he moved purposefully from the kitchen garden to his hothouse to his study, backing up these written inquiries with careful observations of his own.

Yet his health was deteriorating. "I am unwell & must write briefly," he told John Scott, giving up an interesting discussion on the fertilisation of *Lobelia*. He told Hooker and Huxley that the "smallest exertion" now seemed to lead to vomiting, irrespective of whether he attempted a pleasure trip to see friends, made a visit to London to deliver a scientific paper, or stayed safely at home in the bosom of his family. The bustle of Victorian travel, far from providing the invigorating lift to his spirits that used to be the case, merely exhausted him. At times, even reading "makes my head whiz."

This had been going on for some time. When three old comrades from the *Beagle* had arrived at Down House for a weekend visit in October 1862, and Darwin enjoyed a gossipy evening of reminiscences with them, he had felt too ill in the morning to get up to say farewell. He confided to Wallace that such sickness made it impossible to see friends. "I have always suffered from the excitement of talking, but now it has become ludicrous," he said to Hooker. "I talked lately 1½ hours (broken by tea by myself) with my nephew, and I was ill half the night. It is a fearful evil for self and family." Even an affable half an hour with Lyell, his oldest and most congenial companion, could lead to retching. While it is easy enough to see that Darwin may have sometimes used these illnesses as a way of restricting unwanted social commitments, and that his attacks might mirror the ebb and flow of underlying mental tensions, the practical effects could not be ignored. Leonard Darwin—then aged eleven or twelve—recalled the uneasy air at home.

> As a young lad I went up to my father when strolling about the lawn, and he, after, as I believe a kindly word or two, turned away as if quite incapable of carrying on any conversation. Then there suddenly shot through my mind the conviction that he wished he were no longer alive. Must there not have been a strained and weary expression in his face to have produced in these circumstances such an effect on a boy's mind?[74]

In the summer of 1862 his vomiting alternated with bouts of eczema. He got in touch with James Startin, a London expert known for his specialised prescriptions for skin disorders, and liberally applied the expensive "muddy stuff" he acquired to his arms and hands with a camel-hair brush. This apparently did some good, because he sent the prescription on to Hooker. The same tendency to eczema returned in the summer of 1863.

This time, he sought out William Jenner, a physician recently appointed to Queen Victoria's household. Despite the passing authority of his royal connections, Jenner was unable to stop him vomiting. Putting his faith in home remedies, Darwin then kept himself going with doses of Condy's Ozonised Fluid, a Victorian pick-me-up previously recommended for Leonard's scarlet fever. When it ran out, Horace manufactured soda water for him "in our fizzing machine."[75] Pedantically, he began querying every piece of medical advice that arrived in the post. When Jenner prescribed podophyllum (a drastic purgative), he asked Erasmus whether the dosage was completely safe. Then, when Jenner substituted "enormous quantities of chalk, magnesia & carb. of ammonia," Darwin discussed these at length with Fox. In March 1863 he declared he was worsening. The bouts of vomiting extended from four, to six, to twelve days in a row.

Bypassing Jenner temporarily, and with Hooker's and Huxley's approval, he consulted George Busk, already known to him as a natural history author and practising medical man with wide hospital experience. Busk suggested that Darwin might have "waterbrash," a diseased watery secretion of the stomach, for which he enclosed a prescription. Armed with this interesting new thought, Darwin grasped at the high technology of nineteenth-century biological science and sent some vomit on a glass microscope slide to John Goodsir, an acquaintance specialising in microscopical investigations, asking Goodsir to search for pathogenic vegetable spores. He was disappointed to hear that his stomach fluids were just as they should be.[76]

In the end, bad health forced him back to the water-cure. In September 1863 he went to Malvern, at Emma's request, where he spent some weeks as an out-patient, living with his wife and children in a rented villa in the town, as they had done during his first visit in 1849. Horace was still a frail child and Darwin and Emma wanted him to try the therapy as well. "Our youngest boy is a regular invalid with severe indigestion, clearly inherited from me."[77]

Agreeing to visit Malvern was a measure of how ill Darwin felt. Ever since the "odious" time he had had at Ilkley, he had regarded any extended residential treatment at a water-cure as something to be avoided at all costs. His contact with Edward Lane had been lost—Lane had sold Sudbrooke Park and moved out of the water business for a while.[78] Glumly, Darwin considered Malvern more as a punishment than a promised cure. Furthermore, he felt an inherent dislike for returning to

Malvern itself. Painful memories rose of the days in 1851 when his daughter Annie died during Dr. Gully's treatment. He and Emma steeled themselves for an unpleasant time.

They stopped at Erasmus's house on the way. "He went to bed on his arrival," reported Erasmus in concern. "In the morning [he] came down at once to the carriage where we made up a kind of bed as he cannot sit up from giddiness."[79] He was evidently iller than he had been for a long time.

The Malvern cure failed dismally. He had lost his enthusiasm for energetic therapies—the steaming, slapping, rubbing, and wrapping—preferring the more leisurely routines of recent years. Darwin's customary doctor, James Manby Gully, was unwell (a mental breakdown, hinted Fox) and unable to treat him personally.[80] Emma arranged for another physician, James Ayerst, to supervise Darwin, but the prospective patient did not take the change gladly. He felt sure he was being fobbed off with second-best. So Emma persuaded Gully to come over to see Darwin in order to endorse Ayerst's regime. By then Darwin's eczema was too raw and inflamed to bear any water.

In fact, Darwin seems to have had a breakdown. As soon as he arrived at Malvern he suffered a complete collapse. In the space of a few days he became so weak he could scarcely walk in the garden of their villa. At one point, he said, he kept himself alive with doses of brandy. Afterwards, when Francis Darwin was preparing Darwin's letters for publication, Henrietta singled out this Malvern misery as an unforgettable time.

> Two low-water years were 49 & 63. . . . In the summer of 63 he had loss of memory (Mother says I am wrong as to this) & Mother told me, so as to be prepared, that an epileptic fit was to be feared. But Mother wd. not like either of these said in the book.[81]

In the end Darwin was not strong enough to take any further treatment and the visit terminated in disarray. He believed he left Malvern worse than when he arrived. This time, the water-cure "actually harmed" him.

It was an arduous time in other ways that were probably related. Darwin wrote to Fox in alarm about his daughter Annie's grave.

> Emma went yesterday to the church-yard & found the gravestone of our poor child Anne gone. The Sexton declared he remembered it, & searched well for it & came to the conclusion that it has disappeared. He says the churchyard, few years ago, was much altered & we suppose that the stone was then stolen. Now some years ago, you with your usual kindness visited the grave & sent us an account. Can you tell what year this was? I was so ill at the time & Emma hourly

expecting her confinement that I went home & did not see the grave. It is not likely, but will you tell us what you can remember about the kind of stone & where it stood; I think you said there was a little tree planted. We want, of course, to put another stone. I know your great & true kindness will forgive this trouble.[82]

Then Emma found the grave, complete with stone but overgrown and without the balustrade that Darwin thought he remembered once ordering. "A dear and good child," the memorial read—as true to them that day as it had always been. Had she lived, Anne would have been twenty-two years old. Poignantly, Darwin searched his heart and realised the bitterness was gone.

For a moment, at least. While the Darwins were at Malvern, Hooker wrote to tell them that his six-year-old daughter had died. He poured out his distress to Darwin. She was "my very own, the flower of my flock. . . . it will be long before I cease to hear her voice in my ears—or feel her little hand stealing into mine, by the fireside & in the Garden—wherever I go she is there." He knew he would be treated with sympathy by Darwin, who had suffered the same desolation.

I think of you more in my grief than any other friend. Some obstruction of the bowels carried her off after a few hours alarming illness—with all the symptoms of a strangulated hernia.[83]

Each man needed the other to share compassion and fortitude. Religious consolation was impossible for both of them. Only time softened the pain, Darwin told him. "Trust to me that time will do wonders," he soothed, "and without causing forgetfulness of your darling."[84]

*chapter*

7

# INVALID

THE PERIOD of ill health that followed this trip to Malvern was the worst Darwin ever experienced. In a practical sense, it permanently altered his private life. Afterwards, he felt like an old man, although still only in his fifties. His habits became those of an invalid. His perspectives changed, his ambitions waned. He started to look more like an old man—his hair greyed, his face yellowed, his shoulders began to slump. The neat little beard he had begun so proudly in the summer of 1862 grew bushy and unkempt. For three or four years he withdrew from all society, preferring to give the small energies he could muster to his biological work rather than to what he regarded as pointless small talk or society engagements, a preference that may well have been heroic in its dedication to science but was also a welcome convenience that gave him the privacy he desired. Even though Emma remembered him during this cycle of ill health as being "cheerful & affectionate" in the face of his "sufferings," it is likely that he became increasingly querulous. His work tailed off. Understandably, his spirits drooped. "Depressed," he said at one point.

The tyranny of illness became pronounced. He had to adjust. The family had to adjust. His friends had to adjust. One by one, scientific colleagues, publishers, correspondents, and informants all adjusted. If Darwin had deliberately sought to establish iron control over his personal world, he could scarcely have hit upon a more effective device. Illness changed them all, most of all himself.

For one thing, the physical nature of his sickness could not be ignored. On occasion Darwin was confined to bed for two or three weeks, prostrated with retching, colic, giddiness, and fatigue, sometimes barely able to walk. Emma recorded the wreckage in her diary. During the opening weeks of 1864, a month or two after their return from Malvern, her husband was too weak to get out of bed, vomiting into a chamber-pot four or five times a day.[1] "For 25 years," he wrote in 1865, "extreme spasmodic daily & nightly flatulence; occasional vomiting, on two occasions prolonged during months. . . . All fatigue, especially rocking, brings on the head symptoms . . . cannot walk above ½ mile—always tired—conversation or excitement tires me most."[2] At these times he gave up all pretence of leading a normal existence.

For another, Darwin fought back with compulsive determination—a determination that was perhaps part of the problem. He turned his eyes inwards, attempting to establish some form of supremacy over his ailing body, forcing himself to obey a rigid self-imposed discipline, willing his mind to focus on his work, not letting ill health get too much the upper hand, all in an effort to subdue the elemental forces that illness thrust into the daily context. He had a "horror of losing time," said his son Francis, an observation that was reflected in Darwin's involuntary desire to record in his diary every day lost through sickness. Sometimes the self-scrutiny became actively interventionist. He would apply to his forehead a stinging substance from a small bottle kept for the purpose that he claimed was useful for restoring his concentration. In the process, he probably succeeded only in making his illnesses the centre of attention. To relinquish self-discipline was, for him, to admit defeat.

The effort nearly broke him. When he reemerged, he had become the figure that most people subsequently remembered—the frail, grey-bearded invalid of Downe.

## II

At such a distance it is surely impossible to pinpoint the exact grounds of Darwin's medical condition. His physical and mental states probably combined so completely with his familial and intellectual setting and with the consequences of his being the author of the *Origin of Species* that there must be room for doubt whatever is suggested. The multiplicity of his symptoms makes it unlikely that he suffered from a single disease or condition. Nor were the ailments he displayed early in life necessarily part of the same problems experienced later on.[3] His personality, and his current preoccupations, probably contributed in a decided manner.

However, there can be no doubting the troubles he endured. He described these over and over again in letters to friends and relatives, and they were corroborated by family members and his doctors. His symptoms were largely gastro-intestinal, at times severe, sometimes only chronic, often accompanied by exhaustion, skin disorders, and dizziness. It seems possible that there may have been some digestive malfunction at the root of his problems, perhaps (in modern terms) an ulcerated gut, duodenitis, diverticulitis, or gall-bladder problems. It is unlikely from the existing archival records that he had a blood or immunological disorder, or any latent tropical infection picked up during the *Beagle* voyage such as Chagas' disease.[4] Nausea, wind, and colic were his constant accompaniments. "I feel nearly sure that the air is generated somewhere lower down than stomach," he told one doctor, "and as soon as it regurgitates into the stomach the discomfort comes on."[5] Whatever else might have been the result of an over-active imagination, these colicky pains were horribly real. "When the worst attacks were on he seemed almost crushed with agony," wrote Edward Lane, his water-cure doctor. "Of course such attacks as I have spoken of were only occasional—for no constitution could have borne up long under them in their acute phase—but he was never to the last wholly well—never robust."[6]

To these should be added other idiosyncrasies that appear only transiently in the written record but affected his daily routine. Darwin's friends saw how his mind churned and boiled for hours on end underneath the placid and affable outer surface, ceaselessly picking over the threads of his research, pursuing natural history problems in obsessive detail, seemingly trapped in an everlasting game of chess in which he felt he needed to hold all past and future moves in his head before deciding which way to step. Francis Darwin noted his father's intense mental activity, his inability to sleep once he was fired up, his fatigue the following day, and suggested that this teeming intellectual energy was extremely difficult to quench. Darwin's mind evidently raced with ideas in the small hours of the night, dwelling on problems large and small. On one occasion he woke up his son William in order to apologise for a remark that he had magnified out of all proportion with the passing minutes—he could not settle down to sleep until he made amends. Another time, he dressed by moonlight, slipped out of the back door, and visited John Innes in the dark to clarify a troubling point. In the daylight hours, he would dispatch lengthy, almost amusingly convoluted explanatory letters to colleagues to make sure that they did not misunderstand him. Not unexpectedly, he experienced headaches, had difficulties with sleeping, woke early, and occasionally thought he detected slight palpitations of the heart.

Often he felt cold. Cold fingers and toes were apparently a family trait, and Darwin may have suffered from poor circulation and low blood pressure. For years, he was accustomed to take his own pulse on a regular basis, and that of the children, saying with concern that the rate was "feeble," "irregular," and "hesitant." A sluggish circulation could have contributed to the numbness that he sometimes noted in his hands and feet, and giddiness when he stood up. Emma called these fits of "dazzling." His precautionary measures against the cold were simple. He wore the traditional heavy clothes of an English gentleman accustomed to large, draughty rooms in Victorian country houses. According to his account books, these included merino wool undergarments, a high-necked shirt and cravat, "shooting coat & double breasted waistcoat," wool trousers, sometimes a knitted shawl or overcoat on top, and in the evenings a brightly coloured silk dressing-gown and flannel nightshirt. Fires smouldered in the grates throughout the year.

Last but not least, Darwin was addicted to nicotine in the form of snuff. He acknowledged the extent of his dependence, and off and on made half-hearted efforts to restrict his intake. His wife recorded somewhat doubtfully in her old age that she thought he did manage to give it up once for nearly a year before backsliding. He took two sorts, one dubbed an "evil, dark Lundy Foot" by a servant, which made the maids sneeze when they swept the floor.[7] He liked this best. For a long time he kept his snuff-jar in the hall, a little distance away from his study door. The inconvenience of getting up and stopping whatever he was doing would, he hoped, regulate the habit. This gambit never worked. He would invent foolishly pressing reasons to pass through the hall ("the drawing-room fire needs stoking") and, as the children recalled, the clunk of the lid always gave him away.[8]

Another trick involved two flights of stairs: for six months Darwin locked his snuff in the cellar and kept the key in an attic room. Even so, Parslow kept meeting him on the servants' staircase. In the end he resolved that he would take snuff only if he was away from home—"a most satisfactory arrangement for me," said his friend Innes, because Darwin started to call by for a chat much more often.[9] His reliance on snuff probably added to the restless nights. The large pinch that the children remembered him taking before bedtime was not the best way to calm restless thoughts. He drank coffee, too, "at 4:30 every day." But Dr. Andrew Clark made him give it up in the 1870s.[10]

With all these ailments to manage, if not conquer, Darwin's preferred solution was to keep quiet at home. "For the last two years I have led the life of a hermit," he told his old Cambridge University friend Charles Whitley before going to Malvern, "seeing no one & going nowhere; & doing nothing but two or three hours work daily on my good days at natural history."

> I am become that most wretched & despicable object, a confirmed valetudinarian. I have much, very much to be thankful for in life; but everyone has his heavy drawbacks & my own health & even more that of my children is our sore drawback. For years we have had one or other of our children invalids. But I have said enough & more than enough.[11]

It fell to Hooker to ask the obvious question: "Do you actually throw up, or is it retching?" It was both, Darwin replied, but food hardly ever came up. "You ask after my sickness—it rarely comes on till 2 or 3 hours after eating, so that I seldom throw up food, only acid & morbid secretion." Elsewhere he said, "What I vomit [is] intensely acid, slimy (sometimes bitter), corrodes teeth." Doctors puzzled, he added.[12] Hooker did not venture a diagnosis. Nor did Huxley, or any of Darwin's closest medically trained friends. Their medical training had not led them into clinical practice, although they offered sympathy and advice.

Still, there was evidently much more to his illness than vomiting alone. Darwin himself was at a loss to explain his continued afflictions, sometimes saying that he had first gone "wrong" during the *Beagle* voyage and since then had got steadily worse, a kind of perpetual seasickness. He told Edward Lane that his condition might be a result of the bout of sickness he had experienced in Valparaiso. Harriet Martineau said much the same, following Dr. Henry Holland's impression. As the decades wore on, however, Darwin mostly blamed his spinal cord or nervous system, or his "accursed stomach," or just simply "nerves." He tended to think that long periods of intense mental effort would terminate in an attack of illness, and often said that his stomach operated in opposition to his head. This was not such a bad thing. "I know well that my head would have failed years ago, had not my stomach always saved me from a minute's over work." Too much thinking, or too much talking, he declared, would usually end in physical disarray.

But he also said that his work was the only thing that made him forget the pain. When he concentrated on a difficult piece of writing, or an intricate experiment, he became oblivious to everything else, and it was this that made him reluctant to obey doctors' advice to reduce his scientific undertakings. Time spent away from his work was anathema. He would bargain with Emma over their holidays, shaving away a morning here or an afternoon there, or agreeing to go only if he could take some potted plants for experiments or proof sheets to read. When one of his sons asked whether Darwin might take a rest away from home, he replied that "the truth was he was never quite comfortable except when utterly absorbed in his writing." He "dreaded" idleness.[13] Nevertheless, he apparently recognised something of a vicious circle in all this. He wanted to work, and his work helped him rise above his illness, yet it was work that made him ill. He spoke of holidays as if they were a punishment for the enjoyment he found in his work, or as an unavoidable balancing of his physical status quo, a personal accounting system, in which he must sacrifice a few days in order to squeeze more labour out of his fragile frame.

How far all this was linked to the disruptive contents of evolutionary theory, or to his defence of the *Origin of Species,* is hard to say. These factors must have played a leading part, filling his days with obstacles and worries.[14] He must have thought about his wife's religious views, remembering the letter she wrote to him soon after their marriage about her faith, on which he had written, "When I am dead, know that many times I have kissed & cryed over this. C.D." In a more recent letter she had told him how she believed "suffering & illness is meant to help us exalt our minds & to look forward with hope to a future state." This Darwin knew he could not do. How could a good God impose suffering? He could not share her faith or subscribe to the belief that suffering led to redemption.

Even without the religious dilemmas in which he found himself, either at home with Emma or in the larger world with Asa Gray, Kingsley, Wilberforce, and the rest, he was beginning to buckle under the strain. Always, he was writing, writing, writing. He stretched himself further than was realistic in his correspondence. He read the reviews, worrying over every minor point, feeling he should be able to account for each detail of natural selection, for each detail of the encyclopedia of life. People asked him about the existence of God, the origin of living beings, the nature of the human mind, or wrote to abuse him, repeating in personal form much of the Victorian era's larger apprehension about the reshaping of the natural world. Darwin responded to every one.

In these illnesses, Darwin was also a man of his time.[15] Thomas Carlyle, George Eliot, Charles Dickens, Florence Nightingale, George Biddel Airy (the Astronomer Royal), Huxley, Wallace, Spencer, Lewes, Hooker, and plenty of less well-known individuals endured similar digestive complaints, including his brother Erasmus. The "demon of dyspepsia" was an archetypal Victorian preoccupation on which medical and pharmaceutical fortunes could safely be built. Darwin experienced maladies that were widespread and typical of his era.

He was similarly not alone in experiencing crushing bouts of nervous exhaustion. His friend Huxley was one of many individuals whose occasional breakdowns are generally forgotten today. Even so, Darwin scarcely showed the impenetrable misery, hopelessness, mental distortions, or inability to participate in normal existence that are the usual clinical signs of depression. To be sure, his disorders might be considered psychiatric or psychosomatic in a colloquial sense. Grief and guilt surely played their part in his psyche. Fear, too, especially in the way his body would most often fail when he intended making a public appearance, suggesting some deep-seated dread of exposure. His customary reticence may have reflected a wish to avoid getting involved with other people's emotions—reticence and modesty could have been the polite face of dissociation, the spurning of closeness. If Darwin had been a woman, for instance, he might well have been dubbed hysteric, a disorder tied up with a restricted female role in a masculine universe. If he had fallen ill further towards the end of the century he might have been called neurasthenic, a victim of modernity and the hastening pace of life. The terminology he employed, of a "complete breakdown in health," although vague, is probably appropriate enough.

And as several contemporaries remarked, ill health presented a number of advantages for a Victorian intellectual. Bad health excused Darwin from boring evenings at scientific societies and from talking to people he did not wish to meet. It sanctioned his retreat after dinner for some tranquil minutes alone with the newspaper. It allowed him to give the best hours of the day to his work. A weak stomach provided an excellent reason for avoiding social engagements, for it was a much more meaningful excuse to offer than an attack of unmentionable boils. Indefinable stomach pains also created ideal circumstances for restorative mornings at Down House, with Emma or Henrietta in solicitous attendance, and signalled a convenient cut-off point for conversations with guests. These mildly self-indulgent aspects of Darwin's middle life are readily comprehensible. His need for peace and quiet occasionally took the form of genteel indisposition, a pattern of existence characterised by Nietzsche as "freedom from good health." Others of the same social background might have visited Madeira for the winter season or tried water-colour painting in the Highlands.

Emma colluded with these social excuses.

> She used to go and scold him for working too long, and would warn him he was talking too long with guests, etc. This part was very useful to him as it gave him an easy way of leaving the room before he was tired out. He would tell her beforehand that she was to send him away early.[16]

Underneath it all, sickness was by now an integral part of his married life, a time for being the centre of Emma's attention. The benefits worked both ways. Emma was moved by his dependence on her when he was ill, and he expressed his love for her much more freely under sickroom circumstances. "She has been my greatest blessing. . . . She has never failed in the kindest sympathy towards me, and has borne with the utmost patience my frequent complaints from ill-health and discomfort." Although it seems hard to believe that she did not get irritated with him, he praised her sympathetic manner throughout his life.

Perhaps there were advantages for her too. Illness brought him out of his study into the drawing room. When he was ill he spent much more time in Emma's company than usual, listening to her reading or playing the piano, sometimes having his back massaged or his scalp rubbed.[17] As soon as he felt well enough, he would be off back to his work—and it is worth noting that as a consequence Emma and the children customarily saw Darwin only when he was too sick to do anything else except join the family. To some extent this feature of his life partly explains his wife's and children's emphasis on his ill health in their reminiscences.

Darwin, in turn, undoubtedly found bad health an effective way to claim his wife's undivided attention. Her days were otherwise much given to the children, to relatives, and to household and village concerns.[18] "She could hardly endure doing nothing even for a quarter of an hour," reminisced Henrietta.[19] Such fugitive disorders, as novelists like Elizabeth Gaskell recognised, could easily become the primary means of social interaction.

## III

What is certain is that he became worse from the autumn of 1863. He returned to Downe from Malvern with a note from Dr. Gully about a new bathing routine and started cold baths, with Parslow acting as his bathman. But he gave up soon after, and never again bothered with water therapy.

Impatient for progress, he approached Dr. Jenner for more medication. He also turned for advice again to George Busk. This time, Busk thought his problem might be mechanical—the stomach did not push on its contents as rapidly as it ought to do.[20] Busk gave him the name of William Brinton of St. Thomas's Hospital in London, the author of a book on ulcers and stomach diseases. Even so, Darwin did not consult Brinton without first talking it through with Stephen Engleheart ("Spengle"), the doctor for Downe village, whom he saw on a regular basis. "They are in rather a pick of troubles about which Dr. to have," wrote Henrietta to William. "Dr Gully has sent to Dr Jenner, but Spengle strongly recommends a Dr Brinton, so they don't know what to do." Engleheart wondered about the possibility of water-borne infections, a common cause of disease. For a week or two Darwin worried about the state of his drains.

Much of this elaborate procedure revealed just how influential Darwin was becoming. He could afford the best therapeutic help, and the doctors were in their turn keen to offer their services to a celebrated scientific man. Physicians, for example, mostly came to Darwin. All of Darwin's medical consultations took place at Darwin's convenience, in a private, domestic environment, either at his own home, by correspondence, or at the house of his medical adviser. None of his ill health was experienced in a hospital setting, and as was customary for people of his social standing, nursing care was provided by his womenfolk, sometimes for long periods. The practical arrangements for the care of a sick man—the nursing, special diets, sending for prescriptions, and medical attendance—were part of a wife's duty as manager of the Victorian household. Yet in essence, Darwin, as the patient, controlled the medical show. He felt no scruple over consulting three or four physicians simultaneously and was prepared to ignore their advice or turn to another if treatment failed. He kept his village doctor on hand and made supplementary inquiries of medically trained friends or relatives like his brother Erasmus or Henry Holland. At any one moment there could be several well-qualified people giving him advice of one kind or another and recommending different drugs and diets.

In short, Darwin came to treat his illness as if it were a natural history problem and he were setting out on some complex research project, asking everyone he knew for second and third opinions, reading around the subject, experimenting here and there, bombarding colleagues with inquiries, finding out everything he could, and reorganising his hypotheses to accommodate new information. The old adage about doctors making the worst patients seemingly held true for naturalists as well.

And it is entirely possible that these disorders in part became a crutch to him. He may have come to enjoy the medical attention and the rituals of dietary restrictions, and have been stimulated by the pursuit of one therapy after another as he sought an all-embracing cure, rather in the same manner as he relentlessly pursued botanical or zoological facts. He may have come to rely on being able to shelter behind an illness which allowed him to escape commitments, intellectual and domestic alike.

As he worsened, Emma seized her chance and took over completely. Previously little more than a ghost in his working life, she took Darwin firmly in hand and made her presence felt. For a few months in 1864 she transformed Down House into a sanatorium. She discouraged visitors. She or Henrietta dealt with all Darwin's correspondence and sat for dictation at his bedside. She regulated his invalid hours, told him when to get up or to stay in bed, monitored his daily walks, was constantly on hand to read novels to him or to offer a nourishing cup of beef tea. As frequently happens in the domestic arena, illness provided her with an opportunity to pick up the reins of authority. Emma became Darwin's nurse, keeper, and mother, the real ruler of the establishment, conscious of her role in his life, willing to give him the care he needed, and receive his gratitude in return. As well as being the pliable, endlessly supportive person of his recollections, she here showed the strength of her character and a certain satisfaction in taking charge.

"He does not feel the least temptation to disobey orders about working for he feels quite incapable of doing any thing," she told Fox firmly.[21] "One day is a little better & one a little worse," she wrote to Hooker, "but I cannot say that he makes progress at present. He stays in his bed room & gets frequently in & out of bed & occasionally goes down stairs for a very short time, but he can only stand very short visits even of the boys."[22] Horace Darwin relinquished the old *Beagle* telescope so that his father could scan the garden from an upstairs window for any natural history activity. Horace and Leonard obligingly brought him news from the hothouse.

Most of all, Darwin disliked having his letters and his work taken away from him. Without this, he had "nothing" to make his life worth living. So he pleaded with Emma to be allowed to read a little scientific paper here and there, or to make a few notes for some project of his own. He did complete the last pages of the paper he had been composing on the triple fertilisation of *Lythrum* and sent it off to Hooker and the Linnean Society. But Emma would not let him read John Scott's manuscript on experimental plant fertilisation when it arrived in the post. It looked dauntingly technical.

Instead, Emma and Henrietta served as his amanuenses. Many of Darwin's letters during these months are in his ladies' handwriting—large, round, and easily readable. Both women learned to forge his signature. One letter to Daniel Oliver, which was written and signed by Emma, carried at the bottom a note in Darwin's own writing, "Excuse brevity & forgery as I am unwell in bed." A letter to his son William in 1866 similarly ended with Henrietta signing Darwin's signature to which Darwin added the mocking, self-explanatory words, "(miserable forgery)."[23] Several of the fake signatures on other letters carried supplementary remarks from Darwin—"a forgery" or "base forgery." When his women got bored, he would call on any available youngster. He did not hesitate to summon a nephew from a flock of Wedgwoods playing croquet one afternoon in 1866. "Godfrey has been very nice & Papa feels him so tame as to get him to write for him," said Emma.

The pace of his life slowed to that of a plant. "A few words about the stove-plants," he wrote to Hooker. "They do so amuse me. I have crawled to see them two or three times." Occasionally he tinkered with the insect-eating plants that Hooker sent to tempt his intellectual appetite, and Leonard remembered trotting off down the garden to fetch one of these specimens up to the study because—as his father said—he was so "infernally healthy." Confined to his room by sickness on many occasions, he mostly watched twining species that were brought indoors to stand on a warm windowsill. He noticed how the tendrils slowly swept in a circle until they met a suitable object to twist around. The tendrils were "just like fingers."

John Horwood, the gardener from the Rookery, thought otherwise. "I believe, Sir, the tendrils can see, for wherever I put a plant it finds out any stick near enough," he said.[24] Thereafter, Darwin observed twining plants with rising curiosity. Once he struggled out in a storm to see how the wild bryony in his garden hedge fared in the wind and found it riding the gale like a ship with two anchors down, straining at the end of its lines like the *Beagle*. Otherwise, watching the movements of leaves matched his invalid habits admirably. For two summers he observed the tendrils of pea plants circle about in search of something to fasten onto, a serene activity just right for an ailing naturalist with plenty of time on his hands. Botany was a solace to him: "It is all that I am good for; I can just do an hour or two's work, when I can do nothing else," he told Innes.[25] Bit by bit he built up a file of observations on climbing plants that he published in 1865 in the Linnean Society *Journal*.

Emma encouraged William to take part in Darwin's botanical work and keep the necessary momentum going. "Your father wishes the following observation to be made to you, viz. that Meneanthes is now in flower & to be left to your comprehension," she wrote in May 1864.[26] But Darwin missed his exchanges with Hooker. Emma participated by proxy.

Dear Dr Hooker,

I cannot give a very good account of Charles. He has frequent attacks of sickness but recovers from them in a wonderful manner & they are often with very little distress. The stomach retains the food in a surprizing manner which accounts for his not getting thin. His medical men speak confidently of his regaining his former state of health. He desires me to say that your letters always give him the *greatest pleasure* tho' he grudges you the time you employ in them. The partridge's foot has now produced 54 plants which he hopes may stick in your throat.[27]

Almost as bad, in the short term, he was obliged to give up his account books, the other defining aspect of his private life. Darwin did not have sufficent spirit to maintain the double-entry bookkeeping his father had taught him. So Emma cast the family accounts while he attempted to explain to her how to enter his stock market transactions, trust fund records, and mortgage interest receipts. She was accustomed to do her own accounts for household expenditure, listing meat, butter, cheese, candles, clothes, "best tea," and "servant's tea," in a separate book, including a place on each page for "errors," which once or twice ran to over £7 in a single year.

But Darwin was alarmed to see the family's expenditure for 1864, totted up by Emma in August at the end of his financial year. It was way beyond the usual sum. In the end he could bear it no longer and went over the figures himself. "Papa found out the great error in his accounts & also that we are not spending much more, if any more, than for the last few years, all of which cheered him,"[28] Emma said, oblivious to the fiscal distress she had caused. There were many mistakes in her arithmetic. Somehow, Darwin never again found his ill health sufficiently debilitating to relinquish his accounts to another's mathematics.

Hooker obeyed instructions not to communicate, saying that he felt it a positive privation. "I was very glad to get his letter this morning, but he must not try to write to me," he told Emma. He knew better than to argue with a wife who had issued sickroom orders. But he was accustomed to hearing from Darwin once or twice or week, in letters crammed with requests for information, descriptions of projects in hand, books and articles to discuss, and scientific gossip, every one the sign of a man in his active intellectual prime. In March 1863, when no letter from Darwin appeared for eight days, he had asked Lyell to inquire whether Darwin was all right. "Hooker not having heard from you, is growing anxious, and hopes it is because you are corresponding with me and not because of serious ill-health," wrote Lyell obediently.

Darwin felt the ban equally hard. He loved Hooker as a brother and liked to share his life with him. He sympathised with Hooker's administrative problems at the Gardens, respected his judgement, read his botanical texts with admiration, marvelled at the variety of his knowledge, laughed at his temper, and nagged him to look after his health. In a way, Hooker almost was a brother, for Darwin had regarded Henslow

(Hooker's father-in-law) as their shared surrogate father. United by affection and regard, the relationship meant a great deal to both of them, and the death of Hooker's daughter became a deep personal bond. In that respect, Darwin was relieved to hear that the other friends had rallied round to soothe Hooker's despair over the death. Lubbock whisked Hooker away to search for flints in Abbeville as a scientific distraction. From his sickbed, Darwin insisted on sending word. "My dear old friend," he scribbled in a wavery hand. "I must just have pleasure of saying this." Later, they talked about the death more freely. Darwin's scrap of a letter was a silver lining to his cloud, Hooker said.

Sometime soon after this, Hooker inquired whether Darwin might sit for a portrait bust by the sculptor Thomas Woolner. A competent watercolour artist himself, Hooker had a good eye for the fine arts, sculpture, and china. He admired Woolner's work, having previously commissioned from him a bust of his father, William Jackson Hooker, and of John Stevens Henslow. Personally, he collected Wedgwood ware, especially portrait medallions, a hobby-horse that became a joke between him and Darwin. "We are degenerate descendents of old Josiah W. for we have not a bit of pretty ware in the house," Darwin teased him. Still, in November 1863 Darwin asked his sister Susan to rummage around in the old house in Shrewsbury. He sent Hooker an original Wedgwood vase given to Robert Waring Darwin by the first Josiah Wedgwood.

Hooker was uncertain whether Darwin would agree to a sculpture.

> I am very anxious to get Woolner down to take a clay model of your bust for myself, as you kindly promised I might; & I look to Mrs. Darwin to let me know when—he shall cut it in marble at his leisure for me. Such heaps of people want to know what you are like—& the photographs are not pleasing.[29]

Darwin replied he was far too ill to contemplate the upheaval that such a commission would entail.[30] Nevertheless, he accepted a few years later.

Other friends were concerned by his lengthy sickness. "Mr Huxley writes to Papa that I am to treat him like Vivien did Merlin & shut him up in an oak," Emma informed her daughter Henrietta. "He then apologises that Vivien is not a very proper person, but as Papa does not read Tennyson concludes it does not signify."[31] Quite how Huxley meant this remark to be taken remains something of a mystery—it was a strangely inventive friend who could liken Darwin to the proud old magician of *The Idylls of the King,* seduced by sex and flattery into giving up his secrets. Huxley may have regarded himself as another Vivien, prising truths out of a recalcitrant fount of knowledge, or meant only that Emma should keep her husband secluded in the leafy countryside until he was restored to health.

Dr. Jenner's doses of antacids brought relief of a kind. "Hurrah!" the patient shouted in March 1864. "I have been 52 hours without vomiting!!" He wrote to Hooker and Huxley praising Jenner's ability to get him back on his feet. "I shall certainly vote for him for F.R.S. this year," quipped Hooker, putting the thought into action later that year.

The improvement was only temporary. For the next two years Darwin continued much the same, "confined to a living grave," he told Hooker, living on "endless foolish novels which are read aloud to me by my dear womenkind."[32] He said the same to Gray. "I have heard during late 9 months an astounding number of love scenes."[33] He saw hardly anyone, not even Hooker. In time, he felt well enough to make a few botanical observations, asking for plants to be brought indoors to be placed next to his chair or sofa, or walking down to the hothouse for a hour. "The only approach to work which I can do is to look at tendrils & climbers. . . . This does not distress my weakened brain."[34]

## IV

In November 1864, Darwin received the Copley Medal, the Royal Society's highest scientific honour.[35]

The internal politics behind these annual awards were formidable, and every year fellows of the Royal Society would throw themselves into fast and furious negotiations for various candidates. They would haggle and bargain for votes, host soirées and dinner parties, lobby and campaign for their man as if electioneering for a parliamentary seat, perhaps more so, for the Copley Medal signalled an individual's accession to the very pinnacle of British science. Sometimes proposers had to make do with one of the lesser medals for their candidate, prestigious scientific honours in their own right but not quite the same as a Copley. Sometimes they had to admit defeat or wait their turn. Once or twice, a candidate waited rather too long. In 1866, William Henry Harvey died between nomination and voting day. To get Darwin a Copley Medal was a considerable coup for the Darwinian army—an acknowledgement of his intellectual standing and an endorsement of the naturalistic approach to science. It was just the kind of challenge that Hooker and Huxley relished. "Many of us were somewhat doubtful of the result," said Huxley ominously.

Nominations had started in 1862 when John Lubbock and William Carpenter put Darwin's name forward. That was bad timing. The debate over the *Origin of Species* was at its height, and the Royal Society's council doubtless felt that Darwin's proposals were as yet far too controversial to mark with an award. The medal was given to the chemist Thomas Graham. By 1863, Darwin's friends better understood how to work the system. Carpenter backed the proposal with the passage from John Stuart Mill's *Logic* in which Mill praised Darwin. "Dr Carpenter showed me the extract from Mill's Logic which he read when he argued for your having the Copley Medal," reflected Erasmus. "Have you seen it?"[36] The award went to Adam Sedgwick, a popular choice, although he was nominated by Richard Owen almost certainly to block Huxley and the Darwinites. "The numbers were 8 to 10 for Charles, but the Cambridge men mustered very strongly for Sedgwick," Erasmus told Emma. Given these circumstances, it seems entirely possible that Huxley's presence on council, his known support for the *Origin of Species,* and his continuing disagreement with Owen materially impeded Darwin's chances. Sedgwick was sufficiently elderly to warrant prompt attention.

Edward Sabine, the Royal Society's president, nevertheless saw pressure mounting. That same year he warned John Phillips, the Oxford geologist, that the old guard would not be able to squash Huxley so easily next time.

> With all respect to Darwin's great services, and recognising that his recent work on Orchids must be classed amongst these, I cannot see without extreme concern the efforts of a very strong party to obtain the award of the Copley Medal to him expressly on the ground of his conclusions as to the "Origin of Species." . . . We may not have so good an alternative next year.[37]

In 1864, Darwin's supporters changed tactics. Lubbock, Carpenter, and Huxley were obliged to retire from the council after their fixed term of office, and so George Busk agreed to nominate Darwin, seconded by Hugh Falconer. It looks as if Busk and Falconer deliberately downplayed the *Origin of Species* in a more cautiously formulated nomination. This time the omens were promising. William Sharpey and Gabriel Stokes were the joint secretaries, both of whom knew Darwin through the Philosophical Club (the Royal Society's dining club), and the treasurer was William Hallowes Miller, the Cambridge mineralogist who had helped him with *Beagle* rocks and the mathematics of bees' cells. While none of these were close friends or supporters, they were fair-minded respectable men. Hooker was on the council, ready to push things along more subtly than Huxley was ever able. Rival nominees were famous but not so famous as to create a major threat: the chemist A. W. Hofmann (who received the Copley in 1875), M.H.V. Regnault (1869), and Hermann Helmholtz (1873).

Falconer was abroad when the nominations were discussed, so he sent a letter to the Royal Society describing Darwin's achievements. These he divided into five areas—"geology, physical geography, zoology, physiological botany, and genetic biology." Though he was by no means a convert to evolutionary theory, he emphasised the importance of Darwin's *Origin of Species,* calling it "this great essay." Falconer's letter became the basis of the formal statement that Busk put forward.

Such attention to detail paid off. Hooker told Darwin that the affair was managed so cleverly that two "old fogies" voting in the council chamber even asked politely what had Darwin written. The vote went Darwin's way—a majority of twelve. Irrepressibly, Huxley took the credit. "The more ferocious sort had begun to whet their beaks and sharpen their claws." Yet in an odd turn of phrase, the minute book stated that the *Origin of Species* did not form part of the council's deliberations. The public benediction that the Darwinians desired was intentionally withheld.

Scientific politics were at stake here. When he heard the news, Huxley exploded to Hooker. Not only was the omission of the *Origin of Species* a calculated slight on all those like himself who publicly supported Darwin, he declared, it reeked of scientific censorship. "I felt that this would never do." At the award ceremony in November he demanded to know why the *Origin* was excluded. Was the council placing the book on an "index expurgatorius"?[38] He called for the minutes of council to be read. Everyone was shocked by the outburst, but there was truth in Huxley's words. The minutes did imply that the *Origin of Species* had been deliberately omitted. Later that day, in the speech after dinner usually given by the medallist, Lyell (speaking on Darwin's behalf) told the assembled fellows that he supported Huxley's intervention and considered Darwin's *Origin* a lasting achievement.[39] Writing to Darwin the next day he confided, "I said I had been forced to give up my old faith without thoroughly seeing my way to a new one. But I think you would have been satisfied with the length I went."

All this excitement took place without Darwin's presence—another of those key occasions on which his friends fought for him in his absence. Saying he was too ill to go, he asked Busk to accept the medal on his behalf. As with the Oxford British Association meeting, his presence would have inhibited everyone's freedom to fight. Certainly Huxley would not have been so outspoken. "What a pity you can't be there," Erasmus wrote, "and yet if you were it could not be done so well."[40] The medal was a fine thing, Erasmus cynically continued. The gold did not amount to a pair of candlesticks.

Nor could Huxley let it rest there. For weeks afterwards he pestered

Stokes to change the wording in the published report. "What I do protest about is that without the knowledge and consent of Darwin's proposer and seconder, a phrase should have been inserted which compresses the maximum amount of offence into the handiest possible form for Darwin's opponents." Stokes defended his minute book. The bulldog bit harder. Finally, Stokes altered the wording in the printed *Proceedings*. And shortly thereafter, when Huxley published the medal announcement in the *Reader,* he edited out the offending words. Through force of personality he showed that he could play the same game to greater effect.

"I hear from Hooker there has been some row about what Sabine exactly said," inquired Darwin. It was Falconer who relayed the news.

> You will see the President's address—and what he said about you and the Copley award—in this week's "Reader." As it stands, I think you have had very fair measure of acknowledgement. But the passage in the third column, "on the Origin of Species"—has not been given—as it was delivered. The "Reader" has left out a little sentence or two. This will be explained to you hereafter.

The full story came out in letters afterwards. The only one who did not comprehend the high drama was Emma. "I suppose you have heard of Ch getting the Copley Medal from the Royal Society," she wrote to an aunt. "He has been much pleased but I think the pleasantest part was the cordial feeling of his friends on the occasion."[41]

Cordial they certainly were, but also ruthless.

## V

Thoroughly vexed over the Copley Medal affair, Huxley decided to start a private dining club of his own that would bring together a significant bunch of like-minded activists. Over dinner they would catch up on gossip and scheme about science. These men soon became an important informal pressure group, focussing their attentions on the reform of scientific administration and promoting the liberal naturalism associated with Darwin's theory. Huxley's first dinner party included friends like Hooker, Spencer, Tyndall, Lubbock, and Busk, and two new acquaintances, Thomas Archer Hirst, a mathematician, and Edward Frankland, a clever chemist at the Royal Institution. For the next party, he added William Spottiswoode, another mathematician, making with himself nine in all. "I think originally there was some vague notion of associating representatives of each branch of science," he recalled. "At any rate the nine who eventually came together . . . could have managed among us to contribute most of the articles to a scientific encyclopaedia."[42]

They never could agree on a tenth member, nor could they for a while think of a suitable title. Mrs. Busk suggested the X Club, which gave the advantage of committing them to nothing, as Spencer said approvingly. The only rule was that there were to be no rules. Her suggestion also supplied amusing nicknames, such as the Xquisite Lubbock, the Xemplary Busk, the Xalted Huxley, and so on, and allowed invitations for mixed social outings to include Xs and Yvs (Wives). Underneath the humour, the serious aims were obvious. "Beside personal friendship, the bond that united us was devotion to science," recorded Hirst after the first meeting, "pure and free, untrammeled by religious dogmas. Amongst ourselves there is perfect outspokenness, and no doubt opportunities will arise when concerted action on our part may be of service."[43]

Some of the Xs' early discussions would have been absurdly self-important if the members had not so obviously been destined for success. In 1864, for example, when Huxley's *Natural History Review* collapsed in financial disarray the Xs discussed the need for another journal that would adequately represent their views. They toyed with the *Reader* for a little while, each sinking £100 into a fund for establishing this as a broad-minded magazine-style review, staffed by more than thirty leading intellectuals including Ruskin, Maurice, and Kingsley, all under the general editorship of Francis Galton. The scheme worked well enough until Huxley destroyed its credibility with a slashing attack on the Catholic Church that made even the atheists among the Xs shudder.

The *Reader* was to expire in 1867. Not long afterwards, Norman Lockyer, one of its editors, put up the idea of founding a periodical which they would call *Nature,* to be owned and published by Alexander Macmillan, a journal that would provide cultivated readers with an accessible forum for reading about advances in scientific knowledge. Lockyer brought *Nature* into existence in November 1869, fronted by an introduction by Huxley ("as if written by the maddest English scholar," said Darwin indulgently). To command the periodical market was a shrewd tactic in any contested cultural arena but one as yet little exploited in science, and while Lockyer was never a member of the X Club he displayed similarly strong, progressive liberal opinions. Far more than any other science journal of the period, *Nature* was conceived, born, and raised to serve polemic purpose.[44] In the first year of its existence, there were six or seven articles urging Darwin's scheme, two of which were written by Darwin himself. Darwin became a lifelong subscriber, claiming he got a kind of "satisfaction" in reading articles he could not understand.

The Xs also advanced favoured candidates for Royal Society council elections and medals. In 1865 they seriously discussed whether Tyndall should accept the chair of natural philosophy at Oxford. One by one, the members infiltrated every government panel and committee that dealt with scientific affairs. Later, they cultivated the American publisher E. L. Youmans, expecting to win his help in arranging American editions of significant English scientific works. Occasionally they dispensed patronage to young men or visiting naturalists by inviting them to dine. Nonchalantly, they developed an impregnable aura of exclusivity.

Darwin was not a member—he was insufficiently engaged with the cut and thrust of scientific politics—yet the Xs pushed his perspective as an integral part of their own. Without the Xs, Darwin's ideas would never have percolated British culture quite as far as they did. And Darwin did everything he could to help them, even writing a number of articles for the *Natural History Review.* He willingly lent the Xs his name.

In return, the Xs counted on being able to use his influence—and were tireless in asking for it. "Have you any objection to putting your name to Flower's certificate for the Royal Society herewith enclosed," Huxley inquired late in 1864. "Mrs. Darwin perhaps will do me the kindness to send the thing on to Lyell as per enclosed envelope."[45] In similar terms, Hooker casually mentioned, "I have been thinking of Wallace for Gold Medal R.S. but it seems to be half engaged to Dr Lockhardt Clark this year. How would you word Wallace's claims? Will it not be difficult to cite sufficient paper work?"[46] Wallace had to wait his turn for a medal until 1868, when Huxley proposed him for the essay on species in the Linnean Society *Journal.* He waited even longer (until 1893) for fellowship.[47] Curiously, the Xs never invited Wallace to join their company, perhaps believing that he was not in quite the same position to make things happen as each of them was individually. Possibly the Xs decided that getting him a medal—a public sign of approval—was the most effective way to utilise him.

Huxley overheard two scientists in the Athenaeum Club chatting later in the year. "I say, do you know anything about the X Club?" one asked the other. "What do they do?" The reply amused Huxley no end. "Well they govern scientific affairs, and really, on the whole, they don't do it badly."[48]

## VI

Hidden away in Downe, Darwin missed much during these years of illness. "I suppose your destiny is to let your Brain destroy your Body," his cousin Fox put it alarmingly.

As it happened, Darwin's theory was only one of many profound challenges to conventional opinion. His book was necessarily caught up in the transformations of thought and ways of life taking place all over the developed world in the nineteenth century, so much so that he and his *Origin of Species* became part of the transformation itself.

Chief among these was theology. Continued dissent and fragmentation in the established Anglican Church in Britain was reaching a peak even without the push that biological sceptics provided. The comfortable liberal Anglican tradition—the Broad Church—was already in difficulties, with high church, low church, Evangelical, Tractarian, Episcopalian, Arminian, and Calvinist factions arguing over points of doctrine, as sharp a debate in real life as in Trollope's Barchester. Missionary societies, publishing companies, theological colleges, universities, and even individual parishes tended to adhere to a particular party. These internal difficulties reflected wider discontents with the Church of England. The 1851 population census had revealed that only about one-third of the seventeen million people counted in the poll actually attended Anglican parish churches. Despite the difficulties experienced by the census recorders and the fact that religious avocation was not a compulsory question, the statistics still made surprising reading.[49] Of the remainder, some four million people declared themselves nonconformist Protestant dissenters of one kind or another: Methodist, Congregationalist, Unitarian, Quaker, or similar. These dissident non-Anglicans not only objected to traditional theological liturgy but also strongly disapproved of the church's civil power. Broadly speaking, they found a natural home in the Liberal party, the party of moderate reform under the successive leadership of Palmerston, Russell, and Gladstone, each of whom every now and then took a shot at curtailing the church's institutional might. To some, Catholicism appeared increasingly attractive. To others, doubt or scepticism beckoned. In one and the same ten-year period, Francis Galton could declare that science was a valid alternative to all forms of religion and pious John Henry Newman could convert to Rome. At the personal level, men and women queried the evidence for Christianity, the nub of the Victorian crisis of faith. On a larger scale, the Anglican Church, so much a part of the state with its vast wealth, its lands and property, its right to draw income from parish taxes, its parliamentary presence, and its stranglehold over education, looked around and found itself beset by repeated calls for change. While it would be too much to claim that Darwin's work triggered these movements, his *Origin* sowed doubt and intensified doubt where doubt already existed. His book was generally regarded as dangerous. Yet the fluidity and adaptability of religious doctrine during this period made it possible for his theories sometimes to harmonise with contemporary cultural convictions.

In this unsettled atmosphere some Anglican clergymen began to think that liberal theology was going too far. They wondered where the church's stability and authority would lie if scholars kept on pointing out errors in the familiar Bible stories. The simmering crisis over *Essays and Reviews* climaxed in 1864 at the same time as Huxley's and Lyell's books exposed the ancient, animal lineage of humanity. To many it looked as if the desire to explore naturalistic explanations, especially in science and religion, might be undermining the primary basis of spirituality.

Bishop Wilberforce got his ecclesiastical court case against *Essays and Reviews* early that year and the seven authors were formally condemned for their heterodoxy by the church's Convocation. However, the government promptly overruled the church's verdict. Nearly eleven thousand clergymen (almost half the Anglican clergy in the country) then signed a declaration protesting to the Archbishop of Canterbury about this interference in their affairs and asserting that despite the doubts broadcast by *Essays and Reviews,* they at least fully believed in the divine inspiration of the scriptures. A number of devout scientists associated with the Royal Institution prepared a similar petition. Lubbock briefly tried to raise a sceptics' counter-proclamation.[50]

Similarly, John William Colenso's liberal reevaluation of the first five books of the Old Testament was more than most parsons could take. Colenso, an energetic missionary bishop in Natal, regarded the earliest sections of the Bible as little more than a collection of ancient historical documents that had accumulated errors and misreadings over the centuries. Like other biblical commentators such as Renan, Strauss, and Eliot, Colenso took an advanced interpretative line in higher criticism, much of it naturalistic in tone. He was accused of heresy and fled to England to argue his case with the church authorities. Huxley's band of X Clubbers revelled in the ensuing confrontation, invited Colenso to dine, and in 1864 set up a subscription fund to provide a sum of money should he be unceremoniously defrocked—which some of them half hoped might happen. Very soon a petition was circulating in progressive philosophical circles, "a declaration in favour of freedom of opinion & defending the rights of Bp. Colenso," as Erasmus called it. Erasmus and Darwin both signed.[51] Again, Colenso's comparative historical approach must have seemed to conventionally religious readers as if it were based on the same bold disregard for the divine as Darwin's theory of living origins.

And at the Oxford Diocesan Conference in November 1864, Benjamin Disraeli had little trouble identifying the question of the day. Disraeli—then leader of the Conservative party in opposition to the Liberal government—was the guest speaker at the clerics' conference, invited to attend by Wilberforce. He had accepted the invitation with alacrity as part of a campaign to woo Anglican clergymen, every one a potential Tory voter and many of them looking to Disraeli to preserve the system of tithes (the annual sum of money due to clergymen from their parishioners), then under threat from the Liberals. Exotic, ex-radical, and consummate politician, Disraeli was prepared to promise whatever was needed, be it tithes or biblical orthodoxy. The irony of Wilberforce's last public appearance at Oxford did not escape him. "Is man an ape or an angel?" Disraeli asked his clerical audience at the height of his address. "I am on the side of the angels."[52] Within the month, he was in the pages of *Punch* caricatured as a simpering angel, adorned with false wings, political opportunism oozing from every pore. His phrase carried the day. Although this diocesan conference hardly turned the tables on Palmerston, who went on to win the general election of 1865, Disraeli scented Tory victory not too far ahead.

In other areas, too, Darwin let most of the shifts in opinion temporarily pass him by. An Anthropological Society had been founded in 1863 in London, dedicated to the "science of man." This new society was set up by James Hunt in direct opposition to the philanthropic, missionary connections of the older Ethnological Society, a bias that was in part stimulated by the Ethnological's intention to admit women to their meetings, and its abolitionist stance both before and during the American Civil War. By and large, the members of the Ethnological Society believed in monogenism—that all human beings belonged to the same species.

By contrast, Hunt was convinced that the geographical diversity of human beings was a consequence of having emerged from several species—the doctrine of polygenism. Brash, noisy, and masculine, the members of the new society pursued the biology of human difference. They talked of skulls and brain size, buttocks, and primitive civilisations.[53] Huxley, Wallace, and several others interested in human origins joined this extrovert band, at least for a while. They turned up in force to see Colenso come onto the podium in 1865 flatly to contradict the story of Creation and the Deluge on scientific grounds. These "Anthropologicals" acquired an eccentric reputation, helped along by the president's posturing with a gavel topped by a carved human skull to call members to order during lively meetings.

With its emphasis on human anatomy and the origin of mankind, Hunt's society brought the question of human diversity back into focus. Like Louis Agassiz, and Nott and Gliddon before him, Hunt stoked contemporary prejudices with his assertions that four or five modern human types had existed since the earliest times. What was new was that Hunt placed these views in evolutionary context. He demarcated humans into biologically distinct races, in the terminology of the day, and said that these should be regarded as self-contained taxonomic entities, separating black from white. Reproduction between blacks and whites, he claimed, was biologically unnatural—the offspring were said to be sterile or reduced in fertility—and "the analogies are more numerous between the negro and the ape, than between the the European and the ape."[54] Under Hunt's direction, some members of the Anthropological Society not only espoused racist categories of thought but also mostly sided with the South's cause in the war.

These opinions did not go unchallenged. Huxley and Busk took a dim view of Hunt's anatomical assertions in his paper on the Negro, and others reminded Hunt that reproduction took place across all supposed human frontiers. James Horton, an African doctor educated and living in England, icily inquired if fellows of the Anthropological Society had ever seen a black man, for he could not recognise any of his countrymen from Hunt's "prejudiced" and "absurd" descriptions.[55] Even so, there were numerous elements in these programmatic statements that touched a chord. For all his criticism of Hunt's blunders, Huxley also believed in a racial hierarchy and the inferiority of Negroes. Moderate men like Lubbock and Darwin relegated primitives to a stage early in social organisation.[56]

The Anthropological Society also began a programme of translations of foreign texts to promote the polygenist view, including important works by Paul Broca and Carl Vogt. Vogt had proposed separate ape origins for each race of human being (followed by interbreeding and blending) in his *Vorlesungen über den Menschen,* translated by the Anthropologicals as *Lectures on Man* in 1864. Such views were regarded with suspicion by those who believed in a single common ancestor, although Vogt espoused evolution and usually praised Darwin. As Wallace complained to Darwin, "the Anthropologists . . . make the red man descend from the Orang, the black man from the Chimpanzee, or rather the Malay & Orang one ancestor, the Negro & Chimpanzee another."

Darwin was pleased when Wallace made his own standpoint known. He was the first of the Darwinians deliberately to apply natural selection to the emergence of human difference. Everyone else had published his views on various aspects of the evolutionary scheme. Wallace felt he was ready to speak out too. He did so in an article in the *Journal of the Anthropological Society,* in 1864. In this, Wallace proposed that human beings emerged in a single group from apelike ancestors and then rapidly diverged under the impetus of natural selection. In effect, he gave a chronological, developmental account of human origins that united conflicting theories of single or multiple beginnings. First, they were one. Then they were many. "*Most* striking and original and forcible," remarked Darwin. "I wish he had written Lyell's chapters on Man. . . . there is no doubt, in my opinion, on the remarkable genius shown by the paper."[57]

Wallace included a second theme that was unlike anything yet said about human descent. He proposed that at an early point in human evolutionary history, selective pressure must have shifted away from the physical body onto mental processes. Natural selection would then act mainly on the human mind and behaviour, producing the faculty of speech, the art of making weapons, and the division of labour. In this way, human beings would adjust to their local environment and free themselves from the operation of the laws of natural selection. Thoroughly interested, Darwin agreed with much of what Wallace had to say. He too thought human beings had been gradually unshackled from purely biological necessity, although he differed from Wallace in the manner this might have come about.

At the end of the article, Wallace discussed the hierarchy of races projected by his fellow anthropologists. On this, he was the same as the rest. Each group of mankind would not be equal in its fitness to survive, he said. "Improved" races would inevitably "displace the lower and more degraded races," a point on which few of his Victorian readers would have disagreed. Civilised races were destined to increase at the expense of the latter "just as the weeds of Europe overrun North America and Australia."[58] While Wallace, like Darwin, was for the most part a humane and cultivated man, and dispensed with many of the racial prejudices of his contemporaries, he nonetheless endorsed Western cultural superiority and matched it to evolutionary theory.

Impressed, Darwin covered his copy with pencil marks. He wrote to Wallace a few days later, mentioning for the first time his own ideas about sexual selection that were in the future to play an important part in his account of mankind. Courteously, he remonstrated with Wallace. The theory of natural selection was "just as much yours as mine." Neither could have predicted that Wallace's article contained the seeds of what would become the most significant difference between the two men.

Soon after, John Lubbock published his researches into archaeology and Edward Tylor described the progressive development of human societies. Both of them believed Darwin's theories helped them to open up new areas of thought about the lives and minds of early and indigenous peoples.

In *Pre-Historic Times* (1865), Lubbock brought the notion of "prehistory" to the fore—a word that he invented to bridge the gap between the geological past and more modern times. His subdivision of the Stone Age into Neolithic and Palaeolithic periods gave intellectual structure to the theme of human antiquity as described by Lyell, and he connected in chronological sequence the flint-making peoples of Europe to societies that produced pottery and other cultural artefacts. A good all-round naturalist, Lubbock was probably happiest with a pottery shard in his hands, recreating in his mind's eye the men and women who had once used it. He had spent many of his recent holidays sifting through the margins of Scandinavian lakes and bogs in search of these ancient objects, either in buried kitchen middens, among abandoned house piles, or in other recognisable residue of human habitation. His book was widely read and went to five editions in the first year of publication. He and Lyell argued again over the proper attribution of their archaeological data. Lubbock was right in claiming that he had published the information first. The other Darwinians had to smooth things over.[59]

On the other hand, Tylor came to the view that human behaviour itself was subject to evolutionary considerations and could indicate ancestry as surely as any physical attribute. His ideas marked the beginnings of evolution's impact on the social and human sciences. Tylor used well-established disciplines such as linguistics, folklore, mythology, and comparative theology to propose that human thoughts and behaviour patterns could be relics or "survivals" of early cultures just as much as vestigial anatomical organs could be left behind in the evolutionary sequence; as he explained it, the piano was a descendant of the harp. Cultural similarities between disparate groups, he maintained, could be explained on a developmental model comprising independent invention, inheritance, and diffusion. In this he brought the comparative method to a new level of sophistication. From 1865, with his *Researches into the Early History of Mankind and the Development of Civilization,* to 1871, when his *Primitive Culture* was published, he was responsible for securing a place for social and cultural anthropology in the minds of educated British readers. He and Darwin exchanged appreciative letters, and Darwin later consulted him on cultural behaviour and material evidence for the evolution of mankind.[60]

With anthropological books and projects hogging the limelight, it is not suprising that the vexed issues of slavery and race would not go away. Still sick at Downe, in 1865 Darwin learned about Governor Edward John Eyre's suppression of emancipated slaves in Jamaica. The newspapers—and Darwin's friends—split into opposing camps, and Eyre was alternately castigated as a monster of cruelty or admired for his firm action. James Hunt defended Eyre to the point that the Anthropologicals became identified with blatant anti-Negro sentiment, believing in black "barbarism." Kingsley openly approved of Eyre, as did Carlyle, Ruskin, and Tennyson. Darwin was startled to hear that Hooker also supported Eyre. Darwin did not. "You will shriek at me when you hear that I have just subscribed to the Jamaica Committee," he wrote in surprise to the botanist, referring to John Stuart Mill's Jamaica Committee, established to secure Eyre's prosecution for murder.[61] Huxley also supported the committee, collecting subscriptions from his friends for this fighting fund. Darwin had sent £10. "I am glad to hear from Spencer that you are on the right (that is *my*) side in the Jamaica business," Huxley thanked Darwin in November 1866. "It is wonderful how people who commonly act together are divided about it."[62]

Elsewhere, great political events marched forward. The assassination of Abraham Lincoln shook the world. "The noble manner in which our country has borne itself should give you real satisfaction," Gray said soon after. "We appreciate too the good feeling of England in its hearty grief at the murder of Lincoln. Don't talk about our 'hating' you,—nor suppose that we want to rob you of *Canada*—for which nobody cares."[63] The man who was born on the same day as Darwin died amid national mourning on 15 April 1865.

> We continue to be deeply interested on American affairs; indeed I care for nothing else in the Times. How egregiously wrong we English were in thinking that you could not hold the South after conquering it. How well I remember thinking that Slavery would flourish for centuries in your Southern States.[64]

## VII

Darwin's illness had little effect on the way his name was progressively spreading across Europe. The wave of editions and translations taking evolutionism back and forth between Britain and continental Europe and America presented a social and intellectual phenomenon in its own right. Few scientific concepts of the nineteenth century were to experience such recasting, popularisation, negotiation, and consolidation as the body of work associated with the *Origin of Species*. From 1864, when the *Origin* was translated into Russian, Italian, and Dutch, through to the end of the 1870s, by which time it had appeared in Swedish, Danish, Spanish, Hungarian, and Polish editions, and a host of commentaries, criticisms, and supporting texts had been published and translated, the dissemination of evolutionary thought was firmly embedded in individual national and religious contexts.

In the process, there was plenty of room for confusion. Although science in translation travelled across geographical boundaries fairly efficiently for the period, the characteristic language in which it was expressed was seldom well understood outside particular national settings. Natural philosophers did not necessarily share unambiguous procedures and principles of reasoning. Metaphors rarely travelled well, omissions or additions to translations might affect the argument, and an unsupervised preface or addendum could undermine the results.[65] As Darwin already knew to his cost from Clemence Royer's and Heinrich Bronn's early translations of his book, these volumes ought more properly to be regarded as creative reworkings of an author's ideas in which texts were repositioned in another intellectual and social milieu. And always, in the end, when a translated book reached the hands of a reader, no matter how scrupulously it preserved what the author wished to say, the reader might easily perceive messages different from those that the author intended.[66]

After Heinrich Bronn's death in 1862, Darwin arranged to have his books translated into German by Julius Carus, although not before Carus engaged in a proprietorial tussle with Carl Vogt, who also wished to be Darwin's German mouthpiece. Two other possible translators emerged as challengers to Carus in 1866, but were quickly eliminated by him.

Carus was a good choice. His enthusiasm ensured that he did much to spread Darwinism in Germany, and he was the moving force behind the first collected edition of Darwin's works in any language in 1875. Moreover, he was the first commercial writer to capitalise on his relationship with Darwin by asking whether he could produce a "little biographical sketch" including "Birthday, school, and so on. If you should not like it I trust you will tell me quite openly and will not be angry with me." He also translated Huxley's *Man's Place in Nature,* a book that made a lasting impression in the German-speaking states, and then Tylor's anthropological works. Darwin was evidently in safe hands.

A large number of other evolutionary texts appeared in German too. Adolf Meyer translated Wallace's works as they were published, and the philosophical materialist Ludwig Buchner translated Lyell's *Antiquity of Man* in three successive editions between 1864 and 1874. As might be expected, evolutionary thinkers like Vogt, Ernst Haeckel, and others wrote independent texts in their native language. The main scientific publishing firms in Stuttgart, Frankfurt, and Berlin additionally captured a ready market by matching these books, cover for cover, with anti-evolutionary publications, including German translations of Louis Agassiz's attacks on Darwin. There was much debate and creative engagement with these ideas. Perhaps the naturalist Friedrich Rolle was the only one to "swallow the medicine whole," as Darwin expressed it. Rolle wrote several early articles and a book on Darwinism in 1863. Elsewhere, Karl von Baer rejected transmutation outright. Ludwig Rutimeyer, Carl Nägeli, and Oswald Heer agreed more or less with the evolutionary view of nature while dispensing with natural selection and reinstating personal religious commitment. Rudolf Virchow, the most prominent of them all, never accepted Darwinism, but religion played no part in his rejection. Despite the range of response, Darwin seemed cautiously optimistic. In 1868 he let slip to William Preyer that he thought "the support which I receive from Germany is my chief ground for hoping that our views will ultimately prevail."[67]

The picture was broadly repeated in tsarist Russia. Lyell's 1859 British Association address and Huxley's 1860 Royal Institution lecture were translated into Russian early on, prompting S. A. Rachinski to translate the *Origin of Species* in 1864 and the botanist Kliment Timiriazev to write a favourable commentary titled *Charles Darwin and His Theory.* These works were followed by Vladimir Kovalevsky's translation of *Variation Under Domestication* in 1868, by translations of Lubbock and Vogt, and by the publication of evolutionary expositions by noted naturalists like Andrei Beketov and Il'ia Mechnikov. By 1870 it was possible for Russians to read in their own language all of the Western European texts that presented the salient features of the controversy. The dispatch with which Russian scientists took account of these developments was for the most part probably due to the intelligentsia's acceptance of the general phenomenon of evolution some years before the *Origin of Species* was published.[68]

From the outset, however, many Russians considered the Malthusian basis of Darwin's and Wallace's argument relatively unimportant. Darwinism arrived in that country without Malthus's political economy of struggle, changing its costume as it moved across cultural and ideological frontiers. Karl Marx had judged Darwin and Malthus together as part of the British national type. So did Leo Tolstoy, who criticised Darwin for bringing "a fictitious law" into biology. In *Anna Karenina,* Tolstoy's character Levin attacked the moral consequences of Darwin's views so sharply that in real life Timiriazev felt compelled to respond. Notwithstanding this, Kovalevsky and Timiriazev ensured that Darwin was elected to the Imperial Academy of Sciences in 1867, one of the first major scientific awards he received.[69]

These evolutionary thinkers were in the main young, unconventional, patriotic, still relatively fluid in their careers, and united by a dislike of traditional forms of religion, especially Kovalevsky, Vogt, Haeckel, Fritz

Müller, and Buchner. While Vogt had caught the headlines as a radical member of the National Assembly in 1848 and Buchner's hardline social views were expressed in numerous treatises, Ernst Haeckel had first met Darwin's ideas in a biological context. Darwin wrote to them all, carefully establishing the rapport that he felt would lie at the heart of his theory's progress abroad. "He was sometimes troubled how to reply to Monsieur et très honoré Confrère or Hoch verehrter Herr!" explained Francis, amused at his father's caution. He was driven to use "Dear & respected Sir."

Haeckel championed Darwin in the Versammlung Deutscher Wissenschaftler und Aerzte (Assembly of German Scientists and Medical Doctors) and preached his theories in biology lectures at Jena. He contacted Darwin in 1864 by sending him a copy of his book on Radiolaria, minute protoplasmic sea organisms that inhabit elaborate carapaces, telling Darwin that nothing had made such a "powerful impression" on him as the *Origin of Species*.[70] Haeckel became by far the most ardent Darwinian in Germany, influencing a generation of scholars that included Anton Dohrn, Hans Driesch, Hans Spemann, and Richard Goldschmidt. In biology, he believed that Darwin's ideas opened fresh research areas and generated original methods of analysis. In social and political terms, he also set about using Darwin's views to bring about a cultural transformation.[71]

Fritz Müller lived a more obscure life in science than Haeckel but was rather more typical for that reason. He too was a political animal. He regretted the failure of the 1848 revolution, declined a position in the Prussian educational system because of its Christian observance, and emigrated to Brazil, where he spent his final years imprisoned by rebel forces. A dedicated rationalist, he believed in free love, natural history, and Darwin—one of the increasing number of relatively unknown naturalists who were coming to think that Darwin's principles genuinely assisted their work. Müller's name became linked with Darwinism through his small volume on Crustacea, in which he applied evolutionary considerations to the life cycle of prawns and similar species. *Für Darwin* was published in Leipzig in 1864. In it Müller declared that Darwin's theories furnished "the key of intelligibility for the developmental history of the Crustacea."[72]

Darwin was delighted. Müller's book at last delivered the embryological analysis of evolution in action that he had longed to see—the work he had hoped that Huxley, with all his gifts, might have provided. From Down House he wrote Müller an enthusiastic letter. "A man must indeed be a bigot in favour of separate acts of creation if he is not staggered after reading your essay."

> I look at the publication of your essay as one of the greatest honours ever conferred on me. Nothing can be more profound and striking than your observations on development and classification. . . . What an admirable illustration it affords of my whole doctrine![73]

Darwin promptly arranged to have Müller's book translated into English by William Dallas, a capable naturalist who usually prepared the indexes of Darwin's volumes. This was published in 1869 at Darwin's expense by John Murray as *Facts and Arguments for Darwin,* clad in the same green cloth as Darwin's titles. Darwin took on the publisher's risk. "I think you wd. be safe with 750 [copies]," Murray advised him. Once it was out, Müller's words of praise for Darwin's evolutionary system were there for all to see, a "splendid structure which he has raised with such a master-hand," said Müller. This was highly effective publicity.

In France, Darwin felt his situation was still difficult—impenetrable.[74] While Clemence Royer's translation of the *Origin of Species* could hardly be responsible for what Huxley called the "conspiracy of silence," reactions were decidedly muted. Several influential positivists followed Auguste Comte's example by attacking transformism. Catholic opinion, although far less interested in biblical literalism than that of other churches, was for the most part opposed to proposals for the existence of godless, independent natural laws.[75]

More to the point, the centralisation of learning in France probably allowed any rising hostility to Darwin to take root and reemerge as dogma. Despite the efforts of French-speaking naturalists like Alphonse de Candolle, François Pictet, and Édouard Claparède, all based in Switzerland, Darwinism fared poorly in Paris itself, in the Académie des Sciences, the Sorbonne, and the Musée d'histoire naturelle. So when Pierre-Jean-Marie Flourens, the most conspicuous French Academician of the day, declared in 1864 that Darwin's work was deficient in the basic rules of logic, there seemed little reason for Frenchmen to give any more thought to the *Origin of Species.* Flourens has written "a dull little book against me," said Darwin in 1864, encouraging Huxley to give him a drubbing in the *Natural History Review* ("hang the scalp up in your wigwam!" said Huxley). In France, however, Flourens's *Examen du livre de M. Darwin sur l'origine des espèces* carried special weight.

There were discussions about racial biology and brain anatomy from time to time among some Parisian anthropologists in the Société d'Anthropologie. Darwin's book also impinged in a general way on the continued argument between Louis Pasteur and Charles Pouchet about spontaneous generation and the defining features of life. And Lyell's and Huxley's books were translated into French in 1864 and 1868, Spencer's and Haeckel's soon after. Yet during the 1860s only fifteen new scientific books carried the word *transformisme* in their title; in the 1870s, the number crept up to twenty-nine. Undeterred, Darwin counted Jean Louis Armand de Quatrefages, Albert Gaudry, Édouard Claparède, Henri Milne-Edwards, and Louis Charles de Saporta as friends and received qualified approval from Paul Broca, Paul Topinard, and Isidore Geoffroy Saint-Hilaire. Unknown to him, the pioneering psychologist Hippolyte Taine mentioned Darwin's work favourably in 1863 and again afterwards. Otherwise, struggle and natural selection were considered alien concepts by French intellectuals.

Darwin naturally felt there was still much to be done in establishing his views a mere twenty miles across the English Channel. He felt gratified by Quatrefages's efforts to get him elected to the prestigious Académie des Sciences, however unsuccessful these efforts were against Flourens's resistance. In all, Darwin was nominated three times by Quatrefages, supported by the naturalist Henri Milne-Edwards, in 1870, 1872, 1873, and finally with success in 1878. Even then, Darwin was elected on the basis of his botanical work, a snub that intentionally ignored his *Origin of Species* rather as the Royal Society of London had omitted any mention of the *Origin* when awarding Darwin the Copley Medal. "It is curious how nationality influences opinion," he remarked after hearing about the first Académie rejection. "A week hardly passes without my hearing of some naturalist in Germany who supports my views, & often puts an exaggerated value on my works; whilst in France I have not heard of a single zoologist except M. Gaudry (and he only partially) who supports my views."[76] Quatrefages considered Darwin was the only man to have proposed an evolutionary theory that was properly scientific and embraced all aspects of natural knowledge. Generously, Quatrefages said he appreciated the "grandeur" of Darwin's work.[77]

Darwin's book of *Beagle* travels in fact seemed to be the only thing that the French were willing to read. "Though I am so despised by the great guns of the Institute," he protested to Hooker in 1869, "I presume I am rising in estimation amongst the mob, for another man has applied to translate my Journal of Travels.—Here is a boasting note."[78]

As a translated volume, Darwin's *Origin of Species* was plainly dropping into a range of social contexts bursting with their own continuing trends of thought, several of which already included evolutionary ideas. Spanish authors took up the *Origin*'s call in 1869 with a flurry of commentaries on the chemical origin of living beings.[79] In Italy, on the other hand, the intellectual elite already advocated secularism and evolutionary naturalism, to the point where Paolo Mantegazza suggested that science itself should become a religion. As expounded by Italian positivists, Darwin's theory was above all seen as a genealogy of living forms rather than an explanation of the method of change.[80] The *Origin of Species* was partially translated into Italian by Giovanni Canestrini, the pioneering zoologist, in 1864. But it was Filippo de Filippi who, in Turin, really generated discussion with his article "Man and the Monkeys," stimulated by Huxley's book. Filippi accepted all the basic arguments put forward by Huxley but maintained that human morality could not be explained by descent from the animal kingdom.[81] These polemics encouraged Canestrini to publish the complete text of the *Origin* in 1865, followed by an influential book titled *Origine dell'uomo* in 1866. Shortly thereafter, the botanist Federico Delpino began an analysis of Darwin's experimental work on plants, sustained by a non-Darwinian belief in the directed and progressive nature of evolution.

The story was generally the same in Australia, New Zealand, the United States of America, and Canada—each nation divided from Britain by a common language. The *Origin*'s author plugged away at them all. Whether for or against him, or willing to meet him at some point halfway, men and women across the globe began participating in one of the first international scientific debates.

## VIII

Day after day, Darwin spiraled downwards physically. "When you meet Busk, ask him whether any man is better than Jenner for giving life to a worn out poor Devil," he implored Hooker in 1865. His wife, Emma, began to wilt under the strain. "I have taken a little to gardening this summer and have often felt surprised when I was feeling sad enough how cheering a little exertion of that sort is."[82] Keen to have company, she installed an elderly aunt at Down House to play cards with her in the evenings after Darwin had gone to bed. With her ear trumpet and tendency to stand in front of the fire monopolising the newspaper, Aunt Fanny proved an irritating addition to the household.[83] When she went home to Wales, Emma noticed how Darwin "brightened up very much the last day."[84]

Whichever doctor he called in, conventional medicine only partially relieved the miseries. He revived a bit when he came across eccentric Dr. John Chapman, the former editor and owner of the *Westminster Review*, who was closely connected to George Eliot and her circle. Chapman was company of a kind Darwin rarely encountered. Although more or less properly qualified as a medical man, he regarded doctoring as a last-ditch insurance against disaster in the publishing world. Every so often he put a handful of expensive therapeutic gadgets on the market, relying more on medical fads and fashions than on proven practice. Ice-bags for seasickness and nausea were his current venture, and it was an advertisement for these that brought Darwin's letter to his door. Chapman knew everybody: Huxley, Spencer, Lewes, Dickens, Tom Taylor of the *Times,* John Stuart Mill, and Harriet Martineau and her brother James. He was moreover impressed by the *Origin of Species,* saying in his diary that it was "likely to effect an immense mental revolution."[85] Darwin liked him immediately and wrote to him about his symptoms more openly than to any other man, "Tongue crimson in morning . . . evacuation regular & good. Urine scanty," and so on.[86]

Chapman's cures were as ineffective as the rest. He certainly pandered to Darwin's yearning for uncomfortable remedies (or to his desperation) by selling him a set of rubber water-bags of various sizes that were to be chilled and worn next to the spine every day for a hour or two. "Ice is a direct sedative to the spinal cord," Chapman wrote in one of his medical handbooks.[87] Darwin strapped these bags to his lower back three times a day. He seems not to have felt either foolish or gullible about this. Cold, inconvenient, and bulky as they were, at least he could trudge round the Sandwalk at regular intervals. It soon became obvious that "Ice to spine did nothing." But the rubber bags were not wasted. Long afterwards, when William took a boat across the English Channel, his thrifty father recommended that he wear them as a precaution against seasickness.

Then came Henry Bence Jones from St. George's Hospital in London, altogether a more regular kind of physician, who tested Darwin's urine for proteins and uric acid and diagnosed suppressed gout. Jones was a keen disciple of Justus von Liebig's philosophy of chemistry and believed wholeheartedly in analyzing the chemical constituents of the blood at a time when haematology and laboratory researches were not yet an established part of medical therapy. Nutritional imbalances were his forte. He made Darwin stop eating the sugary foods he liked best and advised that acidifying liquid medicines like colchicum and bitter aloes should take the place of sticky puddings. "I have been half starved to death," his patient moaned after a couple of weeks of this, "and am 15 lb lighter, but I have gained in walking power & my vomiting is immensely reduced." It was unlucky, said Francis Darwin, "that so many Drs forbade him sweet things, for which he had a boylike love."

> He often said that the meat of dinner very dull, & the sweets the only part worth. He was not very successful in keeping the vows which he made not to eat sweets; and didn't consider them binding [unless] he

> made them aloud. He often made a vow aloud after breaking a silent vow.[88]

Furthermore, he was forbidden by doctors to eat bacon, continued Francis. "But as it suited him particularly well he never obeyed, and used to laugh at the whole race of Drs for their spite to bacon."

Jones also prescribed exercise "to get the chemistry going." At first he suggested yachting, tipping his hat to the customary medical advice for affluent Victorians. Jones may have recognised Darwin's mental unrest and thought the old-fashioned remedy of a change of scene would work as well as any other. In the event, Darwin stayed at home and began horse-riding instead, buying a steady old gelding called Tommy from a local dealer (the boys tested it out beforehand) and jogging along the country paths for a year or two before taking a bad tumble and deciding he was too old for the venture.

Yet there was no escaping the mournful atmosphere at Down House. Death tugged at the heartstrings when Robert FitzRoy committed suicide in April 1865, cutting his throat with a razor, a ghastly reenactment of his uncle Lord Castlereagh's own suicide. Darwin was taken aback. With a pang of uneasy familiarity, he saw in this last act much of his former captain's impetuous, unstable, and courageous behaviour.[89]

"Ch. was very sorry about FitzRoy—but not much surprised. He remembered him almost insane once in the *Beagle*," said Emma. They had heard intermittently about FitzRoy's deteriorating existence. For years FitzRoy had been aware of the deadly outcomes of his Fuegian philanthropy. The three indigenous Fuegians, whom he had taken to England, Christianised, and reintroduced to Tierra del Fuego during the *Beagle* voyage, had reverted to their aboriginal behaviour. In the 1840s, a Christian mission to Tierra del Fuego had starved to death near the Beagle Channel. Poor organization was to blame. Then Jemmy Button, one of FitzRoy's original three, helped establish another mission in Woollya Sound, which ended in the massacre of every single European. FitzRoy took that news badly. According to the ship's captain who had dropped the missionaries off in Woollya, Jemmy had shouted to the officers from his canoe in perfect English and asked for a pair of braces for his trousers. He went on board and gave them a gift for his old captain.[90] FitzRoy was devastated by Jemmy's apparent treachery (although it was never clear whether Jemmy was directly involved in the massacre).

Furthermore, FitzRoy's work at the Meteorological Office was persistently undervalued. Admittedly, he was a demanding perfectionist. But his attempts to introduce scientifically based weather forecasts and storm warnings for sailors at sea were the constant butt of Victorian satire. "Nature seems to have taken special pleasure in confounding the conjectures of science," jibed the *Times* after another unexpected hurricane disrupted the countryside. FitzRoy took every criticism personally. The wreck of the *Royal Charter* off the coast of Anglesey, with dreadful loss of life, turned his mission into a moral crusade.[91] He argued with Lieutenant Maury in Washington over the meteorological causes of storms and how to predict their arrival. Darwin heard all about it from Bartholomew Sulivan, now an admiral himself, when Sulivan paid an autumn visit to Down House. The two old comrades shook their heads over FitzRoy's depressions and misplaced ardour.

> I never knew in my life so mixed a character. Always much to love & I once loved him sincerely; but so bad a temper & so given to take offence, that I gradually quite lost my love & wished only to keep out of contact with him. Twice he quarreled bitterly with me, without any just provocation on my part. But certainly there was much noble & exalted in his character.[92]

When Darwin heard that Mrs. FitzRoy was left penniless he sent a cheque for £100 (about £5,000 in modern terms), a large sum that probably indicated guilty feelings towards the memory of his captain.[93] He did not feel nearly as sad as he ought to have done. In much the same way as he responded to Henslow's death, he quickly pushed the inconvenience of grief and regret aside.

Old Sir John Lubbock died too, leaving his fortune, title, banking firm, High Elms estate, and village charities to his son John, Darwin's friend. These long-awaited riches burned in Lubbock's pocket, and soon afterwards, he decided to run for Parliament. Darwin regretted the decision, for he felt Lubbock's election as an M.P. would be a loss for science. Nevertheless, he loyally subscribed to the West Kent Liberal Association, discovered a new interest in the Maidstone newspapers, where Lubbock intended fighting for a seat, and told him he would "be very sorry if he succeeded but very sorry if he was beaten."[94] Underneath lay the unspoken assumption that the two shared the same kind of forward-looking politics found in Palmerston and Lord Russell's Liberal party. These were exciting times, with an increasingly split party system giving rise to weak coalition governments beset by confused policies. In 1865, Lubbock hoped to join Palmerston's administration, although whether Palmerston would live that long was a well-aired problem. Over the next five years there were to be three changes of government brought about by deaths and resignations. Darwin looked forward to political discussions with

Lubbock, for he enjoyed after-dinner governance as much as any man of his acquaintance.

Lubbock was beaten. "It made me grieve his taking to politics," said Darwin afterwards, "and though I grieve that he has lost his election, yet I suppose, now that he is once bitten, he will never give up politics, and science is done for. Many men can make fair M.P.s; and how few can work in science like him!"[95]

Lubbock did not achieve electoral success until 1870, thereafter representing West Kent as a Liberal for eighteen years before changing tack on the Irish question and subsequently representing London University for the Unionists. He received a peerage in 1900 and moved to the House of Lords as Baron Avebury, the "banking baronet," renowned for safeguarding archaeological sites like the stone circle at Avebury (from which he took his title) and Stonehenge, and for introducing the concept of bank holidays, national days of leisure. All through this process Darwin wished Lubbock could give more time to science. As an ambitious youth, Lubbock had once confided to him that he hoped to become president of the Royal Society, lord mayor of London, and chancellor of the exchequer, and Darwin reflected that Lubbock was good enough to have achieved any one of these had he been "willing to forgo the other two." Lubbock never produced the innovative biological research of which he was capable. His scientific impact afterwards lay in popular expositions of natural history topics, especially his best-selling *Ants, Bees and Wasps,* and in his parliamentary work and government policy.

The deaths continued. In 1865, William Jackson Hooker died, leaving his son Joseph Hooker the herbarium, library, and directorship of Kew Gardens, a mixed blessing that Hooker tackled resourcefully. In his hands Kew's scientific and imperial role expanded dramatically. And in 1866, Darwin's sisters Catherine and Susan died.

Catherine Darwin went first. Her life had taken a very different turn three years beforehand when, in 1863, she married Charles Langton, the widower of her cousin Charlotte (Emma's sister). Although Langton was by no means a blood relation, and was breaking no ecclesiastical rule by moving so promptly from a Wedgwood to a Darwin, the marriage seemed vaguely indecent to some of the older members of the circle. Mutual loneliness was an obvious motivating factor. Yet Catherine Darwin was never completely well and Langton was so set in his ways that Emma doubted whether they could be happy. The Darwin brothers rearranged the family finances to provide Catherine with a trust fund that would have matched old Dr. Darwin's wishes. They put William Darwin, the next generation's banker, in charge.

"Her life was an abortive one with her high capacities," grieved the combined womenfolk, a story that must surely have been repeated over and over again in the social tapestry of middle-class women's lives.[96] After this death, Langton lived on for another twenty years, outlasting most of his Darwin and Wedgwood brothers-in-law, a tall, benign man who had early on renounced his position as a clergyman apparently after experiencing all the doubts of a culturally engaged Victorian. Although nothing is known about Langton's private thoughts, close contact with Darwin's theories possibly played some part in eroding his faith. Emma scarcely mentioned him in her letters, except to say, "Poor C.L. I suppose it is doubts about future life [that] trouble him." His story was one of the lost tales behind the Darwin and Wedgwood chronicle: always there, never mentioned. One Wedgwood grandson, however, remembered him as an integral member of the pack of great-uncles.

> There were my grandfather [Frank Wedgwood] and his three brothers [Hensleigh, Harry, and Jos], and his brothers-in-law, Charles Darwin, and Charles Langton—all eighty-ish, all grey, all towering, while countless little cousins ran about below them on the floor. They never laughed; they never seemed to be angry. . . . They had read everything, and knew everything, and had their own judgement on everything. The opinion of the world mattered to them not one jot. Their standards were law, and we would as soon have disputed with the Deity.[97]

Susan Darwin died unmarried a few months afterwards, bringing the Shrewsbury era to an end. Erasmus was profoundly upset. Susan was his favourite, the one he considered most like their father in personality, a clever, unsentimental woman, full of an unshakeable belief in home-made jam and the value of conversations about pigs or poultry. Emma used to have the windows cleaned before Susan visited Downe. She would stay with Erasmus in London for the winter season, going to the theatre and exhibitions together, she criticising his housekeeping, he indiscreet about his London cronies, each reassured by the other's predictible attitudes. She was at Erasmus's house in London when she died. "What a very easy thing death is," he wrote to Fanny Wedgwood. Darwin told Lyell (who knew her well) that Susan suffered greatly and that there had been no hope of recovery. The Mount in Shrewsbury was sold, the furniture auctioned off.

## IX

Darwin began to feel slightly better in the spring of 1866. He resumed his writing on the variation of domestic animals and plants and undertook to provide Murray with a fourth edition of the *Origin of Species*. This new edition of the *Origin* took him eight or ten weeks to complete and was published in December 1866. He revised the text thoroughly to include fresh material drawn from his friends' recent publications and answered a number of criticisms posed by reviewers, glad that he kept the reviews and correspondence close to hand.

Soon he felt well enough to venture to the Royal Society for a soirée attended by the young Prince of Wales (later Edward VII). This gala evening was an important event in the society's calendar, hosted by the president, Edward Sabine. Sabine usually held three such soirées during the season, but "Bertie," the royal princeling, had never agreed to come before. His interests took more of a social and sporting turn, although he admired the Oxford and Cambridge professors who had formerly tutored him. Privately, Sabine regretted the untimely death of Prince Albert, an enthusiastic patron of the cultured world, who valued the role of science and technology in Britain rather more than his son, and endorsed the Royal Society's place in the established scheme of things.[98] But the presence of the Prince of Wales as a royal patron was appropriate and would add considerably to the Royal Society's annual celebration of Victorian achievement. As was customary, ladies were not invited. As was customary too, the society's rooms in Burlington House were crammed with exhibits and working demonstrations. Natural philosophers roamed the stalls, familiarising themselves with developments in fields outside their usual domain.[99] At the soirée that Darwin attended there were telegraph cables encrusted with barnacles, photographs of sun-spots, a measuring device used during the Trigonometric Survey of India, and a cannonball retrieved from Fort Sumter. "The President's reception was an evening to be remembered," said J. W. Rogers in a note afterwards.[100] Three fellows of the Royal Society were to be presented to the royal party. Darwin was invited to be one of them—an honour that tickled his vanity but threw Emma and the Down House staff into consternation.

Getting him to London turned out to be an exercise of near-military proportions, occupying Emma, Parslow, Erasmus, Henrietta, and Fanny and Hensleigh Wedgwood for the full space of a week. Emma arranged that the Down House contingent would stay with the Wedgwoods for the occasion, and she asked Fanny to purchase tickets for the ladies to go to *Hamlet* the night before and a Philharmonia concert the night after. Downe never seemed so far away from central London—a mere fifteen or sixteen miles for Emma, a continent for an enfeebled and apprehensive naturalist. The women laughed irreverently at his grumbles. They were determined he would go.

On 27 April 1866 they dusted Darwin off and deposited him at the

Royal Society's front door dressed in his best. Henry Bence Jones welcomed him on the threshold almost as if he were a walking advertisement for medical science. "His Dr. Bence Jones was there and received him with triumph, as well he might, it being his own doing."

The royal presentation went well, although Darwin said that the prince obviously had not a clue who he was. But the most notable part of the evening was of a different order. After so many years of illness, none of his friends recognised him. "In the evening Papa dressed up quite tidy in his dress suit," Emma wrote.

> It felt quite cheering to see him go. He staid more than an hour & saw every one of his old friends, who were delightfully cordial as soon as they found out who he was, which they never did till he told them owing to his beard. General Sabine presented him to the P. of Wales. . . . He made his most respectful bow & the P. muttered some little civility which he cd. not hear. The P. looked very nice & gentlemanlike but utterly uninterested in any of the things & curiosities. Papa was very much pleased with having gone. Yesterday was rather too fatiguing—a long call on Mr Grove, something about forces, then a long call from Sir C. Lyell who was so polite to Lizzie, standing up & talking a little to her.[101]

She repeated the observation in another letter: "He was obliged to name himself to almost all of them, as his beard alters him so much." Her husband returned home from his week in the metropolis dumbfounded by the effects of a long seclusion in the country.

## X

Since Darwin would not go to his acolytes, they began to come to him. For some of his followers, the opportunity to visit Darwin started to resemble a religious pilgrimage. In 1866, Ernst Haeckel made a special trip to England from Jena during which he fulfilled a wish to call at Down House. By now Haeckel was a devoted follower. He praised Darwin in his own books and articles and lectured on *Darwinismus*. One student at Jena, Anton Dohrn, after a few sessions with Haeckel, came to believe that receiving a letter from Darwin was like being granted a "scientific knighthood." Haeckel had just published *Generelle Morphologie der Organismen* (1866), which he dedicated to Goethe, Lamarck, and Darwin, describing them as the preeminent leaders of evolutionary theory. In this book he set out his expansive views on morphology, systematics, the genealogy of mankind, and embryology. He sent it to Darwin.

Not for the first time Darwin wished that Camilla Ludwig was still part of the family economy. He found Haeckel's written German very difficult to understand. He fought his way through Haeckel's definitions of concepts like "ontogeny" and "phylogeny," and the "biogenetic law" that encapsulated parallels between embryology and the evolutionary tree. "Ontogeny recapitulates phylogeny," said Haeckel in densely packed German. Then Darwin worked his way through Haeckel's subdivisions of heredity (conservative or progressive) and adapation (indirect or direct). He struggled past Monism, the philosophical position invented by Haeckel to convey the study of all biological phenomena on mechanistic and naturalistic lines. He contemplated dizzyingly complex illustrations of evolutionary history that showed the whole of living nature emerging from a single ancestral stem. "Your boldness sometimes makes me tremble," he said, "but as Huxley remarked someone must be bold enough to make a beginning in drawing up tables of descent."

> I received a few days ago a sheet of your new work, & have read it with great interest. You confer on my book, the "Origin of Species," the most magnificent eulogium which it has ever received, & I am most truly gratified, but I fear if this part of your work is ever criticized, your reviewer will say that you have spoken much too strongly. . . . I shall feel very curious to read your remaining chapters when published; but it is a terrible evil to me that I cannot read more than one or two pages at a time of German, even when written as clearly as is your book.[102]

Whatever else, the book served as an introduction and Darwin invited Haeckel to visit. He and his household planned the occasion cautiously. Darwin warned his guest in advance not to expect any "lengthy" scientific conversations. He felt a preemptive move in this direction was probably necessary, since Haeckel's book indicated a garrulous propensity to run on. He invited Lubbock to join them for dinner to relieve any social pressure. He asked Hooker and Huxley to take a look at Haeckel first in London and save him the bother of bringing Haeckel up to date with contemporary natural history affairs. He wrote another letter to Lubbock, urgently asking whether perhaps Lubbock could take Haeckel away with him to High Elms for the night.

The visit hardly justified so many precautions. Even though Darwin could scarcely decipher the other man's heavy accent, and Haeckel shouted as if he were on a sailing ship, roaring against the winds and waves, their mutual interests carried them safely through. Haeckel treated Darwin as if he were a god.

> Tall and venerable . . . with the broad shoulders of an Atlas that bore a world of thought: a Jove-like forehead, as we see in Goethe, with a

lofty and broad vault, deeply furrowed by the plough of intellectual work. The tender and friendly eyes were overshadowed by the great roof of the prominent brows. The gentle mouth was framed in a long, silvery beard.[103]

Darwin reported that "I have seldom seen a more pleasant, cordial & frank man."[104] He and Emma called him "a great good-natured boy." Of all the young men Darwin met, this was the one who most behaved like a religious disciple.

Acolytes or no, Darwin's reputation mounted steadily. At the British Association meeting in Nottingham in August 1866, Hooker sang his praises in terms that reverberated through the community. The other Darwinists had stepped aside for this meeting, saying that it was Hooker's turn to push into the "ranks of the enemy." Hooker chose botanical geography as the subject of his address. "Would you believe it, I have in cold blood, accepted an invitation to deliver an evening address on the Darwinian theory at Nottingham. I am utterly disgusted with my bravado," he demurred.[105] He intended to air his views on island floras and the possibility of their former connections to continental landmasses, a topic over which he and Darwin had argued contentedly for years. "I think I know 'Origin' by heart in relation to the subject," he said.

In this address Hooker also provided a sardonic résumé of the larger battles fought over the last six years, particularly alluding to the Oxford British Association meeting of 1860. He created an elaborate metaphor in order to do this, "a parable" he called it, describing his Darwinian friends as if they were a band of missionaries who had encountered "savages" with primitive convictions. Six years ago, he said, the missionaries had attended a gathering of those savages.

> The missionaries attempted to teach them, amongst other matters, the true theory of the moon's motions, and at the first of the gatherings the subject was discussed by them. The presiding Satchem shook his head and spear. The priests . . . attacked the new doctrine, and with fury, their temples were ornamented with symbols of the old creed, and their religious chants and rites were worded and arranged in accordance with it. The medicine men, however, being divided among themselves (as medicine men are apt to be in all countries) some of them sided with the missionaries—many from spite to the priests, but a few, I could see, from conviction—and putting my trust in the latter, I never doubted what the upshot would be.
>
> Upwards of six years elapsed before I was again present at a similar gathering of these tribes; and I then found the presiding Satchem treating the missionaries theory of the moon's motions as an accepted fact, and the people applauding the new creed!

> Do you ask what tribes these were, and where their annual gatherings took place, and when? I will tell you. The first was in 1860, when the Derivative doctrine of species was first brought before the bar of a scientific assembly, and that the British Association at Oxford; and I need not tell those who heard *our* presiding Satchem's address last Wednesday evening that the last was at Nottingham.[106]

With a shimmer of understanding, the audience realised he was talking about them. They were the primitive tribe, at first resistant to new ideas, and now suitably enlightened. Fanny Wedgwood wrote to the family describing Hooker's unexpected artistry. "When the *Sachem* began, for a minute or two we were all mystified & then there came such bursts of applause from the audience first & going on—It was so thoroughly enjoyed amid roars of laughter & noise making a most brilliant conclusion. . . . how I longed for you to hear the noise they made when Charles' name was mentioned. 'Our illustrious countryman'—it made us feel so *grand*."[107]

## XI

Soon afterwards Darwin agreed to have his photograph taken for an album of biographical memoirs written by Edward Walford and published in 1866. He had already elsewhere supplied information to Carus for the first scholarly bibliography of his works, and one or two other authors had put articles about him in dictionaries and encyclopedias of men of the times. Agreeing to Walford's request marked the beginning of a different form of commemorative activity. His face and life were becoming interesting to the public beyond his contribution to biological science.

Darwin posed for the photograph in the studio of Ernest Edwards, afterwards a notable photographer in the United States. This portrait revealed the ravages of a long illness. He looked frail, less confident, more like an invalid than in previous photographs taken only four or five years earlier, a different man from the pictures familiar to his friends. He sat again for Edwards for another album published by Walford in 1868, this time called *Representative Men in Literature Science and Art,* in which the handle of his walking stick can just be glimpsed. In both of these photographs, he gave the awkward impression of not knowing what to do with his hands.

The results did not please. Darwin refrained from ordering any copies for his personal use, although Emma said that she thought Edwards's pictures were very "true to life." Perhaps Darwin had yet to learn to feel comfortable with the idea of pictorial fame.[108] To send personal photographs to friends and relations was one thing. To see oneself in a newspaper or published album was quite another. Unused to self-advertisement, and by nature modest, Darwin may have felt that photography for public consumption was somewhat vulgar unless required by position. Certainly it was not yet clear in those early years of celebrity culture that the public might want to know what he looked like, or that his image would eventually become an integral element in his scientific distinction.

This does not mean to say that he rejected the craze for photographic *cartes de visite* that took off in the late 1860s. On the contrary, he had his own *cartes* made by Maull and Fox, and then by other firms, and swopped them enthusiastically with naturalists. "One likes to have a picture in one's mind of any one about whom one is interested."[109] After visiting Down House, Haeckel sent him a collection of *cartes de visite* of German biologists, to which Darwin replied that he "liked to know what people looked like." Hooker, Dana, Quatrefages, Haeckel, Gray, Lyell, Huxley—he had pictures of them all. And they had him. Much of the friendly coherence of the Darwinian group, almost a family feeling at times, was encouraged by this exchange of photographs. For nineteenth-century communities coming to grips with the advancing technologies of images, the novelty of cheap portrait photography was immense.

Something of the same novelty also eased into the world of print. Although Darwin declined to join Wallace for a double photographic portrait, as Adolf Meyer "coolly proposes,"[110] he agreed to let Carus print his photograph as a frontispiece to the third German edition of the *Origin of Species* (1867), the first opportunity that people who were neither friends nor colleagues had of seeing his face. Previously mostly an empire of the mind, the world of learning was gradually becoming visual and personal. Portrait frontispieces—always a distinguished literary tradition—were giving way to photographic representations of living authors, each picture lending personal authority to his or her volume.[111]

Statues were different again. In 1867, Darwin agreed to Erasmus's suggestion that he should pose for a bust by Thomas Woolner. Erasmus assured him that he would make all the arrangements. This was to be a family piece, not for public display. Momentarily surprised when he heard the news, Hooker reminded Darwin that he had previously made an identical request that had been refused, and asked if he could have a copy made if this new proposal went ahead. "I thought that you had given up all idea about my bust," responded Darwin.

> You cannot be such an ass as to think of a marble bust.—I shall be proud to give you a cast, & surely that will do. The bust is making for Erasmus; & we are fighting here, for Emma votes for a marble

> copy & I maintain it is absurd, & plaister of paris just as good, or any good enough [*sic*].[112]

Because of Darwin's ill health, the sittings did not take place until nearly a year later, in November 1868. When at last artist and model met at Downe, each was pleasantly intrigued by the other. Darwin discovered that Woolner was a large, loose-limbed, muscular giant, a founder member of the Pre-Raphaelite brethren, now sculpting the intellectual and liberal heroes of the era. Woolner's bust of Tennyson, his portrait medallions, and his statues of Bacon and Moses showed his empathy for colossal subjects, and his capacity for conveying the dignity of thought. He had made a portrait medallion of George Warde Norman, one of Darwin's neighbours, a former director of the Bank of England.

And they found much to discuss. Darwin told him of his interest in human expressions, asking for Woolner's professional advice as an artist. When Woolner revealed that he had tried his hand at portrait-painting in Sydney, they exchanged stories about their travels, and Darwin confided a youthful, half-baked plan to emigrate to Australia and search for gold. It turned out that Woolner had painted Philip Parker King in Sydney, commander of the first *Beagle* expedition and father of Midshipman King who shared a cabin with Darwin.[113]

"I shd. have written long ago," Darwin apologised to Hooker,

> but I have been pestered with stupid letters, & am undergoing the purgatory of sitting for hours to Woolner, who, however, is wonderfully pleasant & lightens, as much as man can, the penance.—As far as I can judge he will make a fine Bust, & I tell my wife she will be proud of her old husband.[114]

It fell to Erasmus to convey the brotherly view. "I hope it won't be very hideous which is the most that I expect."[115]

*chapter*
8

# THE BURDEN OF HEREDITY

IN THE MIDST of all this ill health, Darwin devised a remarkable theory of inheritance. The achievement would have been noteworthy enough had he been well. For a sick man to have mustered the concentrated powers of thought that lay behind it was arresting. Much of the time he was literally battling with himself. In the end, he felt he understood where variations came from and how they were transmitted. He placed the idea at the heart of his *Variation of Animals and Plants Under Domestication,* published in 1868, systematically building up the text to support it. He called the theory "pangenesis."[1]

Pangenesis was important to him because it plugged the gap left in the *Origin of Species.* Darwin knew that all his arguments would remain worryingly incomplete until the essential foundation stones of inheritance were located. Huxley told him so; the reviewers of the *Origin* constantly highlighted the point; he came up against the same stumbling block himself time after time. Now that a number of people were seemingly prepared to accept the general thrust of evolutionary ideas, this last gap loomed increasingly large. For natural selection fully to succeed he needed to show that "organic beings . . . have varied largely and the variations have been inherited." Pangenesis was constructed to fill the breach. "It is the facts and views to be hereafter given which have convinced me of the truth of the theory," he wrote in the opening pages of the manuscript for *Variation.*

As Darwin explained it, pangenesis was the highly abstract notion that every tissue, cell, and living part of an organism produced minute, unseen gemmules (or what he sometimes called granules or germs) which carried inheritable characteristics and were transmitted to the offspring via the reproductive process. He was careful to specify that each part of an organism produced only information about itself. There were gemmules for hands and feet, not for whole organisms. Individual gemmules did not contain a complete microscopic blueprint for an entire creature in the way that Herbert Spencer or Carl von Nägeli described. When the gemmules from each parent mixed in the foetus they would produce a unique new individual.

In saying this, Darwin joined a long line of thinkers stretching back to Buffon, Maupertuis, and beyond who believed the visible structure of organisms rested on an invisible arrangement of elementary units and that these units somehow contained a memory that could be passed to the next generation—a "mould," Buffon had called it, more than a hundred years before. Since Buffon's day, the notion of cells had become current, and the organising functions of the cell nucleus were appreciated if not precisely understood. Yet Darwin was unusual in thinking in such realist terms, almost as if gemmules were letters in the postal system, individually conveying a specific packet of information. The ovules, spermatozoa, and pollen grains each "consist of a multitude of germs thrown off from each separate atom of the organism."[2]

Offspring would therefore receive a collection of parental gemmules, some from the mother, some from the father, which mingled and were expressed in different ways in the child. Some gemmules, he proposed, could remain dormant for many generations. Some lost their distinguishing characteristics. Others were routinely carried by all offspring but perhaps only one or another might be expressed, or a certain combination was needed to allow full representation. Some blended. Every child was therefore the fruit of its parents' loins—built up from a mixture of the parents' and grandparents' gemmules coming from either side. Using a metaphor close to his rural heart, Darwin likened this to gardening; a flowerbed could be sprinkled with seeds, "most of which soon germinate, some lie for a period dormant, whilst others perish." His awareness of the longlasting vitality of seeds—their fertility, superabundance, and longevity—gave him confidence that his ideas might be valid.

New as these ideas were to him, they rested in impressions he had been turning over in his mind for decades. All his life Darwin had puzzled over the phenomena of breeding, by which he meant not so much natural selection, although that usually entered his deliberations somewhere or another, but rather the whole complex of biological phenomena connected with reproduction, pedigree, sexual identity, fertility, mating patterns, embryology, and inheritance. Even before Malthus had illuminated his understanding of the natural world, he had been intrigued by animal and plant breeding, seeing it (like his grandfather Erasmus Darwin) as the underlying rationale for nature, the essential moving stimulus that drove each and every living being. "Why such high object generation?" he had asked himself when first opening his earliest notebook on the evolution of species. His answer sprang from the following page: "Generation to adapt & alter the race to changing world."[3] Sex governed all, as he had said in these secret notebooks. As soon as he then hit upon natural selection as a working concept, he plunged into investigations on fertilization, hybridity, inheritance, courting rituals, sexual choice, selfing, and crossing, all of which became highly relevant researches because his new theory turned reproduction into the mechanism responsible for both maintaining and varying living structures. For thirty years, his interest had never wavered. Although it is commonly said that Darwin aimed primarily to explain the meaning of adaptation in living beings, this underestimates the scope of his continuing programme. He was instead a lifelong "generation" theorist.[4] It could legitimately be claimed that the origin of species was not even his primary focus in life when compared with the dedicated attention he lavished on sexual and reproductive concerns. To understand breeding was a fundamental objective.

Darwin's assumptions about the society in which he lived inevitably played a part. These researches into reproduction had always carried a touch of autobiographical interest. On the one hand, he was more anxious than ever before about unmasking the facts of heredity. The possible transmission of ill health from parents to children was an issue of all-consuming worry. His own health was as bad as it ever had been, and the health of his children was constantly present in his thoughts. During this period, he wondered, not for the first time, whether he might be the source of their disorders, and whether his marriage to Emma, his first cousin, had played any significant part in this inherited burden.

And on the other, he lived in a world in which heredity was an obvious organising principle. The upper reaches of Victorian society were, after all, built on the notion of human pedigree and good breeding, not only in the sense that an individual's position in the existing social order depended to a large degree on birth, but also in the heightened emphasis then laid on manners and the cultivation of taste and intellect.[5] Family wealth was transmitted along lines of descent, marriages were contracted according to well-known social codes, and increasing numbers of diseases and disorders were recognised by doctors as hereditary "taints."

Darwin had every reason to muse on good and bad breeding among humans. His personal circle belonged to a close-knit stratum of society, the intellectual aristocracy of the high Victorian era, sympathetic to Mill's idea of a "learned elite" and Carlyle's "aristocracy of talent." Most members of this intellectual elite associated themselves with the rising ideologies of meritocracy, utilitarianism, and personal "character," a Smilesian sense of personal effort and determination under adversity, while for the most part enjoying inherited private incomes and status by birth. Darwin's position as a gentleman was secure. Expansively, he felt free to value gentlemanly qualities in others who might not have been born into favoured families like his own, regarded himself as an egalitarian, applauded merit and industry, promoted civic duty and progress, and appreciated the attributes of refined society. He felt no guilt about being an elitist, and yet he managed most of the time not to be too much of an obvious snob. He made room in his life for the conviction that effort, manners, intellect, and hard work could make a difference.

Assumptions like these could smoothly feed into ideas about family lines of inheritance. Any walk through his village community might reveal local families that had existed on the same spot for generations, the same names on the gravestones beside the church, the same fields, shops, and public houses running in families, the great houses like High Elms passed down from father to son, although these patterns were never quite as stable as pastoral nineteenth-century imagery might suggest. Everywhere he looked he might see sequences of marriage and heredity, whether dynastic unions between royal princes and princesses or the small-scale reinforcements of social identity that took place in his own circle. The family was the primary Victorian institution in which the meaning of individual lives and lines of descent were constructed and transmitted over the generations—the repository of personal history.[6] Embedded as he was in his own family connections, these notions surely seemed a natural basis for his researches into animal and plant inheritance.

And reproductive advice washed around him, not just in medical texts and household manuals but also in the romantic fiction of the period. It would not be going too far to suggest that during these years of ill health the sentimental novels read to him by his wife Emma put mating and heredity in the spotlight.[7] The love plot suddenly mattered, ranging from the tangled nets of courtship and marriage in Eliot's *Middlemarch* or *Daniel Deronda* to the darker writings of Thomas Hardy, who, although an author unread by Darwin, declared himself one of "the earliest acclaimers of the *Origin*." The novels that Darwin best liked to hear were liberally embroidered with squandered fortunes, obscure births, unusual lineages, misplaced inheritances, ancestral maladies, inappropriate marriages, and the transmission of family "temper." In effect, the biological ideas at the heart of evolutionary theory turned human genealogy into palpitating drama. Much of Darwin's interest in good and bad breeding and his sympathetic concern for the travails of pretty heroines consequently reflected the wider preoccupation of his class with marriage, position, manners, wealth, and the fear of congenital ill health. No wonder he insisted on a happy ending.

His attention to the transmission of wealth was significant too. While Victorian aristocrats criticised the law of entail that turned an eldest son into a merely temporary custodian of a landed estate, and country landowners squealed at the erosion of their patrimony, the newly affluent ranks of society claimed an increasing share of the country's financial and political power, sometimes through paper wealth alone. The old order was breaking down. Darwin's interest in the subject was profound, and it probably acquired vivid personal meaning through his own experiences as an investor and a financially supportive husband, brother, and father. He went to elaborate lengths to ensure financial security in the form of an inheritance for his loved ones. The circulation of capital, in this sense, and particularly the accumulation of interest on that capital, could easily serve as a metaphor for the inheritance of characteristics in living beings.[8]

Hereditary disease naturally bothered him too. While he was ill, Darwin lingered on the topic with morbid unease. His own afflictions supplied an alarmingly personalised case study, although his fears about the children having inherited his "wretched constitution" have to be taken in due proportion. Many families experienced at least some comparable congenital disabilities or fears about such. Bad stomachs, melancholia, early senility, hypochondria, stillborn babies, gout, mysterious "fits," and idiocy regularly appear in the medical records of the century. Even so, his children's disorders seemed to him like variants of his own. Perversely, no matter what the children suffered from, Darwin always thought he could identify something of his own failings in each and every illness. "When we hear it said that a man carries in his constitution the seeds of an inherited disease," he was to write in *Variation,* "there is much literal truth in the expression."[9]

Time after time, the topic of inbreeding arose. An undercurrent of worried self-interest ran through his researches into plants and animals, for he was never sure if reproduction between close relatives might inadvertently bequeath to the offspring a series of innate weaknesses, infertility, or a tendency towards disease ("diathesis" in the terminology of the period). In human affairs, consanguineous marriages were at that time prohibited by civil law, prohibitions that extended from blood relatives in the immediate line (fathers and daughters, for example) to relatives by marriage. Even though the prohibition did not include marriages between first cousins, as in the case of Darwin and his wife, Darwin knew there were sufficient medical warnings about such marriages to cause concern. He had carefully studied Alexander Walker's book on intermarriage when first hoping to propose to Emma, a book which addressed "the causes why beauty, health and intellect result from certain unions, and deformity, disease and insanity from others."[10]

Moreover, in England, continued newspaper agitation about repealing the laws prohibiting marriage with a deceased wife's sister rose and fell in the nation's consciousness. He wondered about the marriages between cousins that were so common in his personal circle. No fewer than three of his or Emma's siblings had married cousins, not counting himself, and whose offspring he had an opportunity to observe at Down House on a regular basis. Cousin marriages were commonplace in his wider sector of society, at times seemingly the preferred norm. In fact, Victorian notions about cousin marriages were nearly the opposite of those of the modern day. Given the restrictive social structure of the day and constraints on marriage, cousins were often the only members of the opposite sex whom young men or women could come to know with any intimacy. Such familiarity and shared social background were often regarded as the best possible start to a marriage, as Darwin's own experience showed. The advantages of cousin marriages for retaining land, money, or titles in the family were also well understood. So Darwin was caught in a dilemma that is not easily understandable in modern terms. He did not necessarily regard cousin marriages as a bad thing. Few people, other than a handful of medical authors or social commentators, worried about cousin marriages as a likely source of inbreeding or inherited defect, and even those warnings would have been perceived as mostly applicable to aristocracy or royalty, or to geographically isolated populations in areas of Britain like the Welsh hills or the East Anglian fens, where long-continued intermarriage over many generations was known to result in a number of recognisable congenital physical and mental disorders. Yet he could not help but be anxious about the possible medical consequences of his own actions for his children.

Plants and pangenesis were his salvation. The breeding experiments he performed at Down House during this period were designed in part to test this assumption and—perhaps unsurprisingly—helped him come to the conclusion that the deleterious effects from inbreeding were not nearly as pernicious as supposed.

To this end, he made crosses between individual plants that were closely related, especially back-crossing descendants to their parents, and putting brothers to sisters, techniques much used by Naudin and Gärtner; and he consulted animal breeders and other agriculturalists who knew about the same kind of mating patterns in pedigree animals among which it was important to conserve and emphasise particular bloodlines. In an era when foxhounds or racehorses could be more highly bred than members of the Houses of Parliament, he found plenty of experts to help him on the point. William Tegetmeier carried out similar breeding experiments for him on poultry, pigeons, and even a few turkeys. Darwin studied the results of European specialists, returning again and again to Gärtner's *Versuche und Beobachtungen,* as well as Nägeli's *Botanische Mittheilungen* (1866) and Naudin's *Nouvelles recherches sur l'hybridité dans les végétaux* (1863). All three books were in his library and carried on every page evidence of his attention.

The results were suggestive. He noted that inbreeding was primarily a feature of the domestic kingdom. Farmers would consolidate their stock-lines by putting father to daughter, brother to half sister. Continued inbreeding of this kind between blood relations certainly diminished the "constitutional vigour, size and fertility of the offspring; and occasionally leads to malformations but not necessarily to general deterioration of form or structure."[11] Under the intensive conditions of domestication, with assisted matings and supplementary feeding, even the weakest offspring could survive and reproduce.

> It is unfortunately too notorious that man and various domesticated animals endowed with a wretched constitution, and with a strong hereditary disposition to disease, if not actually ill, are fully capable of procreating their kind.

These situations would not generally arise in the wild. Taking advice from the anthropologist Edward Tylor, Darwin suggested that there must be in nature a form of incest taboo. If by chance there were such matings, natural selection would weed out any malformed offspring.

The surprise came with his realisation that matings between first cousins did not necessarily fall into these deleterious categories. His research into Victorian farming practice showed him that farmers considered cousins sufficiently far apart to present little risk to the viability of the offspring. As often as not these matings were regarded as good procedures for conserving, say, the prime features of champion bulls.

Pangenesis looked to him as if it might supply the answer. Darwin proposed that some limited effects from the environment might become embedded in an individual's constitution and thus be liable to be transmitted, via the gemmules, to the offspring. If two very closely related individuals, who had grown up under rather different external circumstances, were paired, these small differences would make each sufficiently distinct from the other to bear normal offspring.

> There is good reason to believe that by keeping the members of the same family in distinct bodies, especially if exposed to somewhat different conditions of life, and by occasionally crossing these families, the evil results may be much diminished, or quite eliminated.[12]

As he would acknowledge in *Variation*, the point was of "high interest, as bearing on mankind."

Even so, he remained uncertain. The philosophical difficulties and practical consequences of cousin marriages troubled him for years afterwards. There was no other theme in Darwin's science that more clearly reflected the personal origins of his intellectual achievement. He could scarcely have arrived at pangenesis without this attention to his marriage, his children's ill health, and his own sickness.

## II

Significantly, he also thought pangenesis resolved some of the more important criticisms brought up by reviewers of successive editions of the *Origin of Species*. While he was ill, and then while writing the book on *Variation*, he composed his answers.

One of these criticisms went to the nub of the matter. In 1867 the Scottish engineer Fleeming Jenkin pointed out in a long review of the *Origin of Species* in the *North British* that any individual variation, however favourable, could hardly maintain a secure foothold in a large, freely breeding population, at least not in sufficent numbers to produce a new species.[13] Jenkin was one of the most original thinkers of his generation, a friend of Robert Louis Stevenson, and greatly interested in population studies and political economy. Although it took Darwin a full year to discover who had written this review in the *North British*, he took the objection seriously and admitted, "Fleeming Jenkin has given me much trouble, but has been of more real use to me than any other essay or review."

At the centre of the review lay a concrete problem. Calling on archetypal Victorian racial assumptions, Jenkins gave the hypothetical case of a white-skinned sailor shipwrecked on an island inhabited only by black-skinned tribes. Free reproduction between "civilised" and "savage," said Jenkin, would not lead to the preservation and spread of the white man's characteristics. Quite the reverse. The European's supposed advantages would be blended or "swamped" in the larger population.

> Our shipwrecked hero would probably become king; he would kill a great many blacks in the struggle for existence; he would have a great many wives and children . . . but can anyone believe that the whole island will gradually acquire a white, or even a yellow, population, or that the islanders would acquire the energy, courage, ingenuity, patience, self-control, endurance, in virtue of which qualities our hero killed so many of their ancestors, and begot so many children?[14]

Taken aback, Darwin went to enormous lengths to explain in *Variation* that pangenesis would permit the preservation of some favourable variations in a population. Advantageous characteristics, he claimed, did not always disappear through blending.[15]

And pangenesis allowed Darwin to propose that the effects of use and disuse could sometimes be inherited, a feature of the living world he had always hesitated to claim in case he looked too much like the French naturalist Lamarck. Possibly his doubts about cousin marriages partly encouraged him to take this line. Still, Darwin's hesitation was very real. He hated to have his name linked with Lamarck's. For Lamarck's scheme always seemed to him excessively simplistic, with organisms striving to adapt to external conditions and then somehow transmitting these acquired characteristics to the next generation. To his mind, a large gap yawned between the "speculations" of late-eighteenth-century French thought and the steam-driven pragmatism of mid-nineteenth-century British science which he adopted. "I can see nothing in common between the *Origin* & Lamarck," he said. Lamarck's *Philosophie zoologique* was a "wretched book," one from which "I gained nothing."[16] Natural selection, as he viewed it, was a completely different theory, in which an organism's adaptation to its conditions of life was firmly based on statistics and the rules of chance.

Moreover, Darwin disliked the prospect of being linked with the atheistic, progressivist, loosely Lamarckian ideologies of the 1820s and 1830s, doctrines of thought that had haunted his first explorations of evolutionary theory and were now retrospectively associated in people's minds with the political change and upheavals of that period.[17] Although he was not generally a vain man, he was therefore vulnerable to jibes that his work was merely a restatement of Lamarck's. He defended his originality as far as good manners would allow, but he always corrected friends and reviewers if they remarked on the similarities. Time after time, Lyell innocently lumped the two together. Time after time, Darwin rebuked him, and muttered ungraciously to Hooker about Lyell's failings.

But Darwin now wanted to include in his scheme the possibility of the inheritance of some limited acquired changes. Pangenesis gave him the chance to be Lamarckian without any of Lamarck's inner strivings. As he put it, some aspects of the external environment could modify the inheritable gemmules.

> In variations caused by the direct action of changed conditions, of which several instances have been given, certain parts of the body are directly affected by the new conditions, and consequently throw off modified gemmules, which are transmitted to the offspring.[18]

No doubt the whole hypothesis of pangenesis was extremely complicated, he conceded. "But so are the facts."

## III

The awkward terminology did not help. So original in his outlook elsewhere, Darwin's mind would usually go blank when faced with inventing names. Even the relatively straightforward process of naming animals and plants stumped him. He had needed Hooker's help when manufacturing barnacle names in the 1850s, and he relied heavily on experts to identify, name, and classify his *Beagle* collections. In 1841, he had joined the British Association committee charged with establishing rules for scientific nomenclature that any "ignoramus" could follow without anxiety.[19] As a general rule, he believed in the magic of names. To name was to possess, as every contemporary naturalist recognised—a system of controlling diversity in order to comprehend it. The precision of knowledge could be reflected in the precision of names.

This time Darwin consulted his son George, reading mathematics at Trinity College Cambridge, asking George if he could find a classical scholar to provide an appropriate name for his inheritance theory. Darwin wished to convey his belief that the inheritable materials were offshoots of individual body cells. "Do you know any really good Classic who cd. suggest any Greek word expressing cell, & which cd. be united with genesis?" he inquired.[20]

George responded with a series of outlandish Greek terms straight from the armchairs of the junior common room, including "cyttarogenesis," a word-for-word translation of cell-genesis. "Atomo-genesis sounds rather better I think, but an atom is an object which cannot be divided," Darwin commented to Huxley in bemusement. "Perhaps I shall have to stick to Pan." At home, Emma thought "pangenesis" sounded wicked, "like pantheism." Darwin could not tell if she was serious. But he joked to Bates, "The great god Pan has been an immense relief to my mind." The hint of a pagan spirit of nature operating in a non-Christian world was evidently appealing.

Darker waters swirled. Isolated by sickness for so long from his friends and the hurly-burly of scientific life, he had time to dwell on morbid thoughts. He was convinced no one would pay proper attention to his ideas on inheritance. He called pangenesis his beloved child, his baby. He said his child would be stillborn, that he was the only one who loved it. The allusions were disturbingly intense. Sometimes his talk turned feverishly sub-religious in tone. "The poor infant Pangenesis will expire, unblessed & uncussed by the world, but I have faith in a future & better world for the poor dear child!" he told Bates excitedly. "My fear has always been that Pangenesis would be a still-born infant, over whom no one would rejoice or cry," he said to Lyell.[21]

If Darwin ever teetered on the brink of mental disarray, this was the time. He was locked into ill health, endlessly turning the same questions over in his mind, and the pangenesis hypothesis held hidden meanings. Dead or unloved children supplied an unnerving source of metaphor. He felt protective, apprehensive, defiant. His previous intellectual child—natural selection—had nearly died at the same time as his real baby, Charles Waring Darwin. How was this one to fare? The strength of his fatherly emotions blinded him. He looked indulgently on the theory's frailties, just as if it were Henrietta or Horace, sickening quietly in their rooms upstairs. Although his written work was dedicated to the idea of letting the weakest fall by the wayside, the case took a very different hue when it involved an infant of his own.

Furthermore, it had been a long time since he had constructed a completely new theory from scratch. Although he spun interpretative webs around everything he did, large or small, this project was on a higher plane, full of ambitious intentions. He wanted to do it properly. But he was slowed down by age and illness, by his correspondence, his need to keep up with the criticisms of the *Origin of Species*, his continuing writing and research. Stubbornly, he refused to acknowledge that he was tired, full of tensions, not completely free from the sickbed, capable only of a carefully regulated hour or two at full mental pitch. He could not recapture the buoyant inspiration of his early notebooks. He had not just stepped off the *Beagle* in Falmouth, young, alert, and eager.

Moreover, he laid psychological weight on the way pangenesis linked together what he regarded as otherwise inexplicable conundrums. Long ago, the same kind of explanatory embrace had helped convince him of the validity of natural selection. He had no reason to change his mind. He felt that if a single theory of inheritance could explain so much, it might well approximate to the truth. He admitted as much to the botanist George Bentham.

> To my mind the idea has been an immense relief, as I could not endure to keep so many large classes of facts all floating loose in my mind without some thread of connection to tie them together in a tangible method.[22]

To another, Darwin confided, "Pangenesis has very few friends, so let me beg you not to give it up lightly. It may be foolish parental affection, but it has thrown a flood of light on my mind in regard to a great series of complex phenomena."[23] He assured Asa Gray that "at the bottom of my own mind I think it contains a great truth." To Lyell he said that "it will be a somewhat important step in Biology." As time went by, he put on a bolder front. "I fear my dear Pang. will appear bosh to all you sceptics."[24]

As usual, Huxley grasped the essential point. "Genesis is difficult to believe, but Pangenesis is a deuced deal more difficult."

## IV

Darwin produced *The Variation of Animals and Plants Under Domestication* in two substantial volumes in January 1868, having delayed publication for William Dallas to complete the index. Henrietta helped her father correct the proofs, for he was weary with this big book, glad to see it go. In fact the mass of material was so unwieldy that Murray printed the specialist information in smaller type in order to reduce the number of pages, an ugly typological innovation that was not repeated. This time around, Darwin sent fifty presentation copies to the people who had helped him, including John Jenner Weir, Lubbock, Tegetmeier, John Scott, Fritz Müller, James Paget, his children William and Henrietta, and his nephew Edmund Langton, who had fed goldfish and trapped spiders on his behalf. One went to an unidentified colleague, personally inscribed with the words "With very kind regards from his friend and opponent the Author." Others were transported by steamship and camel train to India, Australia, South Africa, and New Zealand. By now Darwin was a global publishing phenomenon. When the second edition came out a few months later, he could afford to be more local in his distribution, giving one to his gardener, Henry Lettington, and another to Camilla Ludwig, the family's former governess, with whom Emma kept up a correspondence.[25] Camilla had translated some difficult German texts on inheritance for him.

Author and publisher also saw to it that *Variation* was issued in French, German, and Russian translations. The Russian one was prepared by Vladimir Kovalevsky, a palaeontologist and brother of Aleksandr Kovalevsky, the rising embryologist. The brothers both admired Darwin. Vladimir was so prompt in his work that his translation was published in St. Petersburg in November 1867, two months before the English first edition. His title showed that even at that late stage Darwin intended to publish in sequence the whole of his original long manuscript on natural selection—the Russian title of *Variation* reads, in translation, *On the Origin of Species, Section 1*.[26] Through this translation, Vladimir Kovalevsky and his wife, Sophie, the mathematician, became personal friends of the Darwin family. At the same time, Aleksandr pursued notably original research in embryology, arguing on Darwinian grounds that the great class of vertebrate organisms originated in the lowly sea-squirt, the transparent, sedentary ascidian.

Despite this apparent early enthusiasm, *Variation* was never popular, selling only five thousand copies in Darwin's lifetime. It was excessively detailed, complained the reviewers. Readers looked in vain for remarks about ape ancestry. Few noticed that Darwin described a number of anomalous human variations, including Julia Pastrana, the "gorilla-woman" known professionally as the "ugliest woman in the world," who had died some ten years beforehand. Wallace had assured Darwin that Julia Pastrana was no artificially constructed circus freak. She was a genuine curiosity, although neither Wallace nor Darwin was able to examine her, and in *Variation,* Darwin cited a third party's report. She possessed "a thick masculine beard and a hairy forehead; she was photographed, and her stuffed skin was exhibited as a show; but what concerns us is that she had in both the upper and lower jaw an irregular double set of teeth, one row being placed within the other. . . . From the redundancy of teeth her mouth projected, and her face had a gorilla-like appearance."[27] Julia Pastrana's career as an exhibit had already been artificially extended after death even without Darwin's discussion. She had married her manager and given birth to a stillborn son as unusual as herself. After her early death she was mummified, dressed, and displayed as an ape-woman by this manager all over Europe, making a bizarre personal contribution to the distaste that surrounded much of the debate on evolution. She reappeared in print among the gooseberries and chickens that were Darwin's primary focus.

Apologetically, Darwin gave Hooker some valuable advice. "Skip the *whole* of vol 1, except the last chapt (& that need only be skimmed) & skip largely in 2nd vol., & then you will say it is a very good book." To Huxley he merely breathed, "Oh Lord, what a blowing up I may receive!"[28]

As for pangenesis, the centrepiece of the argument, Darwin's colleagues were cool. Very few of them understood what he was driving at except Herbert Spencer, who had proposed something similar a few years earlier, and Francis Galton, whose interest in heredity was intense. Wallace adopted it and then dropped it. Much later, in 1880, the naturalist Hugo De Vries took up what he called Darwin's "pangenes" in order to investigate the material basis of heredity, describing his own theories as a modification of Darwin's ideas, and Darwin and De Vries corresponded in the years just before the older man's death.[29] Otherwise, Lyell prevaricated; Huxley quibbled; Henry Holland found it "very tough reading." Soon, Huxley warned Darwin that pangenesis duplicated the discredited work of the eighteenth-century naturalist Charles Bonnet, a disagreeable revelation also mentioned by a reviewer in the *Athenaeum,* who sneered that Darwin's pangenesis was all of a piece with Pouchet's devotion to spontaneous generation. Darwin thought (rightly) that such an unfriendly review could only be by Richard Owen.

One or two naturalists, including Lydia Becker, an intelligent botanist shortly to redirect her energies towards the suffrage movement, wrote to praise pangenesis. But he could sense his theory was not having the impact he desired. Hooker warned him no one would comprehend it, "not one naturalist in 100."[30] To cap it all, Clemence Royer accused him of stealing the scheme from Spencer, slipping the accusation into a new preface to her French *Origin of Species.*

Darwin hardly knew whether to laugh or weep. "I must enjoy myself & tell you about Mad[ll]. C. Royer who translated the *Origin* into French, & for whose 2d. Edit I took infinite trouble," he wrote to Hooker.

> She has now just brought out a 3d. Edit without informing me, so that all the corrections &c in the 4th & 5th English editions are lost. Besides her enormously long & blasphemous preface to 1st Edit, she has added a 2d Preface, abusing me like a pick-pocket for pangenesis, which of course has no relation to the *Origin*—Her motive being, I believe, because I did not employ her to translate "Domestic Animals."[31]

So he was grateful when Galton professed himself a convert. Galton's interest in heredity was already displayed in scientific circles. Three years earlier he had published in *Macmillan's Magazine* a pair of papers on "hereditary talent," arguing that intellectual ability ran in families and employing techniques derived from the Belgian statistical pioneer Adolphe Quetelet to support his argument. These led shortly afterwards to his full-length study *Hereditary Genius* (1869), and culminated in his coining the term "eugenics" and initiating significant developments in the study of heredity and early genetics.

Up to this point the relationship between the two cousins, Darwin and Galton, had never been particularly intimate, although they shared Dr. Erasmus Darwin as a grandfather, tracing their descent from Erasmus Darwin's first and second wives, Mary Howard and Elizabeth Pole respectively. When they were boys, the age gap of thirteen years had been slightly too large for easy comradeship, although Darwin remembered visiting Galton's father, his uncle, once or twice in Birmingham. The youngest of that branch of the family, Galton was a Cambridge graduate and traveller like Darwin. His *Art of Travel* was a characteristically flat-footed account of an expedition to southern Africa, full of tips for campsites and how to wield a machete. This sold well and in 1867 went into a fourth edition. But for some reason Galton was unable to penetrate the inner circles of British science, despite contributing to the Royal Geographical Society and dancing attendance on Huxley, Spottiswood, Tyndall, Spencer, and the rest by editing the *Reader* for a while. He may have hoped to be invited to join the X Club.[32] His views were as starkly naturalistic as the X Clubbers, and he made a point, as they did, of attacking the conservative doctrines of the day. Galton became prominent in December 1871 by publicly criticising the Day of Intercession—a national day of prayer—imposed by the government to give thanks for the restoration of young Prince Albert's health after a dangerous attack of typhoid. Galton questioned the efficacy of such intercesssions, laughing at those who believed in prayer as a remedy for cattle plagues, cholera, or excessive rainfall, and half seriously called for proper statistical tests in which the recovery rates of hospital invalids would be measured, some with prayers and some without.[33]

Yet he was never really part of the Darwinian club as the others were. Perhaps his pedantic manner made for shaky personal relations.[34] Either way, Darwin kept up contact pleasantly. For his part, Galton declared that his cousin's book marked an "epoch in my own mental development." He "devoured" its contents, saying that he assimilated Darwin's arguments as fast as he could read them.[35] He liked to think that he and Darwin shared the same hereditary bent of mind—a proposal that probably seemed more convincing to him than it would have to Darwin.

While Galton's views were generally more sophisticated than usually assumed, there could be no denying the zeal with which he defended the idea that talent was an innate, inheritable quality unaffected by social considerations. Ability "clings" to families, he asserted. Intelligence and ability were biological traits, the product of nature not nurture. In his opening salvos in *Macmillan's Magazine* he listed the genealogy of English judges since the Reformation, a theme that Darwin, with his penchant for reading about Lord Chancellors and famous legal trials, found absorbing. Judges, said Galton, were usually descendants of other judges or other eminent figures in the ancestral tree.

These views were developed in *Hereditary Genius,* a book that can almost be read as a collective autobiography of the masculine Victorian elite. There, in Galton's pages, nearly every member of the British intelligentsia could find data that indicated that his ability ran in his blood and, moreover, was inherited through the male side (although Galton conceded that an able mother could tilt the balance). In truth, Galton assembled some striking materials. In the section dedicated to "Men of Science" many of the people whom Darwin knew, or had read, turned up in print, the family lines of de Candolle, Herschel, Hooker, Humboldt, Geoffroy St. Hilaire, and de Saussure, among others. Darwin was in the book himself, along with his grandfather Erasmus Darwin, Robert Waring Darwin, and a little-known botanical great-uncle, another Robert Waring Darwin (1724–1816). Somewhat disingenuously, Galton did not include his own name in the Darwin entry, saying only that "I could add the names of others of the family who in a lesser but yet decided degree have shown a taste for subjects of natural history."[36]

In all this Galton seemingly ignored, or was not able to conceive, the effects of a hierarchically distributed society based on the advantages of education and wealth, or a culture in which professional openings were customarily purchased or passed on from generation to generation. Galton likened brains to the possession of sporting talent. Training could do only so much. Beyond that point innate ability was needed for high achievement.

However, the book caught Darwin's eye. "I do not think I ever in my whole life read anything more interesting and original," he admitted. He, like Galton, believed that primogeniture was pernicious, both socially and biologically. Like Galton, he was interested in heredity and the biological basis for success. "You have made a convert of an opponent in one sense, for I have always maintained that excepting fools, men did not differ much in intellect, only in zeal and hard work." He moderated this praise by adding, "I still think this is an *eminently* important difference."[37]

Already a supporter of Darwin, Galton read *Variation* with mounting enthusiasm. Pangenesis particularly caught his attention. Never having shown any interest in practical natural history observations before, he arranged for a series of experiments to be made on rabbits housed in the Zoological Gardens of London with the intention of demonstrating the transmission of gemmules. At Galton's request, a curator injected blood from one rabbit into another in the hope that gemmules would be artificially transported from one individual to another, thence to reappear in the next generation. Coat colour would serve as a marker. Blood drawn from the common black-and-white type was injected into a silver-grey, which was then allowed to breed with another silver-grey. Galton expected to see white or black patches emerge in the coats of the offspring. He waited like an anxious father to be called to the birth. Soon finding his paternal visits to the zoo an inconvenience, he asked for the cages to be moved into the garden of his Kensington home.

The cousins held their breath. "F Galton said he was quite sick with anxiety till the rabbits accouchements were over & now one naughty creature eat up her infants & the other has perfectly commonplace ones," reported Emma. "He wishes this expt to be kept quite secret as he means to go on & he thinks he shall be so laughed at."[38]

At last the message came. "Good rabbit news!" Galton cried after a particularly long sequence of injections. Two silver-greys had produced an infant marked with white.

> One of the latest litters has a white forefoot. It was born April 23rd but as we do not disturb the young, the forefoot was not observed till to-day. The little things had huddled together showing their backs & heads and the foot was never suspected. The mother was injected from a grey and white and the father from a black and white. This, recollect, is from a transfusion of only ⅛th part of alien blood in each parent; now, after many unsuccessful experiments, I have greatly improved the method of operation and am beginning on the other youngsters of my stock. Yesterday I operated on 2 who are doing well to-day & who have ⅓rd alien blood in their veins. On Saturday I hope for still greater success, and shall go on at any waste of rabbit life until I get at least ½ alien blood. The experiment is not fair to Pangenesis until I do.[39]

Despite his initial joy, Galton ultimately discovered that such colour variations were common in rabbits. Not a single instance of induced variation occurred in a total of eighty-eight offspring from transfused parents. The gemmules did not work in the way that he thought Darwin described them: "My experiments show that they are not independent residents in the blood." Unsuccessful he may have been, but Galton here firmed up the line of thought that took him towards his influential "ancestral law of heredity" in which he proposed that the inheritable material was passed on to offspring in due proportion from previous generations.[40]

Galton was troubled because he began the work in good faith, intending to prove Darwin right; and he praised pangenesis in *Hereditary Genius* in 1869. Somehow he had unintentionally proved Darwin wrong. Cautiously, he criticised his cousin's theory, although qualifying his remarks by saying that Darwin's gemmules (he called them "pangenes") might be only temporary inhabitants of the blood and that his experiments could have failed to pick them up.

Naturally enough, Darwin wanted Galton to keep these unsatisfactory results to himself. Yet Galton went ahead and published them in *Nature* in 1871, followed by another article in the *Proceedings of the Royal Society*. Darwin objected to having his theory's shortcomings advertised in this fashion among his scientific friends. He published a rebuttal in which he maintained he had not said anything about gemmules being in the blood. Galton was surprised to receive so curt a response. Offended, he backed down, claiming he was acting only as "a loyal member of the flock." In the end, Darwin also backed down. He modified his wording in later editions of *Variation,* admitting in a footnote that he would have expected to find gemmules in the blood, although their presence there was not absolutely necessary to his hypothesis. Pangenesis suddenly seemed much harder to establish than either man anticipated. Only the rabbits benefitted. Darwin agreed to rehouse them at Down House, sending his man Mark up to London to collect them. "The rabbits arrived safe last night & are lively & pretty this morning," he reported.

The setback did not prevent him encouraging others. In 1870 he thanked E. Ray Lankester for saying a few kind words about pangenesis: "I was pleased to see you refer to my much despised child, 'Pangenesis,' who I think will some day, under some better nurse, turn out a fine stripling."[41] And he congratulated John Tyndall for commenting on it in a presidential address to the British Association. "You are a rash man to say a word for Pangenesis, for it has hardly a friend amongst naturalists, yet after long pondering (how true your remarks are on pondering) I feel a deep conviction that Pangenesis will some day be generally accepted."[42]

In the main, *Variation* was noted by contemporaries for its densely packed accounts of horticultural and agricultural practice, its attempt to classify types of variation, and Darwin's useful rounding-up of historic material about early breeds.

One clarification was helpful. In the closing pages Darwin provided an explanation of the crucial difference between variation and selection. In those days, it was not always apparent that these were distinct processes. Indeed, perhaps only Darwin, with the advantage of many years of thinking about the distinction, and a handful of experienced French and German experimentalists were able easily to separate them.[43] Investigators more usually felt there was some form of inbuilt direction in the process of variation—they put back into the evolutionary process the purpose or even the divine guidance that Darwin removed. Asa Gray, in particular, advocated this view, both in print and in private, one result of his conversations with Darwin about design in orchids. In *Variation* Darwin deliberately contradicted Gray's opinion.

> If an architect were to rear a noble and commodious edifice, without the use of cut stone, by selecting from the fragments at the base of a precipice wedge-shaped stones for his arches, elongated stones for his lintels, and flat stones for his roof, we should admire his skill and regard him as the paramount power. Now, the fragments of stones, though indispensable for the architect, bear to the edifice built by him the same relation which the fluctuating variations of each organic being bear to the varied and admirable structures ultimately acquired by its modified descendants. . . . Can it reasonably be maintained that the Creator intentionally ordered, if we use the words in any ordinary sense, that certain fragments of rock should assume certain shapes so that the builder might erect his edifice?[44]

Although the architect or builder would always choose the best stones for building a house, the shapes of the stones themselves were completely random—or rather, their geological production was not related in any causal way to the architect's intention.[45] The final configuration of the house derived only from the architect's ability to utilise local resources, not from any innate adaptive power of the rocks. "However much we may wish it, we can hardly follow Professor Asa Gray in his belief."

## V

Successful or no, his release from *Variation* signalled a marked change in family activities. Darwin felt "very well," as well as he ever did. And after such a long haul at home alone with him and his illnesses, Emma and the girls longed for "dissipation." Leaving Darwin behind in his greenhouse, they took the train up to London and the theatre. Francis and George, both studying at Cambridge, joined them for entertainment; so did Erasmus and the sociable Wedgwood cousins. Soon, Darwin began venturing out for lunch parties in neighbouring country houses, amenable to sitting in the drawing room with the ladies, especially if these included young Lady Lubbock. "Papa adored her as usual," observed Emma after one such lunch.

For the older Darwin offspring, independence started to flower. During this period, Henrietta was to make a short continental tour with her younger sister Bessy, and a longer one in 1870, starting off with her cousin

Edmund Langton and his bride, Lena Massingberd, waiting in Paris for her brother George to come and collect her for the last leg home, although, as Emma said, "Parslow seems quite ready for a trip to France as far as Calais or Boulogne if you like to meet there in preference to this side of the channel." Emma wrote almost daily with advice about clothes, sights to see, and "Papa's health." Bessy went to Germany on her own in 1866.

The names they called each other changed, too. In general, the family was much given to nicknames. No one ever really knew why the children were given their particular baptismal names, apart from William, whose first name ran in the Darwin family. Francis used to joke that "our parents lost their presence of mind at the font and gave us names for which there was neither the excuse of tradition nor of preference on their own part."[46] There seemed little rationale behind the nicknames either. Even so, it had been a long time since George was called Jingo or Leonard answered to Pouter. On her 1870 tour abroad, Henrietta changed her name to Harriot [*sic*], an adjustment that Emma suppressed as soon as possible. Emma began to call Henrietta "Body," perhaps an allusion to the ever-present illnesses; Lizzie opted for Bessy; Darwin moved from "Etty" to "Hen"; and Henrietta organised her brothers and sister to abandon "Papa" in favour of "F.," an abbreviation of Father. Darwin did not like this. When she informed him of the forthcoming change he retorted, "I would as soon be called Dog."[47]

Far away from parental supervision, William Darwin's mind ran on equally independent lines. He wrote home to say he was thinking of getting married. Emma was thoroughly startled: "Do not marry for marrying sake," she exclaimed. "Look at Uncle Frank for a warning." She meant look at Frank Wedgwood's wife, who was idle, dissipated, and expensive. By return of post she advised William to choose a wife only from among family friends. In her eyes a successful marriage depended on the families knowing each other beforehand, no doubt a reflection of her own marital circumstances and other relationships in the close-knit Darwin-Wedgwood clan. She told him she remembered Darwin making a list of the disadvantages and advantages of marriage before he proposed to her, and said she would turn this up when William next made a visit home.[48] William bowed gracefully to maternal pressure and did not venture into matrimony for ten more years. Then he pleased his parents by chosing Sara Sedgwick, the sister of a close friend of the family.

George's success in the final examinations at Cambridge gave Darwin much the same tingle of parental surprise. Francis afterwards said that his father appeared—at least to them—as if he sometimes doubted whether any son of his could truly succeed, a dispiriting state of affairs that propelled the boys into proving themselves over and over again. "We used to laugh at him, and say he would not believe in his sons, because, for instance, he would be a little doubtful about their taking some bit of work for which he did not feel sure that they had knowledge enough."[49] George in particular had made a slow start as an undergraduate. He tried unsuccessfully for an entrance scholarship at St. John's College, Cambridge in 1863. He joined St. John's as an ordinary undergraduate but left shortly afterwards to start again at Trinity, failing the scholarship examination for that college in 1864. He eventually achieved a Foundation scholarship at Trinity in 1866, but "did not display any of that colossal power of work and taking infinite trouble that characterised him afterwards," said a mathematical friend, Lord Moulton.[50] Francis Darwin also failed the scholarship examination at Trinity College when he tried in 1869 (Francis may have sat for the new natural science scholarship at Trinity begun in 1867), and he too entered as an ordinary undergraduate. These ambitious targets and subsequent failures cannot have helped the boys' sense of themselves. Although Darwin must at some level have endorsed the decision for the boys' attempts at scholarships, he seems to have thought of them as very like himself at the same age, characterised by application rather than brilliance. Over the years he became uncertain about their aptitude for any of the traditional occupations of well-connected Victorians, and he convinced himself that each boy's trust fund would be insufficient to allow independent financial existence. The boys would have to get jobs. "Papa is reading a book upon the choice of profession," Emma had remarked at an early point, "which makes him very low as it appears quite impossible to get on in any."[51]

So he was completely delighted when George sent news that he was second in the final mathematical honours list at university, "Second Wrangler" as the Cambridge terminology put it, one of the highest university achievements that a young man could then obtain. Darwin had no inkling that George was mathematically able. To him, it seemed like a transformation. The boy who used to spend his time drawing knights in armour and hunting up the family genealogy had emerged like a butterfly out of a chrysalis.

> I am so pleased. I congratulate you with all my heart and soul. I always said from your early days that such energy, perseverance and talent as yours would be sure to succeed; but I never expected such brilliant success as this. Again and again I congratulate you. But you

have made my hand tremble so I can hardly write. The telegraph came here at eleven. We have written to W. and the boys. God bless you my dear old fellow—may your life so continue.[52]

George's success reverberated through the Darwins' small world. At George's old school at Clapham the mathematics master took the honour as a personal triumph and awarded the boys a half day's holiday. Leonard and Horace, in the top two forms, were heroes of the hour. In London, Francis Galton "crowed v. much because of his theory."[53] And at Down House, congratulatory letters from Darwin's scientific friends poured in, almost as if it were Darwin himself who had excelled all expectations: these letters were much more complimentary, more jubilant, more admiring, than any of the letters pangenesis inspired. Male congratulated male. George's award was precisely the kind of achievement that these men of science understood and valued highly. Huxley inquired which son was it, remembering only a gang of similarly aged boys from his visits. "It is the herald," Darwin replied contentedly.

Thereafter George took a suitably academic line. He was elected to a junior fellowship at Trinity College and contemplated reading for the bar, until his health, previously robust, deteriorated rapidly. At that point he gave up the law and opted to stay at Trinity as a mathematician for the rest of his life.

Then Leonard came second in the entrance examination for Woolwich College, the training school for military engineers. "By Jove how well his perseverance and energy have been rewarded," said his father. Leonard was usually regarded as the duffer of the family. Amused, George dropped him a note. "Bless your soul we're always second we are."

Warming to the renewed sociability, Darwin, Emma, and the girls spent March 1868 in London at Elizabeth Wedgwood's house in Chester Place, dining with Fanny and Hensleigh Wedgwood and Erasmus nearby. They made some new friends during this visit, notably Thomas Henry Farrer (later Lord Farrer) and his wife, Frances, who was a distant relative of Fanny Wedgwood's. The men were united as much by appreciation of Fanny's singing as by their mutual interest in natural selection. Thomas Farrer was a lawyer and high-ranking civil servant at the Marine Board of Trade, a good amateur botanist, with a range of hothouses in the country, and he had many questions to ask Darwin about plants. The two corresponded regularly after this meeting. "What a capital observer you are," said Darwin with pleasure. "A first rate naturalist has been sacrificed, or partly sacrificed, to public life." They became intimate friends in 1873 when Farrer, whose wife died in 1870, married Hensleigh's daughter Effie, another accomplished singer. With webs of relationships like these, Darwin's science was increasingly becoming an extension of his domestic circle.

They also met Frances Power Cobbe, the social reformer and antivivisectionist. "Miss Cobbe was very agreeable," said Emma. "She is very fresh and natural."[54] Eccentric, well-read, sentimental about dogs, loud, and overpowering, she was a type of English countrywoman very familiar to Darwin and members of his family. She persuaded Darwin to read Kant, telling him that he would find much to reflect on in his writings, and asking him bluntly, "are you never going to unite your lines of thought & let us see how metaphysics & physics form one great philosophy?" She was encouragingly avant-garde over evolution but less willing to contemplate an animal foundation for human morals. The human idea of justice is all our own, she insisted. She had written an intelligent commentary on Kant's ethics and could speak perceptively about the problem of how humans come by a conscience. Darwin thought he was not up to the task of reading Kant. However, he read, at her urging, Kant's *Metaphysic of Ethics,* in an early-nineteenth-century translation.

> It has interested me much to see how differently two men may look at the same points, though I fully feel how presumptuous it sounds to put myself even for a moment in the same bracket with Kant;—the one man a great philosopher looking exclusively into his own mind, the other a degraded wretch looking from the outside thro' apes & savages at the moral sense of mankind.[55]

By now Darwin was reluctant to grapple with any of the great European thinkers unless he was chivvied into it or persuaded that he would find something directly useful for his work, much preferring to hear about leading philosophical systems in colorful synopsis from Huxley. During March and April 1869, for instance, he heard a great deal about the finer points of Comte's doctrines. Huxley had launched a violent attack on Comte's positivism in what was to become one of his most famous essays, "On the Physical Basis of Life." Positivism was nothing more than "Catholicism *minus* Christianity," as he derisively called it.[56] The smear so annoyed Vernon Lushington, a leading British Comtean and one of Darwin's London acquaintances, that he sent Darwin an irate letter to pass on to Huxley. Lushington sardonically suggested that Huxley should perhaps read Comte before criticising him quite so roundly. Terse letters were exchanged between the three of them until Huxley had the gall to tell Darwin that he disliked these sorts of fights. "I begin to understand your sufferings over the 'Origin'—a good book is comparable to a piece of meat

& fools are like flies who swarm to it, each for the purpose of depositing & hatching his own particular maggot of an idea."[57] He followed this high-minded sentiment with a sketch of himself as an angry dog bristling at another. "You *must* read Huxley v. Comte," Darwin afterwards exclaimed to Hooker. "He never wrote anything so clever before, & has smashed everybody right & left in grand style. I had a vague wish to read Comte & so had George, but he [Huxley] has entirely cured us of any such vain wish."[58]

Even so, Darwin noted Kant's views on morality and duty, and tucked the comments away for use in his work on mankind.

More to Darwin's liking, he and Frances Power Cobbe discussed the mental powers of dogs, especially their household pets. She was pleased by his affection for the little dog Polly that accompanied the family everywhere, which afterwards struck her as Darwin's only redeeming feature. And he enjoyed Cobbe's self-mocking air. "Though I attended on Saturday a most successful Woman's Rights Meeting," she was to declare in 1870, "I am of opinion that our ancient privilege of talking nonsense even to those we most deeply honour,—is one not to be parted with on any terms!"

After this gregarious London visit, Elizabeth Wedgwood moved to Downe in order to be closer to Emma; they were the last two remaining Wedgwood sisters. There, in Tromer Lodge, she lived out her days, nearly blind, until her death in 1880. Henrietta Darwin remembered her regularly tottering into the drawing room at Down House, followed by her dog, demanding, "Where is Emma?" Parslow did not bother to announce her any more, and Emma fitted up a bedroom so that she could stay the night at any time. The Darwin girls spent hours reading to her aloud. Emma's sisterly devotion sometimes proved a burden to the younger members of the family.

The socialising continued apace, although in June and July, Darwin's health deteriorated again. "Unwell . . . did hardly anything," he noted. In July 1868 he took the family on holiday to the Isle of Wight, renting a house in Freshwater Bay owned by the photographer Julia Margaret Cameron.[59]

The holiday turned out one to remember.[60] Erasmus went too, eager to experience the bohemian artistic resort that Julia Cameron and Alfred Tennyson had between them created; Joseph Hooker rented a hotel room for a week or two nearby, seeking peace and quiet to write a speech for the next meeting of the British Association; and a mixed handful of nieces and nephews arrived for a day or two every so often. Darwin liked Mrs. Cameron, who was exuberant and droll, a gust of fresh air in his sedate existence. She swept them off hither and thither, plunging them into the social whirl associated with her home in Freshwater.

It was hardly a secluded rest. During their six weeks' stay, Darwin visited Tennyson, who lived in Faringford House, not far from the Camerons, talked with Henry Wadsworth Longfellow and his brother-in-law the poet Thomas Appleton, both of whom were visiting Tennyson, and received courtesy calls from each in return, as well as meeting Julia Cameron and her husband on a daily basis. Darwin enjoyed his discussion with Longfellow, which soon turned to Harvard University, where Longfellow was professor of poetry. Darwin praised Gray and Agassiz. "What a set of men you have in Cambridge!" he said to Longfellow. "Both our universities put together cannot furnish the like. Why there is Agassiz—he counts for three!"[61]

Mrs. Cameron took a fancy to Horace Darwin (aged seventeen), often asking him to Dimbola Lodge to help pack and unpack photographs, and on one occasion getting him to pose in one of her allegorical photographs. The Darwin entourage privately smiled at her effusive manner, so different from the self-contained Down House ethos, her flapping silk robes, untidy hair, and capable, dirty hands, stained blue with photographer's inks. She did all her photographic work herself, rushing from the "glass house" that was her studio to the cellar that acted as a darkroom. The menfolk had never met anyone quite like her before. Erasmus was captivated and willingly sat for a photograph.[62] The holiday ended in a transport of mutual affection, with Erasmus calling over the banisters to her, "You have left eight persons deeply in love with you."[63]

It was Julia Cameron who arranged that Darwin should meet Tennyson. She was well aware that an encounter between these flagships of Victorian culture was highly appropriate, each in their own way delving into shadowy realms of thought and transformation, their writings distinguished for the grandeur of their intellects, and each drawing close to the pinnacle of Victorian celebrity. To bring such intellectual lions together, she thought, would be a marvel of social ingenuity. Certainly Mrs. Cameron was convinced that only she could have managed it.[64] But like many other strenuously engineered occasions, the fact of the meeting was more significant than the occasion itself.

Although the two men felt courteously interested in each other, they found it difficult to exchange pleasantries and even harder to engage in real talk. Emily Tennyson made a note in her diary:

> 17 Aug. Faringford. Mr. D called and seemed to be very kindly, unworldly, and agreeable. A. said to him, "Your theory of evolution

does not make against Christianity," and D. answered "No, certainly not."[65]

Emma Darwin, on the other hand, was as excited as any teenage girl. To see in person the poet who fired her imagination was a thrill she remembered for a long while. Merely to be in the same room making inconsequential conversation with Emily Tennyson etched itself into her memory. Tennyson gave her a glass of wine and showed her about the house, flirting gently with Julia Cameron. Ever after, Emma's elderly relatives referred to her "beloved Tennyson." Aunt Fanny Allen mocked him as a second-rate poet, good enough only to write such things as *Locksley Hall.*

This exciting moment had another effect on Emma. She was impressed by her husband. She discovered that she liked to see him being admired for his insights into nature. Even though her loyalty to him had never wavered, this meeting with Tennyson placed her regard for him on a new footing. First the Prince of Wales, and now the Poet Laureate. She felt proud of him. His secularised science, the one thing that might possibly have come between them, was never to be a serious obstacle to their relationship.

The Irish poet William Allingham was visiting Freshwater Bay at the same time. More cynical than most of the people who met Darwin, he described Darwin as "yellow, sickly, very quiet." He noted Darwin's physical frailty and Emma's attentiveness. Disinclined to bow his knee at the shrine of natural selection, he implied that he considered Darwin selfish and self-indulgent. "He has his meals at his own times, sees people or not as he chooses, has invalid's privileges in full, a great help to a studious man. . . . Has been himself called The Missing Link."[66] In his diary he recorded that Emily Tennyson "dislikes Darwin's theory." He and Tennyson afterwards wandered in the garden talking of Christianity. "What I want," declared Tennyson impatiently, "is an assurance of immortality."

During this stay on the Isle of Wight, Mrs. Cameron fell on Darwin as a photographic subject. She had a sharp eye for a passing celebrity, welcoming the cream of Victorian society on their way to see Tennyson, and persuading the most important visitors to sit for a portrait. No matter who they were, they usually found it impossible to refuse. "She thinks it is a great honour to be done by her," said one exhausted guest. "Sitting to her was a serious affair, not to [be] entered lightly upon. . . . she expected much from her sitters."[67] Yet the sessions were an absorbing experience, and many notable figures afterwards said that they were flattered to sit.

Moreover, the results possessed lasting impact. Not yet at the height of her fame, she depicted the giants of Victorian intellectual life as they had never been seen before, creating an indefinable aura that contributed materially to their widening public stature and added to the general preoccupation with national heroes permeating Victorian culture. She subscribed to the "men of genius" school of thought, and her studies of nineteenth-century thinkers like Tennyson, Robert Browning, and John Herschel showed an intense appreciation of manly intellect. She loved to bring out the power of thought in a celebrated man's face; and her greatest talent probably rested in these monumental male biographical studies. Her photographic tableaux were quite the opposite. For these she dressed her sitters in costume and posed them for historical or allegorical purpose, such as "Alethea," or "The gardener's daughter." Housemaids, fishermen, and miscellaneous young visitors like Horace Darwin decked themselves in crowns and robes to depict the death of King Arthur, an illustration for Tennyson's *Idylls of the King*. These sentimental *tableaux vivants* were ridiculed in Cameron's own day. Furthermore, she gave full rein to her special trick of fuzziness: "very daring in style," said the *Photographic News*, reviewing her first exhibition in 1864; "out of focus," complained the *British Journal of Photography*.[68] Such images were in marked contrast to her other more rugged pictures of eminent men.

She took three portrait photographs of Darwin. Two of these show Darwin looking slightly apprehensive, perhaps because Cameron's photographs required ten or fifteen minutes' exposure.[69] In the third and best-known photograph, Darwin sat three-quarter profile, serene and thoughtful. Mrs. Cameron usually posed her male sitters carefully. She made John Herschel fluff up his white hair for effect and draped Robert Browning in a velvet cloak.[70] In her picture of Joseph Hooker, probably taken during this same visit to Freshwater, he leans forward, wrapped in thought. By contrast, Darwin's costume was evidently his own and mostly betrayed his precautions against the cold, especially on holiday on the British south coast in July.

On the whole, Darwin's dress was but a minor part of the composition. Through her use of ceiling light she ensured there was almost nothing in the photograph except Darwin's massive forehead, top-lit to emphasize the vast dome of his skull, the brow creased in thought, eyes sunk under his enormous eyebrows, and his huge, scholarly beard. As in classical statuary, the effect was of a wise and venerable figure. More than anyone else Mrs. Cameron created the visual image of Darwin as a powerful abstract mind.

The beard made the difference. Over the preceding years of ill health, Darwin had let it grow unencumbered, for convenience as much as for anything else. Now it reached to his chest, thick, grey, and luxurious, a reassuringly masculine counterpart to his receding hairline, a traditional sign of male maturity and wisdom, symbolic of all the qualities that Victorians were coming to value in their leaders, both patriarchal and patristic. When Darwin distributed an earlier photograph (taken by William) of himself with this beard, Hooker replied immediately, "Glorified friend! Your photograph tells me where Herbert got his Moses for the fresco in the House of Lords—horns & halo & all. . . . Do pray send me one for Thwaites, who will be enchanted with it. Oliver is calling out too for one."[71] The Darwin boys agreed that their father looked like Moses. Asa Gray echoed the general chorus. "Your photograph with the venerable beard gives the look of your having suffered, and perhaps, from the beard, of having grown older. I hope there is still much work in you—but take it quietly and gently!"[72]

It was a philosopher's beard, as Mrs. Cameron, Hooker, and Gray implied, with strongly religious overtones. Darwin was delighted by this maverick idea: "Do I not look reverent?" he asked his relatives. For himself, he hardly gave the possible symbolism a second thought.[73] He liked having a beard, and he intended keeping it. Still, at some level it probably served as a badge of his intellectual vocation and his increasing age and status. A beard like this symbolized his arrival at the summit of Victorian masculine existence. It also indicated that he was no fresh-faced radical, no dangerously well-groomed Frenchman.[74] The beard moreover disguised his expression, helping him keep his thoughts private and allowing him, if he wished, to be a sage or a prophet.[75] A beard like Darwin's was a visual symbol of the real seat of Victorian power, and one of the most obvious outward manifestations of what Darwin would soon be describing as factors involved in sexual selection among humans.[76] Intuitively, Julia Cameron captured these elements of Darwin's emerging public persona far more eloquently than any other photographer of the middle years of the century.

Underneath Cameron's artistic imperatives, however, lay marked financial necessity. She and her husband were virtually penniless. The couple depended financially on Mrs. Cameron's entrepeneurial endeavours in photography and on the collection of rental fees from holiday houses, as the Darwin family and others (among them Benjamin Jowett, Jenny Lind, and Anne Thackeray, the writer's daughter) were aware. Shrewdly using the proximity of Tennyson as a lure, she just about managed to support herself and her husband. Mrs. Cameron charged Darwin a fee for the photographs and asked him to assign the copyright to her. As with her other studies of prominent men, she went on to publish a number of authorised prints as a commercial venture sold through Colnaghi's, the London fine art firm. Hidden behind impeccable Victorian manners was the inescapable fact that she needed to make money out of Darwin's face.

Darwin obliged. He was interested in photography both as a process and as a sign of the times, and he enjoyed Mrs. Cameron's company. More than this, he liked her photograph better than any other portrait of him, and he wrote her a sentence to that effect. She promptly included the remark as a mechanically reproduced inscription at the bottom of the Colnaghi prints (although some prints exist without it).[77] This was evidently a technique she refined in the future. Tennyson would complain about constantly signing photographs for her but did it anyway. She registered the signed picture of Darwin under copyright later in the year.

At the end, artist and subject were mutually pleased. Darwin paid Julia Cameron £4 7s. for his photograph and other sums later on for various quarter-sized reproductions.[78] These he distributed almost like an autograph until the expense became too much and he sat for another, cheaper *carte de visite* from Elliot and Fry, and other firms after that. Mrs. Cameron moved smoothly into retail action at the next British Association meeting, at which Joseph Hooker was president. "I have between £8 & £9 to hand over to Mrs. Cameron for sale of photographs, chiefly yours," Hooker told Darwin, "but it is far too big for travellers to carry away—I wrote twice to her from Norwich."[79]

"I have got your photograph over my chimney piece, and like it much," Darwin told Hooker in return, "but you look down so sharp on me that I shall never be bold enough to wriggle myself out of any contradiction."[80] In this way, the public became more widely familiar with Darwin's face.

## VI

Back from holiday, Darwin felt as revitalised as his family. "Charles's book is done," said Emma contentedly, "and he is enjoying leisure, tho' he is a very bad hand at that. I wish he could smoke a pipe or ruminate like a cow."[81] Certainly his health was better, an improvement that lasted more or less for another year.

Emma noted his returning air of anticipation. Striding down the garden path after arriving home from the Isle of Wight, he already planned a riot of fresh experiments. Without delay, he invited Hooker to visit and asked him for plant specimens to restock the hothouse. When lumpy parcels wrapped in coconut matting started appearing on the doorstep,

Emma knew that all was right in his world. "You are very good about the lilies," Darwin wrote conspiratorily to Kew. "We want only a few for pots to be brought into the drawing room when in flower. . . . we had not the least intention of begging bulbs from you."[82]

He pushed *Variation* and its poor reviews to the back of his mind.

> The devil take the whole book; & yet now I am at work again, as hard as I am able. It is really a great evil, that from habit I have no pleasure in hardly anything except natural history, for nothing else makes me forget my ever recurrent uncomfortable sensations. But I must not howl anymore, & the critics may say what they like: I did my best & man can do no more. What a splendid pursuit Natural History would be if it was all observing & no writing.[83]

If not plants, then mankind. Before he had finished correcting the proofs of *Variation,* Darwin began turning over in his mind again the links between animals and humans, and evidence for descent from earlier forms of life. For him, the best part of any project always lay in gathering a mass of examples.

In particular he returned to his interest in facial expressions. He felt sure that some human expressions were universal, indicating mankind's single origin. Moreover, he thought the majority of human expressions were also identifiably the same as animal expressions, or at least their origins could be traced in animal movements and emotions, another sign of evolutionary connections. The human grimace of pain would be the same all over the globe, he imagined. It was equally recognisable in dogs. Surely, he asked himself, this revealed the "mental continuities" between animals and humans?

These notions had already supplied him with a pleasant diversion when he was bored with writing *Variation.* Early in 1867, he had distributed a printed questionnaire in which he asked overseas correspondents about the facial expressions and gestures of indigenous peoples. Did all peoples shake their heads to convey a negative, for example? As soon as *Variation* went to press, he arranged for the same questionnaire to be circulated more widely, first by Gray to friends in America, then by Robert Swinhoe in the Far East, and finally in 1868 by the Smithsonian Institution in Washington, there calling it "Queries About Expression for Anthropological Inquiry."[84] Connecting with the Smithsonian Institution was an astute move. This body was famous for its systematic programme of gathering information from all over the world, especially in meteorology and ethnology. Darwin hoped to get information about indigenous peoples from the far west and north of the American continent, perhaps even from Alaska. Puzzlingly, his correspondence system failed in this regard. Only one reply to all his efforts survives.[85]

Darwin began sorting through his private papers until he found his notes on the Fuegians and other ethnic groups encountered during the *Beagle* voyage. He studied his early notebooks again and located his manila folders full of miscellaneous details about the physical and mental properties of human beings. With mounting eagerness he leafed through his collection of printed materials, some relating to human skin colour, others to parasites, or anthropological studies of forest tribes, embryological monstrosities, hairy women, chimpanzee behaviour, religious beliefs, insanity, the human senses, and so forth.

Most eagerly of all, he retrieved his notebook about the Darwin children when they were small, and compared his descriptions of William's and Anne's facial expressions with information gleaned from the wives and mothers among his social circle, and then from observant zookeepers like Abraham Bartlett of the London Zoological Gardens. This sequence of observations on his children had been unusual enough when he first began in the 1840s. He now felt sure that mothers—particularly the ones married to scientists and accustomed to the criteria of science—were as knowledgeable in their own field as geologists or horticulturists in theirs, and could be relied upon to provide accurate observations on the faces that their children made. He was one of the first natural philosophers to make use of this generally unexploited area of expertise, and one of the first to make comparative observations between ape and child.[86]

"Give Mrs Huxley the enclosed," he suggested to Huxley in 1868, "& ask her to look out (for hints) when one of her children is struggling & just going to burst out crying." What did the little Huxley's eyebrows do? "A dear young lady near here, plagued a young child for my sake, till it cried, & saw the eyebrows for a second or two beautifully oblique, just before the torrent of tears began."[87] A human baby seemed to him entirely comparable to a monkey. "When the *Callithrix sciureus* screams violently does it wrinkle up the skin round the eyes like a Baby always does?" he asked Bartlett. He cornered Dr. Engleheart, the village doctor, with similar questions. Engleheart's experience at the local schools stood him in good stead. "Blushing commences quite as early as 5 yrs. as I have two fine little Blushers of this age on show at Chesham School," the doctor told Darwin.

And he threw himself into reconsidering the old favourite of sexual behaviour, human as well as animal. Darwin's notebooks were full of material about mating rituals among birds, insects, mammals, and faraway peoples, and these spilled into observations about beauty and the moral sense in general. As Darwin understood it, marriage was a selective process. "Our aristocracy is handsomer . . . than the middle classes, from having the pick of the women," he said to Wallace, "but oh, what a scheme is primogeniture for destroying natural selection!"[88]

The driving force behind all this lay in Darwin's idea of sexual selection, the mechanism he invented to explain secondary sexual characteristics such as the different colouring of male and female birds or the male peacock's wonderfully ornamented tail. In animal species, he had suggested in the *Origin of Species,* females would mate more readily with males displaying the largest antlers, the brightest colours, the neatest nest, or the most beautiful song, and thereby leave descendants liable to possess the same characteristics. Over the generations such features would build up in a population. Sometimes the attributes might determine the victor in a fight for possession of the female but generally they served no life-preserving adaptive function. They merely increased the chances of mating and thus the number of offspring.

He was convinced that this explained many aspects of human evolution. "Among savages the most powerful men will have the pick of the women, and they will generally leave the most descendants," he mused to Wallace. Strictly speaking, this was not natural selection, since choice was involved. In humans, said Darwin, the choice was exercised by males. The situation was otherwise in the animal kingdom, where he believed females took the decisive role. He was coming to believe that this process generated most of the physical differences between the human sexes and between human races: "every race has its own style of beauty." Superficial attributes like skin colour might easily shift under reproductive preferences like these.

Wallace disagreed. "How can we imagine that an inch in the tail of the peacock . . . would be noticed and preferred by the female?" he asked derisively. Crickets did not stridulate for sexual purposes. Male and female toucans both possessed brightly coloured beaks. Gloriously flecked and spangled male game birds were polygamous, so how could the females possibly exercise choice? In Wallace's opinion, the dull colour of female birds gave them vital protection from predators and hence possessed genuine survival value. By far the better tropical naturalist, he systematically forced Darwin to reexamine his evidence as if he were a connoisseur pointing out the subtlest aspects of a rare piece of china.

Nothing could have suited Darwin better. In his new frame of mind, a good argument like this brought his mind alive again, set the machinery going. All his competitive spirit burst forth. "My difficulty is, why are caterpillars sometimes so beautifully and artistically coloured?" he asked

Wallace in 1867. An abstruse correspondence on caterpillars ensued. He tried to catch Wallace out. "Can butterflies be polygamous?" he probed. What of reindeer horns? Elephant tusks? Cocks' combs? "It is an awful stretcher to believe that a peacock's tail was thus formed; but, believing it, I believe in the same principle somewhat modified applied to man," he asserted. Wallace set out his alternative theory of the colours of caterpillars in the *Westminster Review* in 1867 and explained protective colouration and birds' nests in 1868. "I believe I was the first to give adequate reasons for the rejection of Darwin's theory of brilliant male colouration or marking being due to female choice," he claimed in unruffled fashion in his autobiography.[89]

In short, Darwin felt invigorated, ready at last to confront the pivotal issue he deliberately omitted from the *Origin of Species*. He decided to start a "Man-book." The subject was huge, but very appealing to him. Human variation, geographical diversity, facial expressions, moral sensibilities, inheritance, reproductive behaviour, and sexual selection—in essence, the entire natural history of mankind—at last secured his attention.

None of his friends treated the subject of humans quite as he had hoped. Despite Huxley's continuing writings and lectures, *Man's Place in Nature* mostly served its author's special polemical purposes. Although Lubbock wrote about archaeology and prehistoric societies, he did not explicitly address natural selection. Lyell, for all his reinterpretation of the antiquity of the human species, had done as much as he was able. Spencer primarily interested himself in the development of civilisations. Asa Gray defended the existence of divinely guided variation. Haeckel ran amok with missing links and recapitulation theory. Vogt and Hunt believed that humans emerged from multiple origins. Galton pursued the human intellect with statistics alone.

Other authors less intimate with Darwin also tackled the evolution of mankind in ways that he would rather not have encountered. George Douglas Campbell, the Duke of Argyll, had issued an influential challenge to Darwin's presumed ideas about mankind in his book *The Reign of Law* (1867). Since reviewing Darwin's *Orchids,* Argyll's occasional remarks on evolution had earned him the newspaper epithet "the Darwinian Duke," although he supported the alternative idea of a providential God acting in nature. Huxley irreverently referred to him as the Dukelet. "How can you speak so of a real living Duke?" protested Darwin. For all Darwin's revolutionary ideas, he still felt a stab of awe for the hereditary peerage, strong enough to laugh at himself on occasions. "I have always thought the D. of

Argyll wonderfully clever," he told Hooker, "but as for calling him 'a little beggar' my inherited instinctive feelings wd. declare it was a sin thus to speak of a real old Duke."

Argyll said there must be much more to nature than mere mechanical chance. "Ornament or beauty is in itself a purpose, an object, an end," he wrote in his *Reign of Law,* and he reiterated the point in various articles. To him, progress in the living world was inexplicable except on the assumption that it was planned from the beginning by a divine mind. This kind of creative evolutionism was attractive to many people who felt they could make a compromise by swallowing evolution but rejecting natural selection. Darwin crossly accused Argyll of speaking "absurdly" about "beauty existing independently of any sentient being to appreciate it" and allowed himself a tremor of satisfaction when Wallace "smashed" into Argyll over the issue of bird colours, showing how the iridescence of hummingbirds could be explained by natural selection alone.[90]

Lubbock took on Argyll as well. At the British Association meeting in 1867, Lubbock had argued that primeval mankind lived in a state of "utter barbarism" and had only gradually evolved towards a civilised state. The Duke of Argyll responded in the magazine *Good Words* with a series of refutations that formed the substance of his next book, *Primeval Man,* published in 1869. In this book, Argyll claimed mankind was far more likely to have degenerated from an earlier perfect state than to have risen from the animals. He and Lubbock clashed on the issue at the next British Association meeting. How could early human beings have successfully competed with apes? asked Argyll. Apes would always be better at living in the wild. "Place a naked high-ranking elder of the British Association in the presence of one of M. du Chaillu's gorillas, and behold how short and sharp will be the struggle," tartly observed one of Argyll's supporters.[91]

All the same, their relationship was cautiously well disposed. Argyll admired Darwin's scholarship, saying in his autobiography that he thought "the subject can never go back to where it was before he wrote." And Darwin regarded him with the respect due to an intellectual. Later, when he needed Argyll's political help he was glad they had not irrevocably disagreed.

Only Wallace, as Darwin saw it, was trying to locate human origins in the strict framework of natural selection as originally proposed, and even here Darwin considered that Wallace unjustly spurned his idea of sexual selection. None of these colleagues said what Darwin thought was most needed. The moment was ripe for him to say it himself.

## VII

Months passed. He heard about Hooker's speech as president of the 1868 British Association at Norwich—the speech Hooker had been writing when they all visited the Isle of Wight. "I feel like the parrot which was in the habit of saying in a tone of great contempt after the family prayers were over, 'My God,' " said Hooker.[92] Hooker was the first of the Darwinians to fill the presidential role and took the opportunity to review the reception of evolutionary theory. Wallace reported:

> Darwinianism was in the ascendant at Norwich (I hope you do not dislike the word, for we really *must* use it) and I think it rather disgusted some of the parsons, joined with the amount of *advice* they received from Hooker & Huxley. The worst of it is, that there are no opponents left who know any thing of Nat. Hist. so that there are none of the good discussions we used to have. Vogt told me that the Germans are all becoming converted by your last book.[93]

Then Asa Gray came to stay. Darwin was pleased to clasp his friend's hand at last, a feeling that was reciprocated. Darwin was "entirely fascinating," commented Mrs. Gray that weekend. He was "tall & thin, though broad-framed, & his face shows the marks of suffering and disease. . . . He never stayed long with us at a time, but as soon as he had talked much, said he must go & rest." He had "the sweetest smile, the sweetest voice, the merriest laugh! and so quick, so keen!"[94] Over dinner, she told him about her sister's dog that washed its face like a cat, a story that eventually made its way into Darwin's writings.

Gray magnanimously ignored his differences with Darwin over the Civil War and downplayed Darwin's recent remarks against him in *Variation*. Together they looked over his experimental work. The visiting botanist was silently amused by the simple workbench in the greenhouse and the turned earth in the kitchen garden ready for next year's scientific peas and beans. To Gray, Darwin seemed positively to favour working in humble conditions. Gray went away impressed by his friend's ability to push into unexplored scientific territory armed with only a trowel.

When Gray went to dine at the X Club, he found England's other Darwinists were much more men of the world. Warily, he distanced himself from what he called "the English-materialistic-positivistic line of thought," by which he probably meant Huxley's Whitworth-gun fusillades. At this particular moment Huxley was rampaging on miracles and the existence of the soul. A few months later, he was to coin the word "agnostic" to describe his own position as neither a believer nor a disbeliever, but one who considered himself free to inquire rationally into the basis of knowledge, a philosopher of pure reason, as he liked to say, linking himself with David Hume and Immanuel Kant. The term fitted him well, and although the validity of Huxley's story of its parentage remains uncertain, it caught the attention of the other freethinking, rational doubters in Huxley's ambit and came to signify a particularly active form of scientific rationalism during the final decades of the nineteenth century.[95] "Most of my colleagues were 'ists' of one sort or another . . . so I took thought, and invented what I conceived to be the appropriate title of 'agnostic.' " In his hands agnosticism became as doctrinaire as anything else—a religion of scepticism. Huxley used it as a creed that would place him on a higher moral plane than even bishops and archbishops. All the evidence would nevertheless suggest that Huxley was sincere in his rejection of the charge of outright atheism against himself. He refused to be "a liar." To inquire rigorously into the spiritual domain, he asserted, was a more elevated undertaking than slavishly to believe or disbelieve. "A deep sense of religion is compatible with the entire absence of theology," he had told Charles Kingsley back in 1860. "Pope Huxley," the *Spectator* dubbed him. The label stuck.[96]

Reassuringly, Gray discovered that Lyell was steady in his belief in divine authority and willing to discuss with him some of the religious shortcomings of the evolutionary worldview. Lyell in fact drew support from Gray. "Asa Gray's articles, all of which I have procured, appear to me the ablest, and on the whole, grappling with the subject, both as a naturalist and metaphysician, better than anyone else on either side of the Atlantic." Lyell alone of the inner circle of Darwin's friends adhered to Gray's position on design and theology.[97] The most obvious souvenir of Gray's trip was, however, physical. He returned to Boston with a "venerable white beard" just like Darwin's. Reinforced with this badge of allegiance, he promoted Darwinism with the same dedication as before.

At home, Darwin's life continued steadily, divided between plants in the greenhouse and work on mankind. His wife Emma understood that the domestic arrangements of a man like him should be based on routine. It was enough to keep the windows open, the meals coming, and the sofa ready. Privacy and quietness was all.

In the spring of 1869 he fell off his horse Tommy, hurting his leg badly in the fall, and making him declare that he would not ride again, although it had amused him to plod along the same route every day while noting Tommy's vivid imagination for wolves or highwaymen as they passed and repassed the same heap of grass clippings. The accident put him out of action for weeks. Plaintively, he asked to be towed to the hothouse on a makeshift trolley pulled into position by Horace and a few lads from the stableyard. "I am afraid Ch's nerve will be quite gone whatever animal he rides," Emma told an absent family member. "We have got a Bath chair which is better than the truck on which he rode ignominiously to the hothouse. The boys take him round the sandwalk & he enjoys it after his confinement."

It was in the drawing room, nursing this bad leg, that he seemingly became part of nature himself.

> Yesterday a wasp settled on F.'s face & put its proboscis into his eye to drink the moisture apparently. He got up very quietly from the sofa & stood looking at himself in the glass till the wasp moved.[98]

As Emma said, "A sting in the eyeball wd. have been horrid."

## VIII

Under these circumstances, he found revising the *Origin of Species* for a fifth edition was irksome. "That everlasting *Origin*," he complained to Fox. "I am sick of correcting." To be sure he made several important adjustments. After studying Carl Nägeli's commentary on natural selection, he thought that there were probably some features of living organisms that had no adaptive purpose whatsoever. This was a major concession to make. "I have lately i.e. in new Edit. of Origin been moderating my zeal," he confessed to Wallace, "& attributing much more to mere useless variability.—I did think I wd send you the sheet, but I daresay you wd not care to see it, in which I discuss Nageli's essay on Nat. selection, not affecting characters of no functional importance, & which yet are of high classificatory importance. Hooker is pretty well satisfied with what I have said on this head."[99]

As far as the inheritance of individual variations went, Darwin also admitted that Fleeming Jenkin might be right about blending inheritance. Again, he confided in Wallace.

> F. Jenkin argued in *N. Brit. R.* against single variations ever being perpetuated & has convinced me, though not in quite so broad a manner as here put.—I always thought individual differences more important, but I was blind & thought that single variations might be preserved much oftener than I now see is possible or probable.—I mentioned this in my former note merely because I believed that you had come to similar conclusion, & I like much to be in accord with you.—I believe I was mainly deceived by single variations offering such simple illustrations, as when man selects.[100]

Other issues raised their heads. Friends and reviewers persistently accused him of personifying natural selection. At Wallace's urging, Darwin therefore used for the first time in the *Origin of Species* Spencer's phrase "survival of the fittest," although remarking that the benefits of a change in wording so late in the day could only be limited. "I fully agree with all that you say on the advantages of H. Spencer's excellent expression of the survival of the fittest," he observed. Yet "I doubt whether it [the term 'natural selection'] could be given up, & with all its faults I should be sorry to see the attempt made."

The point had actually come up earlier, and Darwin had accepted Wallace's advice and used the phrase here and there in *Variation.*[101] For this fifth edition of the *Origin of Species* he faced up to the matter more seriously. Wallace had told him that any comparison between artificial and natural selection was liable to be taken literally, and that the word "selection" necessarily implied a selector—the antithesis of what he and Darwin really meant. The objection, said Wallace, "has been made a score of times by your chief opponents, & I have heard it as often stated myself in conversation."

> Now I think this arises almost entirely from your choice of the term "Nat. Selection" & so constantly comparing it in its effects, to Man's selection, and also to your so frequently personifying Nature as "selecting," as "preferring," as "seeking only the good of the species," &c. &c. To the few, this is as clear as daylight, & beautifully suggestive, but to many it is evidently a stumbling block. I wish therefore to suggest to you the possibility of entirely avoiding this source of misconception in your great work, (if not now too late) & also in any future editions of the "Origin," and I think it may be done without difficulty & very effectually by adopting Spencer's term (which he generally uses in preference to Nat. Selection) viz. "Survival of the fittest."[102]

"People will not understand that all such phrases are metaphors," he continued. In private, Wallace went through his own copy of the *Origin* crossing out "natural selection" and inserting the words "survival of the fittest."[103] He never possessed the same emotional commitment to the original phrase as Darwin. Indeed, for Darwin to abandon the word "selection" would completely change his meaning.

Nevertheless, Wallace's opinion mattered to him. Wallace evidently felt that Spencer's phrase properly encapsulated the numerical relationship between death and survival on which evolution rested. It also made explicit the importance of adaptation to circumstances that both of them believed lay at the heart of the process.[104] It was easy to assume some form of improvement in animal structure would result from this process. Some authors were already applying Spencer's words in this way, as illustrated by the early work of Robert Lawson Tait, the Birmingham surgeon, who asked, "Has natural selection by survival of the fittest failed in the case of Man?" in the *Quarterly Journal of Medical Science* for 1869. In actuality, Spencer had defined the phrase partly in relation to Darwin's natural selection. In his *Principles of Biology*, published in two volumes from 1864 to 1867, he stated, "This survival of the fittest, which I have here sought to express in mechanical terms, is that which Mr Darwin has called 'natural selection or the preservation of favoured races in the struggle for life.' "[105]

Yet neither Darwin, Spencer, nor Wallace apparently noticed how far the expression was replete with circular reasoning, nearly self-defining as a philosophical and biological tautology, in which the fit survive and the survivors are fit.[106] Wallace's justification was that evolutionary theory needed to be divorced from the notion of an external selector and this was one way to do it. Unsure how far Darwin would move in this direction, he pressed Lyell to add his weight to the argument. Lyell used "survival of the fittest" almost without thinking in a letter to Darwin in 1869.

Of course, Darwin and Wallace were not social innocents, and each believed in his own form of biological determinism and progress. At root, each felt that adaptation generally meant improvement, that push and shove was an intrinsic feature of nature and that the best individual would be the one that survived. Spencer's words were thus not merely convenient, they were to them for the most part true. Darwin seems to have felt that there was progress in human affairs, while strenuously denying any necessary tendency to advance. Yet there were other issues that apparently bothered him more. Around this time, he made a stark observation to Hooker.

> I quite agree how humiliating the slow progress of man is, but everyone has his own pet horror, and this slow progress or even personal annihilation sinks in my mind to insignificance compared with the idea or rather I presume certainty of the sun some day cooling and we all freezing. To think of the progress of millions of years, with every continent swarming with good and enlightened men, all ending in this, and with probably no fresh start until our planetary system has been again converted into red-hot gas. *Sic transit gloria mundi* with a vengeance.[107]

None of them expected the words to take hold in the way they did. They failed to foresee the social loading that the phrase immediately acquired in human terms, where the "fit" inevitably became associated with the socially successful. Nor did Darwin anticipate the special weight that the publicity surrounding his *Origin of Species* would give these words. Nonetheless, he did not retract. Between them, Spencer, Darwin, and Wallace generated four words that became an integral part of the Victorian frame of mind.[108]

While working on this fifth edition, Darwin also encountered major intellectual problems over the age of the earth. William Thomson (the future Lord Kelvin) had asserted on the basis of experimental physics that the earth was not sufficiently old to have allowed evolution to have taken place. To some extent, Thomson was tilting at Lyell—he had never liked Lyell's endless geological epochs stretching back into eternity. Earlier on, he had attacked Lyell's gradualism and uniformitarianism, saying that geologists ignored the laws of physics at their peril and that the earth was much younger than usually thought, and deliberately initiating a quarrel between physicists and geologists over who had the best claim to study the earth that turned into a boundary dispute between two developing disciplines. In 1866, thoroughly frustrated by what he regarded as pig-headed obtuseness from the Lyellian-Darwinian fraternity, and propelled by anti-evolutionary, Scottish Presbyterian inclinations, Thomson launched a vigorous polemic against the lot of them, stating that 100 million years was all that physics could allow for the earth's entire history. As Darwin noted, Thomson intimated that the earth had had a beginning and would come to a sunless end.

Disturbingly for Lyell and his friends, Thomson's point was supported by Archibald Geikie and James Croll, two good young geologists, and by Thomson's business partner in Glasgow, Fleeming Jenkin, who combined his critique of Darwin's theory of inheritance with this form of geochronological attack. In 1868 Geikie and Croll acknowledged the necessity of fitting geological data into the shorter time supplied by physics. Of course, Lyell could not ignore these arguments against him, and he attempted to answer them in the tenth edition of his *Principles of Geology* (1867–68). Huxley also did his best in an address delivered while president of the Geological Society in 1868. But this was one of Huxley's froth and fury speeches, taking two steps back for every step forward. Darwin felt temporarily deserted when Huxley declared, "If the geological clock is wrong, all the naturalist will have to do is to modify his notions of the rapidity of change accordingly."

That was just what Darwin could not do. In the first edition of the *Origin of Species* he had calculated that the erosion of the Sussex Weald must have taken some 300 million years, a breathtaking length of time that, taken with the rest of the stratigraphical table, provided ample opportu-

Asa Gray, the talented American botanist, was a good friend to Darwin.
(Missouri Botanical Garden Library)

Henry Lettington and William Brooks, Darwin's gardeners, in the grounds of Down House.
(Down House collection, English Heritage Photo Library)

Inside Darwin's greenhouse. Studying plants was a peaceful occupation after publishing the *Origin of Species*.

(*Century Magazine* 1883).

Das
entdeckte Geheimniſs
der
NATUR
im Bau und in der Befruchtung
der
Blumen
von
CHRISTIAN KONRAD SPRENGEL,
Mit 25 Kupfertafeln.
Berlin, 1793,
bei Friedrich Vieweg dem æltern.

The role of insects in carrying pollen from plant to plant was first recognised by Christian Konrad Sprengel in 1793. Darwin praised this book lavishly.

(Courtesy of the Syndics of Cambridge University Library)

Kew Gardens, where Joseph Hooker was director from 1865.
(Author's collection)

Darwin regarded the insect-catching sundews almost as if they were animals.
(C. Darwin, *Insectivorous Plants* 1875. Wellcome Library, London)

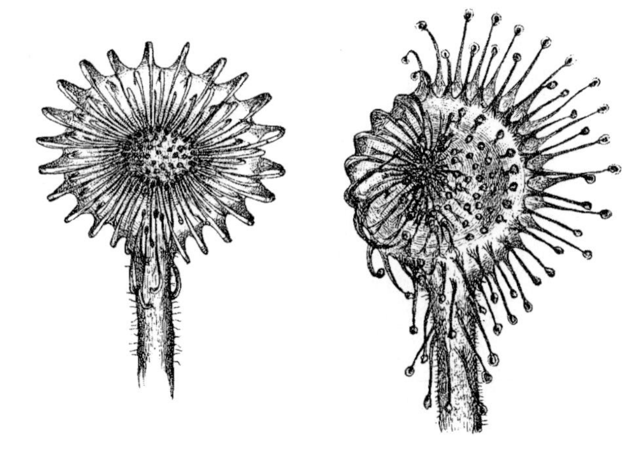

Emma Darwin with baby Charles, whose death from scarlet fever prevented Darwin from attending the Linnean Society meeting at which his and Wallace's theory was first announced. The photograph was probably taken by the Darwins' eldest son, William.

(Courtesy of Randal Keynes)

Henrietta Darwin helped edit her father's books, especially *The Descent of Man*. Darwin's other daughter, Elizabeth, presents a shadowy figure to historians. She is hardly mentioned in the archival record.

(Darwin collection, courtesy of the Syndics of Cambridge University Library)

The family hoped sea bathing would cure Henrietta's debilitating illnesses.

(J. Leech, 1865. Wellcome Library, London)

Darwin on "Tommy" outside Down House. Underneath a copy of this photograph, he wrote, "Hurrah—no letters today!"

(Darwin collection, courtesy of the Syndics of Cambridge University Library)

Erasmus Darwin, Charles's brother, with four of Darwin's five sons; left to right, Horace, Leonard, Francis (with a flute), and William.

(Darwin collection, courtesy of the Syndics of Cambridge University Library)

A LITTLE LECTURE BY PROFESSOR D——N ON THE DEVELOPMENT OF THE HORSE.

Fellow Monkeys,—We commence with the fiery untamed horse-radish, No. 1. We then come to the horse-chesnut, No. 2, whence I need hardly say—the chesnut horse. No. 3. The clothes horse, giving to the modern charger its quality of standing fire. No. 4. The horsey-porsey of childhood, and the origin of the post-horse. No. 5. The progenitor of the high mettled racer (sensation). No. 6. The horse in its mood of "would," which accounts for its obliging disposition. No. 7. The oss. We then come to the horse proper. No. 8. The Louis Cart-horse of the French. No. 9. The Hunter "den Linden" of the Germans. No. 10. The well-known "hors de combat," with others which I shall not insult you by describing (hear hear). No. 11. You will not be surprised to hear is the highly developed race-horse of the future, with jockey and everything complete. I will now retire, feeling a little hoarse myself. (Tumultuous applause, brickbats, &c.)

**MR. BERGH TO THE RESCUE.**

The Defrauded Gorilla. "That *Man* wants to claim my Pedigree. He says he is one of my Descendants."

Mr. Bergh. "Now, Mr. Darwin, how could you insult him so?"

Songs, advertisements, and caricatures took evolutionary theories into the heart of public debate. The cartoons come from *Fun, Harper's Weekly,* and *La Petite Lune.* The advertisement for Merchant's Gargling Oil is published with grateful thanks to A. Walker Bingham. Morris's book *All the Articles of the Darwin Faith* violently attacked the idea of descent from apes.

(British Library; Darwin collection, courtesy of the Syndics of Cambridge University Library; Wellcome Library, London)

Numerous *cartes de visite* were produced as Darwin became increasingly famous.

(Wellcome Library, London)

Thomas Woolner sculpted Darwin's bust. The result can be seen on the lower shelf behind Woolner, in his London studio.

(A. Woolner, *Thomas Woolner, sculptor and poet* 1917)

Darwin became a notorious figurehead for advanced views when his *Descent of Man* (1871) was published. This French cartoon from *La Lune* shows him crashing through circus hoops labelled Superstition, Errors, and Credulity.

(Darwin collection, courtesy of the Syndics of Cambridge University Library)

nity for gradual organic change. But Darwin's calculations were wrong. The actual time was much shorter. "Those confounded millions of years," he had complained to Lyell and deleted the entire example.

So no wonder that "Thomson's views of the recent age of the world have been for some time one of my sorest troubles."[109] The 100 million years that Thomson allowed was not nearly long enough for the exceedingly slow rates of change Darwin envisaged in nature. The fifth edition of the *Origin* bore witness to his discomfort. Rattled, he tried various ways to speed up evolution. He was aware that he was becoming more environmentalist, more Lamarckian, as it were, and producing a poor-spirited compromise. He roped in George, with his Cambridge mathematics, to make alternative calculations, telling him that the age of the earth was the single most intractable point levelled against his theory during his lifetime.

Five years later Darwin was still protesting that Thomson's shortened time-span was "an odious spectre." Some minor relief emerged when Wallace analysed the question afresh, and although Wallace's proposed solution (based on climate changes and tilting ecliptics) was regarded as unworkable by physicists and naturalists alike, Darwin called it "admirably clear & well put." One of the first scientific projects that George Darwin undertook as a mathematical fellow at Cambridge was to rework Thomson's calculations and propose a modification that favoured his father. Although George's relationship with Thomson was close, he warned scholars not to accept all of Thomson's results.[110] Decades of continuing debate over the age of the earth were resolved only with the discovery of radioactivity early in the twentieth century that, broadly speaking, allowed the earth to be as old as evolutionists needed it to be.

In this manner, assailed by doubts, deliberately introducing more and more environmentally induced changes to species, quickening up his adaptive processes, and using the phrase "survival of the fittest," which he neither invented nor admired, Darwin produced his fifth edition of the *Origin of Species*.

Hardly anyone noticed the difference. Only the *Athenaeum* sourly observed that "attention is not acceptance." This edition brought the total number of copies of the *Origin* up to ten thousand, and, perhaps not surprisingly, interest was relatively subdued. Murray said that the booksellers at the November sale purchased only 311 copies of the new edition and eighteen copies of *Variation:* "wh. I hope you will consider not a bad days work." Low sales did not concern Murray overly much considering the long-term harvest he was reaping. Nevertheless Darwin was disgruntled.

Decisively, Darwin said he was selling his French translation rights to Jean Jacques Moulinie. Clemence Royer's third edition was too much for him to stomach. Murray smiled. "You have my best wishes for the putting down of your Parisian blasphemers."

## IX

But it was Wallace who provided the biggest jolt. Although they did not meet much during Darwin's years of illness, they were firm friends, exchanging family news, congratulations, and plenty of natural history information. Of all Darwin's scientific friendships, this relationship with Wallace was the one most dependent on letters—a relationship begun by letter and continued by letter, where they met on the equal ground of pen and paper. Wallace encountered Lyell, Spencer, and Huxley in person much more often. Wallace privately savoured the knowledge that it was Huxley who had nominated him for the Royal Medal in 1868. "Huxley was as kind and genial a friend and companion as Darwin himself." With Huxley, however, Wallace "never got over a feeling of awe and inferiority when discussing any problem in evolution or allied subjects—an inferiority which I did not feel either with Darwin or Sir Charles Lyell."[111]

Wallace's regard for Spencer rested on different grounds. "His wonderful exposition of the fundamental laws and conditions, actions and interactions of the material universe seemed to penetrate . . . deeply into the 'nature of things.' " Wallace would walk over to Spencer's Bayswater boarding-house (where Spencer lived with "rather a commonplace set of people—retired Indian officers and others") to talk through a wide range of current projects, and he seems to have been one of the few men who appreciated Spencer's authentic biological interests, discussing with him recent proposals in the life sciences, such as the origin of flight. He approved of Spencer's egalitarian and atheist principles. Once Spencer turned up to dinner at Huxley's house in evening dress, apparently endorsing the traditional social code they each hoped to abolish. Where could he wear it, if not at the houses of friends, Spencer insisted. "Besides, you will please to observe that I *am* true to principle in that I do *not* wear a white tie!"

It was to Darwin that Wallace revealed his heartbreak when his engagement to be married was abruptly broken; and his happiness when he found a wife some years afterwards. Both men knew how intimately their lives were joined. Every now and then, they expressed these feelings. "Your modesty and candour are very far from new to me," said Darwin. "I hope it is a satisfaction to you to reflect,—& very few things in my life have been more satisfactory to me—that we have never felt any jealousy towards each other, though in one sense rivals. I believe that I can say this of myself with truth, & I am absolutely sure that it is true of you."[112]

Even so, Wallace persisted in taking the lesser role. On one occasion he told Charles Kingsley his views about the resources and stamina that Darwin brought to bear.

> As to C Darwin, I know exactly our relative positions, & my great inferiority to him. I compare myself to a Guerilla chief [*sic*], very well for a skirmish or for a flank movement, & even able to sketch out the plan of a campaign, but reckless of communications & careless about Commissariat;—while Darwin is the great General, who can manoeuvre the largest army, & by attending to his lines of communication with an impregnable base of operations, & forgetting no detail of discipline, arms or supplies, leads on his forces to victory. I feel truly thankful that Darwin had been studying the subject so many years before me, & that I was not left to attempt & to fail, in the great work he has so admirably performed.[113]

Wallace went to Down House in 1868 for a weekend party that included Edward Blyth, recently returned from Calcutta for health reasons, and John Jenner Weir, the ornithologist and naturalist. Darwin chose these guests tactfully. Blyth had been one of the first to recognise Wallace's merits, recommending that Darwin should read Wallace's essays in 1855, and Jenner possessed a fine knowledge of British natural history. "Mr Blyth is a dreadful bore," Bessy wrote to Henrietta afterwards. "He talks incessantly and is always interrupting. Wallace is very pleasant I think. . . . we took Mrs. Wallace to church in the morning and the gentlemen went a walk."[114] Wallace welcomed the intimacy with Darwin that this visit brought.

Early in 1869 he sent Darwin a presentation copy of his *Malay Archipelago,* dedicated to Darwin, "not only as a token of personal esteem and friendship but also to express my deep admiration for his genius and his works." The book was "magnificent," Darwin replied, well worth the labour that Wallace had poured into it. In reading this Darwin was filled with fresh regard for Wallace's scientific thinking. "You make me sometimes feel young again as if I was once again collecting specimens," he assured him. "The dedication is a thing for my children's children to be proud of."[115]

So it was all the more unexpected when Wallace's next publication cut away most of the ground under their combined feet. In April 1869, in a long article in the *Quarterly Review,* Wallace backtracked on his commitment to natural selection. He claimed that natural selection could not account for the mental attributes of modern humans.

Wallace inserted this startling new opinion inside a long and deferential discussion of Lyell's views on species. In the tenth edition of the *Principles of Geology* Lyell at last announced his complete acceptance of Darwin and Wallace's theory—a personal wrench, as his friends noted, and a noteworthy statement for an older man to make, for he could as easily have coasted along to his demise without such public acceptance. Wallace praised Lyell's honesty on the issue. What was more, Lyell completely recast his account of the history of life on earth, admitting that evolution by natural selection provided the only workable hypothesis for naturalists (characteristically mixing up Lamarckism and Darwinism, and making Darwin tetchy).

It was on the grand question of mankind that Wallace felt he could add something more to Lyell's statement. He said he had changed his mind since his 1864 paper on human origins. Wallace now claimed that at some point during mankind's early history, physical evolution had stopped and some higher driving force or spirit had taken over. The human mind alone continued to advance, human societies emerged, cultural imperatives took over, a mental and moral nature became significant, and civilisation took shape. Modern mankind thus escaped the fierce scrutiny of natural selection. The development of human thought freed humanity from the inexorable laws of nature. "Here then, we see the true grandeur and dignity of man. . . . he is, indeed, a being apart, since he is not influenced by the great laws which irresistibly modify all other organic beings." Mankind was composed of a material frame (descended from the apes) and an immaterial spirit (infused by a higher power) that pulled mental and cultural development onwards.

"I hope you have not murdered too completely your own & my child," Darwin exclaimed in horror.[116] Turning over the pages of Wallace's article, he covered the text with pencil marks. "No!!!" he scrawled in the margin, underlining it three times.

"If you had not told me, I should have thought that [your remarks] had been added by someone else. . . . I differ grieviously from you, and I am very sorry for it."

Before this, Darwin had always assumed he and Wallace stood pretty much together. "I was dreadfully disappointed about Man; it seems to me incredibly strange," he confided to Lyell. But Lyell was defensive. "I rather hail Wallace's suggestion that there may be a Supreme Will and Power which may not abdicate its functions of interference, but may guide the forces and laws of Nature."[117]

Bit by bit, Wallace told Darwin of his belief in the existence of spirit forces and the untapped depths of the human mind. There must be something else other than mere matter in this world, Wallace maintained: "whether we call it God, or spirit," it must play an important role in human evolution. Earnestly, he confessed his long-term fascination with séances and spiritualism.

> My opinions on the subject have been modified solely by the consideration of a series of remarkable phenomena, physical & mental, which I have now had every opportunity of fully testing, & which demonstrate the existence of forces & influences not yet recognised by science. This will I know seem to you like some mental hallucination. . . . I am in hopes that you will suspend your judgment for a time till we exhibit some corroborative symptoms of insanity.[118]

Frustrated beyond measure, Darwin wondered whether Wallace might destroy their ten-year-old project with this talk of mysterious forces and powers. It would not perhaps have mattered so much to him if the *Quarterly Review* article had been written by a lesser figure. Nor would he have regarded its publication in a leading literary magazine as insurmountable. But for one of the joint propounders of natural selection publicly to suggest that natural selection should now be modified to incorporate spiritual intervention would surely hand victory to their opponents. The *Morning Post* said it all: "Mr. Wallace's reference . . . to a Creator's will undermines Mr. Darwin's whole hypothesis."[119] Wallace's easy brilliance, so wayward, so innovative, so undisciplined, comprehensively collided with Darwin's innermost beliefs.

Wallace was unrepentant. His life had been transformed by spiritualism when Mrs. Marshall, a well-known medium, tapped out a message from his dead brother Herbert. He now moved easily among the phrenologists, mesmerists, table-rappers, and mediums who provided increasing numbers of Victorians with an alternative to organised religion. These men and women offered him insights into the other world that he thought must exist alongside the material universe. He said that he personally possessed "considerable mesmeric power," although not as good as his brother Herbert. He felt tables move, saw flowers appear, heard "curious musical phenomena."

In giving this rapt attention to spirits, Wallace was fully a man of his time and place. His thoughts always dwelled on the larger metaphysical frame, his inquiries always probed the meaning of existence. Where Huxley wrote of "man's place in nature," Wallace was ultimately to compose a book called *Man's Place in the Universe*. Interest in these forms of pheneomena was then coming to its height, sometimes reflecting a growing curiosity about the way the mind worked, as revealed by hypnotism, mesmerism, prophecy, religious trances and mysticism, and by apparent evidence for miracles and the occasional appearance of ghosts.[120] The invisible operations of electricity and magnetism fell into the same category, especially when similarities between the long-distance tappings of the electric telegraph and the taps of a spirit guide were noted. On the larger scale, interest in spiritualism also represented a yearning for certainty in an age of uncertainty, the very Victorian hope that there was more to this life than material form, a craving probably exacerbated by the doubts raised by the *Origin of Species* and other secularising works. In Lyell's house, for example, there was a quest for greater spiritual knowledge. Arabella Buckley, Lyell's secretary, was an enthusiastic participant in séances, allowing herself to be mesmerised and used as a channel for communication. When Wallace's son Bertie died in 1874, at the age of six, Buckley tried to contact the dead boy on Wallace's behalf. Scarcely able for grief to deal with the messages that she brought him, Wallace discovered his belief in the spirit world was enmeshed with his own personal unhappiness.

Time after time he urged Darwin and other scientists to experience spiritualism without prejudice or cynicism. "I had many opportunities of witnessing some of the more extraordinary phenomena under the most favourable conditions," he insisted, inviting all of them at one time or another to accompany him to a séance. William Carpenter and John Tyndall accepted but were dismissive of what they saw. "It is not lack of logic that I see," complained Tyndall, "but a willingness that I deplore to accept data which are unworthy of your attention." Huxley was less inclined to beat about the bush. "I never cared for gossip in my life, and disembodied gossip, such as these worthy ghosts supply their friends with, is not more interesting to me than any other."[121]

Forced to look elsewhere, Wallace gained support from Robert Chambers. "We have only to enlarge our conception of what is natural, and all will be right," Chambers told him.[122] However, his passions were stirred in 1868 by an attack by G. H. Lewes in a number of the *Pall Mall Gazette* in which Lewes unmasked Mrs. Hayden, a popular medium, as a fraud. In her defence, Wallace set about collecting testimonies and evidence for mediumship, including his own adventure of crouching under a table to verify that an accordion was being played by unseen hands. His Darwinian friends began to wonder if he was becoming something of a liability to the world of professional Victorian science.

Darwin could not believe in any of this. After the jokes died away, he feared that such enthusiasms would lead Wallace astray. It astonished him that an observational naturalist of Wallace's stature could be taken in by what he regarded as obvious fictions. Nor, in his opinion, did the idea of a spiritual world advance the question of human evolution.

> I am very glad you are going to publish all your papers on Nat Selection: I am sure you are right, & that they will do our cause much good. But I groan over Man—you write like a metamorphosed (in retrograde direction) naturalist, & you the author of the best paper that ever appeared in Anth. Review! Eheu Eheu Eheu, Your miserable friend, C. Darwin.[123]

He had waited long enough. He acknowledged to another friend that he felt "taunted with concealing my opinions."[124] Once again Wallace forced him into a full show of speed.

*part three*

# CELEBRITY

*chapter*
9

# SON OF A MONKEY

DARWIN CALCULATED that it took him two years to write *The Descent of Man*—three or maybe four if he included all his preliminary work on sexual selection and facial expressions. Inexorably, the number of pages kept increasing. In the end, the book was published in two thick volumes by John Murray in February 1871. "I shall be well abused," its author murmured in uneasy anticipation.

The book had, in fact, taken Darwin a lifetime to produce. He did much more than simply flesh out his old conviction that humans had evolved from animals. He brought all his accumulated natural history knowledge to bear on the question of human ancestry, all his experience of the human condition as learned from the *Beagle* voyage and from his life as a naturalist, husband, father, and friend. He was at last dealing with what he once called "the highest and most interesting problem for the naturalist."[1] From the start, he perceived his "Man book" as a necessary counterpart to the *Origin of Species*. In it he would deliberately cross the last frontier of the evolutionary doctrine that he and Wallace had set out to establish.

## II

As was customary, he did not work in isolation. What was new and useful, however, was that he was able to take advantage of the shifts in scientific focus brought about in part by his own *Origin of Species* and call extensively on the researches of naturalists and anatomists already operating within the Darwinian scheme, consulting Huxley, Haeckel, Broca, Quatrefages, Claparède, Vogt, Wallace, Galton, Lubbock, and Tylor on many different points. He found specialists to guide him through areas relatively unfamiliar to him, such as the study of insanity or the history of slavery, and was able again to exploit his system of correspondents across the globe, his reviewers, and members of his family circle.

His fame as an author helped. Nearly everyone was willing to assist him. "I read your last letter with very great pleasure," wrote a government agent, William Winwood Reade, from Accra, on the west coast of Africa. "I should consider a letter from *Darwin* a treat anywhere—how much more so out here! I need scarcely say that anything I write to you is fully at your disposal. My only fear is that I cannot send you anything worth having."[2] Reade proved a useful source of anthropological information. And Darwin gave himself time to think about the work of his contemporaries in philosophy, language theory, and cultural history.

Sometimes it was relatively easy to pin down the facts that he wanted. Thomas Woolner, the sculptor, was one source who received a questioning letter. Darwin said he had noticed that Woolner's statue *Puck*, displayed in the Royal Academy's galleries, sported a fine pair of pointed ears and asked if Woolner had ever encountered such in real life. Woolner was sufficiently intrigued to spend a Sunday or two examining monkeys at the zoo and then (more discreetly) studying his human portrait commissions as they sat to him in profile. He sent Darwin a drawing of a human ear which showed a small inward projection on the upper rim, and suggested that this might be the folded remains of a pointed animal-like tip.[3] Darwin was pleased. "The Woolnerian tip is worth anything to me," he replied, and put it in his book straight away.[4] Even a trifle like this, Darwin wrote, indicated close structural links between animals and humans. The time will come, he declared, when it will be thought absurd to believe that the human race and each species of animal were "the work of a separate act of creation."[5]

He was less successful elsewhere. Fired by his research on plant and animal inbreeding for *Variation Under Domestication,* Darwin investigated the issue of cousin marriages through 1869 and 1870, hoping to establish the life expectancy of any children from such marriages. Admittedly, he aimed high. His ambitions for the "Man book" encouraged him to try to get a question about cousin marriages inserted into the 1871 population census, writing increasingly urgent letters to his political friends, John Lubbock, William Farr (at the registrar-general's office), and Thomas Henry Farrer, among others, as the date for printing the census forms drew closer. "I am endeavouring to persuade Mr. Bruce to have inserted in Census [a] query whether in each household the parents are cousins," he told Farrer in May 1870. "I am deeply convinced that this is an important subject: if you can influence any member of government, pray do so. Some few M.P.s will take up the question.—I have given my reasons in a Chapt in 2d. Vol. of my Domestic animals."[6]

Darwin's interest hung on the supposed dangers of long-continued inbreeding among humans. He envisaged that a simple question on the national census—whether the householder was married to his cousin—would allow a correlation between the number of cousin marriages and the number of living offspring. These figures would provide a rough estimate of fertility between close relatives. Over the next few months he managed to persuade Lubbock to put a motion before the House of Commons and sought the backing of William Farr, the medical statistician who had done so much to bring the data collected by the registrar-general's office into useful form. Darwin had previously corresponded with Farr about medical statistics and received from him on loan large volumes of printed reports on the diseases of the nation. Farr indicated that he might agree to Darwin's request. One Sunday in the summer of 1870, said Emma, "the bell rang after lunch & in came Dr Farr about the census. Ch. had been rather done up thinking it was Snow [Julia Wedgwood], but it was wonderful how he revived & enjoyed talking & settling with him."

Darwin could scarcely have hoped to achieve this aim without friends in high places—an aim that reflected his standing in Victorian England. There was at that time no explicit tradition of medical questions on British census forms, or even religious affiliation, although information on the age, occupations, and marital status of every occupant of every household was required. From 1851 the forms had, however, included a question about physical disabilities of sight, hearing, and speech. As it happened, Darwin's request was part of an escalating medical trend, and the 1871 census was due to be extended by Parliament to incorporate a question about "lunatics, imbeciles and idiots," introduced by Farr at the request of eminent doctors. This question lasted for only one census and would be dropped in 1881. Householders were not willing to supply the information. Boldly, Darwin approached the Liberal home secretary, Henry Austin Bruce, afterwards Lord Aberdare, with his own proposal.

Lubbock put the motion on the agenda in July 1870 and read out to the House of Commons a letter from Darwin on the issue, possibly the first time that a biologist's opinions were formally announced in that company. "As you are aware, I have made experiments on the subject during several years," the letter began.

> It is my clear conviction that there is now ample evidence of the existence of a great physiological law, rendering an enquiry with refer-

ence to mankind of much importance. In England & many parts of Europe the marriages of cousins are objected to from their supposed injurious consequences; but this belief rests on no direct evidence. It is therefore manifestly desirable that the belief should either be proved false, or should be confirmed, so that in this latter case the marriages of cousins might be discouraged. If the proper queries are inserted, the returns would show whether married cousins have in their households on the night of the census as many children, as have parents who are not related; & should the number prove fewer, we might safely infer either lessened fertility in the parents, or which is more probable, lessened vitality in the offspring. It is moreover much to be wished that the truth of the often repeated assertion that consanguineous marriages lead to deafness & dumbness, blindness &c, should be ascertained; & all such assertions could be easily tested by the returns from a single census.[7]

The request was turned down, with at least two members of Parliament intimating that that it would be a dangerous precedent to satisfy the curiosity of "speculative philosophers." It is entirely possible that several members of Parliament were themselves married to cousins, as indeed Queen Victoria had been married to hers, and the proposal may well have looked like unnecessary personal intrusion. Darwin's influence had its limits. Lubbock felt personally responsible for the failure. "Do not you think you might get most of what you want by an enquiry at one or two of the largest idiot asylums?" he forlornly ventured.

In actual fact, Darwin did make a stab at contacting the physician John Langdon Down, who worked at the Royal Earlswood Asylum for Idiots in Redhill. Langdon Down (whose career focused on congenital mental disorders and who gave his name to Down's syndrome) had written a short article for *Nature* in 1870 suggesting that inbreeding might be harmful. "A methodical and judicious selection in the marriage of close relations would be of enormous value to the community in the improved race of man that would by that means be obtained," Langdon Down claimed.[8] This was precisely the point that Darwin was attempting to verify with the authority of numbers. The result of his contact with Langdon Down is unknown.

He discovered an unexpected ally in his son George. Ever since Galton's book *Hereditary Genius,* George had been intrigued by Galton's proposals about inherited ability. Such proposals meshed with George's interest in genealogy, and he had been happy to prepare family pedigree charts for Galton. Taking his father's part, George complained bitterly about the unfavourable verdict of the House of Commons. "The tone taken by many members of the House shows how little they are permeated with the idea of the importance of inheritance to the human race."[9]

Nothing came of Darwin's plan, and he was obliged to let his book on mankind go ahead without the statistical data for which he had hoped.

> When the principles of breeding and inheritance are better understood, we shall not hear ignorant members of our legislature rejecting with scorn a plan for ascertaining by an easy method whether or not consanguineous marriages are injurious to man.[10]

Pensively, he continued living out his own life with his cousin Emma, unsure of the burden of heredity he might have imposed on his own children.

## III

His theory and his personal life were by now so closely intertwined that it was becoming difficult for him to maintain scholarly detachment. While preparing *The Descent of Man* he quarrelled irrevocably with a young naturalist of his acquaintance, St. George Mivart.

Mivart was a talented evolutionary biologist who had quickly become a favourite of Huxley's despite the potential for discord that lay in Mivart's unwavering commitment to Catholicism. At first Mivart ignored Huxley's theological taunts, believing they represented, in this instance, a form of rough-and-tumble affection. But in 1869 or so, Mivart parted company with Huxley and the close-knit band of Darwinians, coming to view the group as a dictatorial, self-regarding clique, a powerful brotherhood of older men at the summit of their careers who insisted that acolytes ought to adopt their position and advance the new biology *in toto*. In many ways Mivart read the situation accurately. The inner ring of private clubs and societies which ran scientific London—the X Club, the teaching laboratories and museums in South Kensington, the philosophers and parliamentarians in the Metaphysical Society and Athenaeum Club—were closed to outsiders. The members were influential people who kept a firm grasp on the tiller of scientific progress. Huxley enjoyed his cliques and believed that small groups of "right-minded men" were by far the most effective way to get things done.[11]

Mivart wanted none of this. All through 1869 he published renegade evolutionary articles in the Catholic periodical the *Month* on "difficulties of the theory of natural selection," maintaining that Darwin's ideas could not explain the whole of nature. He dwelled on awkward anatomical cases such as the close resemblance between Australian marsupial "wolves" and European wolves, or the similarities between the eyes of cephalopods and vertebrates. It was hard to explain these similarities as coincidence. "To have been brought about in two independent instances by merely indefinite and minute accidental variations, is an improbability which amounts practically to impossibility," Mivart stated. Like Asa Gray he opted for theological compromise, arguing that there must be some higher guidance in the process of variation that provided an element of design or direction in the evolutionary process. Underneath ran scarcely veiled contempt for the inflexible position of the Darwinians.

Darwin liked Mivart when first introduced to him and welcomed the young man's obvious ability as a natural scientist. He felt bewildered, and then betrayed, by these critical articles, for it seemed to him that Mivart deliberately ignored anatomical points when they did not suit, and that he twisted Darwin's words solely to make the older man look foolish. With sinking spirits, Darwin wondered if Mivart might become another thorn in his side, another Owen. Intemperately, he let his feelings show. He accused Mivart (behind his back) of too much Catholicism, of being overly clever with words as if he were a Jesuit priest in training. When Mivart pointed out the unlikelihood of any intermediate steps in evolution, Darwin snapped back, "If a few fish were extinct, who on earth would have ventured even to conjecture that lungs had originated in a swim-bladder?"[12] From time to time, Mivart wrote conciliatory letters to Darwin stating the high regard he felt for the *Origin of Species*. Darwin regarded the letters as two-faced.

When Mivart pulled his articles together in 1870 for a book called *The Genesis of Species*, published just before Darwin's *Descent of Man*, Darwin felt the facts were being distorted for religious benefit. He covered his copy with bitter remarks. "I utterly deny," "What does this mean," "You cd. not make a greyhound & pug, pouter or fantail thus—it is selection & survival of the fittest." The last straw came when Mivart claimed in print that Darwin had shifted his ground on blending inheritance in the previous edition of the *Origin of Species* merely in order not to lose face. "Not fair," Darwin moaned in the margin.[13]

He found it astonishing that Mivart could still write letters to him. Blindly crashing onwards, Mivart rashly explained, "My first object was to show that the Darwinian theory is untenable, and that natural selection is *not* the origin of species," a point of view that was unlikely to improve relations.[14] Exasperated, Darwin confided in Wallace.

> You will think me a bigot when I say, after studying Mivart, I was never before in my life so convinced of the general (ie. not in detail) truth of the views in the *Origin*. . . . I complained to Mivart that in two cases he quotes only the commencement of sentences by me, and thus modifies my meaning; but I never supposed he would have omitted words. There are other cases of what I consider unfair treatment.

> I conclude with sorrow that though he means to be honourable, he is so bigoted that he cannot act fairly.[15]

Irritated by Mivart's defection, Darwin then came to dislike him. Catholicism was a convenient enemy here, representing to Darwin every outmoded tradition, superstition, and ritual that he felt should be forcibly expelled from modern life. While perfectly prepared to tolerate the low-key formulae of the Church of England, and an admirer of the social values of several of the clergymen he came across, such as Henslow, Innes, and Kingsley, he easily reverted to the unthinking anti-Catholic prejudice of the English middle classes. Nearly everyone Darwin knew regarded Roman Catholicism with distaste or horror. His friends and family could sympathise with him about Mivart's supposedly outrageous Jesuit tricks. In his mind's eye he saw Mivart opposing him not with the arguments of science but with bells and incense.

Faced with Darwin's disapproval, Mivart tried to retain some dignity.

> I herewith close this correspondence & will say nothing even in this letter calculated to annoy you in the least. I am exceedingly sorry to have caused you mortification & I protest, in spite of all you may think, I *have, do and shall* feel more than "friendly" towards you.[16]

For a moment Darwin seemed to be losing his grip. He lost patience with Frances Power Cobbe as well when their paths briefly crossed in Wales during that same summer of 1869. The Darwins and London Wedgwoods went as a family party to stay in Caerdeon in the Barmouth estuary. On the way home after this holiday, Darwin went to Shrewsbury for a last look at the old house and gardens, an occasion he experienced with relative equanimity. At Caerdeon he felt miserable. Hampered by a bad leg, he could scarcely get out on the hills or enjoy the clear Welsh air. "I have been as yet in a very poor way; it seems as soon as the stimulus of mental work stops, my whole strength gives way. As yet I have hardly crawled half a mile from the house, and then have felt fearfully fatigued. It is enough to make one wish oneself quiet in a comfortable tomb."[17]

It was on one of those half-mile crawls that Darwin encountered Miss Cobbe. He was peacefully alone on a hillside when he was spotted. Without any preliminaries, Cobbe bellowed at him over the turf a question about John Stuart Mill's theories on inherited instincts. Startled, but polite, Darwin began a recondite discussion about Mill at the top of his voice, stopping short in embarrassment when a friend came by. Cobbe laughed about "words flying in the air which assuredly those valleys and rocks never heard before" and dubbed the track the "Philosopher's

Path."[18] Unnerved, Darwin chose his walks more carefully in future. Privately he asked William to read up on Mill and tell him what to think.

Cobbe soon afterwards offended him by publishing extracts from one of his letters in the *Echo,* the campaigning newspaper with which she was associated.[19] Darwin had written to Cobbe telling her about his interest in a particular case of miscarried justice in the Bromley region, and mentioning that he had also written a letter in his capacity as a local magistrate to the home secretary (Henry Bruce, Lord Aberdare). Clever editing made him look as if he were publicly criticising the Kent magistrates and offering to set up a subscription fund to "ensure even-handed justice," a situation he hotly denied. Darwin never trusted her again. This odd pair were afterwards permanently estranged over the antivivisection movement when Cobbe attacked Darwin's defence of the use of animals in scientific experiments. After Darwin's death, Emma felt obliged to refuse Cobbe permission to reprint the magistrates' correspondence in her *Autobiography.* She published it anyway.[20]

If Mivart and Cobbe were not enough, there was Wallace still claiming that natural selection did not apply to humans, urging his scientific friends to attend séances, devotedly chasing the spirit world with photographs, heat detectors, and electric recording devices. "I must add that I have just re-read yr article in the Anthropol. Rev. & I *defy* you to upset yr own doctrine," Darwin groaned.[21]

He felt the pressure of alternative stories mounting. The Duke of Argyll's creative evolutionism was gaining ground. Herbert Spencer's *Principles of Biology* and his *Essays: Scientific, Political and Speculative* integrated evolutionary concepts with political, social, and religious ideas already attractive to contemporaries. Galton's critiques of pangenesis were published. Between 1869 and 1870, Darwin's work was reviewed in fifty-two significant journals, some 15 percent of all the reviews he received in his lifetime,[22] and scores of other evolutionary books and pamphlets had been published in the twelve years since the *Origin* had first been issued. Mudie's Circulating Library made a point of distributing many of these, indirectly making it possible for even the most geographically isolated readers to have an opportunity to acquaint themselves with issues of the day.[23]

There was a lot for Darwin to keep in mind, a lot to reformulate and squeeze into shape. Above all, there was the endless problem of propriety. In his "Man book" he was tackling Adam and Eve directly. For twelve years, Victorians had debated whether natural selection could—or should—explain human origins. Many reviewers thought that such matters were not a legitimate area of study. The dawn of humanity was a matter for theologians, they said, not for naturalists. In order to counter this view, Darwin needed to demonstrate beyond doubt that humans were as much a part of nature as any animal.

He began to feel sick with effort and worry. "Pins and needles" kept him from working, he explained to Henry Bence Jones. "Everything has been of late at a stand still with me, for I have not had strength to do hardly anything," he told Hooker. "With respect to my own book, the subject grows so, that I really cannot say when I shall go to press."[24] Perhaps out of desperation, he bought cigars in 1870 as well as snuff and cigarettes, paying nearly £5 to his London tobacconist.[25]

Still, he felt it was time to be frank. "Man still bears in his bodily frame the indelible stamps of his lowly origin," he wrote in his "Man" manuscript.

> The early progenitors of Man were no doubt once covered with hair, both sexes having beards; their ears were pointed and capable of movement; and their bodies were provided with a tail, having the proper muscles. . . . The foot, judging from the condition of the great toe in the foetus, was then prehensile; and our progenitors, no doubt, were arboreal in their habits, frequenting some warm, forest-clad spot. The males were provided with great canine teeth, which served them as formidable weapons.[26]

## IV

While he wrote and worried, his younger children were making their separate ways. Francis (Frank) had followed George to Trinity College Cambridge, first reading mathematics and then turning to natural sciences and graduating in this subject in 1870. The natural science tripos, dating from 1851, had been reformulated in 1861, and it was to be completely restructured in 1871 in tandem with the opening of laboratory and museum facilities, an indication of the gathering pace of high Victorian scientific concerns.[27] At Cambridge, Francis made friends with a number of young evolutionists and physiologists who admired his father's work. After his degree he came into contact with Michael Foster, the charismatic new lecturer in physiology, and turned to medicine. He went to St. George's Medical School in London, taking an M.B. in 1875, but never practised as a doctor.

Next in the family, Leonard had joined the army straight from school and went to Woolwich Military Academy to train as a military engineer. He interested himself particularly in photography and became a useful member of surveying expeditions. Horace, the youngest, whose schooling was constantly interrupted by illness, had gone to a private tutor before entering Trinity like his brothers. He began in 1868 but did not take up his place immediately because of ill health and spent six years acquiring a degree rather than the customary four, graduating in 1874. Leonard's and Horace's entries into these institutions was poignant. Leonard chose the army because he thought himself the stupidest of the children, at one point inquiring of Darwin if a man could hope to develop into a genius after the age of twenty.[28] Horace similarly underestimated himself by thinking he was good only for mechanical occupations. Darwin wrote to Horace's university tutors to say how physically frail the boy was and how cautiously his education must procede. Their father's doubtful attitude probably did little to encourage confidence. "I have been speculating last night what makes a man a discoverer of undiscovered things," Darwin told Horace in 1871, after he passed his first examination at Cambridge. "Many men who are very clever—much cleverer than the discoverers—never originate anything."[29]

Darwin's second daughter, Elizabeth (Bessy), was rather more of a silent entity. She had been an unusual child, given to what Darwin referred to as strange "shivers & makes as many extraordinary grimaces as ever." Her speech was sometimes confused, according to Emma, and her pronunciation peculiar, although the letters she wrote from school and during visits abroad show little sign of this. "She was not good at practical things," said a member of the following generation who was very fond of her, "and she could not have managed her own life without a little help and direction now and then."[30] In her early twenties at this point, she looked likely to remain at home with her parents.

In any case, the children interested and pleased Darwin. "When all or most of you are at home (as, thank Heavens, happens pretty frequently) no party can be, according to my taste, more agreeable, and I wish for no other society," he was to write in his *Autobiography*.[31] The feeling was evidently reciprocated. When the boys came to visit in large sociable parties, accompanied by university friends, with dances, horses, billiards, and bicycle excursions on their united minds, they brought welcome noise and dash to the old house.

Nevertheless, they discovered Darwin's growing fame could be an imposition at times. George, Francis, and Horace mixed with a number of university people who were coming to base their professional trajectories on Darwin's theories. They learned to bow graciously to the pressures of celebrity life at one remove and must sometimes have wondered whether they received social invitations because they were called Darwin, never quite certain if they could make a career on their own, often speculating that colleagues were fishing for an invitation to dine at Down House.

Their father's writings shadowed every conversation. Francis met the *Origin of Species* in his university curriculum, for example, as a practical result of Alfred Newton's and Henry Fawcett's enthusiasm for evolution by natural selection. In the natural science tripos examination in 1871, the year after Francis graduated, Newton asked biology candidates: "What are the objections to the Cuvierian subdivisions of the class Aves? What progress has been made towards a more natural and satisfactory arrangement of the class?" The best answers would include some discussion of Darwin's proposals. Fawcett was explicit (and rather more testing) in the moral sciences exam: "How do you consider that the leading principles of the Darwinian theory stand in relation to the doctrine of Final Causes?"[32]

At Down House, these younger members of the family were obliged to share their Sunday lunch with Darwin's followers. Visitors were sometimes an unintentional source of amusement. Henrietta always found Ernst Haeckel's eccentric turn of phrase funny, especially the time when he commended London banquets with the words "I like a good bit of flesh at a restoration." Emma regarded the guests with a mix of resignation and good humour. "Today we had a thorough Yankee," she remarked in 1871. "He is a sort of jackal of Appleton the publisher, and so amusing we all had great difficulty in avoiding laughing, and did not dare look at each other." At other times, visitors were an encumbrance. Each son and daughter appeared to accept Darwin's theories unquestioningly and was able to contribute to the conversation. Even Emma, who may never have really accepted his views although managing to live with them comfortably enough, was a willing and polite hostess to dedicated evolutionists. As is often the case with the families of the famous, his wife and children became swallowed up in his renown. Deep down, Darwin's sons and daughters were forced to accept that he was not just their father. He belonged to everybody.

Up to a point, Darwin's children were hardly conscious of the consequences of their involvement until they began making their own progress in life. They were, of course, used to assisting Darwin in any number of minor tasks at home. As they became adult, however, Darwin's growing fame transformed what would otherwise be small family chores into useful contributions of which each could be proud. Gradually Darwin drew them into his tactics for the public dissemination of his views.

When George took Bessy to Paris to meet a gang of Wedgwoods for a holiday in 1869, for example, Darwin asked him to make social calls on some of his evolutionary correspondents. Of these, Jean Louis Quatrefages was by far the most helpful to him, although never accepting Darwin's doctrines.[33] "Vous êtes incontestablement le chef de toutes les théories transformistes," he wrote to Darwin in 1869. "J'ai été heureux d'exprimer, publiquement tout l'estime que je porte en vous à l'homme et un savant." Darwin was keen that his French friends should understand the depth of his gratitude. Obediently George sent up his card to Quatrefages ("a tall good looking man with grey whiskers & a kind of beard"), who paid him so much attention that George began to think that "it must have been me after all who wrote the *Origin*."[34]

Then in May 1870, when Darwin and Emma visited Cambridge to see Francis in his final year as an undergraduate at Trinity College, Darwin discovered that he could mingle easily with the young naturalists of the future. He and Emma stayed in the Bull Hotel, admired the spring greenery along the Backs, and lunched in Francis's college rooms. Darwin paid a scientific call on the Cambridge ornithologist Alfred Newton, who had adopted Darwinian theory, and he made sure to talk with the embryologist Frank Balfour, brother of Arthur Balfour, the future prime minister, and already a noted experimentalist. Francis afterwards invited Frank Balfour to Down House for a weekend visit, and Darwin liked him enormously. "A young Mr. Balfour, a friend of my son's, is staying here," he wrote to Galton later that year. "He is very clever & full of zeal for Nat. Hist.—He has been transplanting bits of skin between brown & white Rats, in relation to Pangenesis!"[35] Darwin warmed to him as a friend of his son's and as a good biologist. Family feeling was easily translated into scientific respect and vice versa.

During the same Cambridge visit, Darwin plodded off to see his old professor of geology, Adam Sedgwick, in Trinity College. Sedgwick had lost none of his animation. Well into his eighties, he walked and talked Darwin into the ground while disclosing that he had found it impossible to forgive his former pupil when the *Origin of Species* was published.[36]

> On Monday I saw Sedgwick who was most cordial & kind: in the morning I thought that his mind was enfeebled; in the evening he was brilliant & quite himself. His affection & kindness charmed us all. My visit to him was in one way unfortunate; for after a long sit he proposed to take me to the Museum; & I could not refuse, & in consequence he utterly prostrated me; so that we left Cambridge next morning, & I have not recovered the exhaustion yet. Is it not humiliating to be thus killed by a man of 86, who evidently never dreamed that he was killing me.—As he said to me "Oh I consider you as a mere baby to me."[37]

The two made their peace over the *Origin*. But "Cambridge without dear Henslow was not itself."

The inevitability of these work-based family connections was to emerge similarly in the summer of 1871. Darwin sent George and Francis to the United States for a vacation tour accompanied by a number of introductions to all his American correspondents. The boys established valuable personal contacts for him among the transatlantic Darwinians. Furthermore, in a hotel in San Francisco, two guests saw the name Darwin in the registration book and called on them. One of the men had lunched with Darwin a year or two beforehand at Erasmus's house in London.[38]

## V

While he was writing *The Descent of Man*, Darwin's growing celebrity took a fresh turn. In 1870 he received a letter from Lord Salisbury, the chancellor of Oxford University, awarding him an honorary degree, the D.C.L. (Doctor of Civil Law), the highest public honour that the university could bestow.

He was taken aback. Oxford was the last place from which he might have expected to hear. The university was renowned for its high church religiosity and conservatism in all matters scientific. It, after all, had been the setting for Huxley and Wilberforce's duel. Darwin's old opponent John Phillips still ran the Museum with careful attention. Holman Hunt was about to install his painting called *Light of the World* in Keble College chapel, and Burne-Jones his stained-glass window of Saint Catherine in Christ Church Cathedral. Hunt had "a holy horror of Darwin," exclaimed his friend Edward Lear.[39] At the same time, Charles Lutwidge Dodgson, mathematics tutor at Christ Church and author of *Alice in Wonderland*, was drawing on the dons and internal politics of university life for inspiration. The Duchess in *Alice in Wonderland* was the very image of Bishop Wilberforce. "If everybody minded their own business," she growled, "the world would go round a deal faster than it does."[40] In short, Darwin could not imagine an institution less likely to appreciate the cool rationalism of his work. Nevertheless he did have acquaintances in some of the new science departments, including Henry Acland, professor of medicine, George Rolleston in anatomy, and the younger Benjamin Brodie, professor of chemistry, a freethinker who refused to subscribe to the Thirty-nine Articles.

It turned out that the honour came directly from Lord Salisbury, a well-known doctrinal reactionary. Yet Salisbury was also a cultivated man of the old school, widely read in the humanities, soon to become leader of the Conservative party and ultimately (after Disraeli's death) Queen Victoria's favourite prime minister. In more incidental fashion, he was the uncle of the biologist Frank Balfour at Cambridge and a distant relation by marriage to the Allen side of the Wedgwood family. ("Will George be back for the Hatfield ball?" asked an Allen aunt after hearing of an invitation to the Salisburys' mansion. "Lady Salisbury is recollecting her cousinhood very graciously.") Trying to put Oxford science more firmly on the map, even though he felt that the universities should teach nothing contrary to scripture, Salisbury nominated Darwin, Huxley, and Tyndall for his first batch of honorary degrees. In retrospect, it seems clear that the new rationalism was by now acceptable even to Lord Salisbury.

But the Hebdomadal council—the ruling body of dons—erupted. Edward Pusey, the high church ritualist, took Salisbury's proposal as an outright attack on faith. Angrily, he put forward three alternative nominees. When Benjamin Jowett opposed Pusey's nominees solely to spite him, Henry Acland and Henry Liddon attempted to talk Pusey into accepting the original nominees. "I wish to keep clear of the question whether Darwin's inferences are correct," Acland told Pusey. "It is Darwin's exceeding eminence and his character as a working man that justify and require me to beg you respectfully to pause before bringing about his rejection here."[41]

As it turned out, Pusey said he was not "against Darwin" or even science as such. He occasionally discussed evolutionary matters with Rolleston, the university professor of anatomy, and later delivered a thoughtful sermon reconciling science and revelation. It was Huxley who irritated him. He allowed himself to be persuaded about Darwin as long as Huxley's name was dropped. The whole business was probably a fair reflection of how Huxley and Darwin were individually regarded by the theological establishment of the day—one acceptable at a pinch, the other emphatically not.[42]

Huxley took mischievous delight in his rejection.

> There seems to have been a tremendous shindy in the Hebdomadal board about certain persons who were proposed; and I am told that Pusey came to London to ascertain from a trustworthy friend who were the blackest heretics out of the list proposed—and that he was glad to assent to your being doctored, when he got back—in order to keep out seven devils worse than that first![43]

Darwin's degree was never awarded. Even though it was announced in the *Daily News* on 20 June 1870, Darwin wrote back to Salisbury saying that he was too ill to make the journey to Oxford for the ceremony. Since honorary degrees were awarded only in person, Darwin seemingly took a conscious decision to refuse the honour. After he received Huxley's letter, the circumstances were probably now so fraught in his mind that his immediate reaction was to withdraw, despite Huxley's declaration that "I wish you could have gone to Oxford, not for your sake, but for theirs." Darwin apparently did not wish to be used as a political tool. Altogether more buoyant, Huxley took advantage of the changed situation and slipped a paragraph into *Nature* gloating that Darwin "declined the compliment" from Oxford.

Some of the Oxford dons were not prepared to let Darwin get away so easily. In a previously unnoticed Hebdomadal motion, Liddon proposed that Darwin's degree be conferred *in absentia,* a concession that would require a separate resolution and another vote of council. This took place. The vote was a tie, and according to the rules, not passed. In effect, Darwin's degree was withdrawn and Salisbury was defeated.

Darwin never expressed any regrets about turning the Oxford degree down. But he was at a loss to explain himself. When Bartholomew Sulivan, his old friend from the *Beagle,* wrote to congratulate him on the honour, he could only say, "I shall this autumn publish another book partly on Man, which I daresay many will decry as very wicked.—I could have travelled to Oxford, but could no more have withstood the excitement of a commemoration than I could a ball at Buckingham Palace."[44]

Huxley, on the other hand, relished rejecting the university's Linacre professorship of anatomy when it was offered to him in 1881, and then refused the mastership of University College Oxford, because, as he patronisingly said, of his being "too busy." But he caved in when offered an honorary degree in his own right in 1885. He accepted this with alacrity. "I begin to think I may yet be a bishop," he purred in gratification.

## VI

Meanwhile Darwin struggled onwards with his book. "Many interruptions," he noted in his diary. Judging from the final product, he was trying to do too many different things. In order to show that humans were incontrovertibly members of the animal kingdom, he presented a barrage of information about the natural history of mankind drawn from a wide variety of sources. He also worked his way through the links between the mental faculties of animals and humans. He then discussed language, morals, and music. Most significantly, he gave his views on "sexual selection," an important development in his schemes that accounted, as he thought, for the diverging physiques and behaviour patterns of males and females, animal or human. Towards the end, he argued that this notion of sexual selection could explain the origin of human geographical diversity, perhaps even the foundations of human civilisation itself.[45] The result was a book packed with details that more or less obscured the important points he was trying to make.

He opened the attack by stating that "there is no fundamental difference between man and the higher mammals in their mental faculties." He substantiated this with a series of cameo observations of animal behaviour, ranging from horses that knew the way home to ants that defended their property, chimpanzees that used twigs as implements, bower-birds that admired the beauty of their nests, and cats that dreamed of rabbits in their sleep.[46] The domestic nature of Darwin's observations in this area, the large doses of willing anthropomorphism, his evident delight in traditional country pursuits, and the glimpses he provided of the congenial home life of a Victorian gentleman, "these fairy tales of science," as Frances Power Cobbe was to call them, probably went some of the way towards softening readers before he confronted them with the shock of apes in the family tree.

A large part of his book was dedicated to discussing the animal origins of the faculties that make humans feel fully human—language, reasoning ability, morality, self-consciousness, the religious sense, memory, and imagination. "No one supposes that one of the lower animals reflects whence he comes or whither he goes—what is death or what is life, and so forth. But can we feel sure that an old dog with an excellent memory and some power of imagination, as shewn by his dreams, never reflects on his past pleasures in the chase? and this would be a form of self-consciousness."[47]

Explaining the power of human speech was obviously critical for him, not only because the gift of language was intrinsic to the definition of being human, but also because linguistics and comparative philology then held a leading position in academic scholarship. By the 1870s, there had developed something of an evolutionary swing in the specialist study of linguistics, where ideas about the "descent" of words were generating fruitful insights into the histories of languages.[48] The imagery moved both ways. Lyell had illustrated the value of parallels between languages and the fossil record in his *Antiquity of Man,* and Darwin also mentioned them in various editions of the *Origin of Species,* praising his cousin Hensleigh Wedgwood, F. W. Farrar, and the German philologist August Schleicher for their identification of the telltale vestiges of ancient languages in words of the modern day. Groups of languages could be classified by relationship, just like species, and the presence of rudiments both in languages and species "is remarkable." Darwin was interested to hear by letter that Haeckel had given a copy of the German translation of the *Origin of Species* to Schleicher in 1860, and that Schleicher reconstructed the genealogy of Indo-European languages partly in imitation of Haeckel's evolutionary trees.[49] Schleicher joked to Haeckel that philologists were much better at tracing ancestral connections between words than evolutionists were with animals.

But the genealogy of tongues was somewhat different from the emergence of human speech. Darwin particularly wished to contest Friedrich Max Müller's view that the faculty of language presented an insuperable barrier between animals and humans. Müller had said as much when reviewing a translation of Schleicher's pamphlet on "Darwinism tested by the science of language" for *Nature* in 1870.[50] Darwin had come to believe that the ability to speak must have emerged quite differently, in a gradual fashion from the social vocalisations of apes, and was developed further in early human societies through the imitation of natural sounds.

> As monkeys certainly understand much that is said to them by man, and as in a state of nature they utter signal-cries of danger to their fellows, it does not appear altogether incredible, that some unusually wise ape-like animal should have thought of imitating the growl of a beast of prey, so as to indicate to his fellow monkeys the nature of the expected danger. And this would have been a first step in the formation of language.[51]

Darwin was similarly daring when dealing with religion. Taking his cue from the cultural anthropologist Edward Tylor, he mapped out a comparative evolution of the religious sense, proposing that religious belief was ultimately nothing more than a primitive urge to bestow a cause on otherwise inexplicable natural events. At first, dreams might have given rise to the idea of spirits, as Tylor suggested, or to animism, where plants and animals seem as if they are imbued with spirits. Darwin suggested that these beliefs could easily grow into a conviction about the existence of one or more gods who directed human affairs. As societies advanced in civilisation, ethical values would became attached to such ideas. Polytheism would turn into monotheism. "Strange superstitions and customs" would give way to the "improvement of reason, to science, and our accumulated knowledge." Darwin was careful to separate this instinctive urge to believe from any developing moral feelings. By keeping the two separate he could show the biological roots of both, circumventing critics who might argue that higher moral feelings were bestowed by a single omnipotent deity. In short, he made no secret of his view that he did not believe religion to have any rational foundations at all. Human beings have a biological need to believe, he suggested. Audaciously, he compared religious devotion to the "love of a dog for its master."

He cautiously tried out these views first on his more thoughtful friends and relatives, taking a chance visit to the Wedgwoods in London to consult his niece Julia (Snow) Wedgwood. "F is hard at work on the moral question of man," said Emma in 1870, "& had talks with Snow about defining religious feeling, in which she only admitted love & reverence & left out fear, but owned she was mistaken after all. F is deeply interested in the question & I wish it was over as it absorbs him too much & he had to lie by one day."[52]

As for morality, he could not resist pointing out that the concept was only relative. Long observational experience with household pets, and no doubt with his children as well, told him that living beings had to learn the difference between "good" and "bad" behaviour—the knowledge was not innate. Members of "primitive" societies similarly held very diverse ideas about behaviour. In this he cited the way that some tribes adhere to value systems that shocked Europeans. If honey-bees ever became as intelligent as humans, he continued wickedly, unmarried females would think it a "sacred duty to kill their brothers, and mothers would strive to kill their fertile daughters; and no one would think of interfering."[53]

Of course, Darwin proposed this for effect rather than logical necessity, because he went on to argue that the higher human values emerged and spread only as human civilisation progressed, meaning that duty, self-sacrifice, virtue, altruism, and humanitarianism were acquired fairly late in human history and perhaps not to the same degree by all tribes or groups. Some groups displayed these qualities more than others, he noted; and it is clear that he thought there had been a progressive advance of moral sentiment from ancient societies (such as ancient Rome), which he said were "barbaric," to the polite world that he personally inhabited. "How little the old Romans knew of [sympathy] is shewn by their abhorrent gladiatorial exhibitions. The very idea of humanity, as far as I could observe, was new to most of the Gauchos of the Pampas." In this manner, he kept the English gentry to the front of his mind, and the mind of his readers, as representative of all that was best in nineteenth-century culture. The "higher" values were, for him, self-evidently the values of his own class and nation.

Even the sense of duty was for him biologically based in the social instincts. "The highest stage in moral culture at which we can arrive, is when we recognise that we ought to control our thoughts." To be sure, Darwin praised the intrinsic nobility of this moral feeling, quoting Immanuel Kant. "Duty! Wondrous thought, that workest neither by fond insinuation, flattery nor by any threat . . . whence thy original?"[54] Yet he claimed even the feeling of duty might emerge from animal sources. As he described it, a monkey who voluntarily sacrificed herself for her offspring would not only ensure her children's survival but also supply the next generation with the hereditary "gemmules" that favoured such action again. His social values came into play. Personally, he declared, he would rather be descended from a heroic little monkey that sacrificed her life than from a savage "who delights to torture his enemies, offers up bloody sacrifices, practises infanticide without remorse, treats his wives like slaves, knows no decency, and is haunted by the grossest superstitions."[55]

Plainly, although he rejected the outward trappings of the established Anglican religion, he subscribed wholeheartedly to its underlying values and the presumed onward march of civilisation. Like Wallace, and so many other contemporaries, he believed in the hierarchy of nations.

> Obscure as is the problem of the advance of civilisation, we can at least see that a nation which produced during a lengthened period the greatest number of highly intellectual, energetic, brave, patriotic, and benevolent men, would generally prevail over less favoured nations.[56]

But in truth, he found it difficult to give an actual biological ancestry to humans. Briefly he tracked humans back as far as the Old World monkeys (Catarrhina), saying that the human species must have diverged from the original monkey stock considerably earlier than anthropoid apes, probably at a point close to now-extinct forms of Lemuridae. He further recognised the great apes as humanity's nearest relatives. Darwin knew very little about fossil monkeys and could name only *Dryopithecus,* the largest fossil ape identified in the deposits of Europe at that time. He knew almost as little about fossil mankind, making only a passing reference to the Neanderthal skull, still a disputed fossil. For the second edition of *The Descent of Man* he asked Huxley to fill this gap with an up-to-date essay about fossil finds. He could only guess at possible reasons for human ancestral forms to have abandoned the trees, to lose their hairy covering, and to become bipedal. Nevertheless, he used Haeckel's work in this area to push the primate line back through marsupials, monotremes, reptiles, amphibians, and fishes, ending up at the ascidians, grandfathers of them all. Darwin wrote that Aleksandr Kovalevsky had informed him of his researches into vertebrate ancestry at the Zoological Station in Naples.

> The larvae of Ascidians are related to the Vertebrata, in their manner of development, in the relative position of the nervous system, and in possessing a structure closely like the *chorda dorsalis* of vertebrate animals. It thus appears, if we may rely on embryology, which has always proved the safest guide in classification, that we have at last gained a clue to the source whence the Vertebrata have been derived.[57]

At the end of his discussion of the human family tree he paid tribute to the variety and depth of Haeckel's learning, declaring that if Haeckel had published earlier his *Natürliche Schöpfungsgeschichte* (1868, English translation 1870), in which he discussed the genealogy of mankind, Darwin would not have pursued his own volume on the same subject.

This was a startling ancestry to propose. Yet even the most traditionally minded would see something admirable in Darwin's absolute sincerity. William Darwin Fox regarded his cousin's work with interest. He was not as surprised as he felt he ought to be. It was a curious situation for him—as a country parson—to have a dangerous author as a friend and relative.

> I suppose you are about to prove man is a descendant from Monkeys &c &c. Well, Well!—I shall much enjoy reading it. I have given up that point now. The three main points of difference to my mind—were that Men drink, smoke & thrash their wives—& Beasts do not. . . . I do not think even *you* will persuade me that my ancestors ever were Apes—but we shall see. I have no *religious* scruples about any of these matters. I see my own way clearly thro them—but I see many points I cannot get over, which prevent my going "the whole Hog" with you. . . . Why do not you & Mrs. Darwin run over here, when you have finished your Book—& you can study my little Apes & Apesses.[58]

## VII

At the centre lay Darwin's idea of sexual selection. This was his special contribution to the evolutionary story of mankind, his answer to Wallace, Lyell, and others, and to all the reviewers and critics of the previous twelve years. "I do not intend to assert that sexual selection will account for all the differences between the races," he wrote in his book. Nonetheless, he felt certain that it was "the main agent in forming the races of man." Sexual selection was "the most powerful means of changing the races of man that I know."

In brief, Darwin claimed that human beings were like animals in that they possess many trifling features that are preserved and developed solely because they contribute to reproductive success. Just as peacocks had developed tail feathers to enhance their chances in the mating game, so humans had developed characteristic traits that promoted individual reproductive success. These traits were fluid, changeable, and not directly related to adaptation and survival. But Darwin pushed this claim far beyond the mere acquisition of secondary sexual characteristics. By these means he thought he could also explain the divergent geographical and behavioural attributes of human beings, such as skin colour, hair texture, maternal feelings, bravery, social cohesion, and so forth. Preference for certain skin colours was a good example. Men would chose wives according to localised ideas of beauty, he suggested. The skin colour of a population would gradually shift as a consequence.

Similarly, sexual selection among humans could enhance mental traits such as maternal love, bravery, altruism, obedience, hard work, and the "ingenuity" of any given population; that is, human choice would go to work on the basic animal instincts and push them in particular directions.

> The strongest and most vigorous men—those who could best defend and hunt for their families, and during later times the chiefs or headmen—those who were provided with the best weapons and who possessed the most property, such as a larger number of dogs or other animals, would have succeeded in rearing a greater average number of offspring than would the weaker, poorer and lower members of the same tribes. There can also be no doubt that such men would generally have been able to select the more attractive women. . . . If then the several foregoing propositions be admitted, and I cannot see that they are doubtful, it would be an inexplicable circumstance if the selection of the more attractive women by the more powerful men of each tribe, who would rear on average a greater number of children, did not after the lapse of many generations modify to a certain extent the character of the tribe.[59]

In effect, humanity made itself by producing and preserving differences, a process that broadly mirrored his understanding of artificial selection in which farmers chose traits for "use or ornament," impressing their own taste or judgement on organisms.

He ventured onto thorny ground when he analysed human societies in this way. His naturalism explicitly cast the notion of race into evolutionary and biological terms, reinforcing contemporary ideas of a racial hierarchy that replicated the ranking of animals. And he had no scruple in using the cultural inequalities between populations to substantiate his evolutionary hypothesis. Darwin certainly believed that the moral and cultural principles of his own people, and of his own day, were by far the highest that had emerged in evolutionary history. He believed that biology supported the marriage bond. He believed in innate male intellectual superiority, honed by the selective pressures of eons of hunting and fighting.

> To avoid enemies, or to attack them with success, to capture wild animals, and to invent and fashion weapons, requires the aid of the higher mental faculties, namely, observation, reason, invention, or imagination. These various faculties will thus have been continually

put to the test, and selected during manhood. . . . Thus man has ultimately become superior to woman.[60]

The possibility of female choice among humans hardly ruffled the surface of his argument, although he repeatedly claimed that female choice was the primary motor for sexual selection in animals. Primitive societies, he conceded, may be matriarchal or polygamous. However, he regarded this as an unsophisticated state of affairs, barely one step removed from animals. Advanced human society, to Darwin's mind, was patriarchal, based on what was then assumed about primate behaviour and the so-called "natural" structure of civilised societies. For Darwin, it was self-evident that in civilised regimes men did the choosing. A limited number of women might sometimes be in a position to choose their mate (he was perhaps thinking of heiresses, or royalty, or beautiful heroines in novels). But his vision of mating behaviour was an explicit expression of his class and gender. His personality was evident too. His description of courting practices in *The Descent of Man* gave a romanticised picture of "rustics" at a country fair, "courting and quarrelling over a pretty girl, like birds at one of their places of assemblage." For him, Victorian males set the evolutionary compass.

Try as he might, he could not escape the complications of his work. "I find the man-essay very interesting but very difficult; & the difficulties of the Moral sense have caused me much labour," he told Asa Gray in 1870.[61] He was anxious about breaking new ground in so many different areas. Above all, he wanted to get these notions about sexual selection absolutely right. "Sexual selection has been a tremendous job," he wrote to Wallace. "Fate has ordained that almost every point on which we differ shd. be crowded into this vol."[62]

## VIII

At last he finished and dispatched the manuscript to John Murray, his publisher. Murray flinched a little at the subject matter. Despite his familiarity with Darwin's unorthodox topics and his determination not to let them stand in the way of a successful business relationship, this book on human ancestry rattled his belief in the Bible story rather more than the *Origin of Species* had done. Gingerly, he asked his friend Whitwell Elwin for his opinion and was not surprised at the blast that came back by return of post. Elwin was no longer editor of Murray's *Quarterly Review* but he still possessed the principles of a country clergyman. "It might be intelligible that a man's tail should waste away when he had no longer occasion to wag it," he roared, "though I should have thought that savages would still have found it useful in tropical climates to brush away insects. . . . The arguments in the sheets you have sent me appear to me to be little better than drivel."[63]

Murray partly agreed. Bit by bit, in his spare time in the evenings, the publisher began piecing together a scientific commentary of his own, a modest criticism of Lyell and his associates that he called *Scepticism in Geology,* published in 1877 under the nom de plume "Verifyer," in which he politely, but decisively, disassociated himself from the secular natural history he had successfully placed before the public. Murray was neither a radical nor a conservative in religious affairs, being middle-of-the-road, and his personal dilemma over the age of the earth and "natural development in other branches of natural history" surely reflected at least some of the discomfiture of many of Darwin's more ordinary readers. Insofar as Murray ever let his personal opinions show, this was it. He answered back.

Henrietta Darwin was evidently made of sterner stuff, for she corrected the proofs of *Descent of Man* while she was in the south of France with her cousins Edmund and Lena Langton, scarcely turning a hair at her father's blunt talk about sexual display. In asking her to do this, Darwin relied on her editorial competence. When Thomas Farrer met Henrietta at a social event in London later that winter, he "chastised" her humorously on Darwin's behalf for being out on the town enjoying herself when *The Descent of Man* was not yet published.

Henrietta had first read proofs for her father when she was eighteen and he was producing *Orchids;* and he had increasingly leaned on her during his long illness from 1864 to 1866. All members of the family were accustomed to help with his books in one way or another. Francis remembered how his father would correct proofs first in pencil, and then in ink, getting the younger children to rub out the pencil marks afterwards. Emma sometimes copied manuscripts for him, a point substantiated by one of the few surviving pages of the original manuscript of *The Descent of Man* being in her handwriting.[64] She would read proofs, too, although "chiefly for misprints and to criticise punctuation; & then my father used to dispute with her over commas especially." Henrietta's role as editor grew naturally out of the rest. She was good at it. "He often used to say what a good critic Hen. was, & would sometimes laughingly quote her pencil notes, such as 'this sentence is horrid.' "[65] There is little evidence to suggest that Darwin used her merely as a convenient feminine censor, or as a ready-made moral vigilante, helping him to identify in *The Descent of Man* any hint of nineteenth-century impropriety.

On the contrary, she tightened his prose, wrote comments in the margin, and indicated passages that were hard to understand. These were all tasks he felt unable to ask his men friends to undertake. A friendly appreciation of each other's intellect began to emerge, a mutual sympathy enjoyed by both of them. Strictly demarcated as their intellectual input was, Darwin evidently valued his women for their advice as well as their labour.

He could not stop himself issuing a slew of fatherly instructions.

> My dear H.
>
> Please read the Ch. first *right through* without a pencil in your hand, that you may judge of general scheme; as, also, I particularly wish to know whether parts are extra tedious; but remember that M.S is always *much* more tedious than print.—The object of Ch. is simply comparison of mind in men & animals: in the next chapt. I discuss progress of morals &c. . . . I do not send foot-notes, as I have no copy & they are almost wholly mere authorities.—After reading once right through, the more time you can give up for deep criticism or corrections of style, the more grateful I shall be.—Please make any long corrections on separate slips of paper, leaving narrow blank edge, & pin them to margin of each sheet, so that I can turn each back, & read whilst still attached to its proper page.—This will save me a world of troubles. Heaven only knows what you will think of the whole, for I cannot conjecture.—You are a very good girl indeed to undertake the job. . . . (I fear parts are too like a Sermon: who wd. ever have thought that I shd. turn parson?)[66]

Henrietta must have cut a curious figure abroad, spending the morning correcting her father's account of sexual selection, then putting on a bonnet and shawl to stroll along the promenade in "wicked Monaco," the fashionable gambling resort and centre of the European *beau monde*. This dual experience probably did more to mould her views about human relationships than any other before her marriage. She liked working with her father and felt she understood his arguments. In fact, she surely learned more about men's biological urges than her parents ever expected her to know. It is clear from the few proof sheets that are still in existence that she read the whole manuscript, ranging from the sexual attractiveness of beards to the numerical proportion of the sexes.

"Your corrections & suggestions are *excellent,*" Darwin assured her. "I have adopted the greater number, & I am sure that they are very great improvements.—Some of the transpositions are most just. You have done me real good service; but by Jove how hard you must have worked & how thoroughily [*sic*] you have mastered my M.S. I am pleased with this chapter now that it comes fresh to me." He signed himself "Your affectionate, admiring & obedient Father."[67] Afterwards he gave her a gift of £30 from the profits as acknowledgement of the help he had received. "Several reviewers speak of the lucid, vigorous style, &c. Now I know how much I owe to you in this respect, which includes arrangement, not to mention still more important aids in the reasoning."[68]

Notwithstanding these womanly interventions, Emma Darwin experienced misgivings about the book's subject matter. "I think it will be very interesting, but that I shall dislike it very much as again putting God further off," she sighed to Henrietta. These thoughts were not shared with Darwin. Husband and wife were probably too set in their individual ways for any discussion on the point to have made a difference. They each knew the other's position. Moreover, they both apparently felt easier confiding in Henrietta. Even so, Emma also read the proofs of the *Descent of Man* with a conscientious desire to be helpful. She warned her husband of the dangers of too much anthropomorphism. "F. is putting Polly into his Man book but I doubt whether I shall let it stand," she remarked. Only a wife could be so candid about a favoured example. Polly was Henrietta's dog, a small terrier, as devoted to Darwin as Darwin was to her. "A fond grandfather is not to be trusted," declared Emma robustly.[69]

Shortly afterwards Darwin discovered that his publisher was apprehensive about the subject too. Apes, reproductive behaviour patterns, and human beings in the same book struck John Murray as a recipe for disaster.

> It is with a view to remove any impediments to its general perusal that I wd. call your attention to the passage respecting the proportion of advances made by the two Sexes in Animals. I wd. suggest that it might be toned down—as well as any other sentences liable to the imputation of indelicacy if there be any.[70]

Surprised, Darwin inquired which passages Murray found indelicate. When these were disclosed, he changed them into direct quotations from the original authors. A month later, Murray was back with worries about the title. Darwin's proposal had been simple—"On the Origin of Man." But Murray wanted something less provocative, something more closely related to the contents, more explanatory for intended purchasers. He rejected Darwin's next suggestion, feeling that the word "sexual" could not be used on a title page. "The Descent of Man & Selection according to Sex," would be much better, he proposed, and would "get rid of an objectionable adjective."[71] It was later changed to "in relation to sex."

This time around, Murray prepared a number of special copies for Darwin to present to his friends. These Darwin signed personally, full of warm regard for the men who had come such a long way with him. "I hear you have gone to press, & I look forward with fear & trembling to being crushed under a mountain of facts!" remarked Wallace with a friendly smile.[72]

Despite the worries, the *Descent of Man* was the first of Darwin's titles to make a handsome publishing profit when it was published in February 1871. "I suppose abuse is as good as praise for selling a book," remarked its author. Murray sent a cheque to Darwin for £1,470 with an appreciative nod. "You have produced a book wch. will cause men to prick up what little has been left them of *ears*—& to elevate their eyebrows. . . . it cannot fail, I think, to be much read."[73] For all his misgivings, he was grateful to have this valuable author within his doors. Although other publishing houses were capitalising on the increasingly lucrative evolutionary market, Murray retained the golden goose.

## IX

On the face of it, 1871 was not auspicious for any of Darwin's usual forms of strategic publicity. The Franco-Prussian War, then at its height, seemingly obliterated any prospect of European editions. Even so, he optimistically sent proof sheets to every overseas friend who had expressed a willingness to translate, admitting that "some delay may be advisable."[74]

Astonishingly, in view of the political situation in Prussia, crushing defeats for France at Sedan and Metz, and especially during the "terrible year" of the siege of Paris and the dreadful events around the Commune, the *Descent of Man* went into Dutch, French, German, Russian, and Italian in 1871 and into Swedish, Polish, and Danish shortly thereafter, a testimony to the fortitude of Darwin's colleagues and general interest in evolutionary affairs.

In Britain, comments were muted. Assuredly, reviewers shrank from closing the obvious gap between animals and mankind and objected to descent from "tadpoles." How could mankind become "more crafty than the fox, more constructive than the beaver, more organized in society than the ant or bee?" inquired Sir Alexander Grant frostily in the *Contemporary Review.* Since primitive humans showed no discernible signs of progress it was inconceivable to him that there might be any links between "the backwaters and swamps of the stream of humanity" and cultivated English gentlemen. *Harper's Weekly* complained that "Mr. Darwin insists on presenting Jocko as almost one of ourselves." The *Times* was more emphatic still. "The earliest known examples of Man's most essential characteristics exhibit his faculties in the greatest perfection ever attained. No poetry surpasses Homer."[75]

Other reviewers in other journals picked on the same points. If Darwin's ideas were accepted, said the geologist William Boyd Dawkins in the *Edinburgh Review,* "the constitution of society would be destroyed. . . . Never perhaps in the history of philosophy have such wide generalisations been derived from such a small basis of fact."[76] An anonymous reviewer in the *World,* a New York literary magazine, passed much the same opinion: "Mr. Darwin, like the rest of his atheistic school, evidently rejects with contempt the idea of a spiritual God who creates and sustains the universe."[77] The *Truth Seeker* called the book "hasty" and "fanciful." Another anonymous writer in the Catholic *Tablet* ponderously explained that human beings possessed rationality, a "perfectly distinct faculty from anything to be found in the brutes." The *Spectator*'s reviewer said that "Mr. Darwin has shocked the deepest prejudices and presuppositions" of the English people. A correspondent in the *Guardian* summed them all up by appealing to the direct evidence of the Bible. "Holy Scripture plainly regards man's creation as a totally distinct class of operations from that of lower beings."[78] A columnist in the *San Francisco Newsletter* stooped to a poor joke about Darwin's imaginary son being "not exactly quadrumanous," but "just as handy with his feet as he is footy with his hands."[79]

Few among these countenanced descent from animals. Yet the authors were exceedingly polite about Darwin himself. A reviewer in the *British and Foreign Evangelical Review* praised Darwin's depth of learning. The *Daily Telegraph* referred to his "graceful and conciliatory" prose and "dignified" tone of voice. The *English Independent* suggested that "no loyal servant of the truth will fear the issue of such an appeal." The *New York Express* noted the author's "unassailable integrity and candour," while the *Field* described his "wonderful thoroughness and honest truthfulness." The remarks showed that Darwin's position as a respected man of letters was high. Unknown reviewers in newspapers and periodicals clearly felt that his opinions were worth careful consideration. And the evolutionary debate had by now moved away from the blood-spattered warfare of the early 1860s. Darwin's reputation as an honest man was enhancing the way his volume was being received. Even a widely distributed newspaper like the *Liverpool Leader* could close its eyes and think of England, proposing that *The Descent of Man* was "perfectly consistent with the belief in God the Creator."[80] With a reckless wave, the religious journal the *Nonconformist* wished Darwin "god-speed in his inquiries."

Darwin noted all this in amused surprise. "I think you will be glad to hear, as a proof of the increasing liberality of England, that my book has sold wonderfully," he told Ray Lankester, "and as yet no abuse (though some, no doubt, will come, strong enough), and only contempt even in the poor old *Athenaeum.*" Darwin rather regretted that the *Athenaeum* was losing its anti-Darwinian nerve. An anonymous versifier in the Tory periodical *Blackwood's Magazine* gave up the attack completely and offered a poem about apes to be sung to the traditional country air of *Greensleeves.*[81]

Almost on cue, the Rev. Francis Orpen Morris burst into the open in an offensive little book called *All the Articles of the Darwin Faith* (1875). Morris parodied the Anglican creed by beginning every sentence with the phrase "I believe . . ." and following it with some remark ostensibly drawn from Darwin's *Descent of Man.* Each remark became progressively more insulting.

> *I believe* that although the Mosaic account of creation is borne out by the testimony of the rocks in a most wonderful manner, yet as it does not suit the theory I have taken into my head, it cannot possibly be true, and I do not believe a word of it.
>
> *I believe* that no one who believes in the Bible has any sense or wisdom compared with me.[82]

Priced at one shilling, and attractively packaged in illustrated covers, this tract affirmed the vitality of the evolutionary controversy in Britain's popular marketplace. "Keep as a curiosity of abuse," wrote Darwin across the top of his copy.

By contrast, Wallace was generous to a fault. "Darwin's book on the whole is wonderful!" he told a friend. "There are plenty of points open to criticism, but it is a marvellous contribution to the history of the development of the forms of life."[83] He reviewed it, in a signed article, in the *Academy.* True, he pointed out the places where he disagreed with Darwin, especially their differences over the reasons for protective coloration. He never agreed with sexual selection either. Yet he commented gracefully on Darwin's view of human evolution. Darwin responded appreciatively. "If I had offended you, it would have grieved me more than you would readily believe."

> Your note has given me very great pleasure, chiefly because I was so anxious not to treat you with the least disrespect, and it is so difficult to speak fairly when differing from any one. . . . I care now very little what others say. As for our not quite agreeing, really in such complex subjects, it is almost impossible for two men who arrive independently at their conclusions to agree fully, it would be unnatural for them to do so.[84]

It was probably around this time that Erasmus Darwin wrote to his niece Henrietta, "I think the way he [Wallace] carries on controversy is perfectly beautiful and in future histories of science the Wallace-Darwin episode will form one of the few bright points among rival claimants."[85] To the public, however, evolution usually meant Darwin's theory, not Wallace's. When Wallace went to the British Association meeting in Edinburgh that year he heard Lord Neaves, a well-known wit and song-writer, recite satirical verses on the "Origin of species à la Darwin."[86]

As for himself, Darwin considered that the *Saturday Review* and *Pall Mall Gazette* delivered the most perceptive reviews. He never discovered who was the author of the first, but the second was by John Morley, the literary writer and Liberal politician. All in all, he was "much impressed by the general assent with which my views have been received. . . . everybody is talking about it without being shocked."[87] To a large degree this was surely due to his watchful style of writing and high personal status within British cultural life, a status that he had carefully nursed during the previous decade. There may have been a sense of *déjà vu* for reviewers in rehearsing yet again the controversies that had sprung up when the *Origin of Species* was first published. Apes and angels had been dealt with ten or twelve years before. Faced with a new book about descent in 1871, journalists seemed to find little more to say. They and their readers had become accustomed to the idea of evolution, although not necessarily comfortable with it. Darwin even alluded to the fact in the introduction to *Descent.*

In other areas, too, the times were loosening up. The readership for science was noticeably shifting in focus. In 1872, in an early article written in response to the *Descent of Man,* "Darwinism and Divinity," Leslie Stephen spoke for many of the coming generation by asking, "What possible difference can it make to me whether I am sprung from an ape or an angel?"[88] Stephen proceeded to "give up Noah's Ark," abandoned holy orders, and opted for the genteel life of a well-heeled agnostic, friends with Henry Fawcett, George Meredith, James Russell Lowell, and Charles Eliot Norton.[89] These men were rationalists who advocated religious liberty. As Darwin reflected, "on the whole, the reviewers have been highly favourable."[90]

All except St. George Mivart. Mivart wrote a fierce article for the *Quarterly Review* in 1871 ("a most cutting Review of me"), highlighting the hazards of considering any form of human evolution. This review was one of the most important in Darwin's later career, certainly equal in impact to Wilberforce's attack on the *Origin of Species* that had also been published in the *Quarterly Review* some ten years before. Whitwell Elwin, the *Quarterly Review*'s former editor, was once again the operative force. He had commissioned Mivart to write with the theological difficulties foremost in mind. Although he never met Darwin, and never wanted to, Elwin's effect on Darwin's life through these two reviews was substantial.

Mivart did the job with deadly efficiency. In response, Darwin rolled out his big guns. First, he indulged in a brief but nasty pamphlet war, which satisfied his urge for immediate retaliation. He arranged for the reprinting of an article by Chauncey Wright (already issued in America) that had severely criticised Mivart's 1870 book *Genesis of Species*. This indirect defensive technique had served Darwin well in the past and conveniently allowed him to attack with the words of others while maintaining a reputation for nonconfrontation. But in this case only the Darwinians appreciated the esoteric points Chauncey Wright put forward. Under Darwin's direction, Wright clarified precisely what was, and was not, "Darwinism." The pamphlet was left unread by those people who would be most swayed by Mivart.

Frustrated, Darwin let off steam with a few sharp ripostes in the next edition of the *Origin*. To this sixth edition of the *Origin*, published in 1872, he defiantly added a new chapter expressly directed against Mivart. Here, he seemed to be coming to the end of his tether. He compromised. He defended pangenesis and neutralised natural selection in a manner that allowed considerably more adaptive change in organisms according to use and disuse and the effects of the environment, the most Lamarckian he ever became. It was a cheap edition, intended for mass sales. Darwin had been told how a group of Lancashire workmen were clubbing together to buy a single copy.[91] Impressed, he realised there were more markets to penetrate, more audiences to reach. Yet he felt hemmed in, edgy, and forced to stretch a point. Making these changes bothered him more than usual, and he asked his son William to read the proofs for him. Mivart loomed unpleasantly large in his imagination.

Mostly he watched agog as Huxley savaged Mivart in the *Contemporary Review*. All Huxley's bulldog propensities poured out, and in "Mr. Darwin's Critics" he ruthlessly corrected both Mivart's biology and his interpretation of Catholic doctrine, locating old theological tracts in the university town of St. Andrews (where he was on holiday) to support his cannonade. "How you do smash Mivart's theology: it is almost equal to your article versus Comte," Darwin exclaimed.[92] "The dogs have been barking at [Darwin's] heels too much of late," Huxley explained. Hooker thought Huxley's attack was too cruel and told Darwin so. Darwin replied that he was obviously not so good a Christian as Hooker, "for I did enjoy my revenge." Hooker found it slightly surprising to hear Darwin sneer against Mivart's "bigotry, arrogance, illiberality & many other nice qualities." Even Huxley and Hooker thought better of Mivart than that.

Perhaps the argument might have ended there—distasteful, unpleasant, but final—if Mivart had not then gone on to criticise one of Darwin's sons. In 1873, George Darwin published a short article in the *Contemporary Review* suggesting that divorce should be made easier in cases where cruelty, abuse, or mental disorder became evident. In this George was exploring his developing views on heredity, feeling that quicker and simpler divorce, or easier access to contraception, could prevent traits like criminality or mental deficiency being passed on through the family line. A score of similar papers were published every year in Britain. Yet because it was by a Darwin, George's paper attracted Mivart's attention.

Mivart read the article with undisguised horror. He responded violently, accusing George of ignoring all decency ("hideous sexual criminality . . . unrestrained licentiousness"), and veered close to libel, as Darwin indignantly noted. If Mivart's statements were accepted by readers as true, then George's reputation as a gentleman would be in tatters, Darwin huffed. Father and son consulted desperately together, with the result that George called for a public apology. Mivart reiterated his charge.

Shaken, Darwin looked into the possibility of a lawsuit. "I care little about myself, but Mr. Mivart . . . accused my son George of encouraging profligacy, and this without the least foundation." He slapped Mivart's article in front of all his friends demanding their opinion; and declared to Wallace that "the accusation was a deliberate falsification." Huxley loyally counter-attacked in the *Academy,* ignoring Hooker's warning that it would be much better to send Mivart a private reprimand. The business had, however, gone too far for an apology, even if one was offered, to make any difference. Huxley cut all connections with Mivart, telling Darwin that he was prepared to defend Darwin's son as if he were his own. "I do not think I shall resist telling him how base a man I think him," fulminated Darwin. "You have been, my dear Huxley, most generous in this whole affair."[93]

Cross and powerful, the two united in dislike of a common enemy. A trivial, spiteful incident sealed the sorry episode. When Mivart tried to join the Athenaeum Club under the rule that permitted men of excellence to avoid the usual waiting period, his election was prevented by the Darwinians, X Clubbers to a man. Huxley cast the harshest blow possible by declaring that Mivart's scientific work was not "up to the mark of a Committee election," not as excellent as the rules required. Mivart and his proposer were damned as "brother Jesuits to the backbone."[94]

In fact, the Mivart episode has long fascinated historians for the way it exposed unseen cracks in the Darwinian movement and the heavy emotional investment channelled into it by leading figures like Darwin and

Huxley. It seems more than probable that Darwin was personally wounded by Mivart's defection. For Mivart to reject Darwin's theory, in this regard, was to reject Darwin himself. Darwin never forgave him. On his part, Huxley reacted as if Mivart were criticising the whole of modern science and digging himself ever more deeply into the church's foundations. Both these men felt betrayed. Mivart did not emerge unscathed from the exchange either. Not only was he excommunicated by the Darwinians as a traitor, he was also excommunicated by his own church for his belief in evolution, a sacrificial lamb for each unforgiving camp. Of all the casualties of the Darwinian movement, his was the most pitiable.

Sympathetically, Darwin's friends rallied round. "I am very sorry you are so unwell, & that you allow criticisms to worry you so," wrote Wallace in the summer of 1871. "Remember the noble army of *converts* you have made! & the hosts of the most talented men living who support you wholly."[95]

"Oh Lord, how difficult accuracy is!" Darwin said as letter after letter arrived at Down House disputing his statements.

## X

The main turning point for Darwin and Emma that year was their daughter's marriage in August 1871. Henrietta was the first child of theirs to marry, and the only daughter to do so. Bessy remained a spinster. Henrietta's parents viewed the event with mixed emotions, apprehensive about the rapidity of her courtship. They depended on Erasmus, much closer to the scene of action in London, to tell them whether Richard Litchfield was suitable. It "feels very odd that Hen shd be so intimate with a person of whom I know so little," worried Emma. "I feel quite at ease with him & that he is very nice, but I really have not seen much of him."[96] Erasmus assured them that Litchfield, though penniless, was not a "gold-hunter." The remark was not wasted on her father, who, with George Darwin, was arranging the marriage settlement. George reported that their solicitor recommended a settlement of "£5000 of something of the nature of Debenture stock," and that Darwin should "make the yearly income up to 400£ or 350£, as Litchfield is not a grasping sort of man."[97]

Litchfield was ten years older than Henrietta, trained as a barrister but not practising, who took a minor post in the Ecclesiastical Commission essentially for the pay packet. From 1860 or so he had dedicated his energies to the Working Men's College in London, a philanthropic educational venture of which he was singing master and then principal. The college promoted a sub-socialist, utopian, self-improving vision in which the men were taught by progressive thinkers such as John Ruskin, F. D. Maurice,

Thomas Hughes, and Vernon Lushington, another friend of the Darwins. Henrietta's brothers considered Litchfield a "cool beast," yet they came to like him for all that and ultimately respected his opinion. Henrietta tartly informed George that "you must try to like him for my sake. . . . he seems to be friends with all our sort of people, Spottiswoode, Vincent Thompsons, Lushingtons etc."[98]

She met him while staying with Erasmus Darwin and the Wedgwoods and was primarily drawn by his musical ability. Litchfield was no flag-waving political reformer, but he held liberal views, especially on education in its widest sense. He organised Sunday-afternoon excursions for his working men, during which a group of thirty or forty people, men and women, would walk out of London for a healthy day in the country, returning by train after tea, often singing madrigals and glees. Henrietta was smitten. With his large brown beard, high moral principles, and dedication to duty, he may have seemed like another version of her father. A new life beckoned. She met him in June, became engaged in July, and was married in August.

Before then she indicated a few hesitations in her diary. She regretted that Litchfield did not sweep her up in his arms. "What exquisite joy" she would feel, she wrote, if he had spontaneously appeared at Down House to seek her out one Sunday. She briefly wondered—as all Darwin's children must have wondered—whether he loved her for herself. "I think he must care—it can't be only that he thinks I shd. be a nice sort of person to marry." She regretted that the conventions of the day required that she must not make the first loving advances, a point rammed home by her father's *Descent of Man* as well as the required delicacy of the times. And she recorded her discussions about faith with Hope Wedgwood. She was not very religious, she thought. She said she felt none of Hope's "transcendental emotions."

She took a leaf out of Darwin's book and confided some of these religious shortcomings to Litchfield before her marriage. Arthur J. Munby, a friend of Litchfield's, said that this "petite young woman of 27, with a face not unlike the photographs of her father but very feminine and tender, with bright hazel eyes, and every feature full of life and expression," had told Litchfield that she "did not believe in a personal God."[99] It should perhaps be mentioned that Bessie Darwin also disputed conventional faith. In 1866 she had refused to get confirmed, telling her mother that she believed in neither the Trinity nor the catechism. "Lizzy says she shd feel hypocritical to have anything to do with the Cat & that as she does not believe in the Trinity or in Baptism she does not feel much heart for it," noted Emma. These daughters were not slavish hierophants. Many years later, Henrietta contributed an appreciative word or two about Munby to the *Working Men's College Journal,* although it is clear that even then she had no idea about Munby's unconventional living arrangements with his servant Hannah. Somewhat underestimated by historians, and evidently more thoughtful about her religious position than previously assumed, Henrietta appears in these records as an intelligent, independent, caring young woman. Her parents found it a wrench to let her go.

The impending marriage threw Darwin into "very bad health." As the day drew nearer he cast a morbid gloom over the proceedings. It was Darwin's duty to escort Henrietta to the altar and formally "give her away." But the combination of acknowledging her forthcoming separation from him, the walk up the aisle, and public performance before a God in whom he did not believe was possibly too much. "He could hardly bear the fatigue of being present through the short service," said Francis.

"It was with much exertion that he came to the village church for the wedding," recalled Henrietta many years later.

> Any sort of festivity was quite out of the question, and no friends or relations were invited. But a few of Richard's working men friends managed to find out the day and the hour, and walked the four miles from Orpington Station in order to be present at the ceremony. Great wonder was roused in the mind of our old butler (who was in fact one of the family) as to who these strangers could possibly be, for every face was known in the little village.[100]

After such a start, it was probably inevitable that both Henrietta and Richard Litchfield became ill on their wedding tour in Europe, a situation echoing her parents' marriage and one that proved hard for the newlyweds to handle with equanimity so far away from home. Her silence as to the cause of the trouble may indicate some gynaecological problem. As far as is known, Henrietta never became pregnant, although she certainly at the beginning of her marriage expected children. Darwin's views on the matter were mixed. A year later, when some signs of pregnancy might have been anticipated if all were well, he explained to his cousin William Fox that "Henrietta has no child, & I hope never may; for she is extremely delicate."[101] Henrietta suffered from intermittent collapses in health for the next four or five years, some of which were very disabling, and she and Litchfield moved into a form of relationship that mirrored her previous invalid experiences. "R is a jewel of a nurse," she wrote from Cannes on her honeymoon. "We feel very married each lying sick in our beds as if we'd been at it 30 years like Father and you."[102]

Darwin felt the break dreadfully. "I was a favourite of yours before the time when you can remember," he reminisced sentimentally.

> How well I can call to mind how proud I was when at Shrewsbury, after an absence of a week or fortnight, you would come and sit on my knee, and there you sat for a long time, looking as solemn as a little judge.—Well, it is an awful and astounding fact that you are married; and I shall miss you sadly. But there is no help for that, and I have had my day and a happy life, notwithstanding my stomach; and this I owe almost entirely to our dear old mother, who, as you know well, is as good as twice refined gold. Keep her as an example before your eyes, and then Litchfield will in future years worship and not only love you, as I worship our dear old mother. Farewell my dear Etty. I shall not look at you as a really married woman until you are in your own house. It is the furniture which does the job.[103]

## XI

It was just as well that he had another project to keep him occupied. As soon as *The Descent of Man* was published, he returned to the intriguing theme of facial expression. The subject appealed to him, and had done so for many decades. Admittedly, he had at first intended to include all of his material on the subject in a single chapter in *The Descent of Man,* and he had already collected a great deal of it to that purpose. But it became far too bulky to include. "I have resolved to keep my Essay on Expression in Man & animals for subsequent & separate publication." Now he relished the opportunity to mould his ideas into a separate volume making a sequel to *Descent.*

His delight was obvious. "I feel an exaggerated degree of interest in the subject of expression," he said to Franciscus Donders, in Utrecht, before asking him a complicated physiological question. He loved the sensation of breaking new ground, of uniting disparate fields, and the zest of setting out on a fresh line of inquiry.

Friends and family found the topic just as attractive. People from every corner of Darwin's daily life supplied him with quaint stories about animal expressions. Dr. James Paget knew a terrier that frowned in concentration. Lady Lubbock described the intelligent faces pulled by her pug-dog. Johann Krefft, a museum-keeper in Australia, told him about monkeys throwing temper tantrums like a child, while Alois Humbert saw a hummingbird persistently deceived by flowered wallpaper. On and on the letters flowed, each receiving a place in Darwin's researches. Charles Spence Bate, the dentist and naturalist whom Darwin recalled from his barnacle days, wrote to him with an interesting, though anthropomorphised, account of a old dog who showed "moral courage" while having his teeth extracted.

In turn, Darwin delved into his own and his family's experiences. Pain, of course, was more or less a daily accompaniment. He wrote to Donders about vomiting.

> I had not thought about irritating substances getting into nose while vomiting; but my clear impression is that mere retching causes tears; I will however try to get this point ascertained. When I reflect that in *vomiting* (subject to the above doubt), in violent coughing from choking, in yawning, violent laughter, in the violent downward action of the abdominal muscle as during the evacuation of faeces when constipated, & in your very curious case of the spasms, —that in all these cases, the orb-muscles are strongly and unconsciously contracted; & that at the same time tears often certainly flow, I must think that there is a connection of some kind between these phenomena.[104]

At home, he studied Francis playing the flute, watching his mouth and the muscles straining in his neck.[105] "I have got a good deal of information about the pouting of children of savages, & this makes me wish much for precise details about the pouting of English children," he went on to ask William.

> None of you children ever pouted. I am the more interested, as I fully believe that Pouting is a vestige (an embryonic relic during youth) of a very common expression of the adult anthropomorphous apes when excited in many ways.[106]

Then, after returning from honeymoon, Richard Litchfield tentatively offered his new father-in-law some thoughts on the origins of music and singing. These struck Darwin as "very good." Eager to find something in common, the two men discovered that this book on the expression of the emotions gave them a mutual topic of interest. Litchfield helped Darwin write a section about song emerging as a courting ritual among animals. The ability of music to stir the emotions was something that Darwin could also evidently discuss with Litchfield, and he confided to him that the expressive beauty of Effie Wedgwood's (now Farrer's) voice moved him to tears.[107] Together they criticised Herbert Spencer's theory of the origin of music. Even a son-in-law could be drawn into Darwin's preoccupations.

Young women in Darwin's circle of acquaintance revealed themselves as capable anthropological observers. Margaret Vaughan Williams (Jos and Caroline Wedgwood's middle daughter) helped him with babies' expressions, first describing her own infants and then those of her friends and other family members. "Mary Owen's 3½ yr. old child has a habit of sticking out her lips when she feels shy, but as it is *not* a pout of sulkiness, I do not know if you care about it," she reported.

> She makes no sound. The lips do not seem to become tubular (that is the corners are not drawn together, or hardly). The upper lip is stiffened and projected beyond the lower one, (tho' both stick out to a certain extent) the lips sometimes not quite closed.[108]

Topics like these provided him with easy access into the domestic kingdoms primarily run by women. Darwin was remembered by his nieces for this appreciative attention to their babies' development.

Dogs predictably played a role. Darwin included in his researches his dog Polly, the terrier formerly owned by Henrietta, and Bob, the stable dog. After Henrietta's marriage, Polly adopted Darwin completely. "She has taken it into her head that F. is a very big puppy. She is perfectly devoted to him. . . . She lies upon him whenever she can, and licks his hands so constantly as to be quite troublesome. I have to drag her away at night," declared Emma. This was the dog that slept in a basket by the study fire while Darwin wrote. She appeared in his *Expression of the Emotions* either catching a biscuit on her nose (Darwin thought she was very clever) or as a pictorial example of a "Small dog watching a cat on a table," a copy-book illustration of alertness and attention. Nor were these her first scientific appearances. Darwin was accustomed to claim, with an admiring pat, that her multicoloured coat proved his hypothesis of pangenesis. After she had a bad burn as a puppy, her hair had grown back red instead of white. "My father used to commend her for this tuft of hair as being in accordance with his theory of pangenesis; her father had been a red bull terrier, thus the red hair appearing after the burn showed the presence of latent red gemmules."[109]

Huxley saw the amusing side of this fireside philosophy and scoffed at Polly's elevated place in Darwinian doctrine. He called her the *Ur-hund* (ancestor-dog or idealised type of dog) and sent a drawing of her imaginary evolutionary tree in which equal doses of cat and pig appeared.[110] She was "more remarkable for the beauty of her character than her form," retorted Henrietta defensively.

Bob featured here and there as well, a large dog, full of character. He and Parslow used to sit under the cherry trees every summer, Parslow with a gun to scare the birds, Bob waiting to bark at them.[111] Darwin used him to explain his principle of emotional antithesis, in which individual expressions were said to develop as opposites to an earlier, more basic, emotion. A dog's attitude of submission probably emerged as a deliberate reversal of the signs for aggression.

Similarly, the attitude of dejection was the reverse of the expression of pleasure. Bob was the dog who used to put on a "hot-house face," an attitude of utter despair, when he realised that his master intended visiting the greenhouse rather than striking out on a long country walk.

> This consisted in the head drooping much, the whole body sinking a little and remaining motionless; the ears and tail falling suddenly down, but the tail was by no means wagged. With the falling of the ears and of his great chaps, the eyes became much changed in appearance, and I fancied that they looked less bright. His aspect was that of piteous, hopeless dejection; and it was, as I have said, laughable, as the cause was so slight. . . . It cannot be supposed that he knew that I should understand his expression, and that he could thus soften my heart and make me give up visiting the hot-house.[112]

## XII

For his research into expressions, Darwin made extensive use of photographs and line drawings. Previously, he had little need for illustrations in his investigations or in his written texts, except for a few minimally informative charts and diagrams. This is not to say that he had no visual appreciation or that he failed to think in pictorial terms. Quite the reverse. When he did need illustrations, as in his early work on barnacles or the copiously illustrated *Zoology of the* Beagle, he commissioned good natural history artists and laboured over the accuracy of details. He used diagrams and maps in his geological treatises and happily put pictures of pigeons in *Variation* and other animals and birds in the *Descent of Man*. He paid for George Darwin to have lessons in engraving from George Brettingham Sowerby, a noted natural history artist. His two oldest sons, William and George, made many of the original line drawings for his botanical essays and articles.

Significantly for the expression project, however, Darwin interested himself in photography, the growing art form of the century. In his day, Darwin knew or corresponded with several able photographers, including Dr. George Wallich, whose natural history work initially brought them together, Adolphe Kindermann in Hamburg (Camilla Ludwig, the governess, purchased pictures for him in Kindermann's studio), and Oscar Rejlander, who specialised in photographing emotional expressions. He also acquired a number of portrait images from the Bopp photographic firm in Innsbruck and cabinet cards from Giacomo Brogi in Florence. In the end, his collection ran to around one hundred images. During 1866 he paid out a total of £14 in small sums for photographs, nearly doubling his overall costs for "Science" that year.[113]

This interest stretched seamlessly across his personal and working life. At home, he liked to exchange portrait photographs with other men of science and regularly sat to photographers for this purpose, although not without stating that "of all things in the world, I hate most the bother of sitting for photograph." As occasion demanded he also sent Emma and other family members to the London studios. He encouraged William and Leonard to take informal photographs in and around the house, content to see his sons experience the fun of setting up their paraphernalia and messing about with chemicals. The informality did not extend to himself, however. He was never photographed by the boys in his shirtsleeves or at work in his garden or study. Nevertheless, Leonard took several photographs of his father that subsequently became well known, striving for artistic effects in imitation of Cameron's portraits, and on one occasion depicting Darwin in an armchair on the veranda.

The new medium was an important resource for Darwin. For the first time, visual evidence became helpful for his work in the evolutionary area, contributing in its own way to the transformative moment in the late-nineteenth-century sciences when pictorial representations began to play increasingly exciting (and increasingly problematic) roles in the construction of knowledge. Naturally enough, while working on the *Expression* volume he studied books of art illustrations. At one point he must have asked for one or two of Landseer's typically expressive animal scenes to be photographed for him, for he had copies of these and of a number of Madonnas and female saints in his collection. He mostly concentrated on representational photographs, probably thinking that they were somehow more objective in the way they presented reality, more straightforward documents than even nature itself. But he experienced all the usual problems that scientists encounter in turning artefacts into evidence for theories—problems of attribution and authenticity, of understanding the limits of what the material could tell him, of learning photography's distinctive way of participating in the creative process, much as he had once taught himself how to "see" the geological structures underneath a landscape or the evolutionary connections hidden in pigeon feathers or barnacle valves.[114]

He soon discovered that relatively primitive nineteenth-century techniques—dependent as they were on long exposure times—were unable to capture fleeting facial expressions as he desired.[115] Despite these limitations, he collected a number of photographs from professional portrait studios, from medical and psychiatric institutions, and from individual enthusiasts in England and abroad, including one now attributed to Charles Dodgson, and a series from the Office of the Library of Congress.[116] His collection contained few anthropological photographs, undoubtedly a reflection of the difficulties of photographing in the field, the most notable exception being a picture he possessed of members of the von Ambras family, a "hairy family" regarded as medical curiosities.[117]

Because of the impossibility of recording a flickering expression, Darwin was specially grateful for the research being pursued by Guillaume Duchenne, a French physician who experimented on the activity of muscles. Quite by chance Duchenne had encountered in the Paris hospitals a middle-aged man whose facial nerves were insensitive to pain. Duchenne used him as a human guinea-pig for analysing the contraction and relaxation of different facial muscles. In an unusual series of experiments, he applied electric (galvanic) currents to certain points on this man's cheeks and forehead, rather as a laboratory worker might make a frog's leg contract using the same technique, noting as he did so that the process was like "working with a still irritable cadaver."[118] The man's facial muscles could be galvanised and then kept fixed sufficiently long to take photographs. These pictures were issued in a medical album (not for general circulation) in 1862 as *Mécanisme de la physionomie humaine*. The photographs were unsettling. The juxtaposition of the man's forced expressions and the electrical head-dress that created them made for uncomfortable viewing.

Darwin did not mind. He was impressed by Duchenne's analysis, exactly what he was trying to do himself with less satisfactory results. Duchenne demonstrated that there were no special muscles in human faces dedicated to the higher human emotions, overturning the traditional view established by Charles Bell in 1806. Bell's account had been definitive for much of the century. In that account Bell arranged the "passions" in a system based on pain and pleasure, exertion and relaxation, a system that served as his manifesto for understanding the nervous system, and also exemplified the natural theological view that human expressions were designed by God for the purpose of communicating feelings.[119] There were muscles in the human face, Bell claimed, specially designed for the display of God-given emotions such as morality, shame, and spirituality. Darwin—and others like Pierre Gratiolet—had come to reject such a viewpoint completely.

Darwin then went a step further. When he obtained Duchenne's album, possibly a personal gift from Duchenne (with whom he corresponded in 1871), he showed several of the photographs to people of various ages and sexes to find out what emotion they thought was being expressed. As he hoped, most of the expressions were identified and described in more or less the same terms. Fear, anger, and sorrow were immediately spotted, a point that Darwin would mention in his book to vindicate his opinion that some expressions were universal, recognised the world over. Other photographs were more perplexing. Duchenne's picture of an electrically induced smile confused almost all of Darwin's helpers, most of whom identified the expression as unnatural, possibly malicious. A genuine smile from the same patient was easily recognised as such. Darwin reproduced several of these pictures in *Expression* but not before having his printers mask the galvanic probes in all but one illustration (a photograph of simulated mental distress).

From this and other instances of confusion or misidentification Darwin drew valuable conclusions. The human eye was very discerning, even quixotic. He showed an old picture of religious ecstasy to his cousin Hensleigh and recorded that "Hensleigh W. thinks one side more seraphic than the other."[120] His survey revealed that people recognised expressions only if all the muscular details were exactly right—without crinkled eyelids, a laughing mouth meant nothing. So the eye learns to read faces and stumbles over errors in their syntax.[121]

Fascinated, he hunted out Dr. James Crichton-Browne at Wakefield Asylum, in Yorkshire, for access to photographic records of asylum inmates. Darwin thought that the insane would probably have little control over their mental processes and hence display emotions in a clear, uncomplicated way. By matching the patients' faces to their medical records he could inquire into human rationality and consciousness, issues he had left relatively untouched in *The Descent of Man*. He wondered how consciousness might relate to the emergence of what he called "civilised" behaviour. Was rationality lost by the insane in the same way as it was presumed to be gained by early societies?

James Crichton-Browne was one of the most distinguished psychiatrists of the later nineteenth century, admired for his work on neurological pathology and the classification of mental disorders. Like many of his contemporaries, Crichton-Browne believed there were characteristic "faces" of madness, and, as a keen photographer, he photographed (or arranged to have photographed) all his inmates, labelling each image with the patient's medical diagnosis, invariably one of the "manias" that formed an integral part of Victorian psychiatric classification. Crichton-Browne labelled and sent forty or more photographs of otherwise unidentified asylum patients to Down House, and discussed these and many other points about facial expression with Darwin in letters. His input helped Darwin enormously. "I have been making immense use almost every day of your manuscript," Darwin claimed extravagantly.[122] Tactfully, Darwin did not reproduce any of these psychiatric photographs in his volume except for a single woodcut of a woman's hair bristling like an angry animal. And he made sure to consult other knowledgeable asylum-keepers, such as Patrick Nicol at the Sussex Lunatic Asylum in Haywards Heath. While he hesitated to link mentally disturbed patients directly with primitives, children, or subhuman groups on the evolutionary scale, there can be no denying that he regarded mental activity in all its manifestations as a sure route for mapping the distinguishing biological traits of humanity and the development of nations.

Most usefully of all, Darwin established contact with Oscar Rejlander, the art photographer from Sweden who opened studios in Wolverhampton and then in London, and whose genre studies were appreciated by Prince Albert, Julia Cameron, and Charles Dodgson. This lucky contact lifted Darwin's researches well out of the ordinary and set a precedent for decades of investigation to come. Rejlander's breezy and engaging personality, his enthusiasm for stimulating new projects, and his passion for expressive photography were just the thing to catch Darwin's attention. Irrepressibly, Rejlander threw himself into assisting Darwin in this scientific project.

Rejlander tended to specialise in child character studies. His *Perception, The Young Philosopher,* and *Early Contemplation* were considered uniquely expressive by those in search of unambiguous moral meaning (though overly emphatic for modern taste), and his images of grimy street urchins and a vast composite photograph exhibited in 1857 called *The Two Ways of Life* brought him a degree of fame. Voluble, comically theatrical in his behaviour, and wildly gesticulating, Rejlander would cajole or tease his models until the expression he sought blossomed naturally. The technical difficulties of capturing these in photographs were formidable. Money was always short.

It was Rejlander's genre picture of a vigorously screaming baby that alerted Darwin. This photograph was dubbed *Ginx's Baby* after the title of a popular novel published in 1870 by John Edward Jenkins. In the book the baby interrupted every juncture with screams and yells.[123] Rejlander objected, but the name stuck and soon he was innundated with requests for copies. Sixty thousand prints and 250,000 *cartes de visite* were produced. As he acknowledged, it was not high art but it paid the bills.[124] He hardly liked to say that he had extensively manipulated the image, first of all retouching it, then copying it by hand in chalk in order to rephotograph the chalk drawing alone.[125] It was the rephotographed image that Darwin saw.

To Darwin's eyes, this picture exactly illustrated the information about babies' faces he had tried to elicit from his women friends. He was convinced that very young children cried without tears, and that the characteristic expression of grief gave a square outline to the mouth and created furrows in the cheeks. All he needed was a photograph to demonstrate the point. He approached Rejlander to ask if he could reproduce *Ginx's Baby* in *Expression*. Over the weeks that followed, Rejlander supplied him with a number of other pictures of expressions in children and adults.

The best of it was that Rejlander volunteered to pose himself. Clad in a bohemian dark velvet costume, he struck histrionic attitudes—grief, pleasure, disgust, and so on—and either photographed himself with a time-lapse device or got his wife to aid him. The resulting pictures depended as much on comically exaggerated gesture and body position as on facial expression. On the back of one picture he scribbled in pencil, "My wife insists upon me sending this for you, that your ladies may see that I can put on a more amiable expression." Rejlander's wife posed for a photograph of a sneer (Darwin thought that sneering evolved from the expression of disgust). Gamely, she allowed herself to be reproduced thus in Darwin's volume.

These photographs suited Darwin's purpose. He visited Rejlander in his London studio and maintained friendly contact with him for several years afterwards, glad enough to get him to photograph Polly the dog for *Expression* (reproduced as a line drawing), and personally sitting to Rejlander for a portrait photograph in 1871 that was afterwards reproduced as a line engraving in several magazines. He may have sat to Rejlander again in 1874 when he recorded paying two guineas for a photograph from him. That same year Darwin gave Rejlander a gift of £10 to bolster his declining business—though to little effect.[126] Rejlander died in near penury in 1875.

For all his entertaining histrionics, Rejlander pushed one line of research further than Darwin envisaged. When he learned that Darwin was finding it hard to pinpoint the minute physical differences between laughter and crying—the visual signs were difficult to distinguish—he set about photographing himself sitting next to an enlargement of *Ginx's Baby*, first emulating laughter and then sadness. Using composite techniques, Rejlander printed the two photographs on a single plate for comparison. He had done similar tricks many times before, both as a private joke, in one of which he appeared twice, introducing himself to himself, or in his large composite photograph *Two Ways of Life*.[127] On the back he explained, "Fun only—Here I laughed—ha, ha, ha, violently. In the other I cried e e e. Yet how similar the expression."[128] Darwin found the comparison useful and kept the photograph safely in his collection, perhaps the only copy made and certainly the only one to survive.

"I must have the pleasure of expressing my obligations to Mr. Rejlander for the trouble which he has taken in photographing for me various expressions and gestures," Darwin generously reported in the eventual book.

## XIII

The volume was called *On the Expression of the Emotions in Man and Animals* and was published in November 1872. Ill again, Darwin had to push himself hard to get it written. "I have had a poor time of it of late: rarely having an hour of comfort, except when asleep or immersed in work; & then when that is over I feel dead with fatigue," he told Wallace. "I am now correcting my little book on *Expression;* but it will not be published till November, when of course a copy will be sent to you. I shall now try whether I can occupy myself, without writing anything more on so difficult a subject, as evolution."[129]

It turned out the most successful and readable book he had produced up to that point, selling some nine thousand copies in the first four months, many more than the *Origin of Species* had done in a similar span. "I don't think it is a book to affront anybody," said Emma with obvious satisfaction. "I think it will be generally interesting."

One contributory factor was surely that Darwin included illustrations. The book was notable for presenting some of the first mechanically reproduced photographs of the period as well as a number of line drawings by the zoological illustrators Josef Wolf, Briton Rivière, and Thomas Wood. These helped the book's general interest and popular appeal. Photographic reproductions of *Ginx's Baby* joined pictures of Duchenne's electrified middle-aged man, sulky children, and a number of domestic cats and dogs in action. The illustrations graphically conveyed Darwin's point. Yet even while generating the appearance of technical excellence they had given him untold trouble. Before publication most of the pictures had to be enhanced by hand, or simulated in some way, to show what Darwin required. "The hair is generally smooth on the loins, & this makes the roughness on the back & neck more apparent," he commented on Wolf's drawing of an angry dog. His letters to the artists, in which he naggingly asked for tiny details to be changed, revealed much about the theory of emotional expression that he was trying to establish.

Notwithstanding Emma's comment, Darwin regarded the book as a crucial part of his lifelong evolutionary project. The subject of expression brought his anthropological cycle to a conclusion, seeking to demonstrate a continuum between the mental life of humans and animals. In it, Darwin accepted the commonsensical view that facial expressions and bodily gestures were a primary indication of internal emotional states—that they were the innate and uncontrollable manifestation of what was going on inside. Pain was accompanied by a grimace. Pleasure was accompanied by a smile. Darwin suggested that such expressions must have arisen through the same evolutionary mechanisms throughout the human and animal kingdom. The expressions that pass over human faces were, to him, a daily, living proof of animal ancestry.

Furthermore, in *Expression* he proposed that some habits and learned behaviours could, if advantageous to an animal, be preserved and eventually rendered innate. Gazing into the heart of his original hypothesis of adaptation by natural selection he discovered he must partly concede the environmentalist point in respect to behaviour.[130] Behaviour and biology were inextricably interwoven in ways that natural selection—the anvil on which he tested every theory—was insufficient to explain. Quietly, and without any fanfare, Darwin modified his views, accepting that the inheritance of acquired characteristics needed to be part of his system.

Robert Cooke, the new manager at John Murray's, provided an oasis of calm before publication. He anticipated large sales and was not disappointed. "It may be advisable to get police to defend the house," Cooke sportively declared on 25 November 1872. Murray was astonished. "Your modesty about *Expression* misled me to underestimate its sale."

Darwin was content. His series of evolutionary works was complete. Within the week, the magazine *Fun* issued a cartoon in which a well-dressed, buxom young lady was clasped at the wrist by an apish-looking Darwin who was taking her pulse. "Really, Mr. Darwin," she exclaimed, "say what you like about Man; but I wish you would leave my emotions alone!"

*chapter*
10

# DARWIN IN THE DRAWING ROOM

URING the 1870s, Darwin became the most famous naturalist in the country, "first among the scientific men of England," as Edward Aveling put it, his name inextricably linked with the idea of evolution and with the larger shifts in public opinion gathering pace as the century drew toward a close.[1] In truth he was a reluctant hero, eager to puncture any inflated estimation of his worth among his friends and family. His fame often teetered on the edge of notoriety, as he was nervously aware. Yet he understood that the tide of change carried him along on the unstoppable crest of a wave. His face and his name were becoming known far beyond the confines of academic discussion. The term "Darwinism," almost by default, covered all kinds of evolutionism and unfairly eclipsed the work of Huxley, Wallace, and others. Even his own book was now apparently more discussed than read. "It has been so easy to learn something of the Darwinian theory at second-hand, that few have cared to study it as expounded by its author," Wallace said accurately enough after Darwin's death.[2] Slowly, and somewhat uncomfortably, Darwin recognised that this celebrity, however much he disliked it, worked to his theory's advantage.

As might be expected in an era dedicated to the cultivation of national heroes and heroines, much of Darwin's personal prominence was expressed in characteristically Victorian form. Byron, Dickens, the Duke of Wellington, David Livingstone, and Jenny Lind: all these had risen to fame on the back of escalating expectations of what it might mean to be a celebrity. The notion of evolution by natural selection was, for example, unusual in the way it became part of the richly varied world of nineteenth-century popular culture. Few lay people had direct access to science or scientists, after all. Instead they might encounter scientific knowledge through general nonscientific media like books and magazines, sometimes in a law court or a doctor's surgery, or through pictures, biographies, advertisements, and newspapers. There were plenty of sites for the production and consumption of popular science, too, where evolutionary concepts could also take root, such as natural history museums, horticultural exhibitions, menageries, freak shows, art galleries, agricultural contests, music halls, and fashionable crazes that involved the human psyche like mesmerism.[3]

As far as Darwin was concerned, the dramatic rise in consumer culture had a particular material effect. Although Darwin never encountered his own face on a biscuit tin, as Queen Victoria did, or knowingly allowed his name to endorse household products such as soap or snuff, he did find himself and his theory transformed into various types of commodity. Those people who were conscious of the changing times could, if they wished, buy a pottery statuette of a monkey contemplating a human skull.[4] They might pay to gape at William Henry Johnson, the living "Man-Monkey," at Leicester Square.[5] They could commission an elegant piece of Wedgwood ware decorated by Émile Lessore with cherubs clustering around thc tree of life, hang a Darwinian caricature from *Punch* on their walls, sing a duet at the piano on the "Darwinian Theory," read edifying popular romances such as *Survival of the Fittest* or *The Lancashire Wedding, or Darwin Moralized,* or give their children nursery primers called *What Mr. Darwin Saw.*[6] One freethinking daily almanack produced in 1874 included quotations from Darwin's books in its printed thoughts for each day. Moving abroad, beyond the English coastline, Spanish gourmets could drink a glass of anise from a bottle illustrated with a Darwinian imp. "Science says it is the best—and that's the truth," declared the label.[7] And farmers in upstate New York could medicate their livestock under Darwin's unseeing gaze. The agricultural firm of G. W. Merchant, of Lockport, near Rochester, advertised its gargling oil with a pictorial ape that sang:

*If I am Darwin's Grandpa,*
*It follows, don't you see,*
*That what is good for man & beast*
*Is doubly good for me.*[8]

These diverse commercial products made Darwin and his intellectual achievement tangible to his own generation and the ones that followed him.

For a man like Darwin, fame would, however, take some getting used to. He was a reticent soul, uneasy with anything that demanded public appearance and—while appreciative of approbation from his scientific colleagues—was troubled by the notion of dancing to more popular tunes. Scientists had rarely been figures of interest before. More than this, he did not like the idea of people recognising him or knowing things about him without his knowing them. Public and private become highly problematic categories for society's most conspicuous individuals, as has long been recognised from other celebrity studies. So it was perhaps a relief to him that these first stirrings of celebrity took place in a context where his literal presence was not desired. The books and magazines that ran articles about him, and the evolutionary souvenirs and caricatures that appeared in the marketplace, were relatively abstract phenomena, dissociated from his private life, and for that very reason enterprises with which he could, to some degree, engage. He felt relatively comfortable—and far less emotionally exposed—in the familiar world of print, pictures, and objects.

And yet it was not clear how a scientific man like Darwin might be represented in such a world. Science on the whole possessed few commonly accepted stereotypes. The subject matter of science at that time was esoteric, often quite different from commonsensical notions based in ordinary experience, and then hovering on the brink of shifting from well-known conventional institutional settings like the university lecture room to the more modern spaces of laboratory or clinic. If scientists were to be depicted in public, say in the *Illustrated London News,* it was unclear what might constitute appropriate visual symbols for their activities or what might be the best way to indicate the inner life of the mind. The stereotypes that would come to dominate future images of science and its practitioners—the complicated mechanical equipment lurking in the background, the microscope, white coat, and laboratory bench—did not as yet carry unanimous weight. Such devices would become known only with the growth of science itself accompanied by simpler photography and cheaper means of mass-media reproduction and circulation.[9]

As Darwin rose to fame, cultural commentators therefore came to portray him in books and pictures in more familiar Victorian guises—sometimes as a traveller, collector, or naturalist. Given his life story, it was appropriate to portray Darwin as an explorer, both literally, as in his travels on the *Beagle,* and metaphorically as a man who brought detached, objective vision to new terrains of knowledge. Geographical explorers were given an especially privileged position as truth-tellers in nineteenth-century Britain, not least because they had seen with their own eyes, and Darwin's work was occasionally represented in books and articles as if it were a scientific discovery on the same scale as uncovering the source of the Nile.[10] Darwin found that he could slip effortlessly into this frame of reference, willing to think of his life as a voyage of discovery, one that was initiated by his real voyage, and casting his future autobiography and, in places, the *Origin of Species* in much the same mould. "I have always felt that I owe to the voyage the first real training or education of my mind. . . . Everything about which I thought or read was made to bear directly on what I had seen and was likely to see; and this habit of mind was continued during the five years of the voyage. I feel sure that it was this training which has enabled me to do whatever I have done in science."[11] And it was easy to depict Darwin as a natural history collector, absorbed in the minutiae of nature, building up his achievement specimen by specimen.

But it was as an ape that Darwin was most usually represented—a wonderfully apt and flexible visual imagery that played on the central theme of his writings and marked his passage into the common context of Victorian popular culture.

## II

Evolutionary theory proved an irresistible subject for caricaturists. Of course, caricatures and cartoons in general already supplied Victorians with their sharpest and most developed form of satire, a special form of comment that sanctioned savage public attacks on people in the news, the politicians, royalty, gentry, and social butterflies of the day, combining ridicule, exaggeration, bawdiness, sententiousness, and the deliberate distortion of facts to lampoon victims without mercy. Much of the humour was broad. Much emerged from a puritanical moralism that hated pretension and delighted in exposing what others tried to hide. These comic cuts gave nineteenth-century satirists a lens through which to mirror the times back to contemporary opinion, as had indeed been the case in Britain since the caustic wit of James Gillray had responded to the turbulent years of George III and the American and French revolutions.[12] So it is not too surprising that Victorian satirists leaped on the idea of animal ancestry and evolution as a topical vehicle for commenting on the nation's preoccupations. Moreover, satirical skits about apes fell helter-skelter into the long tradition of identifying animal traits in human beings and puncturing intellectual pretensions the world over. Age-old tropes of metamorphosis and the existence of the beast beneath the human surface—standard topics for political comment—were given fresh meaning that reflected the concerns of the age.

Even so, there was something special about the vigorous combination of advanced pictorial technologies, widening readerships, loosening cultural boundaries, and the publication of *The Descent of Man* and *The Expression of the Emotions* that prompted satirists in the early 1870s to produce an abundant variety of parodies, cartoons, verses, and humorous sketches relating to the general theme of evolutionary progression.

Visual images took centre stage. In Britain, in the era of cheap illustrated magazines like the *Comic Almanack* (founded 1835), *Punch* (1841), *Comic Times* (1850), *Fun* (1861), *Judy* (1867), and *Figaro* (1870), cartoonists played eagerly with the idea of physical transformation, single-mindedly endorsing staple symbolic types that would last well into the twentieth century, turning politicians into unpleasant animals or converting John Bull, the reactionary national icon, into a leg of roast beef, the national dish. Most of these images drew on well-established prejudices that antedated Darwin's writings, sometimes by centuries, and much of their impact rested in these multiple allusions. Evolution in this form meshed comprehensively with contemporary biases, embracing anti-intellectualism, imperialism, slavery, class, race, colour, and political identity, for Darwinism could easily reinforce crude stereotypes and bigotry against Negroes, aborigines, women, Jews, servants, and the Irish.[13] Nevertheless, each artist could only select his subject matter from a repertoire of available topics and print technologies. It was a measure of the publicity accompanying evolutionary writings that they should choose to frame some of their most trenchant observations in terms of Darwin's theory. Outside Britain, talented caricaturists like "Andre Gill" in *La Charivari* (1832), Wilhelm Busch in *Fliegende Blatter* (1844), and Thomas Nast in *Harper's Weekly* (1851), the last memorably documenting the Civil War and etching the elephant and donkey symbols into the American mind, purveyed evolutionism and Darwin along with the ironies of their own culture. In a world of slick and unconstrained popularisation, caricatures and satires about evolutionary theory were a highly usable resource.

Some twenty or thirty printed caricatures about evolutionary theory were produced in Britain from 1865 to 1882, roughly the last two decades of Darwin's lifetime, many more than those of any comparable scientific theory, although, it should be said, not nearly as many as pertained to conspicuous national figures like Disraeli or Gladstone. In particular, the magazine *Punch*, which developed an enduring relationship with Darwinian theory, paid more attention in the long run to Disraeli, chronicling his ascendancy with indulgent contempt, or to the Irish, the last being a specially vehement target during the years of debate over Home Rule. Few of these *Punch* drawings revealed the ferocious bite of earlier artists such as Hogarth, Gillray, or Rowlandson. Nor did they exhibit the cut and thrust of Honoré Daumier in Paris or the patriotic fervour of Thomas Nast, each of whom set out to expose public hypocrisy.[14] The days of vicious campaigning satire had gone—in Mr. Punch's hands at least.[15]

An altogether milder form of topical comment was apparent in British illustrated magazines. This form of attenuated cartoon, in fact, had developed in Britain in 1843 with John Leech's drawings in *Punch,* and was quickly picked up in pictorial journals like Richard Bentley's *Miscellany,* the *Cornhill,* and *Good Words,* and especially in mass-circulation editions of novels, the new commercial phenomenon only made possible by advancing print technologies and best seen in Dickens's works as illustrated by George Cruikshank and Hablot Browne (Phiz). These were domesticated drawings, relying on familiar middle-class values, droll rather than cruel. In the capable hands of John Doyle ("HB") and his son Richard Doyle, followed by those of Sir John Tenniel, Edward Linley Samborne, Ernest Griset, and Charles Keene, cartoons in *Punch* and *Fun* came to express the thoughts of conventionally minded members of the public when faced with the absurdities of modern life. "So you see Mary, baby is descended from a hairy quadruped, with pointed ears and a tail. We all are!" said Jack, in *Punch,* having been reading his wife passages from *The Descent of Man.*

"Speak for yourself Jack! I'm not descended from anything of the kind, I beg to say," retorted Mary. "And baby takes after me, so there!"[16]

*Punch* artists aimed precisely at those readers who would most easily decode and approve its messages. Under the general direction of Shirley Brooks, editor from 1870, and then the humorist Tom Taylor from 1874, their drawings showed well-behaved middle-class men and women like Jack and Mary discussing hairy ancestors beside the fire or attending evening soirées conducted by "Dr. Fossil." They dressed apes as gentlemen or put Darwin in God's celestial chair. None of these images were as shocking or as subversive as they might have been in Paris or Berlin. In a word, *Punch* and *Fun* cartoons mirrored the broad-minded, lazy acquiescence that characterised the complacent, prosperous classes of Britain. With only a few passing grimaces at the follies of science, these pictures softened at least some of the anxieties inherent in evolution by turning them into drawing-room trifles.[17] *Punch* satirists were perhaps creating a new structure of consensus.

So these British cartoons seemingly spoke with the voice of the establishment rather than as its scourge. "At the sight of a monkey scratching himself in the Zoological Gardens, the philosopher might with much propriety observe, 'There but for Natural Selection and the Struggle for Existence sits Charles Darwin,' " said *Punch* prosaically. When Princess Louise, a younger daughter of Queen Victoria, married the Marquis of

Lorne in 1874, *Fun* parodied the wedding procession by including, among the trumpeters and royal guests, the figures of "Dr. Darwin and our distinguished ancestor." In that cartoon, Darwin escorted an ape down the aisle. An earlier representation of Darwin previously appeared in 1871 in the pages of the London edition of *Figaro,* in which Darwin politely invited an ape to contemplate its future in a hand mirror, supported by appropriate quotations from Shakespeare. Marriage, royalty, and Shakespeare could hardly have been more acceptable vehicles for comment. By endorsing the traditional verities of British life, these humorists gave Darwin's theory a relatively unthreatening place in Victorian homes—or at least depicted science as a folly pursued by others and about which respectable men and women need not overly concern themselves.

One significant feature of late-nineteenth-century evolutionary caricatures is the frequency with which Darwin appears in them as the ape itself. His personal facial attributes, such as his beard, the great dome of his skull, and the hairy, beetling eyebrows, were becoming known to the middling sector of society through portrait woodcuts and line engravings reproduced from photographs in the *Illustrated London News* from 1871 onwards. Other pictures of Darwin were available in album-style publications like *Representative Men of Literature, Science and Art* (1868). The most well-known, however, was probably the *Vanity Fair* chromolithograph, published as "Men of the Day, No. 33. Natural Selection" in September 1871, six months after the release of *The Descent of Man.* This *Vanity Fair* caricature was by Carlo Pellegrini under his customary byline of "Ape" (although the Darwin print was in fact unsigned) and was thought by Darwin's relatives accurately to capture his elongated figure and characteristic posture. It belonged to one of Pellegrini's series of character studies that embraced scientists, ecclesiastics, politicians, sportsmen, and other figures of note, each illustration accompanied by a jocular caption and, in the printed part of the magazine, a biographical résumé that invariably made fun of the superficial values of the day. *Vanity Fair* portraits of Huxley and Wilberforce, each squaring up to the other, had been a well-publicised pair a few years beforehand, and Richard Owen ("Old Bones") and Herbert Spencer ("Philosophy") were also in the series. These pictures were not intended to stand alone; they needed to be viewed with the text. They furthermore had an extended life as separate prints. The caricature of Darwin can sometimes be found with a different caption, "You know we all sprang from monkeys?" This variant exists in Darwin's personal collection, and is accompanied by a matching caricature of Owen, retorting, "What a pity you didn't spring a little further." Pellegrini approached Darwin through John Murray for permission to go ahead with the caricature, a precautionary move rarely taken by other Victorian artists, and a practice not usually followed by Pellegrini himself.

Thus recognised, Darwin's facial features were emphasised in every British caricature around the time of *The Descent of Man* and *The Expression of the Emotions,* although to be sure there were one or two in which the artist clearly worked without such information. To some extent, too, the artists consciously played on the iconography of intellectuality. Most caricaturists, then or now, could call on conventional attributes of academic learning such as an absent-minded demeanour, a blackboard, gender, a pair of spectacles, baldness, and so on. "A little lecture by Professor D——n on the development of the Horse" showed Darwin, prolifically bearded, as an absent-minded professor in front of a blackboard, with a handkerchief tumbling out of his trouser pocket in imitation of a monkey's tail. The joke lay in his laborious explanation of the evolution of a horseradish plant into a racehorse through ten nonsensical horsey stages, including a clothes-horse, Louis Quart-horse, and a Hors-de-combat. In 1871, when *The Descent of Man* was being reviewed in American journals, Thomas Nast drew for *Harper's Weekly* a gorilla outside the Society for the Prevention of Cruelty to Animals. The gorilla protested to a recognisable picture of "Mr. Darwin" that his pedigree was being stolen.

Darwin's beard—an eye-catching feature of the commercially reproduced portraits of him—was therefore a bonus for cartoonists. His general hairiness begged to be turned into animal fur. Add a tail, and there was an image that shrieked of apish or monkey ancestors (the general public was never too fussy about which). Artists at the *Hornet* churned out an image of Darwin as a "Venerable Orang-Outang: A contribution to Unnatural History" in March 1871, one month after *Descent* was published. Such a picture of Darwin-as-ape or Darwin-as-monkey readily identified him as the author of the theory in much the same way as a military longboot might have indicated the Duke of Wellington during the Napoleonic period. An undated letter in Darwin's files (probably written in 1871) reflected the point exactly. "Will you permit me to address to you a few lines upon your elaborately evolved theory of man's descent, or ascent (I should rather name it), from the lower creation," asked a Mr. D. Thomas, presumably from Wales.

> Now Sir do not be offended if I just tell you what struck me when I saw your likeness in the Illustrated News on Saturday & not myself only but others with whom I have spoken, the striking resemblance there is to an ape, the thick skull, deep set eyes & hairy face all

> remind one of a fine venerable old Ape & how striking that the person who is most bent upon linking the monkey race to us should so much resemble one in outward form.[18]

The evolutionary tree also began making an appearance in cartoons. After publication of *The Descent of Man,* Darwin was frequently depicted as a monkey swinging from a tree, sometimes labelled the "tree of life." *Punch* caricaturists placed an ape in a tree diligently reading a copy of the *Origin of Species.* The magazine *Figaro* put a hairy Darwin among the branches of "A Darwinian hypothesis," and the Parisian satirical journal *La Petite Lune* dangled him in the guise of a monkey from the "arbre de la science" with an elegant tail draped over his arm.

Few other scientific theories were so readily identifiable in the public prints of the day—perhaps only electricity and telegraphy, the main technological advances of the era, were to have comparable impact on the British public, and these were rarely connected in such an obvious manner to their human originators. A hairy, apish Darwin and a tree became easily recognisable images of evolution—perhaps as recognisable to Victorians as the double helix of DNA is to many people today.

This visual characterisation of trees had subtle but definite effect. Satirists could have called on a number of other possible evolutionary images. Charles Henry Bennett had drawn circular sequences of men changing into cooked geese and Christmas puddings in 1863 and labelled them Darwinian transformations, no doubt a joke on classical notions of metamorphosis and perhaps also on eighteenth-century physiognomic transformations between man and beast as discussed in Johann Caspar Lavater's writings. Bennett, however, seems to have understood the sequence of evolution as a circle rather than as Darwin's branching tree. Circularity as a motif came over just as strongly in Henry Woolf's full-page cartoon in *Harper's Weekly* in 1871 of "The Darwinian Student's After-Dinner Dream," in which knives and forks metamorphosed in stages into the pretty girl the student wishes to marry.[19] Linley Sambourne afterwards portrayed a great spiral from "Chaos" to God's throne in heaven, on which Darwin casually sprawled.[20]

Nineteenth-century satirists hardly ever depicted Wallace or Huxley in similar situations even though they were by far the most prominent men otherwise associated with the gorilla question. One cartoon in *Fun* included Huxley and Spencer (but not Wallace) in a critique of *The Expression of the Emotions,* showing the scientists cowering before an angry hippopotamus protecting her young. This played on the well-advertised birth of a baby hippo in the London Zoological Gardens a month or two before. The caption read, "An open countenance denotes a gentle and good-natured character." Here and elsewhere, it seems as if Wallace was ignored and that it was Darwin's books that were perceived as the primary texts in generating a wholesale transformation in thought. Caricaturists portrayed the theory of evolution as if it were Darwin's alone. Indeed this interweaving of evolutionary imagery and portraits of Darwin probably contributed materially to the sense that evolutionism and Darwinism were one and the same thing.

As for the grizzled old ape himself, Darwin took a friendly interest in collecting the caricatures. "Ah, has *Punch* taken me up?" he said to one passing visitor to Down House, the geologist James Hague. Hague reported the rest of their conversation.

> "I shall get it tomorrow," he said: "I keep all those things. Have you seen me in the *Hornet*?" As I had not seen the number referred to, he asked one of his sons to fetch the paper from upstairs. It contained a grotesque caricature representing a great gorilla having Darwin's head and face, standing by the trunk of a tree with a club in his hand. Darwin showed it off very pleasantly, saying, slowly and with characteristic criticism, "The head is cleverly done, but the gorilla is bad: too much chest; it couldn't be like that."[21]

Still, a dangerous note could be identified here and there. One caricature by the French artist André Gill (L. A. Gosset de Guine) in the Parisian journal *La Lune,* printed in August 1878, extrapolated mercilessly from Darwin's *Descent of Man.* This print made the links between natural selection and the philosophical doctrine of materialism absolutely clear. Under the caption "L'homme descend du singe," Darwin appeared as a monkey at the circus, bursting through a paper hoop labelled "Credulité" and aiming for another marked "Superstitions—Erreurs." The hoops were held by Émile Littré, the medical writer and populariser of Comte, who was repeatedly denounced as the archfiend of French scientific positivism. The threatening implications of Darwinism were thus made obvious to the men and women of the Third Republic, locked in controversy over positivism, republicanism, the Catholic Church, the chemical origin of life, free thought, and the question of progress. While very few Frenchmen or women accepted Darwin's ideas, much preferring their own versions of environmental determinism, Gill evidently wanted to show how Darwin's ideas were being taken up by positivists.

On the other side of the Atlantic, racial issues made an ugly appearance when Sol Eytinge used Darwinism to expose the divisions in American culture. A close personal friend of Charles Dickens, and at one point

Dickens's American illustrator, Eytinge regularly commented on black causes during the most difficult years after the Civil War. In his occasional "Blackville" series for *Harper's Weekly* from 1874, the key period for civil rights issues, he customarily placed Negro families in typically white situations. In January 1879 he sketched an imaginary hometown political debate in which the black chairman asked "wedder Lord Dorwin involved hisself or somebody else" in his theory of evolution. The implication that only Negroes might have descended from apes reflected much contemporary theory, harking back to the views broadcast by Agassiz, James Hunt, and others during the war period who maintained Negroes must be a different and inferior species of mankind.[22] However, the artists at *Harper's* were unsure exactly where they stood on the colour question. Another sketch from a different hand presented "Darwinian Development" with barbed animosity. A rural black couple in a draper's store ask for a *chignon* for the lady, implying impossible social aspirations.[23] In these pictures deep-seated divisions were made plain.

Darwin also made an appearance as a minor character in a vast satirical broadsheet published under the name "Ion" in London in 1873 (with another version following afterwards in 1883). This satire was attributed (probably correctly) to clever, mild-mannered George Holyoake, the leading radical secularist of the period and author of *Half-Hours with Freethinkers*, the survivor of a notorious show trial for blasphemy in 1842. Holyoake had dedicated his life to creating a secular alternative to the established British church, a high-minded liberal movement taken over and dominated by Charles Bradlaugh, the much tougher, hard-nosed, charismatic leader of the National Secular Society.[24] The broadsheet linked ecclesiastical dissent with descent. It was titled *Our National Church* and provided an all-embracing critique on the fragmenting religious beliefs of the nation, depicting rival sects of Broad Church, Low Church, High Church, Dissenters, "No Church," Catholicism, and Science. Up in a corner it included the three priests of scientific naturalism, Darwin, Huxley, and John Tyndall. This complex picture primarily played on James Martineau's widely publicised attempts during the 1870s to unite all clergymen under the single umbrella of a "national" church, and evolutionary theory was merely one of several perceived threats to the theological establishment.

The print showed the dome of St. Paul's Cathedral as a giant umbrella unable to shelter religious traditionalists from the stormy winds of doctrinal unrest. Nonconformists pull the chocks away, atheists rant in a corner, Catholic converts follow a signpost "To Rome," and neither the broad churchmen nor the low churchmen can handle the dome's straining guy-ropes in the gale. A donkey rudely calls, "Let us bray." It was fair comment, said the radical divine Moncure Daniel Conway.[25] Huxley, Tyndall, and the renegade clergyman Bishop Colenso push upwards towards the apish figure of Darwin on a hillside, who calls, "This way to daylight, my sons." The tightly packed text informed readers that over the horizon lay the dawn of an intellectual era which would dispel "the chilling influence of the church."

The second version, usually printed in red and black, was revised to emphasise the evolutionary point. This later version showed an ape carrying the flag of Darwinism, followed by a trail of well-known agnostic philosophers and dissenting clergymen, including Spencer ("Philosophy"), Conway ("We must move on"), Huxley, and Tyndall ("Science"), all aiming for a plinth on which stood Darwin's bust surrounded by a cloud of "Protoplasm."[26] Both versions of the print were in Darwin's personal collection, although it is not known how many others were printed and distributed, or to whom.[27] The artist, whoever he was, considered Darwin and his theory an integral part of the secular, highly politicised world coming into being around him.

Such powerful visual statements propelled the idea of evolution out of the arcane realms of learned societies and literary magazines into the ordinary world of humour, newspapers, and demotic literature. Without Mr. Punch's monkeys and gorillas, *Figaro*'s mirror of nature, and Holyoake's cloud of protoplasm, the transformation in nineteenth-century thought would probably have remained predominantly an elite phenomenon. The full implications of human descent would have taken much longer to sink in. These caricatures were not just a transparent medium of illustration but an actual shaper of contemporary thought, as representative in their own way as any of the fine arts or literary texts of the period. The themes of Darwin's *Descent of Man* were graphically repackaged in a versatile cultural form enjoying wide distribution and popular appeal. The cartoons might appear on the tables of any middle-class home in the country.

"Might we not enter thus a mild protest against Darwin at dinner and Darwin at tea?" inquired a cynical journalist in the *Globe*.[28]

## III

People wanted to see him in the flesh too. An ordinary member of the public could usually hope to catch only the most fleeting glimpse of eminent Victorians. Opportunities to hear their voices were even rarer. Duchesses might be seen rattling past in a coach-and-four or opening a hospital, Gladstone might occasionally appear on the hustings, Tennyson might read to a party of friends, or Dickens might tour the provinces with elaborately staged theatrical readings. Yet before long, Victorian celebrities of one kind or another began making more obvious public showings of themselves, explicitly feeding their audiences' wishes. It hardly mattered that such opportunities were highly structured and limited, little more than a version of the ritual display that has always been an aspect of leadership. A queen and her people, for example, expected both to see and be seen. The ceremonial elements of such occasions were straightforward. A genuinely personal encounter, on the other hand, could create a remarkable thrill, as Browning understood in his famous line "Ah, did you once see Shelley plain?" These sightings were recorded as memorable incidents in the lives of those who came face to face with a star. Many of the people who met Darwin as a private individual consequently could not resist writing about him, even when aware that such comments were a poor return for his hospitable attention. There was an indefinable sense of the public's growing right to sit at the firesides of the famous.

By going to see a celebrity in person, a devotee could in addition satisfy his or her curiosity and bypass or supplement in important ways the whole laborious process of reading. In the nineteenth century, pilgrimages to literary figures began in earnest. Wordsworth was pestered in the Lake District, and snoopers became such a nuisance to Tennyson on the Isle of Wight that he fled to the closely wooded hills of Surrey. "I can't be anonymous . . . by reason of your confounded photographs," he complained to Mrs. Cameron in August 1868.[29] Darwin discovered his position was little different.

Through the 1870s and early 1880s, a stream of sightseers visited or attempted to visit Darwin at Down House. In retrospect, this was quite an undertaking. The journey from London was inconvenient, involving a train journey and then the hire of transport from the nearest railway station. From the outside, too, Down House presented an intimidating appearance to socially unsophisticated callers. The gravelled drive, closed gates, and imposing front door that would be opened by a servant signalled that this was an Englishman's private domain. Relaxed and intimate though the stories were about Darwin's family setting, such pleasantries were only for the people who had an entrée.

Nevertheless, the people came. Several of these visitors Darwin knew by repute or through letters of introduction from friends; some were already acquaintances by correspondence; others became warm friends and visited again and again. The rest were unknown and uninvited, and to Darwin unwelcome. A few were an actual trial. Darwin grumbled with his family about the need to gush over Lady Dorothy Nevill when she arrived to discuss orchids. Still, he regarded his obligation to entertain as an unavoidable duty. Emma Darwin's diaries indicate that although he guarded his privacy jealously, he received many more visitors than his reputation for seclusion suggested.

Interconnections between Darwin's fame, the increasing leverage of his theories, and his shrewd management of personal publicity can be identified here. Many of these visitors regarded a meeting with Darwin as a turning point in their lives. Anton Dohrn, the young Prussian naturalist, longed to visit him during a trip to England in 1867 but was afraid to ask. He had heard so much about Darwin from Haeckel, his teacher and professional colleague at Jena, that he wished literally to sit at Darwin's feet and venerate him. After several years of zoological correspondence, Darwin politely invited him for lunch in 1870. Dohrn described the day as a cherished memory, although noting in a puzzled manner that Darwin did not seem nearly so ill as he expected.[30] The time for private discussion was rationed to one hour. On this occasion, Darwin unashamedly manipulated his reputation for poor health—his time was limited, his strength was weak, his prestige was high. As a result, Dohrn felt he was meeting royalty. He went away glowing with Darwin's endorsement of his biological endeavours. For his own part, Darwin liked this young man who adopted the evolutionary project with such ardour. Both benefitted from the relationship. Darwin became a patron of Dohrn's Zoological Research Station founded in Naples in 1874, and afterwards did a good deal to further the work of the station through gifts and personal recommendations.[31] In return, the station became a leading centre for evolutionary researches—in Dohrn's case, for working on the evolutionary connections between different arthropods and the ancestral origins of vertebrates.

Grown men could crumble in the presence of the god. The American mathematician and philosopher Chauncey Wright paid a call on Darwin in 1872 and burbled about a near-religious experience. Fresh from the natural science tripos at Cambridge University, twenty-two-year-old George Romanes breathlessly recorded that Darwin greeted him with the unexpected words "How glad I am that you are so young!" Romanes became one of Darwin's most dedicated English disciples during the 1870s. John Lubbock said that another admirer of Darwin, whom he brought over to Down House for an hour's courtesy call, burst into tears when safely on the way home. The visitor had been so overcome during the visit that he could not summon up courage to speak to the great man.[32] A few of these young men turned Darwin into a secular saint and Darwinism into a religion.

Occasionally, Darwin was uncertain whether visitors were taking undue advantage. Sometimes acquaintances would arrive with an unknown colleague in tow. If so, Darwin would simply manipulate the elaborate social rituals built up by the Victorian country gentry, presenting himself for viewing only for as long as etiquette required. He would make his excuses and leave, citing his ill health, his need for privacy or for a nap, or the unremitting duties of scientific work, according to the audience. Intimate friends usually conspired to help these plans along. Mary Lyell wrote to an American geologist who proposed a visit reminding her correspondent that although he "might find Mr. Darwin looking well and strong, I should remember his really delicate strength, and not stay too long." Lubbock would solicitously say, "You will I am sure be tired & must not overdo yourself." Guests seldom noticed how selfish these escape mechanisms were, designed to minimise disruption to Darwin alone. Like an old-time patriarch, Darwin granted only what was convenient and more or less ignored the activities among his friends and household that ensured everything would go smoothly.

In this respect, Darwin's entourage worked as if they were a family firm, protecting and supporting their figurehead. Emma Darwin and whichever sons or daughters might be at home were expected to rally round and take a major part in entertaining guests. Either Hooker or Huxley was often invited to ease the proceedings along. Francis Darwin was needed for his languages or to explain botanical experiments, George for his chit-chat about the senior commonrooms of Cambridge University, Leonard for stories of military life, Henrietta for piano recitals. Darwin's household was an integrated corporate enterprise.[33] Emma and Parslow ran the business with practised efficiency.

Inevitably, these social events came to be regarded with a certain weary cynicism within the privacy of the family circle.

> We have been rather overdone with Germans this week. Haeckel came on Tuesday. He was very nice and hearty and affectionate, but he bellowed out his bad English in such a voice that he nearly deafened us. However that was nothing to yesterday when Professor Cohn (quite deaf) and his wife (very pleasing) and a Professor R. came to lunch—anything like the noise they made I never heard. Both visits were short and F. was glad to have seen them.[34]

In this manner, strangers were drawn into the personalised web of Darwin's work, his home, and his family. Making a visit to Down House became an important abstraction in its own right, representing Darwin's status and, it should be said, the intellectual achievement of the visitor. Mischievously, Huxley sent a sketch of someone paying his devotions at the shrine of "Pope Darwin."[35] Pulled as they were into the ambit of

Down House and the family, many of these visitors became Darwin's staunchest defenders.

## IV

The famous came too. Thomas Carlyle was astonished to see Darwin's cheerful demeanour when he visited him in 1875. He had expected an invalid.

> I had not seen him for twenty years. He is a pleasant jolly-minded man with much observation and a clear way of expressing it. Has long been an invalid. I asked him if he thought there was a possibility of men turning back into apes again. He laughed much at this, and came back to it over and over again.[36]

This meeting somewhat eased Carlyle's views on evolution, softening him up sufficiently to declare to Huxley at an Edinburgh dinner, "If my progenitor was an ape I will thank you, Mr. Huxley, to be polite enough not to mention it." Evolution was "rather a humiliating discovery and the less said about it the better."[37] This remark possibly formed the grounds of George Bernard Shaw's anecdote involving a duchess who expresses the same view.

Samuel Smiles arrived in 1876 and was also startled by the noise of crashing illusions. He was taken aback by Darwin's vivacity. "He almost *embraced me*. I staid to lunch with him & his family. He was full of talk—and he, as well as myself, could scarcely *eat* for *speaking*. . . . He went over no end of things. What keen penetration he has. He seems to have an insight into everything. I was almost glad to get away, for his cheeks began to flush; & he is accustomed on such occasions to rush out of the room & take refuge on his sofa in his study."[38] Underneath Smiles's words lay the suspicion of a hint that Darwin might be a bore when he got going. Smiles brought with him a copy of his latest book, *Life of a Scotch Naturalist: Thomas Edward* (1876), another of his tales of heroic self-improvement. Darwin thanked him, saying he had read every one of his biographies with "extreme pleasure."

Hooker, Huxley, and others all testified to Darwin's good humour and lively conversation. At home, on his own terrain, where he was confident that he could control interactions with callers, Darwin appeared to guests as a talkative, hospitable, engaging man, not a sickly hermit. It was his very ordinariness, if anything, that captivated people expecting to meet a sage. "I sometimes feel it is very odd that anyone belonging to me should be making such a noise in the world," Emma told Aunt Fanny Allen.[39]

The more unusual guests naturally etched themselves into the family's collective memory. Emma Darwin wrote to one of the children:

> Do you remember a working man from Australia who rushed in to shake hands with him a year ago and was for going straight off again without another word? We have heard of him again from a Canadian who met him on the road to California, on foot, with nothing on but drawers and shirt, in the pocket of which he carried his pipe and a letter from F. of which he is very proud and shows to everybody.[40]

Some, like the new doctor in Downe, were merely boring. "He sat an hour and a half & expressed so much pleasure at the thought of a chat with Prof D.," said Emma tartly, "that I am afraid the Professor will have to snub him a little to get rid of him."[41]

As time went by, it seemed as though Darwin became like a stately home or historic building briefly opened up for viewing. Henrietta and Richard Litchfield brought a party of Litchfield's working men, accompanied by their wives, to Down on a Sunday excursion in 1873. Darwin greeted the throng with a few words and then left them to "an excellent tea on the lawn, wandering in the garden and singing under the lime-trees."[42] Accommodating these larger groups did not pass without a few troubled thoughts about burglars crossing Darwin's mind. When sixty-seven vagabond boys came for an outing, his friend John Innes outlined for his edification the latest design in house alarm, with "fireworks" if needed.[43] Even in the 1880s, when Darwin was an old man, parties of high-minded enthusiasts might descend at any moment. Leslie Stephen and his walking friends, "the Tramps," recorded how they were entertained by Darwin at Down. On other outings these earnest young men made for John Tyndall at Hindhead and George Meredith at Box Hill.[44] "Of all the eminent men that I have ever seen he is beyond comparison the most attractive to me. There is something almost pathetic in his simplicity and friendliness," said Stephen.[45]

Yet it cut both ways. Darwin and Emma found that they enjoyed easy access to the good and the great. Some visitors were too interesting or too eminent to miss. Moncure Conway, the unorthodox American minister with his own breakaway church in London, the South Place Religious Society, evidently appreciated Darwin's writings and in 1873 preached a sermon entitled "The Pre-Darwinite and Post-Darwinite World." Soon afterwards he was invited to a weekend party at Down House with Charles and Susan Norton, friends of Darwin's who were over from Boston on a visit. It was probably on this occasion that Darwin gave Norton a copy of *The Descent of Man* inscribed in his own hand, "With the affectionate respect of the Author."[46]

The weekend was a great success. Conway was an excellent raconteur with a prodigious memory and had stories to tell about life in Cincinnati, his friendships with the Boston transcendentalists, his part in the abolitionist movement, and a heartbreaking tour he had made of the battlefields in Europe after the Franco-Prussian War. On arriving in England, he had persuaded Lyell and Pengelly to take him to Kent's Cavern to look at extinct bones. "It has been my privilege to know the leading scientific men in America and Europe," he said in his autobiography. "None of them was orthodox, and what could bigotry say against a tree that bore such fruits?"[47]

Conway was prepared to admire Darwin. From his bedroom window he caught sight of his host wandering in the garden early one morning. "His grey head was bent to each bush as if bidding it good morning. And what a head!"[48] Although he did not pursue the parallel, it would not be too fanciful to suggest that Conway saw in that early-morning tenderness something of the risen Christ, whom Mary mistook for a gardener.

George Eliot was another catch. Delicately, Darwin put out social feelers to see if Emma and Henrietta might be introduced to her in 1873. They greatly admired her writings. The general disapproval surrounding Eliot's unmarried life with George Henry Lewes was slowly lifting, and Darwin (who felt kindly towards Lewes, not least because of his encouraging remarks about pangenesis) had called at their London home on his own one Sunday in 1868. In March 1873 he attended one of Eliot and Lewes's Sunday gatherings of literati. At that point he asked if he might bring Emma and the Litchfields. "My wife complains that she has been very badly treated and that I ought to have asked permission for her to call on you with me when we next come to London; but I tell her I have some shreds of modesty" (by which he meant embarrassment at openly requesting an invitation).

Eliot responded the next morning. She was used to this kind of personal sightseeing.

> We shall be very happy to see Mr. and Mrs. Litchfield on any Sunday when it is convenient to them to come to us. Our hours of reception are from ½ past two till six, & the earlier our friends can come to us, the more fully we are able to enjoy conversation with them. Please do not disappoint us in the hope that you will come to us again, & bring Mrs. Darwin with you, the next time you are in town.[49]

She signed herself M. E. Lewes. Henrietta and Richard Litchfield attended a party later that spring. Emma met her some while afterwards, during a visit to Erasmus's house.

## V

The other side of the celebrity coin was that people also wanted to know what he thought. People wrote for his autograph or asked him to describe his theories in his own hand. Rafts of amateur naturalists supplied snippets of information on topics that must have stopped even Darwin in his tracks—a frog inside a lump of coal, a hen that laid eggs with clock faces on them, a hybrid cat-rabbit, beans that grew on the wrong side of the pod in leap years, an avowal that the human soul was really only magnetism. "Dear Sir," one letter began. "I suppose, that you have been very much astonished to receive some months ago a small box from Germany with different fruits of oranges, a specimen of wine and a sort of beans without any letter."[50] But there was little that could astonish Darwin now. Requests for money, jobs, testimonials, photographs, his support for elections to clubs and societies, and the supply of free copies of his books arrived in the post every morning.

If none of these, it was clear that Darwin's correspondents had read his books and found in them people just like themselves who had supplied information. They wished to join in, to participate in the build-up of evidence by reporting their own case histories; and perhaps to make a contribution, however small, to knowledge.

In the decade that followed *The Descent of Man*, Darwin probably wrote around fifteen hundred letters each year and received much the same number in reply. Many of these letters are now missing, particularly those of a routine nature.[51] Only 672 letters survive for the year 1872, and 420 for the year 1876, for example—and these were probably saved because they included useful information for his scientific work. According to Francis Darwin, his father would keep all incoming letters for a period of six or seven years, saying that he never knew when one or another might be useful for an address, and then periodically burn the entire folder when he needed an empty box to start the process over.[52] Nonetheless Darwin paid out the relatively large sums of £37 8s. 6d in 1872 and £38 4s. 11d in 1876 for "stationery, stamps & newspapers," about the same as he paid for "fish and game."[53] At a standard charge of 1d for a penny postage stamp, and 240 pennies in every old pound, Darwin had a substantial undertaking on the go.

Francis Darwin confirmed that his father spent several hours every day dealing with his correspondence. "At 9.30 a.m. he came into the drawing-room for his letters—rejoicing if the post was a light one and being sometimes much worried if it was not. He would then hear any family letters read aloud as he lay on the sofa. The reading aloud, which also included part of a novel, lasted till about half-past ten, when he went back to work till twelve or a quarter past."[54] A number of advertisements and circulars would arrive in the post too, continued Francis, and Darwin would "carry them back to the study having a sort of pride in filling his waste paper basket." Time for writing replies came after lunch and would end at around three in the afternoon. All this suggests that some four or five letters must have left Down House every day.

In the main, Darwin treated correspondents with patience, answering what probably sometimes appeared to him as fatuous questions. One of the few times he showed obvious irritation was in 1880 when Robert Lawson Tait suggested holding a Darwin festival in Birmingham. Tait was a pushy surgeon who constantly asked for favours. "Would it not be better to wait until I am in my grave?" Darwin sharply replied to the proposed festival.[55] Critical comments seemingly arrived by the sackful too, until Darwin wrung his hands. "I should like a society formed so that every one might receive pleasant letters and never answer them," he cried to Huxley.[56]

The logistics of the undertaking were considerable, and on several occasions Darwin toyed with the idea of following Lyell's example and employing a secretary. Lyell's secretary, Arabella Buckley, was a loyal friend to the geologist. She later published natural history works of her own. Darwin was consequently tempted by a proposition put by an anonymous "Miss I." Emma wrote on his behalf to state that "he says he will be glad to experiment for a month." Nothing came of it. Next, he discovered that Wallace was experiencing financial difficulties and wondered if he could—if he should—hire him to edit the second edition of *The Descent of Man.* Wallace was so hard up that he secretly assisted Lyell for five shillings an hour. In the end, Darwin paid his son George to edit *Descent* (Henrietta was ill).

Then Francis Darwin offered to help with the workload, "promising to be as civil as he could wish." Darwin was reluctant to relinquish the task. "When he did let me," recalled Francis, "he used always to say I did the civility well."[57] However, in 1874, Darwin capitulated and employed Francis as his secretary and assistant. Francis liked to describe the moment when Darwin realised he needed assistance. Someone sent him a volume called the *Ingoldsby Letters* dealing with the revision of the Book of Common Prayer, and in acknowledging it Darwin mentioned that his sons often laughed over the book—meaning R. H. Barham's irreverently comic

*Ingoldsby Legends.* That same year Francis married Amy Ruck, the daughter of a family friend from Wales, and came to live in a house in Downe village. Francis walked up the road every day to aid his father with botanical experiments and reply to correspondents. It seems not to have occurred to Francis that Darwin was giving him employment to compensate for his failure to pursue the medical profession for which he was trained.

Darwin's position in the world was becoming big business. With Francis's encouragement he ordered some pre-printed correspondence cards that could be used as a response, especially if a signed *carte de visite* was also slipped into the envelope. Despite this convenience, Francis said that his father scarcely used the printed cards, since he felt anxious if he did not provide a proper handwritten reply. At some point, no doubt with an eye on his duties, Darwin also ordered a rubber stamp of his signature. It was apparently well used. (It survives today, stained with ink, serving as a reminder to autograph hunters that all may not be as it seems.)

His life was carried on through these letters, as was his habit. Sometimes interesting old acquaintances would reappear in an envelope. Thomas Burgess, a long-forgotten shipmate from the *Beagle,* wrote to him in 1875 for a memento. Burgess had been one of the Royal Marines attached to the *Beagle,* and judging from the phonetic spelling, his letter was probably dictated to a poor scribe. He conjured up a time when Darwin's life was energetic and uncomplicated.

> I have often times thought that I should very much like your Portrait as a remembrance of respect during the time I was on Board the Beagle, when she was surveying the South West Coast of Chili and Peru, commanded by Captian Fitzroy, for four years, and you was in connection with the survey on board the Beagle the whole time. For instance, do you remember me calling you upon Deck one night, when the Beagle Lay in Chilomay, to witness, the volcanic eruption of a mountain when I was on duty on the Middle Watch, and you exclaimed, O my God, what a sight, I shall never forget. Another instance, when whe walked eleven miles from the River Santa Cruz, and returning Back, you had forgot your compass and we had to make our way Back without them. Also, do you remember me giving you my water on our returning to the vessel when you was exausted with thirst. I hope I may have said sufficient as to convince you that I am not an imposter,—But one that wishes to have in remembrance of those Happy days I spent with you on Board the Beagle, that his, the presence of your Portrait. I trust you will not think me presumtious in asking for the small favour of respect, Because, I can assure you, I ask from the purest motives.[58]

Messages from the past encouraged Darwin to make his own inquiries about long-lost friends. In 1872 he contacted Sarah Owen from the old Shropshire days, now a widow twice over and apparently living in London. Sarah's younger sister, Fanny, had been Darwin's first love. "No scenes in my whole life pass so frequently or so vividly before my mind as those which relate to happy old days at Woodhouse," he wrote to Sarah sentimentally, forwarding a copy of his *Expression*.

Sarah responded in a long and friendly letter that pleased him. He replied and called her Sarah ("you see how audaciously I begin") as if she were a sister, or a wife. He kept up the correspondence until they ultimately met at Erasmus's house in 1880. The fascination of things that might have been was strong. "Perhaps you would like to see a photograph of me now that I am old," he hesitantly asked.

Most of all, people inquired about his religious views. By now he had privately dispensed with the ambiguities that coloured his writing of the *Origin of Species*. He felt decisive—these were the most godless years of his life. No doubt his imagined grievance about Mivart's behaviour, and the way he still over-reacted to the *Athenaeum*'s constant aggression, soured his perception of religiosity. Yet he also welcomed the gathering pace of change and the first landmarks of a new intellectual topography in which individuals felt able to express varieties of disbelief, a topography which his own book had done so much to open to view. He answered inquiries about his personal faith with a brisk affirmation of his inability to believe, using Huxley's term "agnostic" wherever appropriate. Sometimes sad, sometimes steely, sometimes stately, he told correspondents that on this issue he could not trust the evidence of his own reason because mankind was little more than an elevated monkey. "I can never make up my mind how far an inward conviction that there must be some Creator or First Cause is really trustworthy evidence," he observed to Francis Abbot, a liberal American clergyman and religious writer. The whole concept of the divine was "beyond the scope of man's intellect."

Characteristically, he avoided public statements to this effect, although he fell into a stimulating correspondence with Abbot, the editor of a free-thinking journal, the *Index*. Abbot first contacted him in 1871 wondering if Darwin might supply a few remarks on religion that could be used in a public lecture. Darwin replied promptly and honestly, "I do not feel I have thought deeply enough to justify any publicity." A few months later, Abbot published in the *Index* a clear-headed account of contemporary religious belief ("latitudinarians" were "platitudinarians") that included comments on Darwin's and Tylor's impact on modern thought. He followed this by printing, in an editorial, part of a letter from Darwin under the heading "The coming empire of science: a letter from Mr. Darwin." Darwin was sufficiently interested to subscribe to the *Index* for a number of years. Abbot was a friend of Charles Eliot Norton's, and both were founder members of the Free Religious Association, which owned and published this journal. The association wished to encourage rational humanism, invoking "Truth, Freedom, Progress, Equal Rights, and Brotherly Love."

What Darwin would not do publicly he was sometimes prepared to do privately. When Abbot sent him a copy of his pamphlet *Truths for the Times* (1872), Darwin allowed Abbot to print a recommendation for the book in the *Index,* in which he warmly endorsed Abbot's principles, saying, "I admire them from my inmost heart & agree to almost every word." Abbot was highly gratified. Somewhat at a loss as to what he could offer in return, he gave Darwin a free subscription.

> It was my intention to continue sending the paper indefinitely: for you have paid for it many times over by your kind permission to use your expression concerning the "Truths for the Times." This has been worth a great deal to me; and I cannot tell you how delighted I was by your letter of January 8th, saying that you were pleased with the manner in which I had introduced your name. I can never without deep gratitude remember your generosity to me in this matter.[59]

Darwin evidently kept Abbot's interests at heart. Some time later, during one of those difficult moments that small-circulation, specialist periodicals usually encounter, Darwin sent Abbot a gift of £25. Judging from the comments scribbled in the margins of his copies, he paid attention to the serious-minded articles that Abbot published, especially in 1874, when there was a long discussion of the moral philosophy emerging from *The Expression of the Emotions* and *The Descent of Man.* Darwin wrote to say that although he had "no practice in following abstract and abstruse reasoning," and that he could not see how morality could be "objective and universal," he felt the eulogy was "magnificent."[60] Inexplicably, in 1880 he had William write to request the withdrawal of his endorsement for *Truths for the Times.*

Along with religious probings came questions about the first beginnings of life. Darwin was often asked his opinion about this, and in 1871 had felt moved to write to Hooker on the subject in a more reflective vein than usual.

> It is often said that all the conditions for the first production of a living organism are now present, which could ever have been present. But if (and oh! what a big if!) we could conceive some warm little pond, with all sorts of ammonia and phosphoric salts, light, heat,

> electricity, &c., present, that a proteine [*sic*] compound was chemically formed ready to undergo still more complex changes, at the present day such matter would be instantly devoured or absorbed, which would not have been the case before living creatures were formed.[61]

In other words, even though life might have originated in chains of active proteins, if these simple chains of organic matter were generated today, they would be destroyed before they had a chance to evolve into anything. As Darwin expressed it, that warm little pond was not entirely a fantasy. To him the origin of life was probably a historical phenomenon, unrepeatable, and unrepeated—a direct rebuttal of Lamarck's proposal that spontaneous generation was still continuing in the modern period.

But he remained silent. Ever since Louis Pasteur and Charles Pouchet's argument over the presence of simple organisms in apparently sterile solutions, and his own outburst against Owen's "slime, snot or protoplasm," in the *Athenaeum,* he had of course followed the literature, sometimes for, sometimes against. His own theory of evolution would stand to gain if spontaneous generation was shown to be possible—it would acquire its necessary starting point. Yet it was easy to make rash mistakes. He made this plain to Anton Dohrn, saying that "caution [is] almost the soul of science."

> Pray bear in mind that if a naturalist is once considered, though unjustly, as not quite trustworthy, it takes long years before he can recover his reputation for accuracy.[62]

And to be sure, excited claims by Huxley, Carpenter, and Haeckel that they had found primitive organisms, even naked protoplasm, in the seabed's ooze had been exposed as an embarrassing mistake.[63] Other announcements merely complicated the puzzle of life. Almost alone among practising scientists, John Tyndall put the weight of modern physics behind the germ theory of disease, in which it was contentiously suggested that one germ is always born of another germ, a point of view that immediately ruled out spontaneous generation. But a young researcher called Henry Bastian indicated that there was no real boundary between organic and inorganic substances—"transubstantiation will be nothing to this if it turns out to be true," proclaimed Huxley with a sneer. John Hughes Bennett similarly thought he had evidence for "active molecules." And although cell theory was by now firmly established, the status of bacteria and protozoa as single-celled organisms remained indistinct—Darwin's friend Ferdinand Cohn was attacking this problem head-on.

To onlookers, the interconnections between these ideas and the people who proposed them appeared close—evolutionary theory and the physical basis of life seemed part and parcel of the same sprawling intellectual enigma of scepticism, agnosticism, and materialism. So Huxley's provocative lecture in 1869 on "the physical basis of life" was immediately taken as the Darwinists' declaration of support for spontaneous generation, though Huxley afterwards denied it. When Tyndall succinctly dismissed theology—prayers, cassock, pulpit, and all—during his notorious address at the British Association meeting in Belfast in 1874, it looked as if naturalists were asserting the sole sufficiency of science as a means of comprehending the entire universe. As popular rumour had it, Huxley declared, "Given the molecular forces in a mutton chop, deduce Hamlet and Faust therefrom."[64] Darwin had long been anxious that wavering evolutionists should not be confronted with such starkly impious speculations.

After reading Bastian's enormous two-volume publication on the chemical beginnings of life in 1872, and Wallace's review of it in *Nature*, Darwin therefore wrote to Wallace in an ambivalent frame of mind. Bastian intimated that Lamarck's notion of a constantly replenished source of primitive organisms might be accurate. "I am bewildered and astonished by his statements, but am not convinced. . . . I must have more evidence," said Darwin. Wallace suggested that these rapid transformations of simple matter could quicken evolution to the point where Thomson's warnings about the shortened age of the earth could safely be ignored. Darwin saw the value in this. He would like to see spontaneous generation proved true, he told Wallace, "for it would be a discovery of transcendent importance."[65] For the rest of his life he watched and pondered.

## VI

In 1872, Emma arranged that they should take a house in London for a month in January, an innovation that became a regular habit. That first year they went to Devonshire Street, where Henrietta and her husband joined them. They wanted to keep an eye on her delicate health. The following year, Henrietta had improved enough for her to establish her own married household in Bryanston Street, close to Erasmus. From then on, Darwin and Emma stayed with Henrietta rather than Erasmus or the Hensleigh Wedgwoods. Indeed, Henrietta took on much of the role that Erasmus had earlier played in Darwin's life, providing a safe haven in London and agreeable social connections. Henrietta participated in Litchfield's liberal circles of friends, including F. D. Maurice, John Ruskin, Leslie Stephen, and Arthur Munby, all of whom incidentally appreciated the subtle lure of knowing one of Darwin's daughters. Had Henrietta been anyone else's daughter, it is doubtful whether Munby would have paid her quite so much attention.

> Tuesday, 19 November [1872] . . . to 2 Bryanston Street, and dined with Litchfield. Only his wife & her brother George Darwin. Her father's new book on the expression of emotion was on the table; and Mrs. L. said that often, at home, he used to make her and his other children go out in the garden and look steadfastly at some object, such [as] a particular twig on a tree, to study their faces as they looked. But often, she added, when her father in his rides had stopped to notice the expression of some crying child's face, he found himself saying 'Poor little thing!' and losing his power of observation in his sympathy with the object observed.[66]

During these London months there was plenty of relaxed mingling. "Charles had much rather stay at home, but he knows his place and submits," said Emma approvingly. When added to an annual trip to Southampton to see William at his house, Basset in the New Forest ("a really hideous Victorian villa," said Darwin's granddaughter Gwen Raverat), the days passed agreeably by.

While next in London in 1873, Darwin was disconcerted to discover that Huxley was on the brink of nervous collapse. He was doing too much, too often, at too high a pitch, and for woefully inadequate pay—and Darwin was shocked that he had not noticed before. Sympathetically, the wives suggested to their husbands that a few close friends might club together to give Huxley enough money for a holiday. Henrietta Litchfield said, "My father took eagerly to the idea and became the active promoter of the scheme." His heart was easily fired by Huxley's problems. He longed to do something—anything—to show his affection and high regard. After a conspiratorial talk with Lyell and Hooker, in which they found themselves puzzled how to prevent Huxley taking offence, he wrote letter after letter to mutual friends asking for subscriptions, stashing up the cash in secret until he could present it to Huxley as a cheque. The total came to £2,100, a huge amount for the day (some two years' salary for Huxley) that testified to the respect with which Huxley was regarded.

> If you could have heard what was said [wrote Darwin], or could have read what was, as I believe, our inmost thoughts, you would know that we all feel towards you, as we should to an honoured and much loved brother. I am sure you will return this feeling, and will therefore be glad to give us the opportunity of aiding you in some degree, as this will be a happiness to us to the last day of our lives.[67]

Touched, Huxley accepted the gift as it was intended. The friends, however, took no chances—they knew their man inside out. First, they put

Hooker in charge of him, recognising that it would be a tough job making the patient relax as completely as Andrew Clark, his medical supervisor, recommended. Off the pair went, leaving their wives and families behind, for a walking holiday in the Alps, a favourite occupation from the time when, as energetic young men, Hooker, Huxley, and Tyndall used to hike through the mountains, glacier-spotting and plant-hunting, calling themselves Smith, Brown, and Robinson after the popular Victorian satire on tourism. But they were no Smith, Brown, and Robinson now. Huxley was the leading biologist of the era, Tyndall the leading physicist, and Hooker president of the Royal Society.

Hooker encountered one or two testing moments. At a Swiss railway bookstall, he had to wrestle a book about the miracles at Lourdes out of Huxley's hands in case he lost his temper over Catholicism and Mivart all over again. In France, he rashly let Huxley climb the Pic de Sancy, only to see him feverishly making notes about previously unsuspected glacial action. In a small provincial museum, Huxley thought he saw an unrecorded prehistoric human skeleton and could not rest until he had taken details of the bones. But eventually the unruly ward calmed down. When Hooker left, and Henrietta Huxley came out to join her husband, leaving the children in Emma's care, Huxley proclaimed himself "wonderfully better." He never forgot his friends' kindness in this affair. In one of his most buoyant moods, he dashed off a note to Tyndall telling him he had shaved off his beard as a token of his affection.

## VII

Early in 1873, Francis Galton gave Darwin something different to think about. He asked Darwin to list the special mental qualities that he felt he might possess. Galton was immersed in a survey of the mental attributes of scientific men, intending to identify in quantitative terms the "genius" or talent for science that, as he believed, characterised the British nation. His conclusions would primarily be based on questionnaires circulated among fellows of the Royal Society. His book was published in 1874 as *English Men of Science, Their Nature and Nurture.*

Galton had been collecting these statistics for a long time, although not always so methodically. He was pushed into action by a critique of his *Hereditary Genius* by the Swiss plant geographer Alphonse de Candolle, whose work was admired by Darwin. The links between botany and human heredity were unexpectedly close in de Candolle's mind because he concerned himself with a plant's adaptation to the environment, an approach that enabled him to see many faults in Galton's case for the hereditary transmission of characteristics. Then in 1873 de Candolle produced a biographical history of science in which he delved further into the same issue. "I have hardly ever read anything more original and interesting than your treatment of the causes which favour the development of scientific men," Darwin said appreciatively to de Candolle. Galton decided to look again at the role of upbringing in early development.

Galton was not the first to ask Darwin such questions. By now several people had inquired into what they imagined Darwin's private mental life might be. Some questions originated in simple curiosity about the likes and dislikes of the famous. Others stemmed from a growing interest in what would come to be called developmental psychology, or sought some form of idealised vignette about the scientific "character."[68] Darwin was occasionally asked for his views on school education, for example, and was always ready to repeat his low opinion of Latin and Greek. "I am rather inclined to think that real education does not begin till after school & college days; at least I am sure this was the case with myself."[69] These maverick views did not endear him to many headmasters, and it is worth noting that Darwin's schooldays were not turned into a moralising story for boys, at least not during the nineteenth century when such narratives were commonplace. Nor were his travels on the *Beagle* used as a basis for popular adventure stories, perhaps because the scientific theories that had emerged from the voyage were still highly contested knowledge. Arthur Nicols's account of the history of the earth, written for children and called *The Puzzle of Life* (1877), certainly included descriptions of Darwin's South American fossils but avoided any direct reference to Darwin himself and came to an elevated theological conclusion. There was probably at that time little scope for a suitably Victorian ethos to be drawn from Darwin's life story. David Livingstone's missionary travels in Africa and the search in the Arctic wastes for the missing explorer John Franklin were popular alternatives.

More frequently, he was asked for biographical information, requests that were part of the fashion in Victorian England for real-life stories that might match the abundance of contemporary tropes about character and personality. Samuel Smiles's doctrines of self-help and determination in overcoming difficulties featured widely in this literature. One acquaintance of Darwin's, William Preyer, a physiologist and psychologist in the University of Jena and a friend of Haeckel's and Dohrn's, wrote to him in 1870 to ask whether he could publish a biographical sketch. For some years now Preyer had been teaching *Darwinismus* to biological students in classes numbering up to five hundred. Preyer was alert to the formal honours of science, arranging in 1868 for Darwin to be awarded an honorary degree from the medical school at Bonn. He explained he was particularly interested in Darwin's mental development and asked "which book or books made deep impressions on you in youth."

Flattered at the request, Darwin supplied Preyer with a generously long account of his memories of Edinburgh and Cambridge universities, his love for Henslow, his admiration of the writings of Gilbert White and Humboldt, and his belief that his "education in fact began on board the *Beagle.*" Simple as it was, with this in his hands Preyer had as much to go on as any other biographer of the period. Even with the best will in the world, Darwin never probed very far into his own character and usually relied on comfortable stereotypes that simultaneously conveyed what he regarded as the truth about himself and relatively unexceptional traits that echoed the conventions of his time. He was reserved. These kinds of self-analytic writings were difficult for him. Yet he felt the resulting article was fuller, and truer, than anything hitherto published about him—probably because so much of it came from his own hand.[70] But he laughed at Preyer's description of cart-tracks leading to Down House: "you give . . . too bad a character to our roads or rather lanes; they are rather narrow and rather bad; but when you come to England, I hope that you will prove by visiting me here that they are not impassable."[71]

By responding to inquiries like these, Darwin took an increasingly active part in constructing his public persona and codifying his role in introducing evolution to the world. In 1874, *Nature* began a series of "Scientific Worthies" in which Darwin appeared in an article written by Asa Gray. A similar article was written for the *Judische Literaturblatt* of 1878, edited by Dr. Moritz Rahmer, which featured Lamarck, Etienne Geoffroy Saint-Hilaire, Haeckel, and Darwin as the founding fathers of evolution.[72] Arabella Buckley, Lyell's secretary, wrote *A Short History of Natural Science for Young People* (1876), which included biographical remarks about Darwin and Wallace and their independent construction of evolution by natural selection. "You have crowned Wallace and myself with much honour and glory," remarked Darwin cordially.[73]

Galton's request pushed Darwin into a much more self-analytic framework, and he did his best to respond. The questionnaire was seven pages long, covering family background, health, temperament, religious beliefs, education, and scientific interests. A second column asked for the same information about Darwin's father. Darwin found the personal questions were the hardest. He felt obliged to ask his son George for his opinion to answer one of them. "George thinks this applies to me," he wrote in response to Galton's inquiry about "originality or eccentricity."

> I do not think so—ie. as far as eccentricity. I suppose that I have shown originality in science, as I have made discoveries with regard to common objects.[74]

Others were more straightforward. "I consider that all I have learnt of any value has been self-taught," he declared. Where Galton asked, "Do your scientific tastes appear to have been innate?" Darwin replied they were "certainly innate . . . my innate taste for natural history strongly confirmed and directed by the voyage in the *Beagle.*"[75] To the question "Has the religious creed taught in your youth had any deterrent effect on the freedom of your researches?" he replied, "No." He gave his religious affiliation as "Nominally to Church of England."

As for the rest, "I have filled up the answers as well as I could, but it is simply impossible for me to estimate the degrees."

> *Politics?*—Liberal or radical.
> *Health?*—Good when young—bad for last 33 years.
> *Temperament?*—Somewhat nervous.
> *Energy of body?*—Energy shown by much activity, and whilst I had health, power of resisting fatigue. . . . An early riser in the morning.
> *Energy of mind?*—Shown by rigorous and long-continued work on same subject, as 20 years on the "Origin of Species" and 9 years on *Cirripedia.*
> *Memory?*—Memory very bad for dates, and for learning by rote; but good in retaining a general or vague recollection of many facts.
> *Studiousness?*—Very studious, but not large acquirements.
> *Independence of Judgement?*—I think fairly independent; but I can give no instances. I gave up common religious belief almost independently from my own reflections.
> *Strongly marked mental peculiarities?*—Steadiness, great curiosity about facts and their meaning. Some love of the new and marvellous.

For the final question, he curiously disregarded all his contributions to scientific thought and claimed that his only special talent lay in business and finance.

> *Special talents?*—None, except for business, as evinced by keeping accounts, replies to correspondence, and investing money very well. Very methodical in all my habits.

At the heart of it lay the conviction that his results were brought about by his own "rigorous and long-continued work." Perhaps he needed to believe this. Had Darwin been able to see himself with the eyes of another, he could not have ignored the intellectual resourcefulness he displayed throughout his thinking life, admittedly in untrained form in the earliest pre-*Beagle* days, but then developing through the highly productive years of *On the Origin of Species* and culminating in *The Descent of Man* and *The Expression of the Emotions* in 1871 and 1872. His customary veneer of modesty here allowed him to deflect personal inquiries without the need for any discomforting self-examination.

On another level, however, he may have sensed that the quality of perseverance was indeed the single most important feature of his personality, or at least the feature that had contributed most to his success in science. This is not to say that he regarded diligence as a sure way to scientific achievement, although he did here and there make exactly that kind of assertion and often praised those men who revealed similar qualities. It was more that he identified in himself the ability to persist, and retrospectively turned it into an all-consuming methodology and justification—that this was *his* way of tackling questions. Whereas Huxley or Wallace might have relied on flashes of brilliance, or Hooker on his prodigious memory, it seems possible that Darwin may genuinely have regarded his special talent for science to lie in mental determination. He had given his life to the ceaseless documentation of facts, pushing and pulling at species until everything fell into place. From time to time, he used the expression "It's dogged as does it," drawing on the phrase used by Anthony Trollope in *The Last Chronicle of Barset*.[76]

## VIII

In August 1873 Darwin experienced what Emma called "a fit" in which he temporarily lost his memory and could not move. "Very unwell partial memory loss 12 hours, sinking fits," she scribbled urgently in her diary. Frantically, she consulted Huxley and on his advice sent for Andrew Clark from London, renowned for his sensitive handling of nervous patients. Clark descended the very next day, and would not take a fee for his advice, although Emma wished otherwise. The verdict was reassuring: "there was a great deal of work in him yet." Clark put Darwin on a strict diet, which seemed to work well enough for a while, and prescribed special pills, packed with strychnine and iodine. "Thank God," the invalid sighed to Hooker. "I would far sooner die than lose my mind."

Otherwise, he felt fulfilled, not nearly so ill as usual. Leonard took a pleasant photograph the following year of his father seated in a wicker chair on the veranda, looking tired but benign. This veranda had been built in 1872 after a family holiday in a rented house near Sevenoaks where Emma and Darwin became "acquainted with the charm of outdoor living." They returned to Down House full of architectural visions, going on to construct a wood-framed veranda along the garden frontage of the drawing room, roofed in glass and paved with colourful encaustic tiles. Emma installed cushioned benches for the boys and the armchair for her husband. "So much of all future life was carried on there, it is associated with such happy hours of talk and leisurely loitering, that it seems to us almost like a friend," said Henrietta.

> The fine row of limes to the west sheltered it from the afternoon sun, and we heard the hum of bees and smelt the honey-sweet flowers as we sat there. The flower-beds and the dear old dial, by which in the old days my father regulated the clocks, were in front, and beyond the lawn the field stretching to the south. Polly, too, appreciated it and became a familiar sight, lying curled up on one of the red cushions basking in the sun. After my marriage she adopted my father and trotted after him wherever he went, lying on his sofa on her own rug during working-hours.[77]

Francis recalled those days with similar affection. For him the veranda was redolent with sounds of summer.

> The veranda which was built onto the drawing room was his idea, and gave him much pleasure as he often sat there in his tall Japanese wicker chair shown in Leo's photograph.—It was a pretty veranda with tessellated floor, and wooden posts covered with little Virginia Creeper, and with white clematis trained along the rafters and Adlumia grown from seed springing up in a big flower box in the corner & often decorated with big plants such as Vallota in pots. . . . One sound there was peculiar to Down—I mean the sound of drawing water. In that dry chalky country we depended for drinking-water on a deep well from which it came up cold and pure in buckets. These were raised by a wire rope on a spindle turned by a heavy fly-wheel, and it was the monotonous song of the turning wheel that became so familiar to us.[78]

As the children grew up and moved away, Darwin and Emma would sit here together. Sometimes they would amble across the chalky fields to woods by Cudham and Keston, or sit out in the evenings to catch nightingales in song. "We were out last night nightingaling till 10.30 & returned to bread and cheese. There was a first rate one in the Rookery & we on Green hill heard him perfectly," Emma told William. One favourite walk was to the pretty spot called Green-hill; another to what they called Hangrove, a terrace below Stoneyfield. "There were rabbits in the shaw, and Polly, the little fox-terrier, loved this walk too," said Henrietta. "My father would pace to and fro, and my mother would sometimes sit on the dry chalky bank waiting for him, and be pulled by him up the little steep pitch on the way home."[79]

Much of Charles and Emma's love for each other, undemonstrative and seasoned by more than thirty years of marriage, was tied up in these quiet walks together. Darwin's emotions for his wife were of a piece with his sense of place, his attachment to his garden and the surrounding countryside. Being with her made him complete. As for Emma, although she must at times have found him self-absorbed, withdrawn, and insidiously demanding, these were the times when he was most relaxed, at peace with himself and his world. She became his protectress, straining to conserve his energies. Her commitment to him never wavered.[80]

When they wanted to stay closer to home, they would stroll around the Sandwalk. This land, no more than an acre or so in size, was rented on an annual basis from John Lubbock, who since coming into his inheritance was by far the greatest landowner in the area. Early in 1874 Darwin asked if he might buy the wood from Lubbock, offering some grazing land of his own in return. The transaction caused some unpleasantness between the two friends.[81] Lubbock—with all his riches—squeezed up the price too zealously, saying that Darwin should pay as if the land were suitable for building rather than mere agricultural use. The same accusation could be made of Darwin, a wealthy man who became churlish about a price he could well afford to pay. Francis said his father "was much grieved at Sir John making such a remarkably good bargain out of the Sandwalk which he sold at the highest accommodation kind of price. He used to compare it rather bitterly with Mr. Farrer selling at agricultural price to his neighbours."[82]

But the Sandwalk was his spiritual home. The boys liked to hear the click of his walking stick on the stones as he walked round the path on his own, first the light side, looking over the valley, then the dark side, under the trees. As a businesslike man, he would pile up a mound of flints at the turn of the path and knock one away every time he passed to ensure he made a predetermined number of circuits without having to interrupt his train of thought. Five turns around the path amounted to half a mile or so. The Sandwalk was where he pondered. In this soothing routine, the power of place became preeminent in Darwin's science. It shaped his identity as a thinker.

Meanwhile, the tributes poured in. Some of them he treated humorously, as in 1872 when he accepted an invitation from John Jenner Weir to become patron of a cat show, warning him that "people may refuse to go and admire a lot of atheistical cats."[83] In the same year he was asked by the students of the University of Aberdeen to be their rector; by ancient privilege the undergraduates were allowed annually to elect an honorary leader from public men of the time.[84] He refused, but was tickled to hear that the undergraduates were so determined to have an evolutionist that they asked Huxley, who accepted.

Serious expressions of admiration also proliferated. During the 1870s many honorary memberships and fellowships came Darwin's way. At the end of his life he belonged, in an honorary capacity, to more than sixty learned institutions or bodies world-wide.[85] Some of the diplomas that arrived at Down House were lavish.

Other kinds of honour crept in. In 1873, Karl Marx sent him a copy of *Das Kapital*, inscribing it "Mr. Charles Darwin on the part of his sincere admirer Karl Marx." Darwin was conscious of the compliment being paid him.

> I heartily wish I was more worthy to receive it, by understanding more of the deep & important subject of political economy. Though our studies have been so different, I believe that we both earnestly desire the extension of knowledge & that this in the long run is sure to add to the happiness of mankind.[86]

The book remained in Darwin's library uncut and unopened, and almost certainly unread. The contact between these two remarkable thinkers was, in truth, enigmatic. There is scant evidence for the story that Marx asked if he could dedicate a future edition of *Das Kapital* to Darwin in recognition of Darwin's understanding of struggle in nature. On the contrary, it is much more likely that it was Edward Aveling who asked if he might dedicate one of his books to Darwin, and that this request was refused. The confusion emerged only after Darwin's death, either through Aveling's desire to link Darwinism with his own brand of revolutionary atheism, or because Marx's papers were subsequently mixed up with Aveling's. Either way, Marx consulted Darwin's books when drafting *Das Kapital* and again when working on the second edition. He added an appreciative footnote to the second German edition, the one he sent Darwin, speaking of natural selection as "the history of natural technology, that is, the formation of the organs of plants and animals, which serve as the instruments of production for sustaining their life."[87] As far as is known, the exchange of this presentation copy and the message of thanks was the only formal contact between them.

## IX

If not letters and diplomas, it was spiritualism. Wallace's newest book was out, boldly titled *Miracles and Modern Spiritualism* (1875); celebrated mediums were at work in every metropolitan centre; frauds and sensation-seekers joined hands with sceptics and enthusiasts in darkened rooms around a thousand tables. All of them sought affirmation that there was some form of existence after death. Darwin could no more divorce himself from these issues than from inquiries into his religious beliefs. Robustly, he said he believed in "none of it."

Credulity, he noted in alarm, was already at his front door. He flinched when Wallace told him that all matter was force, and force must be the product of a divine mind. He shook his head over the newspapers and protested at the way mediums exploited the bereaved. Emma's brother Hensleigh Wedgwood had also hastened into the spiritualist movement, attending séances, communing with spirit-guides, and collecting spirit photographs. "Hensleigh in his study, living his separate life among the spirits," complained Emma. She was not far wrong. Hensleigh Wedgwood later served as a founder member of the Society for Psychical Research in the 1880s.

To Darwin and Emma, it seemed as if Hensleigh's mind was nearly turned. "I always feel nowadays that he finds everything so flat & uninteresting except spiritualism," murmured Emma unhappily. "His company is not of the old easy comfortable character it used to be."[88] In 1874 Hensleigh tried to get Huxley involved by sending him photographs of a purported apparition. "I consider fraud impossible," he said, for the entire process had been supervised by himself and his daughter Effie. On the contrary, Huxley could hardly believe the photograph represented reality. He patiently explained that the photographer must have put a second image on the plate inside the camera before taking a picture.[89] Hensleigh refused to believe him.

Nevertheless, these sceptical menfolk wavered between being intrigued and exasperated. Even Francis Galton, an arch-sceptic, attended a number of séances and was "utterly confounded" by the goings-on. Members of the X Club and Metaphysical Society expressed interest in at least some spiritualist phenomena. George Romanes, a committed evolutionist, was intensely caught up in the movement, much to Darwin's disgust, although this was but part and parcel of Romanes's fascination with mental activity as a phenomenon.

At last Darwin and Huxley gave in to Hensleigh's urgings. Somewhat surprisingly, Darwin agreed to attend a séance arranged by his son George at Erasmus Darwin's house in January 1874. It was a highly sociable occasion, uniting sceptics and believers, comprising the Litchfields, Emma and Charles Darwin, Francis Galton, Frederic Myers (the Cambridge classicist and later a prominent member of the Society for Psychical Research), Hensleigh and Fanny Wedgwood with their oldest daughter Julia, George Lewes and Mary Ann Evans (George Eliot), and Erasmus's friend Mrs.

Bowen, as well as the medium hired for the occasion, Charles Williams. Darwin wanted Huxley to be there, and after reiterating that he did not believe in any of it either, Huxley arrived as an anonymous participant, although Williams could hardly have failed to recognise him. There may have been other people there also, because Emma mentioned "séance of 20 persons" in her diary. For her part, Emma was eager to meet the author of *Middlemarch* and took the opportunity to introduce herself as an enthusiastic reader. Darwin and Lewes, too, were as close as they ever were in their intellectual lives, for Lewes had always defended evolutionary theory in his fashion and was about to initiate the journal *Mind,* a publishing venture of mutual interest.

This was to be no ordinary séance but a scientific test, with the medium secured hand and foot, and the room carefully monitored to prevent the legerdemain they all expected. Erasmus and George spent an hour or two beforehand peering under chairs and carpets. Williams—and Williams's spirit-guide John King—were well known to Hensleigh Wedgwood, who regularly attended his séances. Indeed, most of the London mediums must have been wearily familiar with such scientific investigations, for Wallace, Crookes, and the others all insisted on tightly controlled precautions to allay ridicule. But when Williams demanded complete darkness, Lewes and Mary Ann Evans "left in disgust."[90] So did Darwin, although he described the scene afterwards with secondhand aplomb.

> We had grand fun, one afternoon, for George hired a medium, who made the chairs, a flute, a bell, and candlestick, and fiery points jump about in my brother's dining room, in a manner that astounded every one, and took away all their breaths. It was in the dark, but George and Hensleigh Wedgwood held the medium's hands and feet on both sides all the time. I found it so hot and tiring that I went away before all these astounding miracles, or jugglery, took place. How the man could possibly do what was done passes my understanding. I came downstairs, and saw all the chairs, etc. on the table, which had been lifted over the heads of those sitting around it. The Lord have mercy on us all, if we have to believe in such rubbish. F. Galton was there and says it was a good séance.[91]

Henrietta Darwin said, "Mr. Lewes I remember was troublesome and inclined to make jokes and not sit in the dark in silence." Francis later explained that his father regretted the enormous energy people poured into investigating the supernatural when the same effort might be directed into understanding reality. Francis said that Darwin declared "it was all imposture."

On the other hand, Huxley, Galton, Hensleigh Wedgwood, and George Darwin were sufficiently tantalised to arrange another séance a few weeks later. This took place at Hensleigh's house. Huxley described the occasion to Darwin afterwards, with diagrams, explaining how the medium must have produced his effects, such as plucking guitar strings with his mouth, or, as Emma cynically suggested, touching George's hand with his nose. Joyously, Huxley launched himself into exposing Williams as a charlatan. Hot on the trail, he stopped at nothing, teaching himself to snap his toe joints inside his boots in imitation of spirit raps on the table. Thin socks were essential, he said. "My conclusion is that Mr. Williams is a cheat and impostor," he informed Darwin. This was just what his friend wanted to hear. "Now to my mind an enormous weight of evidence would be requisite to make one believe in anything beyond mere trickery."

A little later on, Williams was unmasked as a fraud by George Romanes, who paradoxically continued to give credence to the spirit world. Similarly, Hensleigh Wedgwood doggedly insisted on the independent existence of Williams's spirit-guide, John King, while talking crossly about the "roguery of Williams."[92] With some satisfaction, Darwin told Romanes he should buy a copy of the *Spiritualist,* which was running a long article about the exposure. "Good Heavens what rubbish the whole does seem to be," said Darwin.[93]

But the demise of one medium was not enough to end the curiosity. In 1876, when the most famous medium of them all, the American Henry Slade, was humiliatingly denounced, Wallace testified in court on his behalf. Not a shadow of suspicion darkened Wallace's mind. He said afterwards that Slade had not received a fair trial. In one respect, he was right to think that the scientists had ganged up against Slade. The biologist Ray Lankester, a friend of Huxley's and Darwin's from the South Kensington laboratories, played the central part in Slade's downfall in a letter to the *Times,* itemising the tricks Slade used. For some months beforehand Lankester and Huxley had swopped recipes for fraudulent techniques. Spiritualists were ideal targets for scientists like these—perfect sparring partners—because if nothing else the contest enabled them to bring the full force of their minds into play. Huxley found the exercise wholly invigorating.

This was no time for excessive displays of loyalty to Wallace, thought Darwin. His feelings ran so high that he privately sent Lankester £10, along with a congratulatory letter, to help the case for the prosecution of Slade.[94]

*chapter*
11

# ENGLAND'S GREEN AND PLEASANT LAND

HARLES DARWIN found peace among the plant pots. He knew his likes and dislikes well enough by now. A visit to the hothouse for a talk with his gardeners and then a stroll around the Sandwalk with a dog at his heels were an essential part of his day. All through the writing years he had devotedly kept up a variety of small botanical investigations, telling Emma that these served as his best form of relaxation.

Nevertheless his experimental ambitions had been sidetracked by writing *The Descent of Man* and *The Expression of the Emotions.* Now that he was free of writing, another round of bookish duties had crept in. He produced a revised edition of *Descent* in 1874 and a second edition of *Variation* in 1875, and made minor changes to the sixth edition of the *Origin of Species* (the same edition that had been issued in 1872), the last changes to be incorporated in his lifetime. Darwin paid Henrietta £20 for correcting the proofs of *Variation.* Murray released the amended sixth edition of the *Origin* in 1876 with a title page indicating that 18,000 copies had been printed in England alone since 1859.[1] In the year of Darwin's death the number had increased to 24,000. In these revised books Darwin eased into a more adaptationist frame of mind, suggesting that there was a role in evolutionary theory for the inheritance of some acquired characteristics—the result of his pondering the mechanisms underlying pangenesis, sexual selection, and the expression of emotions. Although he had never categorically excluded behaviourally or environmentally induced adaptations from his writings, he now felt they should play a larger part, telling Wallace that "I think I have underrated . . . the effects of the direct action of the external conditions in producing varieties." As usual, John Murray, and Murray's new business partner Robert Cooke, produced the volumes. Each one sold relatively well.

At the same time, the firm of Smith & Elder, which still held the copyright to his early geological books, encouraged Darwin in 1874 to bring out a revised version of *Coral Reefs,* taking account of James Dana's theories of reef formation and Alexander Agassiz's remarks. They also wished to produce an updated version of *Geological Observations on the Volcanic Islands and Parts of South America* in 1876. These books had been his first scientific writings in the post-*Beagle* years. To him it seemed as if he was turning a full circle. Without hesitation, he agreed. He had been proud of them once, and was still.

If he had paused to consider the matter, Darwin might have quailed at such a punishing self-imposed schedule. Every year from 1860 onwards he had published at least one book, sometimes two, either a new title or a revised edition, and since that date a total of some fifty-five articles serious enough to be listed in the Royal Society Catalogue of Scientific Papers. He persisted in this regime until his death, except for five individual years when he was either too ill or too busy collecting facts. Folder after folder of notes piled up as he worked his way through these publishing obligations. "I have no news about myself, as I am merely slaving over the sickening work of preparing new editions," he moaned gently. "I wish I could get a touch of poor Lyell's feelings, that it was delightful to improve a sentence, like a painter improving a picture."[2]

In between he rewarded himself with a remarkable array of botanical projects. Where he had once used plants simply to explore evolutionary theory and as a pleasant mental diversion, he now reversed the process and made the theory serve as a powerful engine to break into the secrets of physiology. He shifted gears. He desired the real work of day-to-day experimental investigation, not so much the search for so-called "crucial" experiments to which individuals might give their name but the patient probing of nature's mechanisms with no particular way of telling where any line of exploration would end up, the unsung feature of a scientist's life that all researchers know to be authentic. He wanted to pick up problems as he went along, to use the techniques he understood best, to outthink the ingenuity of living organisms.

He was at long last able to give insectivorous plants his full attention, asking Hooker and Gray, and then an army of correspondents, to send specimens to Down so that he could feed them increasingly unusual diets and observe their cellular activities under his microscope. No project in this sense was ever finished. Darwin had waited for ten years or more for this treat. He spent hours studying the leaves and trap mechanisms, still unsure exactly how the plants could digest insects. With Francis, well versed in the new physiology, at his side, he started a major research programme limited only by the *ad hoc* nature of his household resources. He had to write apologetically to Hooker to say that Lettington had destroyed a prime specimen from Kew. "I am very much vexed. . . . I am sure it was an oversight of Lettington's & not carelessness, as he was very proud of the state of *D. capensis.*"

He bought himself a bigger and better microscope, a Smith and Beck compound-lens apparatus, so that he could study the movements of protoplasm inside cells. Hooker surprised him by saying that the accepted wisdom of the day was that plant leaves could not absorb any substances. Absorption was a function confined to roots alone. "Have you tried *Begonia* leaves?" Hooker asked. "Shall I look out for you some plants with hyaline bladdery epidermal cells for you to operate on?" Puzzled, the two old friends consulted William Thiselton Dyer, a younger man who had trained as a physiologist with Foster and Huxley and was assistant director at Kew (soon to become Hooker's son-in-law, too), much closer to the cutting edge of physiological science than either of them could pretend to be. In his spare time Dyer was translating Julius Sachs's important work on plant physiology from German into English. Dyer "would be up to the latest discoveries," Hooker assured Darwin.[3]

And he enjoyed the macabre ingenuity of the plants themselves: the traps, lures, and snares that they devised to capture their prey. The sticky-leaved sundew "is a wonderful plant, or rather a most sagacious animal." A deep-seated appreciation of the sundew's murderous course in life formed the central theme of his inquiries. In this he was helped immeasurably by Edward Frankland of the Royal College of Chemistry, in London, and John Burdon Sanderson, the new professor of physiology at University College London, who both answered his inquiries patiently. When one of Darwin's homespun experiments proved too tricky to handle in his kitchen, Frankland volunteered to macerate *Drosera* leaves in his laboratory and analyse the juices for traces of pepsin and hydrochloric acid, the principal digestive acids of animals.

John Burdon Sanderson had studied under Claude Bernard and then succeeded Michael Foster in London University in 1870.[4] At this point he was beginning his life's work on the fundamental properties of living tissue, and his particular concern was the way muscles work. To that end he specialised in identifying the small electrical changes that took place during contraction and relaxation. He also interested himself in histology and cellular pathology, especially the structure of blood corpuscles. All this was music to Darwin's ears. In June 1873 he contacted Burdon Sanderson (through George Romanes, who worked in the same laboratory) to ask if he would look over his investigations into the cellular activities of *Drosera*. Darwin was still bothered by the process he called protoplasmic "clumping" or "aggregation" inside the hair cells when the leaves were stimulated into feeding. Hooker told him that chlorophyll in plant cells did the same thing when exposed to sunlight. Darwin's new microscope, with the better magnification, told him only that the process remained incomprehensible.

Burdon Sanderson was not at first convinced that the plants could even move—Darwin could tell that he thought he was imagining the whole thing. So Darwin set out to demonstrate the sensitiveness of *Drosera* to both Burdon Sanderson and Huxley, asking them to visit Down House. They peered at one of Darwin's *Drosera* specimens until "Mr. Huxley cried out, 'It *is* moving!' "[5] Burdon Sanderson agreed to make the experiments. Within the month Darwin also persuaded him to investigate the movements of a Venus's fly-trap. Could it be possible, he asked, that the leaves might possess some botanical equivalent to nerves and muscles? So Burdon Sanderson attached his electric probes to individual fly-trap leaves and stimulated them with minute bursts of current. The leaves responded "like animal muscle," he reported at a meeting of the British Association in 1873. It had been a stunning example of laboratory dexterity. Moreover, this announcement was a revelation to the experimental physiologists in the audience who scarcely gave a thought to plants. "Not merely then are the phenomena of digestion in this wonderful plant like those of animals," Hooker said admiringly, "but the phenomena of contractility agree with those of animals also."[6]

Before long, Darwin told Thiselton Dyer he was "in that state in which I would sacrifice friend or foe."

> I fear that you will think me a great bore, but I cannot resist telling you that I have just found out that the leaves of Pinguicula possess a beautifully adapted power of movement. Last night I put on a row of little flies near one edge of two *youngish* leaves; and after 14 hours these edges are beautifully folded over so as to clasp the flies. . . . I have ascertained that bits of certain leaves, for instance spinach, excites so much secretion in Pinguicula, and that the glands absorb matter from the leaves.[7]

The whole business was a sensational story of surface innocence and hidden guile, just the kind of thing Darwin enjoyed in his recreational reading. Who could believe that the pretty common butterwort, with its golden star of leaves, had such a sinister nature? Or that under the quiet surface of a pond there were plants casting sticky nets, like Dictynna in Greek mythology, to trap their prey? Some plants specialised in drownings, others in knocking out victims with anaesthetics, still more in smothering or gluey entanglement.

Intrigued by his evident excitement, Darwin's botanical friends rallied round. Lady Dorothy Nevill sent him an unusual bladderwort from her conservatory, a specimen water-plant that usually drifted on the surface of rivers in South America. Darwin had seen it only in dried form. When he and Francis opened up the bladders they found plenty of animal remains inside. A specimen of *Genilisea ornata* animated them even more. Nothing could escape from the trap it set, for the fly's route to the death chamber was lined with a grid of spikes. Once inside, the victim could not retreat.

> The great solid bladder-like swellings almost on the surface are wonderful objects, but are not the true bladders. These I found on the roots near the surface, and down to a depth of two inches in the sand. . . . I felt confident I should find captured prey. And so I have to my delight in two bladders, with clear proof that they had absorbed food from the decaying mass. For Utricularia is a carrion-feeder, and not strictly carnivorous like Drosera. . . . I have hardly ever enjoyed a day more in my life than I have this day's work; and this I owe to your Ladyship's great kindness.[8]

Messages across the globe rang with the same delighted claims. "Your magnificent present of Aldrovanda has arrived quite safe. . . . You are a good man to give me such pleasure."

Amused by Darwin's enthusiasms, Asa Gray sent a practical joke in the post—an article describing a carnivorous plant in Madagascar that subsisted on humans. Darwin confessed that he "began reading . . . quite gravely." He did not perceive it was a hoax until he came to the woman who was lunch.

Early in 1875, Darwin felt ready to dispatch a manuscript on insectivorous plants to Murray. Inevitably he kept finding new things to observe. "You ask about my book," he said to Hooker in February that year. "All I can say is that I am ready to commit suicide; I thought it was decently written but find so much wants rewriting that it will not be ready to go to printers for two months." In May he was still covering his proof sheets with an inky forest of alterations. Faced with the crisis of placing actual words on paper, he decided that insectivorous plants must possess what he could only call "nervous matter," analogous to animal nerves. He felt unable to describe this nervous matter further without the authority of hundreds of detailed experiments.

Emma despaired of getting him away for the customary break in London.

> F. jibbs a good deal & I am afraid I shall not get him to move, at least not more than for 2 days at Q.A. [Queen Anne Street, Erasmus's house] which he thinks might rest him, & so I dare say it would for a few days. However the proof sheets are coming in at a great rate. . . . F. has had prosperous news today from Dr B. Sanderson. . . . Lady Dorothy Nevill is coming to lunch on Tuesday! It is rather serious. F. will have to be so friendly & adoring (if possible).[9]

The book was nicely illustrated with diagrams by George and Francis and published by Murray in July 1875 under the title *Insectivorous Plants*. It proved far too specialised for a general audience and was not printed again while he lived. To a public accustomed to reading about apes and religious dissent, this Darwin seemed a very different author from the Darwin discussed in the newspapers.

## II

At the same time, Darwin initiated another burst of investigation into orchids. "They are wonderful creatures, these Orchids," he told Bentham. "I sometimes think with a glow of pleasure, when I remember making out some little point in their method of fertilisation."[10]

And he plunged again into a full reevaluation and extension of his earlier work on cross-fertilisation and self-fertilisation in plants. This was perhaps something more than his usual urge to complete projects left dangling many years before. "I cannot endure doing nothing," he told Jenyns in 1877. It was almost as if he feared the moment when his mind might be empty, when his work might be done; and to stave off this abyss constantly found old and new topics to pursue. If not dread of idleness, then dread of decrepitude. He often said that his work made him feel alive, helped his mind sing, was the one thing that blotted out his cares. Although he called himself a "a kind of machine for grinding general laws out of a large collection of facts," the truth was he only felt himself when immersed in some demanding new project.

Francis Darwin helped with the researches. Later, Francis explained that his father had been driven by the idea that cross-fertilised plants would produce offspring that would be more successful in the competition for survival, and any mechanisms developed for ensuring the transfer of pollen from flower to flower therefore gave the species a significant adaptive advantage. This conviction had formed part of Darwin's belief in evolution by natural selection for more than thirty years. The whole of the living kingdom, as he regarded it, was dedicated to ensuring sexual reproduction—man and woman, flower, and beast. Now, in particular, he wished to probe more deeply into aspects of plant fertilisation, heredity, and sterility.

Underneath, it seems entirely possible that Darwin was also becoming anxious about his own family's reproductive future and that his botanical experiments echoed this concern. His seven children were grown up. William, the oldest, was thirty-five in 1874. Horace, the youngest, was twenty-three. Five of these children were as yet unmarried, and the two that were married, Henrietta and Francis, were childless. Even leaving aside any worries he might have felt over hereditary malaise, he probably wondered whether his own family was to prove a biological cul-de-sac. For a man whose intellectual life was structured around reproductive success, it must have been disturbing to consider the possibility of having no issue. His children were not doing what came naturally to most living beings. In biological terms, after all, he was convinced that reproduction kept species actively evolving. Self-fertilising organisms were probably on the road to extinction. Francis Darwin recorded that his father "once remarked to Dr. Norman Moore that one of the things that made him wish to live a few thousand years, was his desire to see the extinction of the Bee-orchis—an end to which he believed its self-fertilising habit was leading."[11] To muse about grandchildren and to investigate the origins of incipient sterility in plants or the advantages of cross-fertilisation were topics likely to resonate at a fundamental level with his private existence.

Once more, he lay in wait in the flower beds to observe bees pushing their way into a nectary, in search of the co-adaptations that had grown up between flowers and their insect pollinators. Once more, he isolated experimental plants under billowing nets of gauze and delicately pollinated them by hand with a camel-hair brush. As the summers passed, he sowed trays of compost with the seeds of particular crosses and counted seedlings. "I have taken every kind of precaution," he assured Gray, telling him how he germinated plants in a pot on his chimney-piece to avoid contamination. Throughout, he intended to demonstrate that the offspring of out-crossing individuals were more vigorous, and more numerous, than the offspring of self-fertilised plants. "Nothing in my life has ever interested me more than the fertilisation of such plants as *Primula* and *Lythrum*, or again *Anacamptis* or *Listera*," he told Hermann Muller in 1878.[12]

Actually, the investigation was tedious in the extreme and revealed Darwin at his most patient, foot-slogging best. "It is remarkable," reminisced Francis, that this work "owed its origin to a chance observation."

> My father had raised two beds of *Linaria vulgaris*—one set being the offspring of cross- and the other of self-fertilisation. These plants were grown for the sake of some observations on inheritance, and not with any view to cross-breeding, and he was astonished to observe that the offspring of self-fertilisation were clearly less vigorous than the others. It seemed incredible to him that this result could be due to a single act of self-fertilisation, and it was only in the following year, when precisely the same result occurred in the case of a similar experiment on inheritance in Carnations, that his attention was "throughly aroused," and that he determined to make a series of experiments specially directed to the question.[13]

"I am experimenting on a very large scale," he confessed to Bentham on another occasion. "I always supposed until lately that no evil effects would be visible until after several generations of self-fertilisation; but now I see that one generation sometimes suffices." If vigour could be lost in the course of one generation, the chances of survival were correspondingly reduced.

Principally he recorded "vigour" by measuring the height of the seedlings. "Lyell, Huxley and Hooker have seen some of my plants, and have been astonished." But Wallace told him, too late, that weight would be a much better criterion than height—the total weight of seed produced by a plant was probably a better indicator of productivity, he said. So Darwin repeated the experiments the following year, this time calculating weights. He germinated the seedlings all over again, he labelled ripening seed-heads and then weighed the seeds in packets, one for each plant, on his old chemical balance. For a change, he tried planting some of the seedlings outside in the autumn (a tough test for the greenhouse varieties) and counted the ones that made it through the winter. After a good hard frost one Christmas morning, he was there in the garden with his notebook taking a roll-call of the survivors. As he hoped, he did ultimately substantiate his point. The crossed offspring were more vigorous than the self-fertilised.

Father and son found these days together rewarding. They sat side by side with their microscopes and plant trays, exchanging comments as necessary. Francis said that Darwin's manner was bright and animated.

> His love of each particular experiment, and his eager zeal not to lose the fruit of it, came out markedly in these crossing experiments—in the elaborate care he took not to make any confusion in putting capsules into wrong trays, &c. &c. I can recall his appearance as he counted seeds under the simple microscope with an alertness not usually characterising such mechanical work as counting. I think he per-

sonified each seed as a small demon trying to elude him by getting into the wrong heap, or jumping away altogether; and this gave to the work the excitement of a game.[14]

In the evenings Darwin relaxed with books about orchids. He stoked up an enjoyable correspondence with tropical experts like Fritz Müller and William Ogle, the latter a medical statistician at the registrar-general's office whom he had met through William Farr. All this he regarded as restful and eminently useful. Friends appreciated the self-effacing manner he showed at these times. They felt able to tease him. Once, when Darwin was wandering about the garden at Down House in the company of William Ogle, he paused to pick a flower and said that it was staggering to have to believe that the beautiful adaptations which it showed were the result of natural selection. To this Dr. Ogle replied, "My dear sir, allow me to advise you to read a book called the *Origin of Species*."

Thomas Henry Farrer, the husband of Effie Wedgwood (Darwin's niece), became a close colleague in these botanical matters, being well read and sufficiently leisured to participate in some of Darwin's inquiries. Darwin enjoyed his company and started accepting invitations to stay at Abinger Hall, Farrer's estate near Dorking, in Surrey. It was nearly the only place he would visit outside his customary circuit of Henrietta, Erasmus, William, and his sister Caroline at Leith Hill Place. This "pleasant, friendly house was now added to the very few places where my father felt enough at ease to pay visits," noted Henrietta. Effie made them welcome, and as often as not Erasmus would run down for a day or two as well, for Effie was his special favourite among Fanny and Hensleigh's adult daughters. She would sing in the evenings for family parties, as "admirable as any concert," said the Darwin brothers enthusiastically. At Abinger, Emma knew her husband would eat, talk, and rest. He "ran riot rather—bribery tart, peaches, grapes &c. He has promised to reform. He has much botanical talk with THF."[15]

Darwin and Farrer's routine was simple enough. They would call at the greenhouses, then walk out onto the "Rough," a patch of wild commonland. Farrer liked these companionable rambles.

> His tall figure, with his broad-brimmed Panama hat and long stick like an alpenstock, sauntering solitary and slow over our favourite walks, is one of the pleasantest of the many associations I have with the place.[16]

Underneath the pleasantries, Farrer also served as a useful foil for some of Darwin's botanical ideas. He helped Darwin understand the complicated adaptations for fertilisation in *Passiflora*, the climbing passion-flower, then a rarity in English gardens. During one summer visit to Abinger, the two men sat pensively in the dusk beside Farrer's specimen, in full flower, waiting to see if any local insect might serve as a fertilising emissary in the absence of humming-birds. They saw "neither humble-bees, nor butterflies, nor any other large insects." On other occasions, they watched the bees on Farrer's summer bedding schemes, noting how the bees concentrated on the yellow nursery varieties and ignored pink or white. At these times, Farrer's admiration for Darwin was limitless. He seemed never to be irritated by Darwin's habit of following one of Farrer's remarks by saying, "Yes; but at one time I made some observations myself on this particular point; and I think you will find, &c. &c."[17]

They also discussed Darwin's theories about climbing plants. Farrer reported to him how the passifloras "seek & find & hold on & pull up like an animal." Living in the Kent countryside, with hop-fields all around, Darwin could not help but notice the twining tendrils that hitched the plants up their wires in the late-spring sunshine. The ornamental climbers on the veranda at Down House caught his attention. Sweet peas and runner beans scrambling over netting were an amusement. "I kept a potted plant, during the night and day, in a well-warmed room to which I was confined by illness," Darwin said, and observed the growing points move with the light, like the hand of a watch. The local countryside was a constant source of inquiry to him. "The number of different kinds of bushes in the Hedge Rows, entwined by traveller's joy & the bryonies, is conspicuous, compared with the hedges of the northern counties," he had noted in anticipation when he first arrived at Down.[18]

Many odd facts emerged. Rather absurdly carrying his wooden measuring stick with him on strolls around the Down House flower beds, Darwin discovered that wisteria could move faster than morning glory. Some species, he saw, climbed using their leaves, others with hooks and latches. Most preferred to move anti-clockwise, except for *Loasa euranti-aca,* which moved first one way and then the other, until it hit something to act as a support. His walking stick even participated in this private source of interest, for it was cut from a hardened, twisted rope of native honeysuckle, a daily reminder of the strange byways along which his theory was taking him. Thereafter, he obtained sensitive species from Farrer, or from the glasshouses at Kew, and set them to work in his greenhouse or study, either climbing or twining or sleeping. He regarded them as tenderly as he did Polly the dog. "F. is much absorbed in *Desmodium gyrans* and went to see it asleep last night. It was dead asleep—all but its little ears which were having most lively games such as he never saw in the daytime," wrote Emma.

Francis came closest to understanding his father's dedication. It was "a kind of gratitude to the flower itself, and a personal love for its delicate form & colour. I seem to remember him gently touching a flower he delighted in."

> It ran through all his relation to natural things—a most keen feeling of their aliveness. Sometimes it came out in abuse & not praise, eg. of some seedlings—"the little beggars are doing just what I don't want them to."—Or the half-provoked, half-admiring way he spoke of the ingenuity of a Mimosa leaf in screwing itself out of a basin of water in which he tried to plunge it.[19]

His volume on climbing plants was issued in 1875, an expanded version of the paper he had published in the Linnean Society *Journal* ten years before. Despite the attention that Darwin lavished on his twitchers, twiners, climbers, and scramblers, these adaptive devices did not catch the public fancy any more than the digestive powers of *Drosera*. The book was one of his slowest sellers.

## III

The old order began to pass. Hooker's wife, Frances, died in 1874, the year after he accepted the presidency of the Royal Society. He told Huxley that he still thought of her as the young girl who he had dreamed about when plant-hunting in the Himalayas. Twenty-five years on, the gap her passing left in his life was enormous. He turned to Darwin for solace, visiting Down for a few days after the funeral, bringing his children with him. "I cannot tell you how depressed I feel at times." Two years later he married Hyacinth Symonds, the widow of Sir William Jardine, a marriage that brought companionship, a mutual enjoyment in science, and then the birth of a baby. The couple caused an unprecedented stir by leaving "the president's baby" with the porter at the Royal Society apartments when Hooker was called to London for scientific engagements.

Lyell died on 22 February 1875, aged seventy-eight. "How completely he revolutionised geology: for I can remember something of pre-Lyellian days," mused Darwin. "I never forget that almost everything which I have done in science I owe to the study of his great works." To Hooker he admitted that the death did not catch him unprepared. Lyell had been fading away ever since Mary Lyell's death from typhoid fever in 1873, and Darwin was glad that his friend went out with his faculties intact. "I dreaded nothing so much as his surviving with impaired mental powers." Among the memories Darwin most wished to keep secure were Lyell's "freedom from all religious bigotry" and his eager interest in political and social advance.

> He was, indeed, a noble man in very many ways; perhaps in none more than in his warm sympathy with the work of others. How vividly I can recall my first conversation with him, and how he astonished me by his interest in what I told him. How grand also was his candour and pure love of truth. Well he is gone, and I feel as if we were all soon to go.[20]

As president of the Royal Society, Hooker arranged for Lyell to be buried in Westminster Abbey, as much a tribute to the man himself as any propaganda for the new order of things. Katherine Lyell, Lyell's sister-in-law, asked if Darwin would join Hooker, Huxley, and others as a pall-bearer. Darwin refused. He said he "dared not," for he would "so likely fail in the midst of ceremony and have my head whirling off my shoulders." Nor did he attend the funeral. According to Emma's diary, he sent Bessy, Francis, and Francis's wife, Amy. It was a sorry end to a remarkable friendship, at root surely reflecting the same form of selfishness, the same self-indulgence, that had tainted his response to John Stevens Henslow's death, the other man who had made him who he was and who had given him just as much dedicated assistance as Lyell. Darwin could be heartless in cutting himself off from those to whom he owed the greatest debts. Perhaps he thought that to go to Lyell's funeral—to make himself ill for Lyell's sake—would serve no useful purpose. No wonder he suddenly felt "old & helpless."

A passing problem lay in finding appropriate words for the Abbey tombstone. Katherine Lyell drafted two proposals that Hooker thought were far too religious ("I have fainted away twice," he said cynically). Darwin agreed. "They sound to me like truckling to the parsons or to Westminster Abbey."[21] Together they devised a form of words that played to Lyell's strengths rather than what they secretly had come to regard as his private failings. Yet the epitaph was not as fulsome as it might have been. To them he had perhaps outlived his usefulness.

> Throughout a long and laborious life he sought the means of deciphering the fragmentary records of the earth's history in the patient investigation of the present order of Nature, enlarging the boundaries of knowledge and leaving on scientific thought an enduring influence.

## IV

These ruminative gardening days were interrupted by the political problems of science. Darwin's position as a biologist, as a national figure, and as a man increasingly engaged in physiological researches pulled him into the last great medical controversy of the century—vivisection.

Always a delicate problem, and fairly well regulated in Britain through a code of practice set out in 1871, the issue of vivisection raised fundamental social, moral, and professional issues that collided with public opinion during the 1870s and 1880s.[22] Fears about cruelty to animals were only a part of it. British experimental physiologists faced extreme hostility, not just from the public, but also from practical medical men and clinicians who at that point saw no need to adjust their finely honed diagnostic skills to accommodate laboratory knowledge. Only a few hospitals at that time provided facilities for medical research. The great public institutions were primarily charities for the relief of the sick, run by a board of governors who considered it no part of their business to establish expensive laboratories for the academic investigation of disease. In Britain, University College Hospital, St. Bartholomew's, Guy's, and St. Thomas's hospitals were the exception. Similarly, the universities found it difficult to understand the new desire for laboratories to study blood, muscles, or nerves, although Huxley had made a promising start in South Kensington. A doctor with experimental leanings would therefore tend to equip a room in his house as a small laboratory oriented toward microscopy or chemistry according to his tastes. As a result, physiologists in Britain struggled to establish the value of their researches in the slipstream of spectacular advances made elsewhere in Europe. Furthermore, well-publicised appointments to university chairs of men like Burdon Sanderson, Edward Sharpey-Schafer, and Michael Foster called into question matters of medical authority—why should medicine, hitherto dominated by the Royal Colleges, take seriously university professors whose appointments and research interests lay outside the clinical system? Barely hidden questions of power and authority fuelled the debate, fanned by the political activities of the Royal Society for the Prevention of Cruelty to Animals (RSPCA) and public outrage against the use of living animals for research.

When Burdon Sanderson published a *Handbook for the Physiological Laboratory* for students in 1873, his readers were made uncomfortably aware of the methods adopted by rising young experimentalists. The following year, at a medical meeting in Norwich, the British Medical Association allowed a particularly ill-chosen experimental demonstration of the effects of absinthe on a dog. The session ended in uproar and criminal proceedings were brought against the French neurologist Valentin Magnan for cruelty.[23]

Victorians were alternately stirred, chastised, annoyed, and appalled by reports of such practices that appeared in the press. Led by Frances Power Cobbe, on the one hand, the founder in 1875 of the Victoria Street Society for the Protection of Animals Liable to Vivisection, and Richard

Holt Hutton, editor of the *Spectator,* on the other, public revulsion against laboratory experiments on living animals gathered pace. But there was also accumulating unease about what was claimed to be science's heartless, value-free agenda of inquiry, fears that were displayed in Wilkie Collins's unusual novel *Heart and Science,* in which the villain's callous behaviour was attributed to his vivisectionist researches. The main argument of the antivivisectionists was that experiments on living animals were cruel, useless, and immoral. Cobbe called them "tortures." Human beings, she argued, had moral duties towards other creatures, creatures that also possessed the capacity to suffer. Perhaps she felt that the wanton exploitation of animals reflected something of the political and social vacuum in which women then existed. The link with feminist history has always been strong. Beyond that, antivivisectionists freely expressed their misgivings about the growing ascendancy of science and medicine. It seemed to many that experimentalists pursued their researches without any legal or moral obligations to society or to the higher world of the divine. The campaign groups that emerged were steadfast in their opposition. When Darwin tried to describe Hutton's inflexibility over the issue of animal experiments, he said in awe, "He seems to be a kind of female Miss Cobbe."

The issue involved Darwin directly as a figurehead for advanced modes of thought. At the same time, however, he was disgusted by cruelty to animals of any kind. As a local magistrate he sometimes came across cases of cruelty to farm animals and was inexorable in imposing fines and punishment. In 1853 he had waged a private vendetta against a Mr. Ainslie in the village for cruelty to his carthorses, sending for an officer of the RSPCA and threatening to "have him up before a magistrate & his ploughman also." From time to time, he would jump out of his carriage to remonstrate with coach drivers using whips or spurs excessively.[24] He sacked a Down House employee who had left the family horse standing in its harness for hours. This was more than mere sentiment. Darwin dimly remembered the screams of Brazilian slaves during his voyaging days. A profound dislike of inhumane treatment of any vulnerable being strongly coloured his perspective, and he said that the thought of laboratory animals being made to feel pain simply in order to satisfy human curiosity was "abhorrent" to him. In a little-noted passage in *The Descent of Man* he wrote that "every one has heard of the dog suffering under vivisection who licked the hand of the operator; this man, unless he had a heart of stone, must have felt remorse to the last hour of his life."[25]

In these views Darwin was supported by Emma and Henrietta. Some ten years beforehand, in 1863, Darwin and Emma had written a joint letter to the *Bromley Record,* the local newspaper, protesting about the use of steel traps for vermin. This letter was accompanied by a line drawing made at Darwin's request of a rabbit's paw fractured by a sprung trap. The letter was republished, minus the picture, in the *Gardeners' Chronicle* under Darwin's initials alone, and again as a separate pamphlet that Emma sent around to friends and relatives in order to raise funds for the RSPCA to investigate humane traps. Darwin recorded payment for the cost of distributing the "cruelty pamphlet," and his name was listed for several years beside an announcement in the RSPCA magazine *Animal World* of his intention to give a monetary prize to the inventor of an alternative, more humane, device. The prize was still listed in 1876. With the antivivisection debate in full flood, Emma drew on that published text for another letter that she apparently sent under her own name to the *Times* or *Spectator,* although it seems not to have been published.[26]

So at home, antipathy to vivisection ran high. When George Romanes came for a visit, Darwin warned him, "When in the presence of my ladies do not talk about experiments on animals." Yet Darwin also believed physiology to be "one of the greatest of sciences, sure . . . greatly to benefit mankind." His devotion to science—his belief in it as the way forward—ensured that he pledged wholehearted support to the ideals of pure research. His personal solution to the dilemma was less clear. He believed that experimental animals should be rendered completely "insensible." Then again, as early as 1871, he told Edwin Ray Lankester that although he felt vivisection was essential for the progress of knowledge it should not be performed for "mere damnable and detestable curiosity. It is a subject which makes me sick with horror, so I will not say another word about it, else I shall not sleep tonight."

He did not follow his own advice. He refused to sign the antivivisection petition that Henrietta presented to him in January 1875 and wrote her a long letter explaining his reasons. These were mainly pragmatic. The proposed licensing system seemed to him likely to be overly restrictive. Furthermore, "I do not see who is to determine whether any particular man should receive one." The traditional country sports of English gentlemen, he added, were far crueller than a properly executed experiment. Most of all, he feared a deleterious effect on research.

> If stringent laws are passed . . . the result will assuredly be that physiology, which has been until within the last few years at a standstill in England, will languish or quite cease. It will then be carried on solely on the Continent; and there will be so many the fewer workers on this grand subject, and this I should greatly regret. . . . No doubt the names of doctors will have great weight with the House of Com-

> mons; but very many practitioners neither know nor care anything about the progress of knowledge. I cannot at present see my way to sign any petition, without hearing what physiologists thought would be its effect, and then judging for myself. I certainly could not sign the paper sent me by Miss Cobbe, with its monstrous (as it seems to me) attack on Virchow.[27]

A few months later, Lord Hartismere was persuaded by Cobbe and other antivivisectionists to put a bill forward to Parliament to regulate physiological laboratories. Hartismere's bill proposed that vivisection be confined to premises that were annually registered with the home secretary, and that animals be properly anaesthetised. Curare was singled out as not suitable for the purpose.

Darwin was dismayed by the terms of this bill, not least that private scholars would need a personal license to work in their own homes. With deadly efficiency, he set about drawing up a counter-petition to protect "the science of physiology as well as animals." His physiological friends were astonished to see him roused to such intense political action. Astonished and gratified. Perhaps only someone with Darwin's Olympian presence could have united such a disparate band at this juncture. His petition hit his friends' letterboxes and they signed it in droves.

The activity increased during his customary visit to London. Although the response to his petition was gratifying, Darwin decided that he would be better advised to get a blocking bill presented in Parliament. He lobbied all his senior political acquaintances. Richard Litchfield drafted a proposed bill for him, which Darwin sent for approval to Burdon Sanderson and other leading physiologists, each and every one personally known to him. Hooker signed as president of the Royal Society, and Darwin wrote to advise Lord Derby that the bill would be introduced to Parliament by Lyon Playfair (as Lubbock recommended). Darwin made sure to secure prestigious backing in both houses, including Lord Cardwell and the Earl of Shaftesbury. With a turn of speed and purpose that may have surprised even himself, he deliberately used his status to advance his cause.

He did the work well. The opposing bills from science and the public (with more common ground between them than might be at first be supposed) caused such a stir that a Royal Commission was appointed to investigate the entire issue. The commission was put under the direction of Lord Cardwell, a senior Liberal statesman who was often brought in by Gladstone for seemingly impossible problems of arbitration. Not surprisingly, Darwin was called to give evidence to this commission. "F. went to the Vivisection Commission at two," wrote Emma.

> Lord Cardwell came to the door to receive him and he was treated like a duke. They only wanted him to repeat what he had said in his letter (a sort of confession of faith about the claims of physiology and the duty of humanity) and he had hardly a word to add, so that it was over in ten minutes, Lord C. coming to the door and thanking him. It was a great compliment to his opinion, wanting to have it put upon the minutes.[28]

Darwin was involved again and again until the year of his death. The authority of his presence in this debate and his unimpeachable qualities as a witness made him more visible than he had been for years. He felt fighting spirit flood back into his aging veins. Fame was not always a burden.

## V

At last his son William persuaded him to sit for an oil portrait. The process of persuasion had taken ages. William first made the suggestion in 1872 after the family had visited Southampton, confiding to Henrietta, "I mean to have a portrait done of Father by Watts, unless anybody can persuade me that it will be a failure probably either from Watts not taking to this kind of subject, or being ill, &c. &c. &c. The expense will not be more than £500 and if F. jibbs, I am game to pay it. . . . Please keep quite dark."[29]

Darwin kept finding objections. "He often talked laughingly of the small worth of portraits," remembered Francis with regret, "and said that a photograph was worth any number of pictures, as if he were blind to the artistic quality in a painted portrait. But this was generally said in his attempts to persuade us to give up the idea of having his portrait painted, an operation very irksome to him."[30] The root cause of these objections was probably the feeling that many people share about portraiture—anxiety about being thought vain, and the time that it would take. Underneath, he may have sensed that the process of having a portrait painted was a form of intense personal appraisal. He really did not like the idea of anyone penetrating too closely. He had placed Woolner's bust in an unobtrusive corner of the dining room and routinely deflected any comments by laughing about it in a dismissive fashion. "He used to point with scorn at the conventional way in which the head in a great many busts such as the one in the dining room was attached at the neck," said William, a trifle sadly.[31]

Patiently William parried all his father's objections. At the end of 1874, Darwin agreed. For unknown reasons, the artist Frederick Watts was not approached, even though he was then working on his "Hall of

Fame," the portraits of eminent figures that he ultimately presented to the National Portrait Gallery. Instead, the commission went to Walter William Ouless. The portrait was completed in March 1875. William intended it to remain in the family. As it happened, the picture was distributed rather more widely than this. Ouless exhibited it at the Royal Academy in 1875. It was subsequently engraved by Paul Rajon and published in the *Illustrated London News* in 1877. A little while later Ouless made a copy in oils for Christ's College Cambridge, Darwin's old college.

Ouless stayed at Down House for the sittings for a few days in February 1875, Darwin grumbling about time that could better have been spent in his hothouse and Emma wondering how to entertain the artist between his painting engagements. Eventually, she hit on an idea which she revealed to Leonard, away with the army in New Zealand.

> Mr Ouless is painting F. and me! He cannot nearly fill up his time with F. so it was a convenient time for me to sit. Both portraits are unutterable as yet; but he puts in the youth and beauty at the very last.[32]

She never liked the results. Despite Francis's and William's praising the painting of Darwin, with Francis regarding it "the finest representation of my father that has been produced," Emma felt it did not depict the man she knew. No trace of Ouless's portrait of Emma has ever been found. When Rajon came down in 1877 to show his engraving, she said, "We expect M. Rajon today with the picture & a proof sheet of etching, which I expect will exaggerate the faults of the picture viz. roughness & dismality." Three days later she was still cantankerous.

> M. Rajon came down on Thursday bringing the picture & a proof sheet of the etching. They all admired it but I rather dislike etchings & don't like the picture; so it was not likely to please me.[33]

Darwin felt ambivalent too. Writing to Hooker after the painting was completed, he reflected, "I look a very venerable, acute melancholy old dog,—whether I really look so I do not know."[34]

Silently, he and Emma removed this painting from their immediate surroundings. When their friend and neighbour Alice Bonham Carter wondered if it might be displayed at the Bromley School of Science during an evening *conversazione* to encourage local naturalists and scientists, they withheld consent. "Alice wanted us to send Ouless' picture," Emma told William, "but F. cd not stand exhibiting himself in that way."[35]

The following spring he and Emma went to London to visit Henrietta as usual. This time Darwin stated a wish to see Hooker sitting in the presidential chair at the Royal Society. The two Cambridge sons, George and Horace, came up to London to join him.

> G. and Horace came on Wed. to go to the Royal Soc. soirée and F. went with them! There was such a crowd he only cd. behave like a crowned head shake hands & before he cd. enter on any talk somebody else came up. Many of them he did not know, & one was so affectionate he was ashamed to ask & parted from him with the greatest effusion. I don't think it will be worth while his going again. However it did him no harm.[36]

Darwin took a different view. "Tell Hooker I feel greatly aggrieved by him. I went to the Royal Society to see him for once in the chair of the Royal to admire his dignity and enjoy it, and lo and behold, he was not there."

## VI

As time went by, Darwin occasionally thought about writing down a few recollections of his life. There was no pressure to do so, he said; no ambition to seal his past in aspic; and, as he saw it, no particular need to search for self-justificatory causes and effects. Others might regard him as a hero, or, as Emma said, a crowned head of science, but he felt no such conviction.

Yet there were many things that encouraged him to dwell on long-gone days. A little while previously, in the autumn of 1875, he had been asked for some biographical reminiscences by Ernst von Hesse-Wartegg, a German exile in Paris. Hesse-Wartegg had been one of the few reviewers who spoke favourably about *Insectivorous Plants*, and Darwin consequently felt kindly towards him. Hesse-Wartegg inquired if Darwin might supply information about the development of his mind and character for an article in a Leipzig journal called the *Pioneer*, which would then be syndicated to the German encyclopedias.[37] The older man usually disliked dealing with these requests. Still, Hesse-Wartegg's letter sparked his imagination. Like Huxley, he fell into the understandable trap of thinking that the most accurate account would be the one written by himself.[38]

Odd details of the past also teased his memory. He sympathised with Katherine Lyell, doing her utmost to memorialise Lyell in a volume of life and letters. Pensively, he gathered up the letters he had received from his geological friend over the decades and sent them to Katherine to use in her book, an enduring friendship now ignominiously reduced to two brown paper packages. "I hope that I may die before my mind fails to a sensible extent," he caught himself thinking. He reminded William Darwin Fox of their longstanding intimacy. Every morning, as he turned over his botanical notes, many of them written at Maer Hall or Shrewsbury when he was just married, others during long-forgotten holidays at Elizabeth's and Charlotte's homes in Hartfield, he contemplated his past and sensed time slipping away. So many loved ones dead, so many changes in science.

However, it was family news that provided the real push. In the spring of 1876, Francis and Amy announced that they were to have a child. Darwin and Emma rejoiced: the baby would be their first grandchild. They had worried a little about Henrietta's continuing childlessness, and in an unintentionally poignant note written after Francis and Amy's announcement, Henrietta partly acknowledged the gap herself. "I feel as long as I have Father & you it does instead of our having children & makes our lives quite full, for you are the dearest Father & Mother that ever anyone had."[39] Apart from Francis and Henrietta, none of the others had edged towards marriage, not even William at the advanced age of thirty-seven.

The thought of the impending baby pleased Darwin greatly. He had loved his own children dearly. So when a few peaceful weeks sailed into view during a visit to Hensleigh and Fanny Wedgwood's country house in May 1876, he began a short autobiographical memoir, opening it with the hope that it would amuse his children and their children. "I know that it would have interested me greatly to have read even so short and dull a sketch of the mind of my grandfather written by himself, and what he thought and did and how he worked," he confided.[40]

Over succeeding months he added paragraphs here and there as things occurred to him, and in May 1881 he conscientiously filled up the gap from August 1876. One important addition was fourteen pages of reminiscences of his father, Dr. Robert Darwin of Shrewsbury. Darwin very much wanted to write this section not merely to honour his father and to make sure that successive generations would be put in touch with their history but also to resolve his own relationship with the Shrewsbury past. "I do not think any one could love a father much more than I did mine, and I do not believe three or four days ever pass without my still thinking of him," he had revealed to Hooker with unaccustomed frankness when Hooker's father died. There was more than a little self-deception in those words. In his autobiography Darwin turned his father into a bully. He said he owed his father nothing in mental ability. "My father's mind was not scientific, and he did not try to generalise his knowledge under general laws. . . . I do not think that I gained much from him intellectually." Moreover, he evidently thought his father had misjudged him. He still smarted at the outburst that he noted in his autobiography, "You care for nothing but shooting, dogs and rat-catching, and you will be a disgrace to yourself and all your family," exactly remembering the words and the mortification he had felt but neither the year nor the surrounding context. When Darwin's sister Caroline read these passages after Darwin's death she was upset to realise that even in old age he had no idea how much he had been loved by their father.[41] Perhaps she failed to see—as Darwin also failed to see—that he was distancing himself from his father by insisting that he had always been independent, standing alone. As for the female side of his ancestry, Darwin ignored it completely. Darwin, seemingly like many Victorian intellectuals, regarded the development of his mind within a predominantly patrilineal descent.[42]

Once he was into the swing of writing about himself, Darwin found the work an agreeable way to pass the summer, the more especially because he did not need to look anything up in learned journals or check the accuracy of his facts. He made a number of minor slips, as might be expected in an informal family memoir written at the end of a full and varied life. Anecdotes and reminiscences flowed. He included stories about the famous people he had met, just the kind of thing that people were already recording about him, and laughed at Lyell for relishing the same high-society entertainments that he dismissed. He took no care over his style, for the pages were not designed to be read by anyone other than members of the family. He did not bother to pursue episodes that escaped his memory, letting his narrative trot on in a straightforward line toward his achievements. As a result, the work embodied many of the charms and oddities of an English gentleman at home: polite, considerate, shy, and amusingly self-deprecating; and yet unable or unwilling to delve too far below the surface. With a modest smile Darwin deflected every difficult question that a more demanding, self-analytical author might have asked of himself.

He was clear about the piece not being intended for publication. Writing to Julius Carus in 1879, he said, "I have never even dreamed of publishing my own auto-biography."[43] While this remark seems hard fully to accept in retrospect, it seems fair to say that Darwin at the time of writing believed—or wanted to believe—that he was directing his remarks only to friends and family. The skimpy explanations of his actions that he offered, the comfortable expectation that he would be believed, the avoidance of difficult memories or troubling emotions, all suggest that he approached the project informally, in a relaxed frame of mind. Nevertheless, in choosing which memories to record in words, in selecting the anecdotes, he was constructing himself in the shape in which he wished to know himself and to be known by.[44]

Towards the end, he did try to estimate his character, reflecting on the spread of issues that Galton's questionnaire and other inquirers had set before him. But he was not introspective by nature. He possibly felt a mixture of distaste and apprehension about people rummaging around in his life, even if this should be himself, and unwilling to open up to scrutiny. More important, he evidently felt, were words of support to his sons, whose aptitude, he believed, lay, like his own, in perseverance and little else.

Above all he seems to have found it inappropriate or difficult to reveal his innermost feelings. He did not attempt anything like the analysis and score-settling that usually accompany the autobiographical genre. He acknowledged his debts to Lyell and to Henslow with grace but very little fervour. His voice was apologetic, humble, accepting at face value his own and others' motives, unquestioning, even-tempered, and conversational, an unfolding of the pleasantly unassuming persona that would ultimately lie at the centre of the Darwin legend. This was not just the superficial language of gentlemanly respectability. To him, raw emotions were probably too intense for written words. The only exception was his private writing about his dead daughter Annie, pages that were not meant for any eyes other than his own and Emma's.

But in many cases the rawness had evidently dissipated with the years. In writing these "recollections" he felt none of the exuberance of his early letters from Cambridge or Shrewsbury, none of the verve of his *Beagle* correspondence. This does not mean to say that he dismissed the passions he had once experienced, especially the passion for natural history. In general, he looked back on his enthusiasms with indulgent bemusement, full of affability and fondness, almost as if he were another man completely. Perhaps he did believe he was another man. In this memoir he emphasised his intellectual achievements, his books, his contribution to the advance of knowledge, tending to undervalue his creative involvement in the process. He seemed to say that science made him, rather than he made the science. In a disconcerting turn of phrase, he said he wrote "as if I were a dead man in another world looking back at my own life," and his spare, unadorned sentences did convey something of this bleak neutrality. "Nor have I found this difficult, for life is nearly over with me," he continued.

He mentioned Emma only to praise her. He scarcely mentioned his children, except for recollecting Anne's death in 1851 ("we have suffered only one very severe grief") and the fond memory of childhoods outgrown. "When you were very young, it was my delight to play with you all, and I think with a sigh that such days can never return."[45] He said that his life after his marriage was merely the story of writing his various books. He spoke about himself not as a person, living and growing, but as a series of publications, an author.[46]

In the one place where he did describe the changes he had noticed in his personality, his children afterwards went to considerable lengths to discount the self-evaluation. Darwin stated that his aesthetic sense had gradually deadened over the years. He did not thrill to beautiful scenery or to a piece of music as he used to.

> Formerly, pictures gave me considerable, and music very great delight. But for many years now I cannot endure to read a line of poetry; I have tried lately to read Shakespeare, and found it so intolerably dull that it nauseated me. . . . My mind seems to have become a kind of machine for grinding general laws out of a large collection of facts, but why this should have caused the atrophy of that part of the brain alone, on which the higher tastes depend I cannot conceive.[47]

His honesty about this bothered the rest of the family. It was as if Darwin was denying his sensitivity to nature, almost turning his back on his special gifts. One by one, after his death, members of the next generation pointed out counter-examples, where Darwin had enjoyed a scenic view or an evening of music. "As regards his imagination," said Leonard Darwin defensively, "I think that scenery, the beauty of flowers, and music and novels were sufficient to satisfy it. I remember he once said to me with a smile that he believed he could write a poem on Drosera, on which he was then working."[48] There was no real need for the family so diligently to readjust the record because the cultural norms of that time were easily able to accommodate a lack of artistic taste in the great and the good. There has always been a place among the English upper classes for philistinism, even among the intelligentsia. More was at stake. Francis evidently felt that Darwin's disregard for aesthetics was unworthy of the flame of intellect and said so when he came to write his father's *Life and Letters*. Unanimously, the children rejected their father's own view of himself as a deadened, anaesthetic man.[49]

Despite this occasional bleakness, Darwin's writing was characteristic. He dwelled sentimentally on his Cambridge days. "My time was wasted, as far as the academical studies were concerned. . . . we sometimes drank too much, with jolly singing and playing at cards afterwards." He told his *Beagle* story with an appreciative understanding of the way it transformed his life. A censorious edge crept into his account of FitzRoy. He praised Henslow too much, probably guilty about not having thanked him adequately in life. He took a valedictory swipe at Richard Owen and explained, to his own satisfaction at least, his decision never to engage in controversy. He was silent about his charitable work in the village, remarking only that he wished he had done more. He refrained from describing his income—and made sure to disparage his youthful extravagance.

In particular, he emphasised throughout his own personal effort, presenting himself as a man committed to the ethos of self-help, a part of his identity that evidently drew on the ideals of self-determination, "character," and enterprise more generally encountered in Samuel Smiles's books. As he had told Galton, he valued this quality highly. He believed everything he had attained was the result of his own industry. Looking back, he reckoned he learned nothing at school; nothing from his father, who considered him "a very ordinary boy"; nothing from two universities except that which was performed under his own steam. Everything accomplished on the voyage was through his own hard work. He claimed that he had listened to Dr. Robert Grant's eloquent advocacy of Lamarck "without any effect on my mind." His grandfather's views produced "no effect on me." There was a decidedly self-congratulatory element to this. He could not even believe that the subject of evolution had been "in the air." The black-and-scarlet beetle he had seen on holiday at Plas Edwards, in Wales, aged ten, was recalled more clearly than his dying mother. All this conveyed—whether intentionally or not—his belief in personal creativity and autonomy. Like all memorialists, he pieced together his own view of himself, remembering only those things that endorsed his particular inner picture, and he moulded his memories to the hidden conventions and assumptions that shape any individual's manner of thinking about his or her life.[50] Rather astutely, he painted himself as developing from a good-hearted numbskull, forever "surprised" and "astonished," into an unlikely prince.[51] This was the unquestioning individuality of the age of laissez-faire, a kind of capitalist autobiography where by skill and personal effort the author's losses were turned into gains. It was full of masculine assumptions about the world of work and positive intellectual activity.[52] His moments of uncertainty and the paths left untaken were discarded.

Furthermore, Darwin smoothed out the turbulent path of his early evolutionary speculations and turned them into a steady march from facts to theory. To omit, or to forget, the intellectual electricity of his London years, his pride and excitement in his theory, his despair on reading Wallace's letter, and the steady support given to him by others was to rewrite his past. The old Darwin forgot how the young Darwin had returned from the *Beagle* voyage hard-headed and full of drive. He did not tell his sons about scientific courage or competitiveness, or the ecstasy of ideas.[53] He avoided talking about sparks that explode into insights. Even his account of the moment of creative inspiration that had made him into what he was sounded flat and full of future hard work.

> Fifteen months after I had begun my systematic enquiry, I happened to read for amusement Malthus on population, and being well-prepared to appreciate the struggle for existence . . . it at once struck me that under these circumstances favourable variations would tend to be preserved, and unfavourable ones to be destroyed. The result of this would be the formation of new species. Here, then, I had at last got a theory by which to work.[54]

His affable nature, his wish to be kind, his feeling that he had a gilded life, his love for his family: all these obscured the very real shocks he had encountered. As others have remarked about the genre of autobiography, the author depends not so much on the art of recollecting as on the art of forgetting.[55] He nowhere acknowledged that competitive steely element that drove him on. Huxley afterwards believed Darwin's greatest attributes were a "clear rapid intelligence, a great memory, a vivid imagination," which were subordinated to "his love of truth."[56] Darwin was incapable of seeing himself as others saw him. In an oddly engaging manner, he remained a stranger to himself.

This blurring of the past softened his account of the *Origin of Species* and the unrelenting dedication with which he had ensured its world-wide consideration. Throwing all memory to the winds, he offered a pleasant encomium on his reviewers.

> I have almost always been treated honestly by my reviewers, passing over those without scientific knowledge as not worthy of notice. My views have often been grossly misrepresented, bitterly opposed and ridiculed, but this has generally been done, as I believe, in good faith.[57]

## VII

In this autobiography Darwin expressed startlingly harsh views on Christianity. Like many Victorian thinkers finally coming to terms with their loss of faith, he blamed his increasing doubts on the absence of any rational proof for God's existence. No "sane man" would believe in miracles, he said; the Gospels were demonstrably not literal accounts of the past; comparative studies of the Hindu, Mohammedan, and Buddhist faiths, along with scholarly descriptions of primitive animism and spirit worship, showed that Christianity could not be regarded as a divine, monotheistic revelation. Formerly, he said, he had believed in a personal deity. He remembered standing in the Brazilian forest and his "conviction that there is more in man than the mere breath of his body." Disbelief had only slowly "crept" over him. In fact, his account is distinctive in the annals of autobiographical writing for its lack of any pivotal moment of loss. Darwin had no epiphany. When he composed the *Origin of Species* he retained some religious views. But now, as he wrote, he said not even the grandest scenes would cause such feelings to swell.

To him, the God of the Christians was cruel.

> I can indeed hardly see how anyone ought to wish Christianity to be true; for if so, the plain language of the text seems to show that the men who do not believe, and this would include my father, brother and almost all my best friends, will be everlastingly punished. And this is a damnable doctrine.[58]

Perhaps it was for self-protection that he did not dwell on the moral or ethical dilemmas that had beset his notebooks. He did not mention his conscience. There was little here to guide his sons on the right way to live a moral life in the newly secularised world that he had helped to create, no sense of personal transformation or rebellion either, where one set of views was finally jettisoned in favour of another. He did not struggle to find meaning in his loss of faith, seemingly accepting it as an inevitable feature of the life of a scientist. No other experiences, he implied, not even the loss of faith, could match those he had encountered in science.[59]

As for the evidence presented by a deep internal conviction that God must exist, he admitted that he could not trust his own mind to reason properly on the issue, knowing that his faculties were "developed from a mind as low as that possessed by the lowest animal." A dog might as well reason on the mind of Newton, he had once said. More forcefully, he put the same words into his father's mouth, recounting Dr. Darwin's story of an elderly female patient who had remonstrated with him. The old woman believed in God without ever asking herself why she believed. "Doctor, I know that sugar is sweet in my mouth, and I know that my Redeemer liveth."

All these were opinions that Darwin repeated here and there in letters during the 1870s, sometimes using the same turn of phrase. They were evidently strongly felt. "Science has nothing to do with Christ," he told Nicolai von Mengden, a Russian biologist, in a letter written in 1879, "except in so far as the habit of scientific research makes a man cautious in admitting evidence. For myself, I do not believe there has ever been any Revelation."[60] As it turned out, Emma Darwin had once suggested to him that his scientific approach might make him question the validity of religion. Darwin seems to have concurred in the truth of this opinion. His niece Julia Wedgwood recollected that when he was writing *The Descent of Man* he told her that "the habit of looking for one kind of meaning I suppose deadens the perception of another." Julia admired Darwin's reticence.

> If he had ever said a word that was either on the side of, or on the side against, what we mean by religion, he could not have taken the place he has done. His books would have been in either case more interesting to a good many people, but no one could have felt, as everyone must feel now, that they are a manifestation of science in its absolute purity. This gives them a coolness & repose, unlike any other written in the last thirty years.[61]

Living out for himself the archetypal Victorian crisis of faith, Darwin perhaps recognised that he had lost the last vestiges of faith when he discovered that biology provided him with the answers he most desired. In the end, in his autobiography, he asserted that religious belief was little more than inherited instinct, akin to a monkey's fear of a snake.

Emma Darwin's position in relation to these views was complex. Her husband probably wrote these words with her in mind, concerned not to offend or disturb, sensitive to her fear that his lack of belief would separate them in the hereafter, while trying to be true to his commitments. Yet he may have been almost too sensitive. Emma's faith was gradually ebbing as she grew older. "She kept a sorrowful wish to believe more, and I know it was an abiding sadness to her that her faith was less vivid than it had been in her youth," said Henrietta.[62] She had developed a broadminded tolerance for her husband's opinions. Nevertheless when the family was considering publishing parts of this autobiography after his death, she asked for the sentence about the monkey's dread of a snake to be omitted, telling Francis that "your father's opinion that *all* morality has grown up by evolution is painful to me." This particular remark, she felt, ran a real risk of offending believers, especially those believers who knew and loved Darwin as a man. "I should wish if possible to avoid giving pain to your father's religious friends who are deeply attached to him, and I picture to myself the way that sentence would strike them, even those so liberal as Ellen Tollett and Laura [Forster, one of Henrietta's friends], much more Admiral Sullivan, Aunt Caroline, &c, and even the old servants." The family politics that accompanied her request, and her wish to omit one or two other religious comments that she felt were "not worthy of his mind," became a significant episode in sanitising Darwin's religious beliefs in the years immediately after his death.[63] In the event, the whole section was diplomatically omitted from the first printing of Darwin's autobiography.[64]

Darwin closed the religious part of his autobiography with a statement of honest ignorance. "I cannot pretend to throw the least light on such abstruse problems. The mystery of the beginning of all things is insoluble by us; and I for one must be content to remain an Agnostic."[65]

As the summer blazed away and the garden wilted, Darwin tended to write during the early afternoon and spent the remains of the day lying on the grass under the lime trees, lazily drifting in and out of his memories. When it was done, Emma said, "F. has finished his Autobiography and I find it very interesting, but another person who did not know beforehand so many of the things would find it more so."[66]

## VIII

All along, he worked quietly to promote his sons' careers. Each of the boys, in one way or another, displayed particular aspects of Darwin's own character and abilities. William took after his father's secret talent for finance, as well as helping Darwin with some of his botanical research. His abiding virtue was reliability. George and Francis opted for science, one theoretical, the other experimental. Leonard travelled the world as a military engineer before standing (unsuccessfully) for Parliament as a Liberal and ultimately becoming president of the Eugenics Society, while Horace eventually satisfied his lifelong fascination with machinery by founding what would become the leading scientific instrument company in Britain, whose fortunes rose with the Cambridge school of physics emerging in the Cavendish Laboratory.[67] Horace graduated from Trinity College Cambridge in 1874.

Francis was indispensable to Darwin at home, both as a secretary and as an experimental botanist. A literate, accomplished man, he felt himself wholly unsuited to the profession of medicine in which he had qualified. He was the son that the others thought most understood his father's feelings for nature—although Henrietta considered George most like Darwin in his power of work and the "warmth and width of his affections."

Darwin was proud of Francis's achievements. He was a second pair of hands and eyes to him. At Down House, Darwin identified interesting botanical problems for Francis to solve, monitored his experiments, recommended his researches to editors of scientific journals, eased him into his web of eminent correspondents, and relayed his results to Hooker and other leading botanists. As soon as was decently possible, in 1875, Darwin nominated Francis and George Romanes to become fellows of the Linnean Society, and he encouraged his son to produce his first botanical papers on unresolved questions emerging from his own work on insectivorous plants. He longed to see Francis an established naturalist and did everything he could to help him on his way.

He was consequently outraged when the Royal Society rejected an early paper of Francis's on the absorptive filaments of the teasel, almost as if the Royal had rejected one of his own investigations. "The wicked R.S. has declined printing Fr's teazle paper," Emma reported, "which has vexed F. 10 times more than Fr."[68] Truculently Darwin sent Francis's paper to a lesser journal for publication, wrote a laudatory notice about it for *Nature,* and told Ferdinand Cohn, "I have reason to know that some of our leading men of Science disbelieve in my son's statements & this has mortified me not a little."[69] With an observant sigh, Emma said that Darwin deliberately sent the letter to *Nature* in order "to spite the Royal Soc." Guiltily, Darwin began to wonder if his reputation would make it harder for Francis to establish his credentials. Would his old adversaries attack his son, a much more vulnerable target than himself?

Much of the same dedicated paternal push came in his dealings with George. George Darwin's early work touched on the social sciences, for he advocated marriage reform, and in an important article in 1875 he examined the purported ill effects of cousin marriages. As in Francis's case, these studies grew out of his father's theory of evolution, even an unintentionally amusing piece on clothes for *Macmillan's Magazine* in 1872 in which he discussed "survivals" in dress, such as hatbands and tailcoats, from the standpoint of continuous adaptation and selection, spiced with a dash of Tylor's anthropology.

George began his study of cousin marriages and congenital disorders with Darwin's *Descent of Man* very much in mind. A text by Alfred Huth, *The Marriage of Near Kin,* published in 1875, could well have been the stimulating factor, for Huth examined marriage and incest from an anthropological point of view, calling on Charles Darwin's book for support and advocating the idea of putting a question about cousin marriages on the population census form. Like many of his generation, George felt disturbed over the apparent weakening, or degeneration, of the human race, a common enough response to his father's application of the idea of natural selection to mankind, that also reflected longlived trends in social and economic thought. George decided to examine the question of cousin marriages by using statistics. He gathered these statistics from *Burke's Landed Gentry* and the announcement columns of the *Pall Mall Gazette,* and from the records of lunatic asylums, making contact with two of Darwin's medical correspondents, the psychiatrists James Crichton-Browne and Henry Maudsley. Taking a tip from his father, and Francis Galton before him, he also circulated questionnaires, but discovered too late that people were reluctant to divulge to a stranger their children's mental problems.

George concluded there was some statistical evidence for slightly lowered vitality among the offspring of first cousins, but no evidence for an excessively high death rate in infancy. How he regarded this information in the knowledge that his own parents were first cousins, and that two other sets of uncles and aunts were first cousins, is unknown. It seems likely that he was here peering into his own heredity, searching for his place in the scheme of nature, and was perhaps relieved to find that his results suggested only a few minor incapacities for himself and his siblings. His Darwinian bad stomach was not going to be the death of him, for example. The dangers of inbreeding could be outweighed, he said in his article, by differing features of upbringing between related parents.

> This is in striking accordance with some unpublished experiments of my father, Mr. Charles Darwin, on the in-and-in breeding [*sic*] of plants; for he has found that in-bred plants, when allowed enough space and good soil, frequently show little or no deterioration, whilst when placed in competition with another plant, they frequently perish or are much stunted.[70]

These two sons, George and Francis, plainly began their careers by defending and expanding on their father's principles. In George's next paper he tackled William Thomson over the age of the earth and its implications for evolutionary theory. He questioned Thomson's figures, suggesting that the earth's axis of rotation was subject to fluctuations large enough to cast doubt on Thomson's conclusions. "It's rather like a pea meeting a cannon ball to oppose him, but I feel tolerably safe at present, & if I am right it will be so much the greater triumph," he wrote home in 1876. These fluctuations meant that the earth might be old enough after all for gradual biological evolution to have taken place. The paper was read before the Royal Society in 1876, was published in the society's *Philosophical Transactions* in 1877, and was followed by George's nomination for fellowship. Darwin nearly burst with pride. "Oh Lord what a set of sons I have, all doing wonders." Thomson acknowledged the accuracy of George's mathematics. This line of work, and a growing friendship with Thomson, was to occupy George for the greater part of his scientific life.

More than happy with these developments, Darwin and Emma looked forward to the coming grandchild. Tragedy struck in September 1876 when Francis's wife, Amy, died two days after childbirth, "a most dreadful blow," said Darwin, who had known her since she was a girl. "I think she was the most gentle & sweet creature I ever knew." He and Francis were with her when she died, a horrible death of convulsions followed by kidney failure, probably puerperal fever, a kind of septic shock. He told William, "It is the most dreadful thing which has ever happened, worse than poor Annie's death, though not so grievous to me. I cannot bear to think of the future."

> God knows what will become of poor Frank, his life will be a miserable wreck. He is too young to care for the Baby, which must be brought here, & I trust in God we may persuade him to come here & not to live in his house surrounded by memorials of her. No Father ever had better children than we have & you are one of the best of all.—God bless you.—I hope you keep pretty well. Tell us always about yourself.[71]

He wrote the same miserable message to Leonard, adding only that "Frank seemed quite bewildered and dazed." He turned to Hooker. "My dear old Friend I know that you will forgive me pouring out my grief."

This experience of death and suffering struck hard at the heart of the family. It was not in any sense the good death beloved of sentimental writers. Amy had "suffered greatly." Paradoxically, Darwin was more familiar than Emma with the physical realities of dying bodies, having been at the deathbeds of all three of his children and now Amy. He bore up under these experiences with a fortitude that belies the common assumption that he usually recoiled from emotional involvement.[72] Even though he must surely have preferred to avoid such traumatic personal events, he could face them when necessary. This time, it was he who gave his wife and family the emotional support they craved.

Francis was desperately unhappy, and he and the newborn baby, a boy they called Bernard, moved up the road to Down House, the only sensible step for him to take in such a situation. His wife was buried in Wales, in her father's village, and Francis spent some weeks there before returning to close up Down Lodge, the marital home. "I felt I *couldn't* bear A.'s loss," said Emma, finding her sympathetic feelings for her son nearly "intolerable." Emma arranged nurses and cots at Down House while Darwin tried to keep Francis preoccupied with plants. A sympathetic father, he hoped that Francis might be able to work himself out of his distress.

Live and work with them he did. Domestic chaos reigned for a while. Emma opened up the old nursery wing and turned the billiard room into a sitting room for Francis, next door to his father's study. A girl called Mary Anne ("Nanna") came up from the village to serve as a nursemaid to Bernard, and an admiring circle of female staff quickly gathered round. Fortunately for all, Bernard was fat, placid, and healthy, little trouble for a large and well-trimmed household despite the sad circumstances of his arrival. As he grew up, he became the natural centre of attention. Darwin was besotted. Emma told Henrietta, "Your father is taking a good deal to the baby. We think he (the baby) is a sort of Grand Lama, he is so solemn."

This time around Emma knew how to relax and enjoy the experience of another baby in the house. Her husband was equally transformed. He did not catalogue or observe Bernard, as he had done with his own children. He merely adored at the shrine, along with the others. As time softened the blow of Amy's death, Bernard brought a great deal of simple joy into their lives. "The baby is quite a prize article in point of fat and healthiness and may become handsome, though far from it now," said Emma. "He has a pretty mouth and expression, and is particularly amused at his grandfather's face. I am surprised at his making out the expression from such a mass of beard." The real casualty was Francis. He yearned for physical contact. "How often, when a man, I have wished when my father was behind my chair, that he would pass his hand over my hair, as he used to do when I was a boy."[73]

Soon afterwards, Darwin published the results of his work on cross-fertilisation, wistfully accepting the truth of his suspicion that "there are not many persons who are interested about the fertilization of flowers."[74] At the end of the year his second edition of *Orchids* came out.

## IX

Letters continued to roll in, alternately useful or bizarre. A German editor called Otto Zacharias inquired if he could translate George's article on cousin marriages into German. The resulting pamphlet sold rather well, reported Zacharias a few years later.

Zacharias also asked if he could use Darwin's name on a new journal he wished to start in Germany, called *Darwinia*. The plan came to naught, and Zacharias bowed out in 1877 when Ernst Krause founded *Kosmos* with Darwin's endorsement. Such journals, as they all recognised, played a fundamental role in distributing evolutionary ideas. The story of *Nature*'s conception in 1869 was prime evidence of the value of having a tightly controlled, well-distributed mouthpiece, and it seems that first Zacharias and then Krause intended to fill the same broad-based cultural niche in German science. There was a gap in the market, to be sure. One of the first journals to take up Darwin's views in Germany had been the weekly magazine *Ausland* ("Abroad"), a heady mix of biology and society, pumped up with a stream of articles from Haeckel and other evolutionists. During the Franco-Prussian War, the editor, Friedrich von Hellwald, claimed Darwinism as proof that warfare between nations was a natural law, a standard view of the time that did not prevent his journal's expiry a few years afterwards.

Krause's *Kosmos* conveniently filled the hole. Krause explicitly based the journal on the "theory of evolution in connection with Charles Darwin and Ernst Haeckel." He favoured Haeckel's theory of monism, and made its columns notable for high-level discussions on heredity. Furthermore, it was in the pages of *Kosmos* that the bitter argument between Haeckel and Rudolf Virchow over the political meaning of Darwinism in Germany was played out. Virchow—who never liked Darwin's proposals despite a mild interest in transformism—vehemently attacked natural selection on political grounds at the Munich meeting of German naturalists and physicians in 1877.[75] He denounced the ideological roots of Darwin's theory, calling natural selection a dangerous fantasy of individualistic self-betterment. If this was adopted by democrats, as well as radical socialists, he said, it might contribute to the destabilisation of the German state. Haeckel countered by pointing out that Darwinism (as he saw it) was strictly hierarchical and "aristocratic" because of the driving force of survival of the fittest.[76] In Bismarck's Germany, with north and south forcibly united across widening political and social rifts, the threat of destabilisation was real enough. The journal *Kreuz Zeitung* blamed the theory of descent for the "treasonable" assassination attempt on Emperor Wilhelm by the social democrat Emil Hodel. Haeckel responded to Virchow with his *Freedom in Science and Teaching,* translated into English in 1879 with a rousing preface by Huxley. With this fertile context in which to plant a journal, Krause issued nineteen volumes of *Kosmos* from 1877 to 1886.

Darwin also received letters incongruously revealing his high place in other people's minds. During a correspondence with a gardener at a lunatic asylum he once discovered, secreted into an envelope, a note from one of the patients, in which the writer claimed he was wrongfully confined and begged to be saved. Like a character out of *The Woman in White,* Darwin tried to have the man released, only to find that the inmate wrote letters like that all the time. Elsewhere, an unknown child was named in his honour in Hamburg. A correspondent asked his opinion on the possibility of a flying machine powered by birds. A man living in Yorkshire told him, "I have two Alligators now about 3 feet long, which I keep in the mill ponds. I have good opportunity for noting their habits should you wish to know about them."[77]

Many of these transactions were a source of gentle amusement. In February 1877, Darwin received as a birthday tribute an enormous album of photographs of German naturalists, sumptuously bound. This, conceived as a mark of respect, had been organised by Haeckel, Otto Zacharias, and William Preyer, and forwarded to Down House by Emil Rade from Munster. More than 150 men of science presented their photographs and signatures. "It is by far the greatest honour which I have ever received," Darwin wrote gratefully to Haeckel. In the same postbag, a cascade of letters arrived from naturalists who had been omitted from the volume, each hastening to assure Darwin of their undying devotion and modern attitude to biology.

Within the week, another huge parcel arrived from the Netherlands. Emma noted the contents.

> F. was expecting about this time (his birthday today) an album containing photos of German men of science, when yesterday arrived a most gorgeous purple velvet & silver Dutch album of the same sort with 219 portraits—some of youths, some girls & some fat women, I suppose any one who subscribed. However it shews a v. different state of feeling about him. You wd. not get boys & fat women in England to subscribe & send him their photos as a mark of respect.[78]

Fame manifested itself in other ways as well. William Gladstone, leader of the Liberal party, came to call with a group of John Lubbock's weekend guests in March 1877. The male members of Lubbock's house party, comprising Gladstone, Huxley, Lyon Playfair, and John Morley, walked over to see Darwin at Down House on Sunday 11 March.[79] Morley had reviewed *The Descent of Man* in the *Pall Mall Gazette* and was as keen as the rest to meet its author.

Lubbock knew the visit would be acceptable. Darwin quite plainly regarded it as "an honour," and mentioned it subsequently with considerable awe. Gladstone was less overcome but interested to meet the naturalist nonetheless. He wrote in his diary, "Called on & saw Mr Darwin, whose appearance is pleasing and remarkable. . . . conversation with Mr Morley, Prof Huxley & others."[80] John Morley cynically recorded an alternative view. He said Gladstone settled down in a chair at Down House and for two hours bored them by reading out loud the proofs of his latest Turkish pamphlet. Darwin told Charles Eliot Norton about it with barely suppressed excitement.

> Our quiet, however, was broken a couple of days ago by Gladstone calling here.—I never saw him before & was much pleased with him: I expected a stern, overwhelming sort of man, but found him as soft

> & smooth as butter, & very pleasant. He asked me whether I thought that the United States would hereafter play a much greater part in the history of the world than Europe. I said that I thought it would, but why he asked me, I cannot conceive & I said that he ought to be able to form a far better opinion,—but what that was he did not at all let out.[81]

Darwin and Gladstone exchanged a few letters on colour perception in infants after this, a topic that wisely avoided theological matters but one that intrigued them as a possible clue to what the primal state of the human senses might be.[82] Gladstone never swerved from the strictest possible line of orthodox Christianity. A few years later he sent Darwin one of his essays on Homer, pointing out Homer's accurate observations of expressive movements. Darwin replied obsequiously, "Although you are so kind as to tell me not to acknowledge the receipt of your Essay, in which you show how wonderfully Homer distinguished different kinds of movement, yet I must beg permission to thank you for this honour."[83]

Ever afterwards, Darwin and Emma felt they had experienced high-level politics at first hand. Lifelong liberals, they were predisposed towards Gladstone's policies and followed his and Lubbock's speeches conscientiously, discussing the newspapers and pamphlets in the evenings by the fire until Gladstone's position on the Home Rule for Ireland question drew their ire. But to have entertained him in their drawing room made all the difference. After Darwin's death, Emma became an ardent Unionist.

Inevitably, with a baby in the house, Darwin's mind also turned back to his old notes on infant behaviour. In April 1877, when Bernard was six or seven months old ("such a little duck," said Francis indulgently), Darwin published extracts from his diary of observations on child development as a "Biographical Sketch of an Infant" in the new psychological journal *Mind.* Hippolyte Taine had put forward an essay in a previous number in which Taine described his daughter's development during the first eighteen months of her life, attending particularly to language acquisition. Darwin thought his own observations extended some of the remarks Taine made. He sent the notes to the editor, George Croom Robertson. "If you do not think fit, as is very likely, will you please return it to me."[84]

Once published, Darwin's observations were perceived as a valuable contribution to the emerging field of developmental psychology. Like Taine, and like Tiedemann a century before him, the material provided data for contemporary theories of individuality, consciousness, and the will, as well as contributing to the understanding of emerging language skills and intellect. Darwin's methodology—his careful watching and recording—was a sensitive research instrument.[85] That the observations were made by such a renowned naturalist no doubt helped. William Preyer's book-length study, *The Soul of a Child* (1882), drew on Darwin's and Taine's researches and was widely regarded as an important early work in modern child psychology. Yet Darwin was baffled that a study of a baby, bereft of any obvious interpretative remarks, should be so popular when his plant books were such slow sellers. His *Mind* article was soon translated into German and French and generated a substantial postbag.

X

Without pausing for breath, Darwin and Francis pushed on with a volume called *The Different Forms of Flowers on Plants of the Same Species*, published in July 1877. The book was dedicated to Asa Gray, "as a small tribute of respect and affection."

This book on flowers was another highly technical treatise in which Darwin juxtaposed his previously published thoughts on the question with new researches. "Plants are splendid for making one believe in Natural Selection," he told Huxley around now, "as will and consciousness are excluded." His achievement was to reveal the results of different "marriages" in plants, for example the primrose with its thrum and pin-headed flowers. If two flowers of exactly the same kind mated, the offspring were fewer in number and often displayed reduced fertility. "No little discovery of mine ever gave me so much pleasure as the making out the meaning of heterostyled flowers. The results of crossing such flowers in an illegitimate manner, I believe to be very important as bearing on the sterility of hybrids; although these results have been noticed by only a few persons."

The metaphor of marriage was very real to him, as to Linnaeus before him.[86] Darwin viewed the sexual lives of these plants as if they were flesh-and-blood humans, prone to all the marital mistakes and inappropriate yearnings of romantic fiction. He wrote of "illegitimate unions" between flowers, and of the poor quality of the offspring, reflecting in his imagery all the undercurrents of his own and George's work on first-cousin marriages. Working away in his greenhouse forcibly creating trays of bastards and infertile degenerates, he initiated a wide variety of plant matings that would have made any visiting clergyman blush. He harangued his botanical friends on the question, blind to the fact that only experts would be able to understand the complex issues involved. He read the newest books on plant crossing, hybridisation, and inheritance, taking up some of August Weismann's work on heredity with enthusiasm. Here, he thought, were functional causes for the incipient sterility sometimes observed between varieties of the same species. Here, in fact, was divergence in

New ideas about prehistory stimulated artists
to depict early mankind as brutal and primitive.
"Primeval man" by Louis Figuier, 1867.

(Natural History Museum, London)

Ernst Haeckel, the German biologist and evolutionist, was an ardent disciple.

(E. Haeckel, *Italienfahrt* 1921)

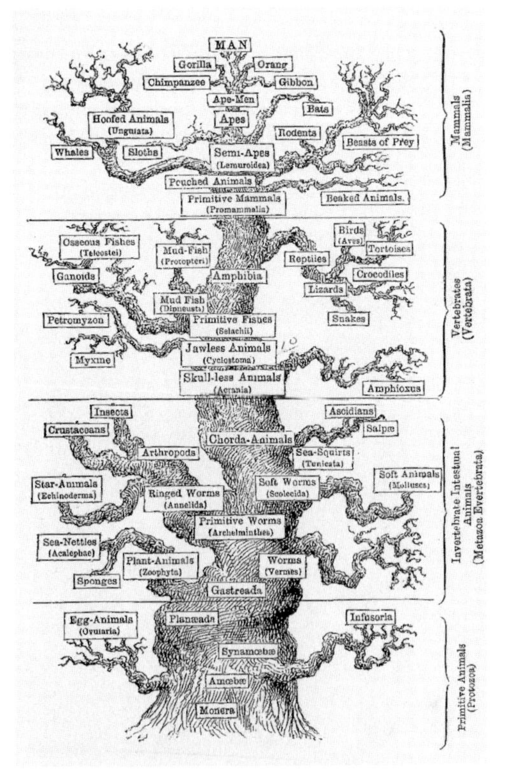


Haeckel was one of the first to construct evolutionary trees.

(E. Haekel, *The Evolution of Man* 1887.
Wellcome Library, London)

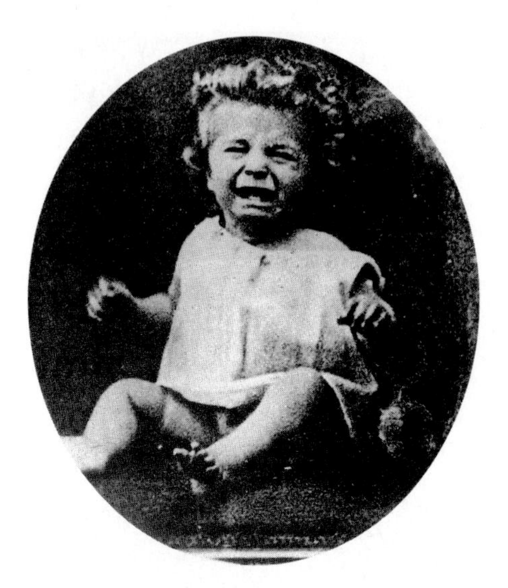

Technical advances in photography made it possible for Darwin to study the evolution of facial expressions. This photograph was dubbed "Ginx's Baby" after a popular novel.

(C. Darwin, *Expression of the Emotions in Man and Animals* 1872. Wellcome Library, London)

The exuberant photographer Oscar Rejlander sent Darwin several self-portrait studies of expressive behaviour. In the double photograph Rejlander compares Ginx's Baby with laughter on the left and crying on the right.

(C. Darwin, *Expression of the Emotions in Man and Animals* 1872. Wellcome Library, London; and Darwin collection, courtesy of the Syndics of Cambridge University Library)

Notes in Henrietta Darwin's hand on Joseph Wolf's drawing ask for adjustments to emphasise the dog's raised hackles.

(Darwin collection, courtesy of the Syndics of Cambridge University Library)

Darwin thought facial expressions would be more intense in the insane. James Crichton-Browne supplied Darwin with photographs of the inmates of Wakefield asylum and discussed his theories with him.

(Darwin collection, courtesy of the Syndics of Cambridge University Library)

Guillaume Duchenne gave Darwin photographs of his galvanic experiments showing "frozen" muscle contractions.

(C. Darwin, *Expression of the Emotions in Man and Animals* 1872. Wellcome Library, London)

As part of his investigations into human variability and heredity, Francis Galton took George Darwin's fingerprints.

(Galton Archive, University College London)

The Darwin family commissioned a portrait from William Ouless in 1875, engraved by Paul Rajon. Darwin said he looked like "a very venerable, acute melancholy old dog."
(Wellcome Library, London)

This stuffed monkey, dressed in "academicals," was dangled from the ceiling of the Senate House when Darwin received an honorary degree at Cambridge.
(By permission of the Master and Fellows of Christ's College, Cambridge)

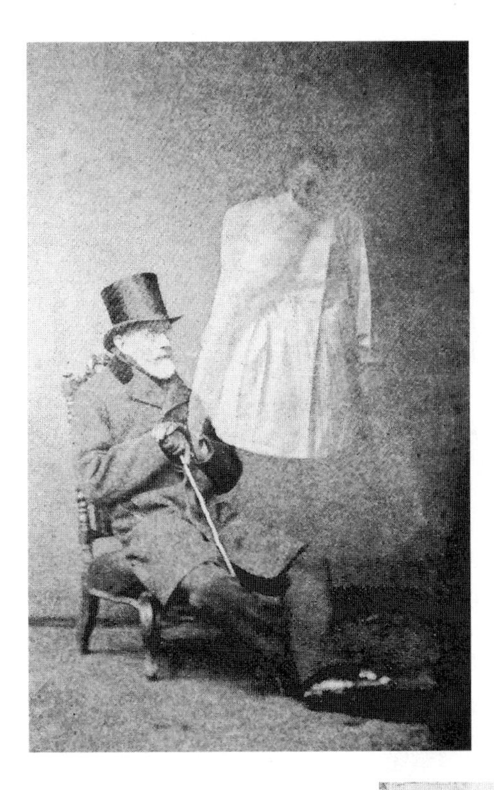

Evolutionary theory was only one of many popular new topics of inquiry for Victorians. Emma's brother Hensleigh Wedgwood was a dedicated spiritualist. He sent this photograph of himself in the presence of a spirit manifestation to Huxley.
(Imperial College Archives)

T. H. Huxley's daughter Marian sketched Darwin during a visit to Down House in 1878. Her husband, the artist John Collier, later made a formal portrait for the Linnean Society.
(National Portrait Gallery, London)

Two of the sensitive plants that Darwin investigated during his last years showing the leaves "asleep."

(Darwin collection, courtesy of the Syndics of Cambridge University Library)

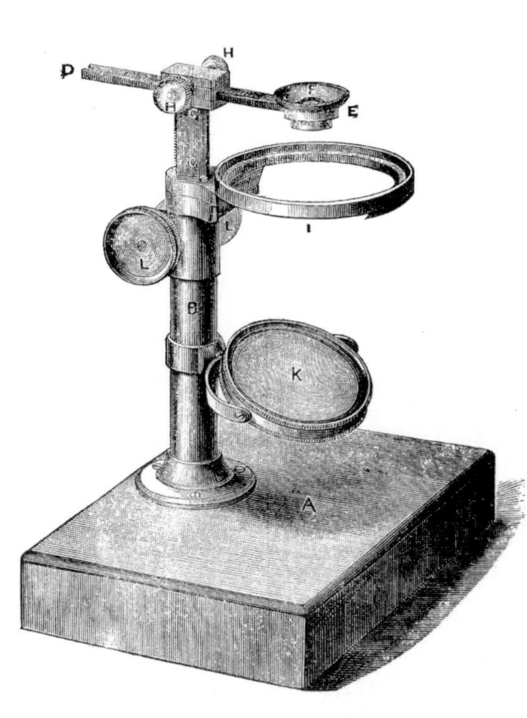


The simple microscope that Darwin preferred.

(R. Beck, *Achromatic Microscopes* 1865. Wellcome Library, London)

The Down House staff in the garden in the late 1870s. Darwin's first grandchild, Bernard, is probably on the pony. William Jackson, the family butler after Joseph Parslow retired, wears the pale-coloured coat.

(Darwin collection, courtesy of the Syndics of Cambridge University Library)

One of the last portraits of Darwin.

(Darwin collection, courtesy of the Syndics of Cambridge University Library)

Darwin's final investigations were on earthworms. The cartoon is titled "Man is but a worm." The spiral shows the evolution of human beings from 'chaos' on bottom left, to Darwin, in God's chair, in the centre.

(*Punch's Almanack* 1881. Wellcome Library, London)

action, an answer at last to Huxley's complaint that natural selection could never select for sterility. Putting it at its simplest, Darwin thought "illegitimate" seedlings were almost like hybrids between members of a single species.

## XI

A frisson of scandal briefly fluttered on the horizon. For a moment he got embroiled in the notorious obscenity trial of Charles Bradlaugh and Annie Besant, the first a prominent atheist and author, the second popularly thought to be a freethinker in sexual matters. The impending trial alternately thrilled and agitated the nation, a Victorian counterpart to the legal commotion nearly a hundred years later surrounding D. H. Lawrence's *Lady Chatterley's Lover.* In a sixpenny pamphlet issued early in 1877, Besant and Bradlaugh had described the perils of over-population and recommended various methods of contraception. The pamphlet was a revised version of Charles Knowlton's *Fruits of Philosophy or the Private Companion of Young Married Couples,* a book that had already been available for several decades. This time around, in Besant and Bradlaugh's hands, the pamphlet addressed a completely new audience, issuing dire Malthusian warnings about degeneration, dissipation, and "unrestrained gratification of the reproductive instinct." By explaining the means of contraception to the masses they hoped to avert these calamities.

The resulting scandal brought together the most learned minds of the Victorian legal system. In April 1877, Besant and Bradlaugh were arrested and charged with obscenity for their "dirty filthy book."[87] In June, Bradlaugh wrote from gaol to Darwin asking if he would testify in their support. Bradlaugh had every reason to believe that Darwin—who was a well-known secular thinker, author of *The Descent of Man,* and a prominent advocate for Malthusian views in nature—would be likely to defend the rational application of natural selection to mankind and verify the pamphlet's views on overcrowding.[88] He did not know Darwin except by repute, although he may have hoped that Annie Besant's connection with Moncure Conway, an acquaintance of Darwin's, might help his request along.

Bradlaugh hoped in vain. Darwin responded immediately, declaring he would have nothing to do with either a subpoena or an obscenity trial. If called to court, he would vigorously testify against Bradlaugh and Besant's case.

> I have not seen the book in question but for notices in the newspaper. I suppose that it refers to means to prevent conception. If so I should be forced to express in court a very decided opinion in opposition to

> you & Mrs. Besant. . . . I believe that any such practices would in time lead to unsound women & would destroy chastity, on which the family bond depends; & the weakening of this bond would be the greatest of all possible evils to mankind.[89]

Here, Darwin made it plain that he believed that civilised societies best advanced by childbirth taking place only within the respectable boundaries of marriage—a point of view that had also been the gist of Malthus's original remarks. Like Malthus, Darwin disparaged contraception, which he regarded as an impediment to natural processes. He thought easy access to contraception would lead to unfettered sexual activity outside marriage, which in turn would introduce licentiousness and vice, inadequate care of children, financial insecurity, death, and disease. "If it were universally known that the birth of children could be prevented, and this were not thought immoral by married persons, would there not be great danger of extreme profligacy amongst unmarried women?" he wrote in a concerned manner to George Arthur Gaskell, an advocate of birth control.

Like Malthus, too, he seemingly believed contraception should not be used within marriage either. His ten children were proof of this view. "I am strongly opposed to all such views & plans," he reiterated in 1878 when asked to defend another contraceptive publication.[90] He was humane enough, however, to feel something of a dilemma over how far society might be justified in intervening in the human reproductive process—he did not wish for humans to exist in a complete state of nature and undoubtedly remained aware that fertility has social consequences. From time to time, he queried the rationale for charitable measures to relieve the sick and the poor, saying that such intervention tended to preserve the unfit. But he also stated that civilised human beings were civilised precisely because they looked after their weak and needy. "The evil which would follow by checking benevolence and sympathy in not fostering the weak and diseased would be greater than by allowing them to survive and then to procreate."[91]

After such a reply, it was hard for Bradlaugh to continue to ask for Darwin's presence in court. As expected, Bradlaugh and Besant were found guilty, although the sentence was relatively light, just a fine and a pledge to withdraw the book from circulation. But the personal damage was done. Annie Besant's life spiralled rapidly downwards thereafter. Her estranged husband initiated a court case to remove their daughter from her care, itself a turning point in legal history. She converted to Madame Blavatsky's mystical doctrines of reincarnation and theosophy and became involved with Edward Aveling on the *National Reformer.* Towards the end, she turned her eyes to India. Bradlaugh won a seat in Parliament that he never occupied because he refused to swear an oath of loyalty to the monarch.

Coming straight from his work on "legitimate" and "illegitimate" unions among plants, his private life a tranquil pool of liberal-conservative social principles accompanied by a vision of an advancing British society built on moral and financial rectitude, Darwin's response to Bradlaugh made it clear that he had never believed more strongly in family values.

*chapter*
12

# HOME IS THE SAILOR

DARWIN SPENT the next few years staring into the past. Discarding his other obligations in the summer of 1877, he started digging for earthworms, harking back to a project he had begun even before he was married. For a long time now, he had taken a tender interest in worms. Bit by bit, this developed into his last major undertaking in natural history.

He believed that earthworms were vastly underestimated agents in transforming the earth's surface by bringing up tilth from below, creating fertile mould, and causing stones to sink. "It may be doubted whether there are many other animals which have played so important a part in the history of the world, as have these lowly organised creatures."[1] When he had first arrived at Down House in 1842 he had recorded the types of agricultural treatments on particular fields with a view to seeing which surface changes over a number of years might be attributed to worm action. That investigation had grown out of the early articles on worm activity that he had presented to the Geological Society in 1837 and in itself had indicated that he intended putting down roots and staying permanently in the vicinity. He recorded similar data when visiting Emma's father's estate and those of his uncles, and from time to time he would inquire of them how various fields were getting along. Occasionally, he asked obliging friends and relatives to investigate worm-casts for him on summer mornings.[2] (Amy Ruck, Francis's wife, had first endeared herself to Darwin through her willingness to measure the depth of Welsh worm-holes.) In the main, Darwin hoped to ascertain how far, and how quickly, dressings of whitish marl, or cinders, or stable manure sank beneath the surface; and how rapidly large stones might become entombed by soil.

These changes he attributed to worms' constantly bringing fine tilth up to the surface in their droppings, gradually covering any surface items and transforming the upper layers of the earth. Worms were making history, he thought grandiloquently. They were archaeologists in reverse, burying the face of nature.

It was a perfect project for a man with his circle of landed connections. Perfect, too, for a slow-moving country gentleman with a penchant for leisurely, appreciative walks at home and on the estates of friends and relatives. He and Emma wandered out in the evenings to the chalk fields along the top of the neighbouring downs, to Cudham woods, Keston, and all about, so that he could prod the ground with his stick, turn over stones, and scan the land for worm activity.

It was an occupation that Emma found attractive too, reminiscent of the relaxed days of their early marriage. "F. was made very happy by finding two very old stones at the bottom of the field," she observed while they were staying at Caroline's house (Leith Hill Place) in June 1877. These stones were ancient slabs from a limekiln, five feet long, fairly sunk into the ground. Caroline's gardeners lifted them for Darwin the following day. Underneath, the fine black mould of worms was obvious. It was a long job and Emma worried whether Darwin was taking proper precautions. "F. has had great sport with the stones, but I thought he would have a sunstroke."[3]

Stealthily, earthworms crawled into his high days and holidays. He asked William to look out for worm-casts during a trip to Malvern. He counted on George for the same when on holiday in Madeira. On rainy days Francis Galton counted dead worms for him on the tarmac paths in Hyde Park ("one for every two-and-a-half paces"). And he sent inquiry after inquiry to George Romanes asking for definitions of animal intelligence that might be stretched to include the limited responses of these simple creatures. "I tried to observe what passed in my own mind when I did the work of a worm," he said engagingly.

> If I come across a professed metaphysician, I will ask him to give me a more technical definition, with a few big words about the abstract, the concrete, the absolute, and the infinite; but seriously I should be grateful for any suggestions, for it will hardly do to assume that every fool knows what "intelligent" means.[4]

There was fun to be had. One year, during his and Emma's annual pilgimage to William in Southampton, Darwin impulsively decided to hunt for worms at Stonehenge. He and Emma took the train to Salisbury armed with a spade and a piece of Lady Lubbock's notepaper on which there was an engraved picture of the stone circle: Lubbock was a prominent parliamentary advocate for archaeological preservation long before such sites were regarded as national treasures. At that time there was access for all, even for scientists with digging on their minds. Darwin grubbed around the base of the fallen stones while Emma admired the scenery.

> We loitered about and had a great deal of talk with an agreeable old soldier placed there by Sir Ed. Antrobus (owner) who was keeping guard and reading a devout book, with specs on. He was quite agreeable to any amount of digging, but sometimes visitors came who were troublesome, and once a man came with a sledge-hammer who was very difficult to manage. . . . They did not find much good about the worms, who seem to be very idle out there. Mrs Cutting gave us a gorgeous lunch and plenty of Apollinaris water. We drove back a lower way, very pretty by the river and rich valleys and close under Old Sarum—very striking.[5]

His curiosity whetted, Darwin also pursued worms at Thomas Farrer's property in Surrey. Luckily for him, in 1877 one of Farrer's estate hands accidentally turned up the remains of a Roman villa in a ploughed field, and Farrer invited Darwin to come and watch the excavations. Five-foot-deep trenches going down through time to the Roman period were just what Darwin wanted. He spent his evenings with Farrer in the pits, after the men had gone home, measuring the depth of the black worm tilth and the intervening layers of stones and earth, all the way down to the broken mortar and tessellated pavement at the bottom with undisturbed subsoil underneath. Calculating backwards through the centuries, he computed how much earth had been deposited by worms since the pavement had lain on the surface, a variant of the technique he had often used in working out the age of geological deposits. Even so, he was cautious, no doubt remembering the miscalculations he had made when working out the age of the Weald. He asked George to go through the figures. The net result was that he estimated that the ruins were covered at the rate of one inch every twelve years. Worm activity alone, Darwin told Farrer triumphantly, had submerged the villa so completely that no one knew it was there. Farrer agreed to keep an eye out for any new earthworm movement under the Roman pavement for the next six weeks or so. "I cannot remember a more delightful week than the last," Darwin wrote in thanks.

These home-based activities suited him exactly—slow, undemanding, and sufficiently interesting to keep him absorbed. In retrospect, it is plain that he had become utterly fixed in his own home to the extent that even his scientific imagination could not help but turn inwards towards it. Having exploited his garden, transformed his house into a factory for theories, and turned his growing children into facts, he now looked downwards, into the very earth he walked on. His science in this regard had come to be absolutely of the place, an expression of himself, of his personality and aspirations, now quite distinct from any comparable researches undertaken in biology by friends such as Huxley or Hooker. Whereas Hooker's botanical endeavours were at heart fixed in a colonial, taxonomic enterprise, in which he evidently found inspiration and motivation from worldwide imperial motifs, Darwin had progressively reduced his gaze from large to small, making his own surroundings the centrepiece of his intellectual project. Whereas Huxley, or Anton Dohrn, might study annelids and ascidians as indications of the ancestral form of the first possible vertebrate in evolutionary history, and make sustained laboratory inquiries into annelid embryology and physiology, Darwin wished to know how many worms—equally members of the annelid family—were working away under his lawn. To be sure, this was the classic natural history approach that had characterised his work for decades. But it was also the distinctive feature that he brought to his individual scientific activities.

Every now and then, worms gave way to other concerns. William announced that he wanted to marry Sara Sedgwick, the younger sister of Susan Norton, the wife of Darwin's friend Charles Eliot Norton of Massachusetts. Darwin and Emma liked the Nortons very much, and Sara suited William well. For some reason the engagement was to be kept secret, at least for a month or two, and Darwin smiled to see how smoothly Emma could lie when required. "You ought to have seen your mother," he said conspiratorially to Henrietta. "She looked as if she had committed a murder, and told a fib about Sara going back to America with the most innocent face." The couple married in November 1877, a lasting, although childless, union.

That same November, Darwin was awarded an honorary degree of Doctor of Laws (LL.D.) from Cambridge University, his alma mater. George was the first to tell him the news, writing directly from Trinity College with the result of the vote. Darwin accepted the honour with alacrity—he valued it highly, probably more than any other in his lifetime. As George told it, the voting was much more straightforward than the bungled affair at Oxford. Edward Atkinson, the vice-chancellor of Cambridge University, was a proponent of scientific advance and proud to announce both the opening of the Cavendish Laboratory for physics and Darwin's election in the same year. The resolution passed easily.

The degree was conferred at a special ceremony held for Darwin alone on Saturday 17 November 1877 in the Senate House in Cambridge. At first Emma doubted whether Darwin would cope with the ceremony, a two-hour haul topped off with a procession. He would be decked out in heavy red robes, followed by a deal of bowing and smiling to be got through. She underestimated her man. He was perfectly capable of effort when it was needed. "I thought he would be overcome, but he was quite stout and smiling."[6]

However, they were astounded by the noise. The Senate House was crammed with members of the university and their ladies, as the *Cambridge Chronicle* noted. Darwin's sons were there too, George, Horace, and Francis. The gallery throbbed with undergraduates cheering and shouting. As soon as Darwin appeared in his robes, a stuffed monkey was dangled from the ceiling dressed in a miniature degree gown and mortar-board.

> His appearance was the signal for a flattering ovation and partial silence had scarcely ensued when cheers were given for "the primeval man." . . . A couple of cords stretched from either gallery now became visible and presently there were affixed to them the effigy of a monkey in Academicals and an object to illustrate the "missing link" hung over Darwin's head.[7]

When the words could be heard above the rumpus, the Latin oration conferring his degree was elegantly incomprehensible. Darwin asked George to send him the Cambridge newspapers afterwards in case a translation appeared. All his works in biology and geology were mentioned, albeit in Latin, from "all that flies, or swims, or creeps" to the "tiny tendrils of the vine." The weekend closed with a sociable lunch hosted by George in his rooms in Trinity. "J. W. Clarke not pleasant," Emma said of the zoology department's superintendent, "but he did me a good turn as I followed his lead in tasting Gallantine which is v. superior. . . . Now we are expecting F. Galton & Fawcett for an evening call & we mean to wind up by Trinity chapel—which I can't believe F. will be up to. F. has been thoroughly pleased by the cordiality of everyone."[8] Whether by design or oversight, they missed the Sunday-morning sermon in Great St. Mary's in which Dr. Wordsworth, the university preacher, reminded his congregation to approach the facts of nature in a properly reverential spirit.

The irony of a university full of ordained fellows and teachers honouring Darwin did not go completely unnoticed. An editorial in the *Rock*, an Anglican journal, commented, "No doubt the affair has its ludicrous side, though a believer will scarcely regard the honour paid to the apostle of evolution as by any means a laughing matter." The *Manchester Weekly Times and Examiner* was more urbane.

> Mr. Darwin has made considerable havoc in the groundwork of Christianity. . . . I can only wonder how the Cambridge Doctors of Divinity and official guardians of orthodoxy will manage to square accounts with Mr. Darwin. Perhaps it is believed that a niche may be found for him in some sequestered corner of the Christian fabric.[9]

By separating the man from his theory it evidently proved possible for Cambridge University, and then nearly all elements of the British establishment over the next several decades, to greet the new regime of evolutionary biology with relative equanimity. Even the Church of England acknowledged Darwin's personal integrity and quest for truth. Perhaps it was merely traditional British hypocrisy, or the spirit of compromise and adaptability in the face of what increasingly appeared to be a scientific truth. Nevertheless, by heaping personal honours on Darwin's head, by praising his individual qualities of probity, commitment, decency, and modesty, the nation could glory in celebration without having to make any judgement on evolution at all.

One secondary consequence of this honorary degree was that the Cambridge Philosophical Society requested a portrait of Darwin in his LL.D. robes. Darwin had been a member of the society since Henslow's day and it was where his first published pamphlet, cooked up by Henslow from remarks taken from his *Beagle* letters, had been read to other members. Forty years on, he felt privileged to comply. Subscriptions from members and other Cambridge graduates quickly amounted to some £400, and the society commissioned William Blake Richmond, shortly to become professor of fine art at Oxford (the son of the artist George Richmond, who had drawn portraits of Darwin and Emma at the time of their marriage), to take the portrait. Richmond lived in Sevenoaks, only a short ride from Downe. Darwin sat to him in June 1879, although not without making a fuss. George Darwin had to hire the scarlet LL.D. robes at short notice from a Cambridge outfitter and bring them home to Down House; and he had to hire them again, a few weeks later, to send by post to Richmond's studio for the finishing touches. Artist and model talked prosaically. "Once I asked Mr. Darwin which of the years of a child's life were the most subject to incubative impressions," said Richmond. The answer was "Without doubt, the first three."[10]

Emma greatly disliked the painting, and was glad to see it go to Cambridge. When she saw it for a second time in the early 1880s, it was just as bad as she remembered. "We went to see the red picture & I thought it quite horrid, so fierce & so dirty. However it is under a glass & v. high up so nobody can see it."[11]

## II

Imperturbably, he settled into a steady routine, a stalwart of the village community, taking on much of the identity of a country squire.[12] He mingled with other landowners, clergymen, farmers, and neighbouring worthies, each of them an integral part of the cross-section of society traditionally most concerned with village affairs. Despite its closeness to the metropolis, Downe was still essentially rural in structure, its surface hardly ruffled by shifting social demarcations or any rush of incomers with new occupations or ideas. Darwin carried out a number of minor philanthropic activities in and around the village, primarily acting as treasurer of the Coal and Clothing Club from 1848 to 1869, and establishing the Downe Friendly Society (also called the Downe Benefit Society) in 1850 and serving as its treasurer for thirty years. He kept the accounts of both organisations and dealt with the official paperwork and government regulations for such societies. He was furthermore a founding trustee for the Bromley Savings Bank and encouraged small investors among the Downe community by arranging savings accounts for them with this organisation.[13] Francis and Leonard Darwin continued the same form of paternalistic, fiscal encouragement by founding in 1879 a "Penny bank" at Downe, for men, women, and children.

On the female side, as was customary for her age and position in society, Emma would "visit" a few women on a regular basis with food, cast-off clothing, and medical advice, primarily concentrating on members of the Coal and Clothing Club whose efforts to save small sums of money declared them to be, in the terminology of the time, the deserving poor. Her letters to Henrietta and Bessy were sprinkled with comments about her cases, and her account books reveal that she backed up these charitable intentions with constant small sums of cash, a sixpence here, three shillings there. In 1870 her annual total for "gifts" was £27 5s. 4d, about the same amount spent on household candles. For some forty years she also issued "penny bread-tickets" to those who rang the doorbell at Down House. These tickets were recoverable at the village bakery.[14] Darwin could be prickly if he thought his wife's generosity over the tickets was being exploited. "Wonderful charitable people the Darwins were," said John Lewis, the village carpenter in an interview after Darwin's death.

> Used to give away penny tickets for bread on the baker. I've given away thousands and thousands. And very good to the poor for blankets and coal and money till they got run on. One man used to brag

in the pubs that he could live without work—till Mr. Darwin heard of it.[15]

Emma poured most of her energies into the Sunday school, where she and her daughters, and other young ladies of the district, voluntarily taught children the rudiments of reading and writing. As part of this project, Emma ran a little lending library, and in later years she promoted a Sunday reading room for adults. All of these activities, however, were nothing more nor less than what was expected of them. Neither Emma nor her husband went so far as to sit on a charity hospital board, which would require them to issue admission notes to hospital, nor did the Darwin womenfolk attend the Bromley Workhouse on a philanthropic basis (as did the Bonham Carters). They performed their chosen civic duties with a sense of satisfaction, not making any public show about commitment.

The Coal and Clothing Club lay closest to Darwin's bookkeeping heart. He and the vicar, the Rev. John Innes, had early on set up this system, in which a few pennies were collected from the villagers every month and larger donations obtained from the local gentry. Darwin gave £5 every year. The money was spent on winter necessities. Members would run up an account in the village shop that was paid off at intervals by Darwin, as the treasurer, against subscriptions. The intentions of this "club" were characteristically Victorian in their encouragement of self-help and self-discipline in expenditure. Darwin, for instance, held the purse-strings and could easily exercise power over those who were unable to bring sober commonsense to their saving and shopping. He continued to run this club with other local gentlemen after Innes left to make his home in Scotland.

The Friendly (or Benefit) Society served a somewhat different purpose in that it provided insurance cover against loss of work, illness, or funeral charges. Similar small insurance societies sprang up all through the nineteenth century as a form of group protection, matched on the larger scale by the trade unions and cooperative societies, and developing in response to the challenges of urban life, industrialisation, and rural uncertainty. All Darwin's instincts told him it was his duty to initiate this kind of local scheme in which self-help and financial independence could be promoted, a kind of monetary natural selection in which his advice might confer the advantage of prudence on his flock. He believed the Downe villagers were unlikely to take such steps themselves. He and Innes had sought advice from John Stevens Henslow, who had been a pioneer in local insurance arrangements. "I have succeeded in persuading our Clodhoppers to be enrolled in a Club," Darwin declared energetically in 1850.[16]

Darwin did not like challenges to his authority in running these organisations. The Downe Friendly Society teetered in 1877 when two or three independently minded members wanted to recoup their investment as a lump sum. Because of the nature of the insurance policy, which was calculated on the expected number of contributions over future years, this would have meant Darwin's dissolving the society at a loss to other members. Annoyed, he called on the dead hand of bureaucracy to prevent insubordination. Armed with official documents from the Registrar of Friendly Societies, he accosted his miscreants in the village pub.

> Last night I gave the club a long harangue, which I think produced some effect; at least it acted like a bomb-shell for all the members seem to have quarrelled for the next two hours. I do not think there is the least chance of the dissolution of the club. I had much satisfaction in reading aloud the penal clause.[17]

Darwin won, but not before he had printed and distributed a single-sided sheet containing extracts from the Friendly Societies Act of 1875 and an amending Act of 1876.[18] He arranged with the registrar to withdraw £290 from the National Debt Office to distribute among the membership as an interim cash bonus—a sweetener. The remainder stayed in the bank.

He participated in the parish's educational activities, too, sitting on the "vestry" for a number of years, the supervisory body for local schooling. Darwin joined Lubbock, Innes, and other parish dignitaries in setting up the "National" nondenominational day school, whose large single room was subsequently made available as an evening reading room for working men. Innes and Darwin performed well together on these projects. The two were very different—the vicar and the scientist, the Tory and the Liberal—but liked each other all the same. Their tendency to take opposite views became a source of humour. On Innes's final visit to Down, before going to Scotland, Darwin declared, "Brodie Innes and I have been fast friends for thirty years, and we never thoroughly agreed on any subject but once, and then we stared hard at each other, and thought one of us must be very ill."[19]

"We often differed," Darwin wrote to him afterwards, "but you are one of those rare mortals, from which one can differ & yet feel no shade of animosity—& that is a thing which I shd. feel very proud of, if anyone could say of me." Innes similarly appreciated his connection with Darwin. When he settled in Morayshire, home of the kirk and rabid sectarianism, he mischievously enjoyed the stunned silence that would greet his casual remark that Charles Darwin was one of his closest friends. "Dear me! If some of your naturalists, and my ritualist friends were to hear us two saying civil things to each other, they would say the weather was going to change, or Paris to be relieved, both which I wish might happen."[20]

After Innes left, Darwin sometimes wrote to him about the curates appointed by proxy in his place. Innes, as the holder of the church living, had the privilege of naming his ecclesiastical replacements, but this was always tempered by the church's centralised administrative system. Often he accepted references at face value. Hence Mr. Robinson, the curate who unwisely took a fancy to a housemaid, had to go. The next one was little better. These to-ings and fro-ings gave Darwin rather a prejudice against young curates and the bishops who appointed them. He would apparently repeat with grim satisfaction Harry Wedgwood's adage "A bench of bishops is the devil's flower garden."[21] The curates became so unsatisfactory at one point that Innes offered to hand over his right of appointment to Darwin, momentarily forgetting that most clergymen would consider the author of the *Origin of Species* the last man in England appropriate to oversee a church living. The letters between them on these issues reveal many of the internal workings of a small rural parish.[22] Darwin and Innes shared the same sense of resignation when faced with human folly or the byzantine rules of church bureaucracy. Innes, Fox, Henslow, and Langton were a breed of unassuming clergymen whom Darwin respected.

These parish activities came to an abrupt halt when Darwin argued with George Ffinden, the vicar from 1871. Ffinden came from the sanctimonious, soulful end of church doctrine with a distinct aura of ritualism and tractarianism about him, someone who would have been much better placed in an Oxford college than a country living whose inhabitants were relatively unmoved by either Anglican church or Nonconformist chapel. Ffinden had evidently been warned that his parish included the great scientist. "I confess that, perhaps, I am a bit sour over Darwin and his works. You see, I'm a Churchman first and foremost. He never came to church, and it was such a bad business for the parish, a bad example."[23] Ffinden started boldly by putting high church teachings on the school curriculum. Darwin thought that a few prayers before lessons were "acceptable," but not reciting the entire Thirty-nine Articles in chorus.

Then Ffinden interfered with the evening use of the schoolroom as an adult reading and social club, one of Emma's pet schemes. Irritated, Darwin applied by letter to the education inspectorate in London to get independent approval of the evening use of the room. He sent the official reply, with a stiff little note, to the parish council. A succession of tit-for-tat retaliations ensued, the very stuff that drives provincial life onwards. Ffinden considered that his position as minister to the souls of the parish was usurped and was curt to Darwin when they next met. Darwin resigned from the school committee in response. Then Ffinden refused to acknowledge him when they met in the street. Shortly afterwards, John Lubbock, who also sat on on the school committee, took Darwin's part and threatened to build a new reading room, outside Ffinden's jurisdiction. Emma flounced off to worship at Keston, rarely returning to the flint-built church in Downe. There is some anecdotal evidence that Lubbock also stopped attending Downe church because of Ffinden's interference with the Lubbocks' customary village patronage. Emma certainly believed Ffinden treated the Down House butler, Joseph Parslow, very unkindly at his wife's funeral in Downe in 1881, as cheap retaliation against all things Darwinian. "Did I tell you of Mrs Parslow's death. Poor P takes it very patiently & has rather benefitted than otherwise by Mr Ff's brutal behaviour at the funeral . . . he thinks the ArchB ought to be informed."

Towards the end of the 1870s, Darwin urged his son Francis to join the school committee. Francis tried to oust Ffinden from the committee, perhaps successfully, because Ffinden shortly afterwards resigned, although he remained as vicar to the parish until after Darwin's death. Darwin sent £25 for the church restoration fund and wiped his hands of any further contact. "I remember his giving me a subscription for the church and the house-restoration or building," said Ffinden later.

> "Of course," he told me, "I don't believe in this at all."
> "I don't suppose you do," I said to him. Quite candid on both sides.[24]

## III

Early in 1878 the Darwins completed an extension to Down House intended for Francis and his son, Bernard. The building work had actually been going on for some time, for they had engaged the architect, William Cecil Marshall (who became a good friend), two years earlier, soon after Amy's death. Darwin spent £1,000 on the rooms and furniture. These new works were supposed to provide a billiard room (Francis occupied the old billiard room) and another bedroom and dressing room upstairs. But the family changed their minds as soon as the new rooms were available. The sons suggested that Darwin should take the downstairs room as a bigger and better study. He moved into the new room in February.

Conservative in his old age, Darwin made the new study look as much like the old as possible. He put the pictures in the same place over the mantelpiece, the armchair and table where he was accustomed to find them. However, even he liked the novelty. He said he was pleased with the couch that Emma purchased, the large tables for his plant pots, the clean northwestern light. With hardly a nod of regret he sold his special billiard table, and he hung a photograph of Emma by his armchair, an act of homage that apparently pleased her. Francis took over his father's original study, which with the former billiard room created a small suite of rooms for him. Gaily, the old people experimented with a telephone which Leonard brought home one day from Chatham, probably a military model for use in the field rather than one of Bell's magnetic-coil units.

> Leo had brought a telephone but it is rather a disappointing creature—They did not bring a long enough wire so that we kept hearing the talking outside the room in the natural way, & I could not hear Leo's voice at all (hardly) tho' I cd. Elinor's very well. Aunt Eliz was m. interested but could not hear v. well.[25]

In a more general sense, Darwin's paternalistic role in the external community of Downe was repeated indoors. Behind the green baize door another world ran parallel to Darwin's. In 1878 there were over a million domestic servants in Britain, making service the second-largest occupation after agricultural work. A household the size of Darwin's would usually have included a butler, who might double as a manservant, one or more footmen (in livery), a coachman or groom (in livery), one or two housemaids, a cook, a lady's maid, a nursery maid, a parlourmaid, a laundry maid, and gardeners, stablehands, and governesses as required. The 1861 census return for Down House listed twelve servants, a little over 2 percent of the total village population of 496. The 1871 census listed only six or seven because the family was away from home on that day. Among the employees living on Darwin's property were William Brooks, the head gardener (aged seventy in the 1871 census), who occupied a cottage in the grounds with his wife and child; Samuel Jones, a coachman, originally from Llandudno; and George Bridger, a footman engaged from Emma's sister when Elizabeth's and Charlotte's homes in Hartfield were sold. By the end of the 1870s, John (a coachman), Fred (a groom), and James (a footman) had joined them. These, and other staff, stayed with the Darwins for years, sometimes all their lives, growing old together. Moffat, the senior footman, arrived in 1858 and remained for twenty years. Most of the staff lived free of charge in the house, their meals and uniforms included, with a small monthly or quarterly cash payment on top. Darwin's total salary costs for servants hovered around £100 per annum, a sum that was about half his butcher's bill for any single year.

The settled nature of the garden staff revealed the steady employment patterns traditionally fostered by the landed gentry. William Brooks remained at Down House as head gardener until 1872, when he retired at the age of seventy-one. Henry Lettington, the second gardener, aged forty-eight in the 1871 census, who married Brooks's daughter, took over the older man's position and stayed on past Darwin's death. Mr. and Mrs. Brooks may have continued to live in the gardener's cottage at Down with their daughter and son-in-law: the records are unclear. There were occasional tremors in this regime. At some point in the 1870s, Darwin's sons persuaded him to engage an active young Scotsman for the garden. "The two old servants were dreadfully bustled by [the Scotsman], and I well remember their flushed faces after the first morning's digging in the serious Scotch manner," said Francis.[26] Considerately, Darwin let the new man go. Lettington was ultimately replaced by Tommy Price, of whom Emma said "his work is not worth 1s. a day."

In small establishments like this the butler reigned supreme, his exalted status emphasised by the fact he wore civilian dress, not livery. Among other duties, Joseph Parslow looked after the wine and casks and the family silver, and took overall responsibility for the dining room, which included cleaning the best tableware and wineglasses himself. The silver was a fair responsibility, valued at £179 12s. 6d in the 1840s (around £10,000 in modern terms), including the Darwins' wedding gifts and Emma's bracelet of "gold, diamond and emerald." Parslow kept his butler's pantry locked, and possibly also slept there. His wife, Eliza (a dressmaker), and their children lived down the road in a tied cottage while he was in service. His daily life was inextricably interwoven with that of the household. One of Parslow's daughters worked as a maid at the house, probably Anne Parslow, aged nineteen when listed in the 1861 census. Bernard Darwin's nurse Mary Anne married Parslow's son Arthur in 1881.

Parslow did much more than guard the silver. As several guests mentioned, he was nearly a member of the family. He was Darwin's valet, accustomed to helping him with all his personal requirements. Over the years he performed as nurse and bathman, sometimes cradling Darwin like a baby while he was sick. "About thirty years ago many's a time when I was helping to nurse him, I've thought he would die in my arms," he said after Darwin's death.[27] He was natural history assistant and trusted messenger. He travelled with the family, young and old, on holidays, to school, and for excursions; taught the boys to shoot; bought cows and sold horses for his master; played billiards; and told the lads to take their boots off in the hall. More than this, he was patriotic and public-spirited, joining in with village life. He signed up for the voluntary militia in Bromley before shifting, in the company of Darwin's oldest boys, to drill with the High Elms Rifle Corps, a volunteer corps set up by the Lubbock family. He served as overseer to the Downe Friendly Society, which probably meant that he collected the subscriptions. Darwin paid for him to join the West Kent Liberal Association. His life was the mirror image of his master's, the other half of a Victorian gentleman's existence. He knew all the houses that the family visited, the names as familiar to him as they were to Darwin—High Elms, Hopedean, Leith Hill Place, Queen Anne Street, Abinger Hall, Kew, and Bassett. One special friend of his was John Lubbock's butler at High Elms. When Lubbock was elected to Parliament, Darwin gave Parslow the afternoon off so that he could join the High Elms staff in celebrating victory in the servants' hall. More often than he liked to remember, Parslow drove the horse and cart to Kew Gardens to fetch plants or took the train to deliver manuscripts to John Murray in Albermarle Street. "No man ever had a truer affection for the whole family than Parslow," said Francis.

Parslow retired from Darwin's service in 1875 with an annual pension of £50. He moved to the "Home cottage" and afterwards occasionally came up to the house for a day's duties when the family were short-staffed or required rooks to be shot. He was fond of the infant Bernard, who called him "Pa," and sometimes invited Bernard to visit the cottage on his own for tea. In return, when Bernard was given a Swiss doll, he christened it Parslow. "He is devoted to it, feeding it with titbits, and having it to look on at his bath," said Francis. In retirement, Parslow's attention turned to his vegetable garden: first prize for potatoes in the village show, runner beans two years after. He attended Darwin's funeral and outlived him by sixteen years. He died in 1898 and was buried in Downe churchyard.

William Jackson took Parslow's place, neither smart, efficient, nor tidy—a little man with red cheeks and wispy side whiskers, looking much like the groom he once was, a distinct character also much loved by Bernard as he grew up. Jackson considered himself fully one of the family. While waiting at table he would pay attention to what was being said and sometimes contribute his own remarks. Bernard once asked Emma what the play *Electra* was about and whether it was nice. Emma said it was very nice. "What is it about?" asked Bernard, still a very little boy. "About a woman who murdered her mother," said Emma, whereupon Jackson doubled up with laughter.[28]

Jackson liked having a small boy around and often used to make things for him, constructing a sentry box in the orchard and then some miniature walking sticks, labelled by Bernard according to the houses he used to visit with his father Francis. Bernard also watched with admiration as Jackson constructed a scale model of Down House out of cork and glue, his masterpiece, on which he was engaged for years. Jackson's treatment of the bow windows and the glass in the roof of the veranda struck Bernard as "the high water mark of human ingenuity."[29] When Bernard learned to ride on a donkey, aged three, it was Jackson, Jones, and Fred who plodded round the paths and Sandwalk with him. And it was Jackson, Jones, and a French nursemaid called Pauline who taught him cricket on the lawn—that is, until Harriet, "sonsy, pink and handsome," and with a loud laugh, took over, lobbing a tennis ball to Bernard along the upstairs corridor with a little red fire engine serving as the wicket.

According to these members of staff, Darwin was a considerate employer, inexhaustibly polite, but one who shied away from taking disciplinary action. Mrs. Evans the cook seemed incapable of organising things in the kitchen as efficiently as he privately desired, yet he could not bring himself to remonstrate. Certain areas of household expenditure were slack. Darwin resented the annoying discovery that a miller had cheated him for years in supplying flour. One gardener sold a cow and pocketed the money without anybody knowing until Darwin did the end-of-year accounts. Francis, who recorded these details, said he "used to regret he didn't look after things like his father used to." In particular, Darwin avoided change in his domestic arrangements, disliking those occasions when he had to sack an employee—another instance of his innate antipathy to involving himself in potentially unpleasant circumstances. The servants that were satisfactory often stayed for a long time. Parslow remarked that it was a contented household.

> He was a very social, nice sort of gentleman, very joking and jolly indeed, a good husband and a good father and a most excellent master. Even his footmen would stay with him as long as five years. They would rather stay with him than take a higher salary somewhere else. The cook came there while young and stayed till his death—nearly thirty years.[30]

Although these men and women shared Darwin's life so fully, his work remained a mystery to them. John Lubbock once asked Lettington, the gardener, about Darwin's health.

> "Oh!" he said, "my poor master has been very sadly. I often wish he had something to do. He moons about in the garden, and I have seen him stand doing nothing before a flower for ten minutes at a time. If he only had something to do I really believe he would be better.[31]

Bernard hit this note of bemusement neatly when he recalled a new nurse's remark. "It was a pity that Mr. Darwin had not something to do like Mr Thackeray; she had seen him watching an ant-heap for a whole hour."[32]

## IV

By now Darwin was far more prosperous than he had ever expected to be. "He used often to say that what he was really proud of was the money he had saved," recorded Francis. "He also took much pleasure in the money he made by his books."[33] The income from Darwin's books was now relatively high, every new edition and translation bringing a cheque of one or two hundred pounds from Murray. When Darwin totted this up in 1881, he noted that the "total receipts from earnings on books" during his lifetime came to £10,248 (close to half a million in modern terms). This sum was large for science, although perhaps not for creative literature as a whole. Anthony Trollope reckoned that his novels, down to 1879, brought him some £70,000.[34]

From about 1878 or so, Darwin took to dividing up his annual surplus between the children, a plan borrowed from a local friend, George Norman. He did not make an equal division. The figures in 1881 were £474 for each son, £316 for either daughter. This reflected his alternative arrangements for the girls, based on trust funds that provided them with an independent income. After Henrietta's marriage he continued paying her a personal allowance of £100 a year.

Darwin's main source of financial comfort was his investment income, based on some shrewd capital purchases over the years. He rarely took risks. Much of the capital in his hands was held in a number of different trust funds for Emma or the individual children, and he diligently reinvested interest payments back into each. The remainder was his own, inherited directly from his parents, a proportion of which was still in the form of loans and mortgages. Darwin also owned Beesby Farm, in Lincolnshire, land that he rented out and that returned a regular income. He and Erasmus, and their brother-in-law Charles Langton, who was noted in the family for his good eye for an investment, cautiously moved the remaining capital around. Darwin spoke with admiration of the way Langton doubled his first wife's fortune through canny investment. Bit by bit, the older generation drew William, the banker, into the process, making him a trustee for his brothers' and sisters' settlements, and executor for his parents, uncles, and aunts.

The last third of the century was a good time for investors prepared to put money into the nation. The joint-stock banking system was stable, the expanding empire filled government coffers, inflation presented no threat, and through its gilt-edged securities the treasury guaranteed a fixed return on loans and investment.[35] Darwin consequently bought into the Exchequer loan and consolidated stock (consols). Giving advice to his sister

Caroline in the last months of his life he strongly advised her to buy "ordinary Government 3 percent Stock," reinforcing his advice by adding, "My father used to say that everyone ought to hold some of this stock." He looked overseas, and towards import and export companies, at times holding shares in Pennsylvania Oil, the London Docks, Massachusetts Railway, and the East and West India Docks. "What do you think about Southampton Docks for your mother's £1000?" he asked William. "I think that the Leeds Corporation or the Leicester Corporation wd. be best, & next the Canada Bonds." He avoided gold and cotton, both unstable after the Civil War years in America.

Although lucky windfalls from railway companies in mid-century had helped him create an extensive railway portfolio, he had sold off most of these stocks by the 1870s in order to reinvest more safely elsewhere. In 1860, for example, he received dividends of £8,000 from the North Western and £7,800 from the Great Northern, part of which he promptly used to buy £5,000 of government consols as security for William. He put £10,150 of this windfall into the Lancaster and Carlisle Railway, and (by finding another £340) added £990 to his North Eastern Preference Berwick railway debentures. From time to time, he bought a stake in railway companies that were moving towards profitable mergers or were building a new line, such as the London and North Western, or the North Eastern, the latter a notoriously aggressive company that preyed on weaker businesses. Here his views on the struggle for existence among animals and plants interlocked exactly with political economy—the united biological and social world that Spencer thought ran according to the "law of railway morals."

Darwin found room for some remunerative local activities as well. He and Lubbock were principal share-holders in the Mid-Kent Railway Company and the South Eastern Railway, which during the 1860s and 1870s opened up a tangle of short-lived railway lines in the area around Beckenham and Bromley where they both lived. In 1855, Darwin and Lubbock had backed a line called the Mid Kent and North Kent Junction, running for less than five miles from Lewisham to Beckenham, and for a period Beckenham was the nearest station to Darwin's house. Then he and Lubbock financed a side line from Bromley to St. Mary Cray in 1858, which was sold at a profit to the London, Chatham and Dover Company. Darwin and Lubbock supported the South Eastern Railway, which gradually absorbed other lines, and aided the development of Farnborough and Orpington stations. Later they interested themselves in the new stations in London at Blackfriars Bridge and Ludgate Hill.[36]

Dividends could be handsome. In 1863, Darwin had calculated that he was worth £122,131.[37] In 1881, the year before his death, William estimated Darwin's capital as £282,000 (nearly £13 million in modern terms). This included the residue of his brother Erasmus's estate. "Did you ever expect to be worth over ¼ of a million?" inquired his son. Further to this, Emma possessed £24,000 invested independently on her behalf.[38]

On the other side of the coin, his household expenses were modest—even frugal. Darwin's account books show an annual expenditure of about three to four thousand pounds, a relatively small sum for an establishment of some twenty people living on an estate of twenty acres. These account books were his private works of art, on which he lavished an inordinate amount of time and attention. Everything was listed, everything was totalled and tabulated. At a glance he could observe whether one year was more expensive than another in coal, beer, stables, books, medicines, holidays, or charitable donations. Emma's accounts were kept separately, but he totalled them up just as efficiently. At the bottom of each column was a place for recording errors, the minute adjustments that to him made all the difference. Excluding the profit from his books, his annual personal income sometimes reached £6,000 or £7,000. Each year he invested, out of savings, £2,000, £3,000, or £4,000. He did not have many outgoings beyond household expenses. Income tax and other obligatory payments such as fire insurance and the Poor Rate were low in the nineteenth century, for him no more than £111 total in 1873. He paid pensions to the longest-serving servants, like Parslow, and was generous to former governesses, who, like the old ladies in Elizabeth Gaskell's *Cranford,* were usually reduced to watchful poverty, including Brodie, the children's first nurse, to whom Emma wrote regularly until her death in 1873. Darwin paid £30 a year to the former governess Miss Pugh so that she could take an annual holiday from the asylum where she lived.

In all this he took personal satisfaction. Understanding and regulating the domestic economy of his house was as delicate a piece of work as any of his natural history projects, and the results were as intricate and judiciously balanced as a patch of natural woodland. Darwin was not an outright Victorian patriarch, financially dominant, top of the tree. Instead, he brought with him a perceptive, sympathetic understanding of a man's place at the head of the household. His objectives were to protect, to provide, to be private and attached to his home. All the same, he expected to oversee the budgets of his wife and children, to be consulted, and if necessary to give his permission.[39] "I think they were a model couple in the way in which the machinery of their joint life worked," said Francis in full sentimental flow.

> My mother was always consulted about everything, but the ultimate decision of anything important really rested with my father; & my mother always seemed perfectly happy that it should be so. On the other hand he was glad to be managed in all little things. He often said in fun that the woman was the real master in a house. He considered her a sort of conscience in small things & matters of small etiquette.[40]

His thrifty nature was a family joke. The children warned each other about his air of worried abstraction when the annual accounting came along. Every September they waited for forecasts of poverty. The "workhouse season," they called it. "Thank God, you'll have bread & cheese," he once said to the boys, and Francis remembered being small enough to take him literally.

The steady hum of these financial cogs was unexpectedly altered in 1878 when Anthony Rich, an appropriately named solicitor, informed Darwin that he wished to leave him his fortune as a token of his admiration.[41] Mr. Rich was apparently sincere in his desire to honour Darwin, in the same way as a devoted parishioner might bequeath assets to a local church, or a graduate to his college. Thoroughly taken by surprise, Darwin tried to persuade Mr. Rich otherwise. He was not an elderly institution in need of an endowment fund. Moreover, it occurred to him that the letter might be a practical joke or a fraud, or that he would somehow be drawn into public embarrassment. He was wary of being bound by the ties of generosity. So Huxley jumped on a train on Darwin's behalf to take a look at Rich in his home town of Worthing, and returned to say that he was a wholly respectable man, interested in science, liberal, decent, and childless. He was a Cambridge graduate who occasionally published little-read works on classical antiquities. In turn, the solicitor was so impressed by Huxley that he put him into his will as well, leaving him in 1881 the house that had excited Huxley's admiration, informally valued by Darwin as worth £3,000.[42]

When Darwin and Rich subsequently met, one intent on giving, the other on refusing, Darwin discovered that Rich would not change his mind. In the end, he accepted the money on the grounds (as Rich pointed out) that he had five sons to provide for. By regarding the sum as a legacy for his boys, Darwin managed to justify the business to himself. Thereafter he scrupulously performed the duties into which this unusual situation pushed him, corresponding with Rich, courteously inquiring about his welfare, drawing him into his researches, and keeping him up to date with the boys' progress, especially as it related to scientific developments. Rich replied at length, taking an informed interest in evolutionary matters. Little else is known about him or his bequest.

The rest of the family regarded it as a wonderfully unexpected bonus. Emma could hardly believe that her frugal husband had agreed. "We shall be *disgustingly* rich & F. is thinking of reckless extravagance in the matter of Chutnee & Bananas, & Bessy & I have been running riot in bulbs & flowers. Leo is scheming a new tennis court."[43] If Rich had hoped for public recognition, however, it was not forthcoming. Darwin regarded the bequest entirely as a private affair, as did the sons who were the beneficiaries. It was hardly mentioned in their recollections of Darwin's life. Darwin never saw the money himself. Rich died in 1891, well after he did. The boys apparently inherited from this bequest several thousand pounds apiece.

## V

Having Francis at home made all the difference to Darwin's botanical researches. From 1876 or so, Francis carried out a wide range of microscopical inquiries, looking into the agglutination ("clumping") that his father claimed to see in *Drosera* cells as well as pursuing observations on protoplasmic streaming. The two worked closely together on the movements of plants and on insectivorous species.

In the summer of 1878, and again in 1879 and 1881, Francis went to Würzburg, near Frankfurt in Germany, for a few months to study with the experimental botanist Julius Sachs and learn the craftmanship that then made Sachs's laboratory the leading centre for plant physiology in the world. To spend time in Sachs's laboratory was a key requirement for ambitious young botanists. In fact, the dissemination of Sachs's techniques to Cambridge marked the beginnings of plant physiology in England, and the story was repeated in every European capital where botany was intensively studied. Francis went intending to investigate the growth patterns of roots and shoots. He came back with his botanical horizons extended in every direction.

In particular, Francis's imagination was fired by Sachs's lavishly equipped laboratory—an enviable procession of the latest instrumental designs, a pageant of rotating drums, dials, electrical terminals, smoked cylinders, microtomes for cutting wafer-thin slices, high-resolution microscopes, and ingenious devices for amplifying and measuring plant growth and movement. Several of these instruments were adapted by Sachs from animal physiology researches and could be encountered only in his laboratory. Francis was deeply impressed. He felt that Sachs was more inventive than even Darwin, especially in the manner that he turned plants into active participants in experiments by attaching the living organism to whatever form of machinery was required and letting it inscribe, or self-record, its own movements on a chart. In this way each plant wrote its personal autograph.[44] This focus on instruments—a fundamental shift in late-nineteenth-century science—transformed the living organism, with all its individual foibles and attributes, into a standardised object of research.

Thus inspired, Francis injected a new sense of purpose into Darwin's researches. He wanted to have instruments like Sachs's, urged his father to use rigorous experimental controls, and initiated wider-ranging investigations into cellular structure and chemical activity than Darwin would have been inclined to pursue on his own. It seems likely that the expensive Hartaack microscope listed in Darwin's effects at his death was chosen and brought back to Down House by Francis. In 1881, Darwin asked Haeckel's advice about a Zeiss instrument, again probably for Francis. In short, Francis became his father's experimental eyes and hands, bringing him fresh ways of thinking about plant functions. He propelled Darwin into new areas, most notably the means by which plants perceive stimuli, a field that Sachs also investigated.

One area of research was to establish that every growing shoot described tiny circles or spirals in the air. Darwin thought these "circumnutations" formed the basis of all other plant movements, with each additional twist or turn being an adaptation to the plant's special circumstances of life, the handwriting that distinguished one moving species from another. The evolutionary aspect excited them both. Darwin and Francis made extensive use of grass seedlings (grass coleoptiles) in this research, a choice of experimental organism copied from Sachs, and as significant in its own way as the use of flowering peas to Mendel or fruit flies to Thomas Hunt Morgan. Day by day, they measured trays of seedlings stretching towards the window. These measurements confirmed, in a manner acceptable to science, that young plants always grow towards the light and could correct their orientation in a matter of hours if needed.

They also inquired into what it was that helped plants recognise the vertical, and the rapidity with which they sensed and responded to stimuli, such as touch. A passion-flower tendril could move within twenty-five seconds of stimulation. It was this phenomenon, more than any other, that made them doubt Sachs's opinion that movement originated in the unequal growth on either side of the tendril. Good-humouredly, Darwin regarded shoots and tendrils almost as if they were earthworms, with tiny brains in their tips.

It is impossible not to be struck with the resemblance between . . . the movements of plants and many of the actions performed unconsciously by the lower animals.[45]

Another area of research was to identify what it was that made roots move downwards. In Darwin's day it was assumed that gravity acted mechanically on the entire root (radicle), pulling it downwards by its own weight. Certainly Sachs said so in his authoritative textbook *Lehrbuch der Botanik* (1868, English translation 1875), and this phenomenon was part of what Francis had studied in Würzburg. If the root was placed horizontally on the ground, the cells on the lower side were thought to become turgid and heavier, pulling the root down into the earth. Darwin discounted this mechanistic explanation and believed he could show that the tip of the root was the active agent. When the tip was cut off, roots lost their sensitivity and remained horizontal. Further experiments led him to suggest that the tip transmitted an unknown internal "stimulus" to the rest of the root, which then responded. Now known as tropisms, these stimuli were almost beyond the bounds of the nineteenth-century botanical imagination. For one thing, plants were not thought to be actively responsive agents like animals, and for another, Darwin's proposals were made some years before the concept of animal and then plant hormones was fully articulated. Interpreting plant movements as responses to stimuli was unusual. Darwin probably arrived at this position by thinking of roots as somewhat like simple animals acting according to their unconscious instincts. Sachs "will swear and curse when he finds out he has missed sensitiveness of apex," he told Francis with satisfaction.

Yet there could be no denying that Francis felt cramped by the lack of laboratory facilities at Downe. Resourcefully, he used lamp-black to darken thin pieces of glass and paper in the fashion he had learned at Würzburg. Using a humdrum collection of thread and blobs of wax, he attached these black papers to root tips to record their movements. It was a poor imitation of Sachs's technique, and the blackened papers that remain in the archive show only faint and incomprehensible traces of the plant roots that once wriggled across them. Darwin was grateful, no doubt about that. But he spoiled the professional effect by adding glass beads as weights or needles from Emma's sewing box.[46]

Few of these *ad hoc* experiments could be easily repeated. Darwin was working in the old-fashioned style of a gentleman-amateur, a tradition that had served him well in the past and was still common enough in the last third of the nineteenth century, even in the case of physics; well-financed individuals might experiment with air pumps when at home in the country or keep enormous electrical batteries in their cellars for scientific investigations. Many of these independent experimenters regarded their activities as an important feature of intellectual integrity. To be paid for scientific work still suggested the possibility of being indebted to an employer. At the very least the scientific data emerging from paid work was not the personal property of the scientist. Nor was a project necessarily of his own choosing.[47] If a man could afford it, he would tend to pursue independent research and therefore "own" his own results.

But Darwin stood on a cusp. Cultural changes within science during the 1870s were shifting the emphasis on to endeavours performed in specialised, usually metropolitan settings, where the findings could be witnessed by others, be repeated time after time, and emerge as validated facts.[48] In actuality, repetition was honoured mainly in the breach, since experimental procedures involved a degree of subjectivity and more incomplete data than most practitioners liked to admit. Nevertheless, these changes created groups of individuals who shared the same criteria of judgement about experimental results, and soon developed into distinctive subcultures with their own rules, technical practices, and languages. The spaces, or places, in which properly authenticated science could be done were therefore shifting. Darwin's study and greenhouse looked increasingly amateurish, increasingly inappropriate. Huxley's innovative programme of laboratory training in the Science Schools at South Kensington had headed the charge, in which a cohort of late Victorian naturalists learned high standards of laboratory technique, including Edwin Ray Lankester, William Rutherford, William Thiselton Dyer, Frank Bower, and H. G. Wells. Wells regarded his time in Huxley's top-floor domain as the most educational year in his life.[49] Further afield, Foster pushed ahead in London and Cambridge, and Hooker and Thiselton Dyer busied themselves in creating a physiological laboratory at Kew, funded by T. J. Phillips-Jodrell and opened in 1876. Darwin wrote to Hooker recommending the purchase of "good instruments" for the advancement of science. He said that if the Jodrell Laboratory owned a centrifuge or heliostat, he would want to use it.

Obstinately, Darwin refused to buy any of the expensive equipment for which Francis yearned. Afterwards Francis said, "I have always felt it to be a curious fact, that he who has altered the face of Biological Science, and is in this respect the chief of the moderns, should have written and worked in so essentially a non-modern spirit and manner." At Down House they made do with a miscellaneous collection of improvised gadgets and measuring devices. None of them would have passed muster on Sachs's laboratory benches. "If any one had looked at his tools, &c. lying on the table," admitted Francis, "he would have been struck by an air of simpleness, make-shift, and oddness."[50] Darwin's seven-foot wooden measuring rule was calibrated by the village carpenter. He measured very small objects with a piece of card marked up from an ancient book that turned out to be wrong. He used the battered old chemical balance from his Shrewsbury days, and he altered his Smith and Beck microscope to suit himself without hesitation, evidently treating instruments as nonchalantly as the scientific books he would tear apart for convenient reading. When he needed a black box, or a turntable, he would find something close to what he wanted and get it darkened with shoe-blacking, as John Lewis, the carpenter, recalled.

> One day Mr. Darwin told me to make him six mahogany boxes, three inch square inside, and one side perforated zinc. I brought them up, and he bid me put them on six wee Dutch clocks and take off the hands. Then they were fixed so that the boxes went round and round with the clockwork. And he sowed seeds in them. I never could make out what it was for, but one day he says to me, "I've gained my object." But I never knew what that was.[51]

Darwin trusted these rough and ready instruments implicitly, and Francis doubted if he even noticed that they were a probable source of error. His one virtue, said his son, was that he used his equipment consistently. Any mistakes were the same mistakes throughout.

Nevertheless, Francis felt pride in his father's ingenuity. This pride contributed in large degree to the admiring imagery that was to grow up around Darwin's simple approach to investigative work, for Francis made much of it in his edition of his father's *Life and Letters,* but it was further bolstered by a general undercurrent of admiration for the string-and-sealing wax approach to scientific research. The Russian botanist Kliment Timiryazev, who had done a great deal to translate and promote Darwin's views in Russia, on visiting Down House was impressed by seeing only "a series of pots with sundew turfs; each one of them was partitioned off into two halves by a tin plate: the leaves of one received meat, the leaves of the other were left without meat products."[52]

Alphonse de Candolle, who visited Darwin from Geneva in 1882, also admired these unsophisticated tricks of the trade. De Candolle plainly thought that the essence of Darwin's approach was to ask very direct questions, the simpler the better.

> He was not one of those who would construct a whole palace to lodge a laboratory. I sought out the greenhouse in which so many admirable experiments had been made on hybrids. It contained nothing but a vine.[53]

# VI

On his seventieth birthday, in February 1879, Darwin's attention turned again to the past. His grandfather came back to haunt him in ways he could never have imagined.

This started innocently enough with a congratulatory number of the journal *Kosmos,* a *Festschrift* to celebrate Darwin's seventieth birthday, published by Ernst Krause in Germany in February 1879. The special issue included an article written by Krause on the life of Dr. Erasmus Darwin, Darwin's paternal grandfather and an early proponent of evolution, titled "A Contribution to the History of Descent-Theory." Krause praised grandfather and grandson, although he drew too close an intellectual connection for Darwin's complete ease of mind.

Nevertheless Darwin was intrigued by Krause's article and decided to have it translated into English. He proposed adding his own "preliminary notice" of Erasmus Darwin to the translation. In part he was buoyed up by completing his autobiographical sketch, in which he had touched on the first Erasmus Darwin, and in part he thought that succeeding generations of Darwins would be as interested as he was in a fuller account of their ancestor. Moreover, an essay of his own provided an opportunity to correct inaccuracies arising from Krause's reliance on what Darwin called Anna Seward's overblown and "sarcastic" memoir, and the "calumnies published by Mrs. Schimmelpenninck." These female contemporaries of Dr. Erasmus Darwin had not told the tale in the manner that Darwin heard it from his own father. He resolved to go through the family papers and supplement Krause's essay. Before long, he wrote to Francis Galton for advice. Galton shared the same grandfather and retained his own family papers. Together the two evolutionists began to explore their own ancestry.

As it turned out, Krause's essay was too long for Murray to contemplate publishing in its entirety, especially if Darwin's own notice was added. Krause's part (translated by William Dallas) was consequently cut down mercilessly by Darwin. "F. keeps quite firm about employing Murray," Emma informed the sons early in 1879, "but he is quite willing that any of you shd undertake the ornamental part of the affair."

> He has got materials to make a pleasant preface—one a very capital letter of Dr D to Miss Howard before they were married—which shows him to be quite a human creature. Things always grow in his hands so that I expect it will be quite a small book & not a mere pamphlet.[54]

One of the sons was in his element. George Darwin had a fondness for family history that he usually suppressed, feeling that his brothers were prone to tease. This romantic urge would take flight with any theme that went back into the mists of time. He liked playing about with ancient dialects or the pedigrees of words, "adored Roman roads" and Walter Scott's novels, and, in time, took up archery and real tennis (the original indoor form of the game), half imagining himself "a character in medieval history."[55] That summer George was inflamed anew with heraldic inscriptions and genealogical charts. Pedigree talk dominated. "All your astronomical work is a mere insignificant joke compared with your Darwin discoveries," chuckled his father in June. "Oh good Lord that we should be descended from a Steward of the Peverel; but what in the name of heaven does this mean?"[56] Leonard's humour was broader. Making play with the lexicographical conventions of genealogists, he quipped: "[William?] Darwin had a son whom he named after himself, leaving out the bracket and the ?"

George commissioned a formal pedigree to be made by an American acquaintance, a Colonel Chester who had an enthusiasm for such researches. The chart went back two hundred years.[57] Darwin roused himself to a similar pitch, poring over the old family papers bequeathed by his father. He was amused by the ancient wills that left only one shilling to descendants.

Henrietta and Leonard pitched in too. Relaxed about the content and still in autobiographical mode from writing his own life story, Darwin dotted his text with personal comments and memories. "Henrietta. Is this too egoistic to include?" he jotted at the top of one page.[58] As a result, Henrietta went through his text with a red pencil, excising passages containing mentions of living individuals who could be identified (all the Galton sisters, for instance), any overly personal judgements, and family finances, including a sentence recording Erasmus Darwin's avowed desire to acquire wealth. A mature woman with opinions of her own, she thought it was improper that her father should discuss money, always the most interesting yet most inviolable barrier in polite Victorian society. Moreover, she also deleted a footnote in which Darwin commented approvingly on Erasmus Darwin's religious scepticism. She had coped well enough with editing her father's ideas about sexual relations in biology; now she censored the family story. Darwin accepted her deletion. Leonard added his own suggestions. "Humbug," he wrote beside Darwin's long, self-absorbed passage about the qualities of a man of science.

Still, he had enjoyed himself. The writing of this, the only biographical study that he ever attempted, was undertaken entirely for personal reasons. Even so, he found that it clarified some of his views about the role of speculation and originality in science. In thinking about Erasmus Darwin he necessarily pondered his own contribution to biology, wondering what made a man a discoverer; and with Galton's opinions on heredity fresh in his mind he reflected on the traits that linked the men of the Darwin family line together as thinkers and innovators. There was much in his grandfather's personality that was mirrored in his own, and in places, Darwin seemed almost unsure about whom he was writing. He could have been speaking of himself when he praised Erasmus Darwin's "indifference to fame." More ambiguously, he did not know whether to criticise or admire his grandfather's inexhaustible zest for speculation—presumably the flash of self-recognition was strong. In the end he criticised Erasmus Darwin's "overpowering tendency to theorise and generalise," saying that the "vividness of his imagination seems to have been one of his preeminent characteristics," while adding the backhanded tribute that "his remarks . . . on the value of experiments and the use of hypotheses show that he had the true spirit of the philosopher." Darwin's pronounced sense of his own individuality made him declare that "unfettered speculation" was worthless unless it was kept closely in check by observation.[59] These internal tensions came to a head in speaking of Erasmus Darwin's most philosophical treatise, the *Zoonomia,* the book in which Dr. Darwin had proposed a system of evolution, and on which Charles Darwin had drawn extensively during the early days of his own riotous speculations on evolution. Darwin dismissively wrote, "I fear that his speculations on this subject cannot be held to have much value." For the last, decisive time, he broke away from his grandfather and from his past.

The biography was published in November 1879. Darwin blundered thoughtlessly when he came to send the material to Murray for publication. Coincidentally, in May 1879, Samuel Butler had published an essay on the history of evolutionary theory which included a comparison between old Erasmus Darwin and his grandson Charles, called *Evolution Old and New, or the Theories of Buffon, Dr. Erasmus Darwin and Lamarck Compared with That of Mr. Charles Darwin.* Butler carried a noteworthy name himself, being the grandson of the Dr. Butler who had been Darwin's headmaster at Shrewsbury School and a descendant of the seventeenth-century Samuel Butler who wrote *Hudibras.* In 1872 this Butler had published the novel *Erewhon,* a vigorous satire on Victorian life, drawing on evolutionary science for a literary set-piece in which a utopian race of human beings abolishes all machinery because it begins to evolve mental ability. Later on, Butler's *The Way of All Flesh* (published posthumously) was an unrelentingly bitter saga of the degradation of a family line. In his nonfiction he was one of the first seriously to compare different evolutionary theories.

Butler regarded natural selection with refreshing dislike. Although he endorsed evolution as a general idea, and as a youth had admired the *Origin of Species,* he had come to feel increasing distaste for Charles Darwin's principles and thought there was much to commend in earlier schemes that relied on the internal drive of organisms to power evolutionary change, such as Lamarck's or Erasmus Darwin's proposals. His book *Evolution Old and New* was aimed partly at reinstating these older schemes. Like Wallace and Spencer he hoped to reshape evolutionary theory and use it as a means of describing the progress of the human mind. But unlike them, he kicked against the mechanism of natural selection, emerging as a leading figure in the anti-Darwinian backlash that was to take shape in the 1880s and 1890s.[60]

Unsurprisingly, Darwin did not think much of Butler's book. He sent it to Krause, along with a note indicating that he thought it had little merit, "without anything new having been added." Krause inserted a critical remark about it into his shortened essay on Dr. Darwin. More than this, Krause apparently copied a few useful phrases to help his argument along, and paraphrased another. When *Erasmus Darwin* was published in English by Murray in November 1879, Darwin omitted to mention that the original German essay had been shortened, changed, and updated. In reality, Darwin had begun to say some of this in his draft for the preface, but Krause said he did not wish to advertise the fact that Darwin had abridged his work. In response Darwin deleted the remarks from the proofs. Now, in final printed form, Darwin's preface made it sound as if the book version of the article had been published before the Butler volume and the criticisms of Butler had been present in Krause's original essay. Neither was true. Butler was outraged, accused Darwin of publishing a falsehood, and in the letter pages of the *Athenaeum* demanded a public apology.[61] As George Darwin afterwards put it, Butler claimed that "this book was fraudulently antedated & was intended as a covert attack on him."[62] A few weeks later, Butler accused Darwin of plagiarism as well.

As in his disagreement with Mivart, Darwin was simultaneously angry at being wrong-footed and unsure how to proceed. He never expected a little book about his grandfather to inspire such venom. Nor was he accustomed to meeting someone who was prepared to fight back. Nevertheless, his first impulse was to explain his position. He drafted a letter for the *Athenaeum* and sent it up to London for Henrietta's and Richard Litchfield's advice, while Emma notified the rest of the family.

> F. got so anxious to get his answer to Butler's attack off his mind that he sent John up with it to R & Hen for their approval. They sent it down again with 2 very sensible letters from R & Hen warmly dissuading him from taking any notice. Their arguments were so good that I (& F) quite changed our opinions. . . . F has ended up referring the matter to Huxley, & he will be glad if he decides against publishing an answer. Meanwhile Dallas [the translator] wants to answer the attack & I have no doubt Krause will also, so Butler will be in the atmosphere of hot water which will be his delight.[63]

Friends and relatives offered conflicting advice. All the sons were involved, as were Richard Litchfield, Huxley, and Leslie Stephen, the last by virtue of his experience as a man of letters, who recommended Darwin to "take no notice of Butler whatever." Erasmus called it a "boggle," reporting to Julia Wedgwood that Darwin's riposte was "suppressed by the Litches" and that he hoped her new book would be more successful than "our attempt to glorify the Darwins." Letters hummed across Bloomsbury, Kent, Cambridge, and down to Southampton.

> Today came Leslie Stephens' award [verdict] which will be sent to you in due course. It is decidedly to do nothing; which I think will please some of you very much & satisfy everyone but steady old Leo. F. is quite satisfied. Huxley alluded to the matter in rather a poor joke as "the Butler spilling some more dirty water out of his pantry."

In actuality, silence was the worst course to follow, as Francis Darwin subsequently realised.[64] Darwin was in the wrong and should have apologised. Butler convinced himself he was the victim of a conspiracy, virulently attacking again and again. Both men took irreparable offence.

Like the Mivart affair, too, this quarrel involved a combination of personal, intellectual, and public concerns. The two men fenced with different weapons. Butler made direct use of the power of the press in order to make his accusations public. Darwin activated his subterranean network of private influence. And even if the quarrel had been promptly resolved, the underlying message was hard for Darwin to accept: he could not always expect his own way; a handful of younger evolutionists were coming to regard the theory of natural selection as inadequate; Darwinism was perceived, at least by some, as a growing orthodoxy that needed to be reevaluated. Darwin himself, moreover, might increasingly be considered the triumphant master of a clique, too grandiose to contemplate alternative opinions, too elevated to apologise, too quick to use his influence to trample on rivals without a second thought. Like Mivart, Butler at heart objected to the rising sycophantic tide of acquiescence.

The reactions of Darwin's confidants were also entirely predictable. By now all Darwin had to do was to wind up Huxley and watch him jump into action. In April 1880, Huxley deliberately knocked Butler on the head, although only in passing, in a rousing speech at the Royal Institution called "On the Coming of Age of the *Origin of Species*," an allusion to the twenty-first anniversary of the book's publication. In this title Huxley intended to convey the dual meaning of both chronological and intellectual maturity. His talk was a statement that the *Origin*'s place in science was assured.

Darwin appreciated Huxley's gesture. But he was not satisfied. Frustrated by his family's advice to remain silent, he secretly encouraged Frank Balfour, the embryologist, to translate Krause's letter of explanation and send it along to *Nature*. This was published in January 1881. Nor did he object when George Romanes went too far with a cutting review of Butler's next book, *Unconscious Memory.* Taking a personal interest, Anthony Rich joined in.

> Who the Devil's Mr. Butler? When he can say, as you can say, that he has opened out to the knowledge of mankind a new field of Science, and has cultivated it through a life time with marvellous skill and industry, till all the most able men of every country have come to acknowledge the work as one of the greatest and most important discoveries of the age, while the worker is regarded with respect and honour by all who know him—well, when he can say that it will be time enough for you to pay attention to anything he may say. P.S. My lawn is the Paradise of earth-worms.[65]

The controversy had brought deeper currents of change to the surface. The body of thought generally labelled Darwinism was entering another phase of existence, easing out of the grip of its first protagonists. Darwin and Butler each took his resentment of the other to the grave.[66]

## VII

Nor did Darwin immediately get his way in another battle of wills. Horace Darwin, his youngest son, wanted to marry Ida Farrer, Thomas Farrer's daughter by his first wife, Fanny. The two had been courting for some time. Unexpectedly, Farrer forbade the marriage, saying that Horace was an invalid and held so few prospects of professional advancement that Ida could not possibly lead the kind of married life that he, as her father, desired for her.

Darwin was dismayed by Farrer's negative response. He tried hard not to be offended, for it was insulting to realise that although he might be an intimate friend, and admired for his intellect, he was deemed unsuitable to join the family as a father-in-law. Letters between him and Farrer became tense as Darwin stated Horace's case. Tetchily, he inquired of his solicitor, the pragmatic William Hacon, how much he could afford to settle on his youngest son. Hacon mentioned a sum that he thought might induce Farrer to look more kindly on the marriage. Eventually Darwin asked Farrer for a personal interview, at which he explained Horace's situation. Emma put the plan of campaign succinctly in a letter to Henrietta.

> . . . especially his turn for mechanical invention, which is his profession tho' not a profitable one; also Dr. C's opinion that he was so likely to get well as life goes on, & that it was suppressed gout. Also how well off he wd. be, which is a matter of some consequence when you are not likely to make money.

Farrer relented in the autumn of 1879, and Horace and Ida married early in 1880. The couple went to live in Cambridge.

Eager for a diversion, the remaining family set out in August 1879 for a month's holiday in the Lake District, travelling by train with Henrietta, Francis, and two-year-old Bernard in a private railway carriage hired for the occasion. Darwin complained about the extravagance but Emma said "we shd. have been quite killed but for the saloon carriage." Admitting defeat, he jettisoned all pretence at invalidism and enjoyed the journey with "the freshness of a boy," greatly savouring the vista of Morecambe Bay as the train snaked around the edge, one of the most dramatic combinations of railway engineering and coastal scenery in the United Kingdom.[67] They stayed in a hotel in Coniston owned by Victor Marshall, the cousin of their architect at Down House, with views over the lake and gardens. Everyone remembered this holiday with pleasure. Darwin was more relaxed than the family had known him for a long time. On a day's excursion to Grasmere, Henrietta said, "I shall never forget my father's enthusiastic delight, jumping up from his seat in the carriage to see better at every striking moment." Darwin recognised the unusual situation himself. "The scenery gave me more pleasure than I thought my soul, or whatever remains of it, was capable of feeling," he told Romanes afterwards. Jovially, he wrote to Marshall to express his thanks, and volunteered to send him a young oak tree from Down House grown from an unusual specimen in his father's garden at Shrewsbury. He agreed to return to the Lake District to Ullswater for another visit in 1881.

On that first trip, Darwin made the acquaintance of John Ruskin, the critic and author, who lived in seclusion at Brantwood on the shore of Lake Coniston. The two had much to talk about, not least the friends they held in common. Richard Litchfield knew Ruskin well from their time as teachers together at the Working Men's College. Darwin listened sympathetically while Ruskin confided that his mind was becoming "clouded." A few days later Darwin paid a return call during which Ruskin took him into a back room to show him his collection of J. M. Turner's watercolours. Politely, Darwin murmured some appreciative words. Afterwards, he confessed "he could make out absolutely nothing of what Mr. Ruskin saw in them."[68] Victor Marshall reported that Ruskin subsequently spoke about this meeting in high good humour, mentioning Darwin's "deep & tender interest about the brightly coloured hinder half of certain monkeys."[69]

Early the following year, Darwin's children arranged a surprise for him. They gave him a fur coat to keep out the winter cold, "a gift planned because we thought it in vain to expect him to be so extravagant as to give himself one."[70] Francis effected the delivery, leaving the coat surreptitiously in Darwin's study during his afternoon rest.

> I think the coat exploded very well. I left it on the study table furry side out and letter on the top at 3, so that he would find it at 4 when he started his walk. Jackson was 2nd conspirator, with a broad grin and the coat over his arm peeping thro' the green baize door while I saw the coast clear in the study. You will see from father's delightful letter to us how much pleased he was. He was quite affected and had tears in his eyes when he came out to see me, and said something like what dear good children you all are. I think it does very well being long and loose.[71]

Darwin said he would not be himself if he did not protest that "you have all been shamefully extravagant to spend so much money over your old father, however deeply you may have pleased him."

Darwin published *The Power of Movement in Plants* in November 1880, arranging for a German translation by Julius Carus to be released early in 1881. This was an unexpectedly controversial book. The views on root movements that he set out were energetically contested by Julius Sachs, who sneered at Darwin's suggestion that the tip of the root might be compared to the brain of a simple organism and declared that Darwin's home-based experimental techniques were laughably defective. Furthermore, Sachs's botanical colleagues and co-workers were unable to replicate Darwin's results—the customary criterion of a valid experiment. The message from Würzburg was that Darwin must have inadvertently maimed the living roots in some way, and that Darwin's practical skills were weak. This hit Darwin where it hurt most. In retrospect, Sachs's criticism undoubtedly rested on the growing gap between laboratory and country-house styles of investigation, underpinned by a hefty dose of institutional pride and nationalist feeling. For Sachs, skill lay not just in the experimenters' hands. Skill could be present only in the proper place, that is, in the laboratory—his laboratory.[72] Had Darwin but known it, he was ultimately more in the right than Sachs, for his results prefigured the line of research that was to lead to the identification of plant "hormones," and the "stimulus" that he thought was transmitted from the root tip was ultimately isolated and named auxin by Firtiz von Went in 1928, himself following some of Darwin's experiments. And yet Sachs was justified in his attack on Darwin's informal and unstructured experimental procedures. The centre of attention for physiological research was shifting decisively, and permanently, into the laboratory.

Half-heartedly Darwin made a few defensive replies. But his mind was not in it. He did not return to the subject in any direct form, leaving Francis to pursue the argument in subsequent editions.

## VIII

More and more, he enjoyed the company of worms. He spent a good deal of time thinking about them during 1880 and 1881.

Darwin made a thorough survey of annelid behaviour and activities, potting up worms from his flower beds as if they were plants, and keeping them in his study for observations, once again transforming his house and garden into nature's observatory. He and Francis would creep downstairs at night to see what the worms were up to in their flowerpots, or stealthily move through the moonlight to spy on them under the lime trees or at work ejecting casts on the surface of the Sandwalk. In the daytime, if it was fine, Darwin dug for burrows. He discovered that worms lined their holes with leaves and that they often plugged them at the top as well. One morning he removed 227 leaves out of a series of burrows, seventy of which came from the row of lime trees near the house. If the weather turned rainy, so much the better for his researches. He tracked the worms' trails across the damp paths, wondering how far their ambitions had taken them. Unfailingly curious, he pursued what he called "fool's experiments" by asking Francis to play his bassoon close to worms in pots to see if they detected sound. He blew a whistle, breathed tobacco fumes, and waved a red-hot poker over them. Finally, with an embarrassed laugh, he put them (in their pots) on top of the Broadwood piano and requested Emma to play the keys loudly. He "has taken to training earthworms," she said resignedly, "but does not make much progress, as they can neither see nor hear. They are, however, amusing and spend hours in seizing hold of the edge of a cabbage leaf and trying to pull it into their holes. They give such tugs they shake the whole leaf."

Darwin was convinced that worms possessed a modicum of intelligence. This was displayed most evidently in their feeding patterns. He fed his potted specimens as assiduously as he had once fed insectivorous plants, supplying them with leaves from many different kinds of tree, shrub, and vegetable, some of which came from tropical species grown only at Kew and would never have been previously encountered by Kentish worms. They liked cabbage best of all. He went on to observe their dexterity in pulling unevenly shaped leaves into their burrows. Each worm was capable of finding either the pointed end of a leaf or the stalk, or could locate the mid-point of a long thin leaf and pull it so that it folded into two halves. Standing under the Scotch firs for several nights in a row, accompanied by the dim light of a hurricane lamp, he and Francis ascertained that worms enterprisingly coped even with pine needles. Darwin surmised that this was a purely mechanical adaptation for pulling food into the burrow. To test his hypothesis he fed worms in pots with tiny paper triangles or diamonds, presenting either the sharp or flat edge to each questing worm in the evenings, no doubt wondering if he was quite right in the head as he carefully cut out shapes from Bernard's drawing paper. "I am becoming more doubtful about the intelligence of worms," he wrote to Galton about this. "The worst job is that they will do their work in such a slovenly manner when kept in pots, and I am beyond measure perplexed to judge how far such observations are trustworthy."

Darwin probably initiated some practical measurements of worm activity in the ground as well, although the archival record is unclear. At some point, a flat, heavy stone, originally a surveyor's bench-stone, was placed on the lawn in the Down House garden, serving as a device for indicating the rate at which worms made it sink. This particular experiment was predominantly Horace's, and involved Horace in constructing a metal recording instrument to measure the descent of the stone, even though the idea was plainly drawn from Darwin's work. It remains uncertain how much of the plan Darwin laid out beforehand. Horace wrote an article about the sinking of the stone in 1900. Another stone was similarly placed in his sister Caroline's garden at Leith Hill.[73]

A gently melancholic air suffused this work. Darwin's life had slowed down to the pace of his subjects. He mused on the vast reaches of time, of days when Roman villas or prehistoric stones stood upright on the earth's surface, of things long gone, and eternity. He knew he was aging, and that the close of his life could not be very far away. Sometimes when Lettington turned over the compost heap for him and he saw countless wormholes in the dark earth, he momentarily glimpsed his grave. The thought did not worry him unduly. Echoing a remark of his grandfather's that he had included in his book on *Erasmus Darwin*, he said he had "no fear of death, after such a life," and he considered himself fulfilled. He contemplated the deaths of Emma's sister Elizabeth Wedgwood and her brother Josiah Wedgwood in 1880 without any obvious distress. Many friends remarked on the serenity of the last years of his life. Darwin became at the end what he had always been in his heart, almost part of nature himself, a man with time to lean on a spade and think, a gardener.

He found time to give financial support to needy causes. He was upset to hear that the Brazilian home of his friend Fritz Müller had been flooded, and that Müller's apparatus and books might have been destroyed. "I have long looked on him as the best observer in the world," he cried to Ernst Krause, and offered Müller money to replace his scientific losses. His ability to empathise with personal scientific disasters like these was always remarked by others. He said he could imagine his own distress should something similar happen to him.

His thoughts also rested on Anton Dohrn, the director of the Naples Zoological Research Station. Zoologists from Germany and Britain travelled every summer to the Stazione to rent laboratory space and investigate the invertebrate fauna of the Bay of Naples. Darwin was interested in the Stazione's progress, even though he must have sighed at Dohrn's incessant requests. Would Darwin donate money for the library or sit on the supervisory board? Could he supply favourable remarks for the report of the Stazione's activities or send a photograph of himself so that a local sculptor could create a plaster bust for the grand saloon? Here Darwin was witnessing Dohrn's single-handed construction of a research school—an intellectual and practical scientific tradition—that was to dominate zoology in coming decades. He seemed not to mind the way that Dohrn exploited him. In fact, Dohrn's determined transformation of Darwin into an iconic father figure was one of the simplest and most potent devices to ensure the coherence of a clearly defined interest group. The Stazione came to represent one of the first and most influential centres of embryological research, almost the only laboratory in the world at that time to research Darwinism in action.[74] Perhaps Darwin found it comforting after the Butler debacle. It was a curious form of Darwinism nonetheless. Dohrn proposed that ancestral vertebrates emerged not from ascidians as Darwin and others popularly supposed but from annelid worms whose digestive and nervous systems must have reversed places in development. "May I venture to caution you not to extend too far the degradation principle," suggested Darwin with disquiet.[75] All his experience of earthworms indicated that they were simple because they were primitive, not because they had degenerated from a more complex form.

Despite the differences, when Darwin was awarded a large sum of money as the Buffon Prize of the Linnean Society of Turin, he asked Dohrn if he might welcome some piece of equipment for the Stazione up to the value of £100 (Dohrn asked that the gift be used to start a travel bursary scheme). In similar vein, Darwin supported the Cambridge University Zoological Museum. He also sent money to help the author Grant Allen over a bad time, and contributed to William Boyd Dawkins's plan to explore limestone caverns in Yorkshire, via the Settle Cave Exploration Fund. Sometimes he let himself be gulled. When a German palaeontologist called Leopold Wurtenberger requested a loan, Darwin replied that he did not lend money and enclosed a cheque as a present.

He had time, too, to consider good works. In 1881, he approached Hooker with an idea that had been gestating for several years. He wanted to pay for the Kew botanists to publish a revised version of Ernst Gottlieb Steudel's *Nomenclator,* an index to the names given to plants. One edition had been published in 1840, and since then the list had been kept up at Kew by handwritten additions. This wish was heightened by his abiding respect for Hooker and the work done under his jurisdiction at Kew. In fact, he wanted to follow Lyell's example and bequeath sums to all his favourite sciences. Lyell had left money in his will to advance geological research, and Darwin had told Lyell before his death that he would do the same if only "he had fewer sons." Anthony Rich's bequest probably helped to loosen his purse. In any event, Darwin notified Benjamin Daydon Jackson, of the Linnean Society, that "it was his intention to devote a considerable sum of money annually . . . in aid or furtherance of some work or works of practical utility to biological science."

> On the occasion of my last visit to him [wrote John Judd, the geologist] . . . he dwelt in the most touching manner on the fact that he owed so much happiness and fame to the natural-history sciences which had been the solace of what might have been a painful existence;—and he begged me, if I knew of any research which could be aided by a grant of a few hundreds of pounds, to let him know, as it would be a delight to him to feel he was helping in promoting the progress of science. He informed me at the same time he was making the same suggestion to Sir Joseph Hooker and Professor Huxley with respect to Botany and Zoology respectively. I was much impressed by the earnestness, and indeed, deep emotion, with which he spoke of his indebtedness to Science, and his desire to promote its interests.[76]

Darwin's desire to expand Steudel's *Nomenclator* did not materialise in the form he anticipated. Backed by his money, the project turned into the *Index Kewensis,* a huge enterprise edited by Benjamin Daydon Jackson and the Kew herbarium staff, listing all known plant names with the written sources in which those names were first given. When it was finished, it included 375,000 entries.[77] It was not published until 1892–95. Darwin's sponsorship was recorded on the title page, and the proper title was *Nomenclator Botanicus Darwinius.*

The vigilant generosity extended to obscure workers as well. From 1878, Darwin supported James Torbitt, a wine merchant and grocer in Belfast, who was engaged in breeding experiments to produce an infection-proof race of potato. Ever since the Irish famine of 1845, Darwin had followed the continuing uncertainty over the causative agent of potato blight, remembering his own failed attempt with Henslow to grow uninfected stock from the *Beagle* specimens he had collected on the island of Chiloé. Darwin busied himself in Torbitt's project, writing to Farrer and other friends in high places to commend the investigation. In 1881 he sent the equivalent of a research grant to Torbitt. "I have the pleasure to enclose a cheque for £100," he told him. "If you receive a government grant I ought to be repaid."

Most of all he was disturbed to hear that Wallace was in desperate financial straits. Wallace was unsure where to turn next, with no permanent job and rapidly looming poverty. Alarmed by the news, Darwin mobilised all his influential contacts. He hoped to get a government pension awarded to Wallace despite the continuing confusion whether scientists were eligible to receive them.[78]

Of course, Darwin regretted what he deemed Wallace's lack of caution in scientific affairs. These regrets were freely shared by many of their scientific contemporaries. When Hooker had come back from the British Association meeting in Glasgow in 1876, he had snorted crossly to Darwin, "Wallace as Presiding Spiritualist made a black ending to a scientific meeting." And Wallace was always in some sort of financial trouble, usually of his own making. Whatever investment he made, the company's shares were bound to drop in value. Architects and builders cheated him. He rashly accepted a challenge to prove the roundness of the earth (and won), but went to court when the wager remained unpaid and libellous accounts began circulating against him. Wallace learned too late that justice does not come cheap. In 1880 he sold his house at Grays, Essex, and purchased a cottage in Godalming, hoping to live on the slender earnings from his publications. But for all that the two evolutionists disagreed over natural selection, spiritualism, mankind, sexual selection, animal colours, and plant distribution, Darwin wished "most heartily" for Wallace's comfort and happiness.

As with Butler and Farrer, not everything went according to plan. Hooker did not wish to support a scheme for Wallace's pension and needed to be persuaded otherwise. He believed Wallace had "lost caste" and would embarrass the scientific cause, although he was mollified by Wallace's pathbreaking study of evolutionary biogeography, *Island Life,* published with a handsome dedication to Hooker in 1880. This book became the foundation of high-level research into animal and plant distribution. Darwin moved in turn to Lubbock, to Huxley, and then to his old sparring partner on the evolution of mankind the Duke of Argyll. They all thought the request would have more power if made by Darwin alone. Nonetheless, a memorial describing Wallace's "lifelong scientific labour" drafted by Arabella Buckley and Darwin, adjusted by Huxley, and signed by Darwin and other eminent figures, including Argyll, arrived on Gladstone's desk in January 1881.[79]

"I hardly ever wished so much for anything in my life as for its success," Darwin told Arabella Buckley. He put a good deal of effort into writing letters for this ambitious political project, one that proved more demanding than many of his natural history researches. "It has been an awful grind—I mean so many letters." In the end, influenced on the one hand by the Duke of Argyll's approval and on the other by Darwin's personal stature and the public nature of his and Wallace's scientific achievement, Gladstone awarded Wallace a pension of £200 a year—a modest sum but manna from heaven for Wallace, "a very joyful surprise . . . very great relief from anxiety for the rest of my life."[80] To Darwin he wrote, "There is no one living to whose kindness in such a matter I could feel myself indebted with so much pleasure and satisfaction."[81]

Darwin called it a "splendid" outcome. Whatever mixture of feelings was motivating him, he recognised that it was both his pleasure and a duty to save Wallace. Their interlocking lives required it. Now, their dual story could be safely closed—he had done his best for his companion in the pages of history. Truly, as Wallace implied, no one but Darwin could have succeeded in initiating and gaining agreement for the award. The link between them was as asymmetric yet as honorable and as extraordinary as ever.

Two months later, Gladstone came back with a request of his own: would Darwin serve as a trustee of the British Museum? He refused. "I am much obliged for the honour which you have proposed to me, & this I should have gladly accepted, had my strength been sufficient for anything like regular attendance at the meetings of the Trustees. But as this is not the case, I think that it is right on my part to decline the honour." This exchange was probably the source of a later tale that Gladstone offered

Darwin a knighthood that was turned down. It seems that Darwin was never offered any such knighthood.

Otherwise, as Darwin pottered about in his garden, he easily forgot how famous he was. Occasionally people might glimpse a tall thin figure in a dark hat and cloak in one of the surrounding lanes. "An elderly man was walking slowly down," said Wallis Nash, a temporary neighbour.

> Seeing me he turned aside and stood as I moved along the road, with his back to me, studying the face of the chalk quarry in the hill, from which the road material of chalk and flints had been dug. The action was that of a shy and nervous man, and I looked curiously at him as I passed. I saw, in side view, a slender and somewhat bowed man, with a "drawn face," heavy white eyebrows and beard, under a soft black hat. He wore black clothes and a cape, with a grey plaid shawl wrapped round his shoulders. There was something familiar in the general outline, and I wondered if I had met that man. Suddenly recognition came to me. It was the pictures of the author of the "Origin of Species" I had in mind, the original of which I had passed, looking at the chalk quarry on the road to Down.[82]

Or visitors would be welcomed into a booklined study by a courteous old man. Edward Aveling, who brought Ludwig Buchner to Down House in September 1881, said their host was very retiring. More unlikely houseguests than these could scarcely be imagined. The two radical social philosophers were attending the Congress of the International Federation of Freethinkers in London. Buchner was widely reputed to be the fiercest materialist in Europe; Aveling was a proclaimed atheist. Only a few months before, Darwin had written to decline Aveling's request to dedicate *The Student's Darwin* to him, saying that the atheistic portions took his views "to a greater length than seems to me safe."[83]

The lunch party could hardly have been more incongruous either. For moral support the Darwins had also invited John Brodie Innes, their old vicar. Yet the occasion was congenial. After eating, the men retreated to Darwin's study, and there, "amidst the smoke of cigarettes, with his books looking down upon us, his plants for experiments hard by, we fell to talking."[84] Aveling urgently asked Darwin if he was an atheist. He preferred the word "agnostic," he replied. "Agnostic was but Atheist writ respectable," responded Aveling, "and Atheist was only Agnostic writ aggressive." The guests pressed Darwin to consider his role in spreading free thought—every freethinker should proclaim the truth "abroad from the housetops!" Towards the end, they cordially settled on the insufficiency of Christianity. "I never gave up Christianity until I was forty years of age," claimed Darwin. "It is not supported by evidence." Impressed by

Darwin's obvious sincerity, Aveling published an excitable description of this interview in 1883, after Darwin's death, calling it *The Religious Views of Charles Darwin*. His article upset the remaining members of the family.

Other visitors to Darwin's study were struck by the evident signs of his mind at work. Wallis Nash recalled that "the walls were covered to the ceiling with well-worn books, which overflowed into the passages and landings on the upper floor." Plant pots stood on the table and window-ledges, a fire burned in the grate, a woollen shawl hung on the high-backed leather chair. During the last few years of Darwin's life, several men and women came to venerate this room almost as if it were his mind itself—a theatre of memory and intelligence that could be revered either with or without his presence. The photograph that Leonard took of the old study just before Darwin vacated it for the new conveyed something of the same sense of an abstract mind at work. In the photograph Darwin was not even in the room.

As he continued to retreat into domestic privacy, Darwin found it all the more perplexing when he was recognised by strangers. A passing incident in February 1881 was a revelation to him. That month Darwin went to London to hear John Burdon Sanderson speak on insectivorous plants at the Royal Institution. Sanderson and he had a friendly word or two beforehand in a back room, and as Darwin entered the auditorium from the front, a burst of applause rippled around the audience. Darwin looked behind him to see who had come in. He took several minutes to realise that it was he who was being applauded.[85]

Nevertheless, he felt himself sinking. "What I shall do with my few remaining years of life I can hardly tell," he wrote to Wallace while visiting Ullswater. "I have everything to make me happy and contented, but life has become very wearisome to me."[86]

In August 1881, he was invited to be a public figure again, this time to make a guest appearance at the International Medical Congress to be held in London. There was more than just celebrity involved here, although that was probably the primary reason that he was asked. Darwin had recently endorsed vivisection again, in a letter to the *Times* in which he put the case for "the incalculable benefits which will hereafter be derived from physiology," a major strike for science. This evoked a spate of rejoinders from Frances Power Cobbe, Richard Hutton, Lord Shaftesbury, and George Jesse, as well as a flurry of pamphlets, reprints, and even a leader in the *Times* itself.[87] Suddenly Darwin looked like a highly appropriate figure to parade before the medical luminaries due to attend the congress.

This congress was by far the biggest medical convention ever held, drawing more than three thousand participants, including Lord Lister, Rudolf Virchow, Louis Pasteur, Robert Koch, and Jean-Martin Charcot. Darwin could not but be involved. Sir James Paget and Sir William Gull asked him to the opening reception to meet the Prince of Wales and the "chieftains of science." Paget invited him to lunch and Gull asked him to dine, both of them evidently with a view to complimenting their guests with a glimpse of England's own star turn. Erasmus remarked that his brother was "clearly becoming a fashionable as well as scientific swell." Indeed, at the reception he seems to have upstaged most of the other dignitaries by shaking hands like royalty with scores of foreign visitors, many of whom he already knew through correspondence, finding a word or two to say to everyone. Huxley was amused by the fuss. He too asked to be introduced to Darwin, bowing horizontally over his friend's hand with a flourish. The chaff between them could not disguise Darwin's international renown. For the meal he was seated at the top table.

> There was an immense crowd of all the greatest scientific swells and much delay and I was half dead before luncheon began. I sat down opposite the Prince and between Virchow and Donders who both spoke bad English incessantly and this completed the killing.[88]

Many of the speakers went on to defend vivisection in their addresses to the congress, especially Virchow, and the delegates adopted a resolution supporting the use of vivisection with humanitarian control. Darwin did not attend any of the events after that first reception, presumably having served the purpose merely by being present. He therefore did not feature in the remarkable composite photograph of the delegates taken by Herbert Barraud; nor did he appear in the enormous oil painting of Angela Burdett-Coutts's garden party, at which Europe's greatest living scientists were depicted standing around somewhat awkwardly in conversational groups on the grass.[89] Most of his friends, correspondents, and enemies were there. For one last time, Darwin's absence was palpable.

He continued to support the vivisectionist cause. A few months later he sent money to help defend David Ferrier, an experimentalist who was unexpectedly prosecuted for vivisecting without a licence. This notorious court case was initiated by Frances Power Cobbe and was regarded by Darwin and others as jeopardising Ferrier's important work on the functions of the brain. A Science Defence Fund was subsequently set up to assist physiologists confronted by similar circumstances, and Thomas Lauder Brunton asked Darwin if he would serve as first president of this fund. Although Darwin declined, he kept abreast of developments.

## IX

Despite all these manifest signs of public approbation, the melancholy persisted. "My life is like clock-work, working away at what little I can do more in science," he said. He was in the grip of a vision of time as powerful and as bleak as anything in Victorian culture. A pensive air underpinned the portrait of him made by John Collier in July 1881.

The commission was undertaken at the request of the Linnean Society. To the extent that he ever welcomed such commitments, Darwin was content to agree because he felt loyal towards the Linnean, the framework for his identity as a biologist. He seems not to have given a moment's thought to the possibility that the Linnean Society, as perhaps the Cambridge Philosophical Society before, was interested not simply in commemorating his long association with the body but also in appropriating his image as cultural capital. The purpose of this picture—the place where it was to hang and the message it was to convey to assembled fellows—was an essential part of the project.[90]

Moreover, Darwin liked "Jack" Collier. He was Huxley's son-in-law, a professed unbeliever and a member of the well-financed, forward-looking, artistic London set. Collier had married Huxley's second daughter, Marian ("Mady), who herself drew an attractive pencil sketch of Darwin in 1878.[91] The portrait to be painted by her husband was financed by subscriptions from the Linnean fellows.

Collier visited Down House twice for sittings, bringing Marian with him. He eschewed any scientific props, portraying Darwin unencumbered by any signs of earthly, practical existence, a man whose claim to greatness rested in his intellect. He posed him as if he were about to set off on his daily walk, dressed in his outdoor cape, hat in hand, a simple, stately, benevolent figure gazing into the complexities of nature and seeing further than most. His white hair, ample beard, and furrowed brow gave him the appearance of a saint, even of a deity. The younger members of the family were pleased. Collier had created an icon of solitude and wisdom. But as was becoming customary, it did not appeal to Emma, who commented that the "likeness is so indefinite."

Emma's dealings with Marian Huxley were another matter. Marian was one of the new breed of women, confident in her own abilities and the future.

> Marian . . . does not make her fortune here. She ought to have been improved by marrying so nice a man—her manner is so indifferent & wanting in respect that Bessy can hardly bring herself to call her

> Marian. My only aim is to be just civil enough to prevent her Jack finding out how little we like her. e.g. she sat at 5 o'clock tea yesterday reading her newspaper without even looking off to drink her tea. It is clear we need never invite her again.[92]

The sittings (the "standings") tired Darwin, so Jackson the butler took over. He modelled for Collier, standing on a box dressed in Darwin's cape to get the height and drape. He chatted brightly to Collier. "Do you know Sir, I believe some people don't think Mr Darwin good looking; but we can't see that at all." At least three copies were made of this portrait, including one for the Darwin family that was afterwards given to the National Portrait Gallery in London.[93] The original hangs in the Linnean Society. "I shall be proud some day to see myself suspended at the Linnean Society," said Darwin as he congratulated the artist.

## X

Darwin's brother Erasmus died in August 1881, a sad time for them all. Erasmus had played a central part in Darwin's life—witty, affectionate, and hospitable, a generous host in London, a favourite guest at weekend parties, on family holidays, and outings, an incorrigible gossip whose letters provided the cement of the family circle. "Caroline had a horrid story about a bilberry tart too dreadful to repeat and I am afraid they all laughed instead of sympathising," he would report whimsically. He wrote about his daily life in the metropolis, his invalidism, his casual detachment.

One by one, his relatives asked each other whether Erasmus had died a lonely death. They all knew he was not a Christian believer. "It isn't my nature to look at the best side of things," he had said in 1878, a self-judgement that none of his relatives recognised. His attitude to life was more engaging, warmer than that. They were pleased at least that he had not faced death alone. His and Darwin's remaining sister, Caroline, had struggled up to town to sit with him for a few days before returning home. Then his dearest friend, Fanny Wedgwood, probably his one true love, came and stayed till the end, accompanied by her daughter Effie and Henrietta Litchfield. The love that the younger set held for him was moving in its simplicity. William Darwin said he seemed "much more than an uncle." Indeed, it is possible that they found in Erasmus an attractive alternative to their father, the other side of Darwin's personality, a man with endless time and loving curiosity about their developing lives, someone who was not incessantly preoccupied with work. Henrietta chose to live close to Erasmus and was with him more or less to the end. His wry humour lived on, making fun of himself and the family claims to fame. He had been the instigator of a private petition that did the rounds of Darwins and Wedgwoods for signature, requesting Effie Farrer to write letters that someone was able to read. Or he had amused them with frivolous plans to spend the winter in Algeria, "utterly averse to anything beyond sitting in the sun on my balcony"—frivolous only because they knew he would never get up sufficient energy to go there. He was alert to his brother Charles's achievement. "I wrote you a very pretty note on Saturday & tore it up on Sunday," he told Effie just before his death, "and perhaps this will be torn up tomorrow but if it survives that will be proof that it is the fittest."

At Erasmus's death, Darwin recalled the "touching patience & sweetness of his nature." Writing to Hooker, he said, "He always appeared to me the most pleasant and clearest headed man, whom I have ever known. London will seem a strange place to me without his presence." Shortly afterwards, he was pained to see Thomas Carlyle's dismissive remarks about Erasmus, in which Carlyle patronisingly suggested that he had been "doomed to silence and patient idleness." Julia Wedgwood rose to defend Erasmus in the *Spectator.* At home, Darwin pulled out the manuscript copy of his autobiography to add the comment that Carlyle's account had, in his opinion, "little truth and no merit."[94]

The burial was at Downe. Darwin refused to have the village vicar, George Ffinden, officiate and shipped in his cousin John Wedgwood, the same cousin who had conducted the marriage service for him and Emma long ago at Maer. This was the last time Darwin attended church except for his own funeral. Francis remembered his look most distinctly, as he stood in the churchyard caught in an unseasonal scattering of snow, wrapped in his long black cloak, "with a grave look of sad reverie." It was 1 September 1881, "bitter cold."

Although undated, it seems likely that the final photographs of Darwin date from this sadly meditative period. A photographer from the firm of Elliot and Fry came to Down House in person, probably Clarence E. Fry, the senior photographic partner. This was a change from Darwin's usual practice of visiting a London studio as he had done several times in the late 1870s for new *cartes de visite.* Darwin posed outside, on the veranda, dressed in his cape and soft hat. Three studies, perhaps more, were the result. He looked sombre. Almost all that could be seen was his beard, his hat, and his eyes. The man as a physical presence had almost disappeared. All that was left was his mental powers.

He published his final book, *The Formation of Vegetable Mould Through the Action of Worms,* in October 1881. "The subject may appear an insignificant one," he wrote in the preface, half-defensive. But the principles behind the study were the primary creed of his scientific life—"small agencies and their accumulated effects." He believed that the natural world was the result of constantly repeated small and accumulative actions, a lesson he had first learned when reading Lyell's *Principles of Geology* on board the *Beagle* and had put to work ever since. His interpretation of South American geology had been based on Lyell's vision of little-and-often, and his theory of coral reefs too, each polyp building on the skeletons of other polyps, every individual contributing its remains to the growing reef. Most notably, he had applied the idea of gradual accumulative change to the origin of species, believing that the preservation of a constant procession of minor adaptations in individuals would lead to the transformation of living beings. His work on barnacles, plants, and pigeons all supported the point. No one, not even Lyell himself, or any of Darwin's closest friends and supporters, accepted as ardently as Darwin that the book of nature was about the accumulative powers of the small.

For this volume he called personally at Murray's with the manuscript in his hands, uncertain whether the firm would find it worth publishing. Murray recalled the conversation with a certain humour. "Here is a work which has occupied me for many years and interested me much. I fear the subject of it will not attract the public, but will you publish it for me?" Darwin apparently said. "It always gives me great pleasure and hope to hear an author speak of his work thus. What is the subject?" Murray replied. "Earthworms," said Darwin.[95]

Despite his fears, it was by far his most popular volume, selling widely from the day it was published, in greater numbers and at a faster speed than even *The Expression of the Emotions* had done. A day or two after publication Murray exclaimed, "3500 Worms!!"[96] The subject matter captured the general imagination. For months afterwards, Darwin received letters from people wishing to add to his quaint and curious enterprise. "I am driven almost frantic by the number of letters about worms; but amidst much rubbish there are some good facts & suggestions," he sighed.[97] A number of contemporaries saw the pathos in Darwin's magnificent obsession. "I must own I had always looked on worms as amongst the most helpless and unintelligent members of the creation; and am amazed to find that they have a domestic life and public duties!" said Hooker. One caricature by Linley Sambourne in *Punch* depicted Darwin as the creator of life, contemplating the evolutionary cycle from worm right around again to worm, portraying mankind's inevitable return to the earth that bred him. Sambourne sent this picture and his respects to Darwin. Leslie Stephen praised the book for Darwin's tenderness towards insignificant creatures. Moncure Conway took a theological view, speaking of "our fellow worms" in the *Index* of 1881, declaring that the worm is the resurrection and the life.[98]

## XI

Darwin slowed down considerably over the following months. Family matters occupied his mind. In December, Horace and Ida presented him with a second grandchild, this one to be called Erasmus. Although the news was welcome, Darwin felt too fatigued to travel to Cambridge to see the baby, much preferring to idle on the sofa. He and William took some time to sort out the family's financial affairs, and around now he put his will into final order, making sure to leave a personal gift of £1,000 each to Hooker and Huxley, "as a slight memorial of my lifelong affection and respect." Somewhat surprisingly, he did not do the same for Wallace.

His weak pulse induced Emma to call in Dr. Andrew Clark, "but he did not take a serious view of it." Privately, Clark warned Emma that Darwin showed signs of encroaching weakness of the heart. This was made clear by a frightening incident at George Romanes's house. During a visit to London early in December 1881, Darwin experienced a sudden spasm on Romanes's doorstep, identified by him afterwards as heart pains of the kind that came to plague him. Again, Clark examined him. "Dr. A. Clark finds that my heart is perfectly right, & that the pain & rapid intermittent pain, must have been only some indirect mischief," he reported with relief.

Five-year-old Bernard was an agreeable distraction. Darwin and Bernard referred to each other familiarly as "Baba" and "Abbadubba," and each looked forward to their daily conversations, the young and old meeting on common ground. Francis appreciated the intimacy between them.

> His love and goodness towards Bernard "Abbadubba" as he called him were great. He often used to say he never saw such a contented child as Dubba; and often spoke of the pleasure it was to him to see his little face opposite to him at dinner; they used to talk about liking brown sugar better than white &c., the result being "We always agree don't we Abbadubba"—Dubba had so many adorers that my father didn't get him to himself much; one ceremony that took place every day was showing "Baba" how big a bit of chocolate he had got. This was at 3 o'clock when my father was lying down upstairs—and Dubba used to run from the chocolate box across to the sofa.[99]

Bernard's high spirits reverberated through the family. He joined his grandfather for walks around the garden, drew soldiers as quietly as he could on the floor in Darwin's study, jumped down the stairs in threes and fours, and held Darwin's hand in self-important splendour on the lawn when the Coal Club band came for its annual performance. He danced with excitement on top of the garden wall when a traction engine trundled round the bend in the road. "They had a delightful expedition all the way to Cudham," said Emma that winter. "He talked all the way when he was not singing, and had to be put down whenever they came to a frozen puddle to stamp upon it." She told Henrietta that "Bernard is at the stage of wanting endless stories or rather talkings, & I find myself very soon run dry—tho' he is very merciful in wanting the same story 40 times."[100] Encouraged by Darwin, he raced his tricycle up and down the long gravel path while the old man timed him on his pocket watch. The sliding board for the stairs, the board that the previous generation had used on rainy days when they were children in the 1850s, mysteriously reemerged from the cupboard where Parslow had stowed it.

In the evenings, when Bernard had gone to bed, Emma would play the piano for Darwin, especially when Henrietta and her husband were visiting. Sometimes Francis, Richard Litchfield, and Emma would play a trio together, "a little tootling" she called it. These trios were engagingly amateur in performance, Litchfield taking the violin part on his concertina, and Frank the violoncello on his bassoon, "and thus they played a great many of the Mozart and Haydn trios and slow movements out of Beethoven."[101]

Darwin continued to plan experiments and interest himself in the house and grounds. Late in 1881 he purchased a strip of land behind the hothouse to add to the garden. He intended to extend the orchard and lay down a tennis court, although a makeshift grass court was probably already marked out. The sons remembered that Darwin would sometimes walk down that way *en route* to the Sandwalk and knock back stray tennis balls with the handle of his stick. They said he was predictable in his complaints about the "idiotic" system of tennis scoring, "maintaining it was all affectation marking 15, 30, 40 instead of 1, 2, 3." Pausing to contemplate the freshly exposed back of this garden wall, Darwin also mapped out in his head a plan for a new boiler-room for the hothouse and discussed with Francis the possibility of a building without windows that appears to have been meant as a dark-room for growing experimental plants. This was not put into use until after his death.

Tweaked into idle new avenues of thought, he reopened a handful of unfinished botanical investigations. One of these involved the condition he called bloom. This bloom could be seen in the powdery complexion of fresh-picked fruit, say a plum, which Darwin believed was an adaptation to reduce the effects of dehydration. "It is a really pretty sight to put a pod of the common pea, or a raspberry, into water," he said enthusiastically. "They appear as if encased in thin glass." He thought bloom was related in an evolutionary manner to the waxy covering that protected plants from salt spray, mud, and rain. So he pestered William Thiselton Dyer at Kew for specimens of seaside species and heartlessly stripped the leaves of their surface coverings, burned holes in them by watering them in the sunshine, immersed them in salt water, and inflicted any other "devastation" he could devise.

From time to time he looked inquiringly at the other end of plants. He immersed roots in carbonate of ammonia to examine the rate of cellular uptake. "My dear Mr. Vines," he wrote understandingly to a new contact at Cambridge, the botanist Sydney Vines, "I fear that you will be utterly tired of me and my roots." Or he tried to identify the reasons for the daily opening and closing of leaves and flowers. He liked to refer to the closing movement as "sleep," although he soon decided that it was actually an adaptive response related to self-preservation from light. He put together some cumbersome wooden gadgets with rotating arms to catch them at it. At night he placed a succession of valuable pot-plants on the lawn with their leaves tied together so that they could not move. "We have killed or badly injured a multitude of plants," he said remorsefully. He sacrificed a favourite *Oxalis sensitiva* to this work before guiltily remembering that it came from Kew.

These idiosyncratic projects diverted his attention from his declining health. Even so, he mused on the life he had spent. Leonard Darwin recorded one moment when this feeling almost lay solidly in the air.

> My father, my sister and I were walking . . . on a beautiful sunny evening when the charm of the quiet scenery was, I am sure, affecting his mind. At all events, in reply to something which my sister had said, he declared that if he had to live his life over again he would make it a rule to let no day pass without reading a few lines of poetry. Then he quietly added that he wished he had "not let his mind go to rot so."[102]

Emma often spent the early afternoons and evenings reading aloud to him. Mudie's Circulating Library kept them well supplied with fiction and memoirs, and a letter of hers written in 1879 indicates that they kept relatively abreast of contemporary literature. After lunch, Darwin rested and smoked a cigarette while she read to him from the text of the day.

> We have begun to read aloud "The Europeans" a very pleasant contrast to a very painful but powerful novel of Trollope's "An eye for an eye" which we have just finished. Mr James is rather too subtle for my taste & I often have to stop to consider what he means.

Otherwise, she would throw two games of backgammon with him, a nightly ritual maintained through thick and thin. The board was bound in leather to look as if it were a large book, and labelled on the spine "History of North America." Darwin remained competitive to the end. His desire to win even at backgammon was almost comic. "Bang your bones" or "Confound the woman" he would exclaim if things were going badly with him. "Pray give our very kind remembrances to Mrs. Gray," he once wrote to Asa Gray. "I know that she likes to hear men boasting, it refreshes them so much. Now the tally with my wife in backgammon stands thus: she, poor creature, has won only 2490 games, whilst I have won, hurrah hurrah 2795 games!"[103]

During these listless days, Emma modified her routine to suit him best. It worried her that Darwin was becoming frail, taking only an egg for an early supper with Bernard, and retiring soon afterwards to bed. "I am fairly well," he wrote to Wallace, "but always feel half dead with fatigue." Francis was quick to catch the elegaic mood. "It was one of his many bits of clocklike regularity that he might be heard blowing his nose with a very loud sound at 10.30 every night as he undressed in his study which he used as a dressing room. His long bright coloured dressing gown was a familiar sight as he went slowly up to bed with his slow tired step."[104] Darwin took to wearing pince-nez for reading, although these kept getting lost in his waistcoat. He never went deaf, said the sons. Emma stayed close, apprehensive that any day might be his last. "Looked out of window," she recorded in her diary on one particularly bad morning, as if this was an achievement of note. When she tended him at night he said, "It is almost worth while to be sick to be nursed by you."

> During my father's last years her whole day was planned out to suit him [reported Henrietta], to be ready for reading aloud to him, to go his walks with him, and to be constantly at hand to alleviate his daily discomforts. . . . My mother would, when her strength and the weather allowed, go with him round the "sand-walk."[105]

He relapsed in February and March 1882 with heart pains. Emma told Henrietta, "The 2 walks he took brought on faintness & some pain about the heart—so we were greatly rejoiced when Dr Clark examined him carefully & pronounced the heart quite sound. He attributes all this discomfort to eczema which will not come out." Yet he depended on Emma completely, unwilling to walk far from the house in case the pains, or "a fit of his dazzling," would seize him, unable to tolerate visitors except extremely close family friends. Instead, he loitered with his wife on the veranda or, as the spring began, sat with her in the orchard, admiring the crocuses and listening to birdsong.

Four doctors were in attendance during these months, an unnecessarily large number but characteristic of Darwin's lifelong multiple relationship with the medical profession. Andrew Clark came to see him from London, although the Darwins disliked interrupting his busy practice to call him down. Clark put him into the combined hands of Norman Moore of St. Bartholomew's Hospital and C. H. Allfrey of St. Mary Cray, a nearby village. Shortly after, Dr. Walter Moxon, a physician from Guy's Hospital, was brought in. Darwin's mind was absolutely clear, his sense of humour intact. On one occasion when Emma told him off for interrupting her, and then asked Bernard rather too abruptly to ring the servant's bell, he laughed. "Yes, look sharp about it, Mammy is not to be trifled with when she is in this humour, I can tell you." Here and there, he made some small observation in the garden or greenhouse, his mind still dwelling on roots and leaves. Two days before his death he fretted in the house until he was allowed to note the results of one of Francis's experiments on bloom.

But he was fading. He died on the afternoon of 19 April 1882, after sinking very low for two or three days beforehand and suffering what Emma called "fatal attack" at midnight on the 18th. There was no deathbed conversion, no famous last words.[106] "I am not the least afraid to die," he apparently murmured to Emma. "Remember what a good wife you have been."[107] Allfrey signed the death certificate giving "Angina Pectoris Syncope" as the cause of death, the gradual ceasing of the heart. He was seventy-three.

## XII

Before the day was through, John Lewis creaked up the road with a coffin on his cart and laid Darwin out, as the old country tradition required. Emma knew that Darwin expected to be buried in Downe churchyard, alongside Erasmus and the Darwins' two dead babies, "the sweetest place on earth," he had once remarked to Hooker. He no doubt imagined that Emma would be buried there in the future. He had made his life, and his books, on that spot. But science claimed him, as it always had. News of his death went directly from Francis Darwin that evening to Huxley, Hooker, Galton, and the rest. The next day Galton hurried to William Spottiswoode, the president of the Royal Society, to request that Darwin should be interred in Westminster Abbey, as a fitting memorial for a great man.

Dying was the most political thing Darwin could have done. As Huxley and others were aware, to bury him in Westminster Abbey would celebrate both the man and the naturalistic, law-governed science that he, and each member of the Darwinian circle, had striven, in his way, to establish. Such an accolade suited Huxley down to the ground. His affection and admiration for his friend meshed seamlessly with elevated regard for modern science. Before long Lubbock was able to send to the dean of Westminster a document stating that "it would be acceptable to a very large number of our countrymen of all classes and opinions that our illustrious countryman, Mr. Darwin, should be buried in Westminster Abbey." Twenty members of Parliament had signed it. Lubbock was confident that Dr. Bradley (the dean), and Gladstone (the prime minister) would give their consent.

Hardly given time to grieve, Darwin's family were initially reluctant to accept this request, conscious not only of Darwin's wishes but also sensitive to the charge that they might be seeking for him the publicity that he had avoided in life. "It gave us all a pang not to have him rest quietly by Eras.," Emma said to Fanny Wedgwood. "But William felt strongly, and on reflection I did also, that his gracious and grateful nature wd. have wished to accept the acknowledgement of what he had done." Several people in the village expressed reluctance for other reasons. The landlord of the George Inn thought a grave in the village would have helped local business by bringing sightseers. "There was great disappointment in Down that he was not buried there," agreed Parslow. "He loved the place and we think he would rather have rested there had he been consulted."[108] Nonetheless the family accepted.

The funeral at Westminster Abbey was held on Wednesday 26 April, a week after Darwin's death. The ceremony was overwhelming in size and content, a chance for science to show its strength as well as render formal acknowledgement of Darwin's place in history. The original Kent coffin was changed for a sumptuous velvet-draped affair. Philosophers, scientists, naturalists, admirals, museum superintendents, and civic dignitaries attended, "together with a host of lay celebrities," all keen to pay their last respects to the man and the achievement. There were a few absences—Gladstone was busy in Downing Street ("off to Windsor at 5. Grand dinner in the Waterloo gallery"), and the Archbishop of Canterbury was indisposed.[109] It is not known how far these may have been tactical.

Even so, among the ceremonial trappings the domestic character of Darwin's work briefly showed its face. Completely unself-conscious, his eldest son, William, felt a draught on his bald head after removing his hat. "So he put his black gloves to balance on the top of his skull, and sat like that all through the service with the eyes of the nation upon him."[110] Incognito among the "dense throng of mourners, amongst whom were men whose names are as household words in European scientific circles," were his servants Parslow and Jackson, and Mrs. Evans the cook.

"Happy is the man that findeth wisdom and getteth understanding," sang the choir from Proverbs 3. Two dukes and an earl were among the men who carried his coffin to the grave—the Duke of Argyll, the Duke of Devonshire, and the Earl of Derby—accompanied by the American ambassador, J. Russell Lowell, and William Spottiswoode, representing English science as president of the Royal Society. The other pall-bearers were equally eminent, the four men whose lives had intertwined so intimately with his own and whose rise to public position had matched the growing acceptance of his theories, Hooker, Wallace, Huxley, and Lubbock. Far away in America, Asa Gray could only write to Julia Wedgwood to say Darwin's death was "like the annihilation of a good bit of what is left of my own life."

Darwin's grave was in the nave, near Sir John Herschel and Isaac Newton, an honourable place but not as close to Lyell as Emma had hoped. The stone was inscribed with his name and dates, and only later did the Royal Society organise a fund to add a bronze tablet that described his contribution to science. As the *Pall Mall Gazette* declared, well-wishers believed they were laying there "the greatest Englishman since Newton," one who had given "the same stir, the same direction to all that is most characteristic in the intellectual energy of the nineteenth century, as did Locke and Newton in the eighteenth." No one, said the *Times,* has "wielded a power over men and their intelligences more complete than that which for the last twenty-three years has emanated from a simple country house in Kent."[111]

Emma remained at home. Unexpectedly, Francis had to put down Darwin's dog Polly a few days afterwards, and he buried her under the Kentish Beauty apple tree in the orchard.

And Darwin himself slipped into legend.

# *Notes*

Books and articles are cited in author-date form, and a full listing of these is in the Bibliography. Some standard sources are abbreviated as short titles, e.g., *Correspondence, Life and Letters,* etc. The full reference is given below. These are also listed alphabetically in the Bibliography, once under the abbreviated title and again under the author's or editor's name.

Darwin's books and papers are for the most part held in the Manuscripts Room, Cambridge University Library. The collection is subdivided and catalogued in various ways. Since this volume was begun, an extensive programme of renumbering has taken place and new items have been acquired, in particular the Down House Manuscripts on deposit from English Heritage. For clarity I have converted old call numbers into new numbers. Some discrepancies may, however, arise. The abbreviation DAR is the prefix for all Darwin items. Where other material from Cambridge University Library is cited I have tried to make the distinction clear. In general, I have cited published sources for letters. Otherwise, the recipient or sender and the conjectured date of unpublished letters are given with the name of their repository. A complete listing of Darwin's correspondence is given in *Calendar* (see below).

## *Abbreviations*

*Autobiography:* Nora Barlow, ed. 1958. *The autobiography of Charles Darwin, 1809–1882, with original omissions restored.* London: Collins.

*Calendar:* Frederick H. Burkhardt, Sydney Smith, et al., eds. 1994. *Calendar of the correspondence of Charles Darwin,* rev. ed. Cambridge: Cambridge University Press.

*Collected papers:* Paul H. Barrett, ed. 1977. *The collected papers of Charles Darwin.* 2 vols. Chicago: University of Chicago Press.

*Correspondence:* Frederick H. Burkhardt, Sydney Smith, et al., eds. 1983–2001. *The correspondence of Charles Darwin.* Vols. 1–12 (1821–64). Cambridge: Cambridge University Press.

*DAR:* Darwin manuscript collection, Cambridge University Library.

*Descent:* Charles R. Darwin. 1871. *The descent of man and selection in relation to sex.* 2 vols. Facsimile ed. with an introduction by John T. Bonner and R. M. May. Princeton, N.J.: Princeton University Press, 1981.

*Earthworms:* Charles R. Darwin. 1881. *The formation of vegetable mould through the action of worms, with observations on their habits.* London.
*Emma Darwin:* Henrietta E. Litchfield, ed. 1904. *Emma Darwin, wife of Charles Darwin: a century of family letters.* 2 vols. Cambridge: Privately printed.
*Expression:* Charles R. Darwin. 1872. *The expression of the emotions in man and animals.* Reprint ed. with an introduction, afterword, and commentaries by Paul Ekman. London: HarperCollins, 1998.
*Journal:* Gavin De Beer, ed. 1959. Darwin's Journal. *Bulletin of the British Museum (Natural History) Historical Series* 2:1–21.
*Journal of researches:* Charles R. Darwin. 1845. *Journal of researches.* 2nd ed. London. Reprinted as *The voyage of the* Beagle, edited by H. G. Cannon. London: J. M. Dent, 1959.
*Life and Letters:* Francis Darwin, ed. 1887. *The life and letters of Charles Darwin.* 3 vols. London.
*More letters:* Francis Darwin and A. C. Seward, eds. 1903. *More letters of Charles Darwin: a record of his work in a series of hitherto unpublished letters.* 2 vols. London: John Murray.
*Natural Selection:* Robert C. Stauffer, ed. 1975. *Charles Darwin's* Natural Selection, *being the second part of his big species book written from 1856 to 1858.* Cambridge: Cambridge University Press.
*Notebooks:* Paul H. Barrett et al., eds. 1987. *Charles Darwin's notebooks, 1836–1844: geology, transmutation of species, metaphysical enquiries.* Cambridge: Cambridge University Press.
*Orchids:* Charles R. Darwin. 1877c. *The various contrivances by which orchids are fertilised by insects.* 2nd ed. Revised with a new foreword by Michael Ghiselin. Chicago: University of Chicago Press, 1984.
*Origin:* Charles R. Darwin. 1859. *On the origin of species by means of natural selection, or the preservation of favoured races in the struggle for life.* London. Facsimile edition with an introduction by Ernst Mayr. Cambridge, Mass.: Harvard University Press, 1964.
*Variation:* Charles R. Darwin. 1868. *The variation of animals and plants under domestication.* 2 vols. London. Facsimile edition with new foreword by Harriet Ritvo. Baltimore: Johns Hopkins University Press.

## CHAPTER 1: STORMY WATERS

1. *Once a Week,* 21 January 1860, p. 67. Victorian Britain is characterised in Houghton 1957, Briggs 1959, Appleman et al. 1959, Himmlefarb 1959, Burn 1964, Burrow 1966, Perkin 1969, Hobsbawn 1975, Hilton 1988, Collini 1991, Hoppen 1998, and Cannadine 1999.
2. *Life and Letters* 1:318. See also Stecher 1961.
3. Described in Wedgwood and Wedgwood 1980 and partly in Arbuckle 1983. See also Annan 1955, Cannon 1964, and L. Stone and Stone 1986.
4. *Life and Letters* 2:288.
5. Wedgwood and Wedgwood 1980, p. 334. On Darwin's place in the rural community, see Howarth and Howarth 1933, Atkins 1974, Moore 1985a, and Neve 1993. More generally on the bourgeois experience, see Gay 1984–88 and F. Thompson 1988.
6. Darwin's long manuscript is transcribed in Stauffer 1975. The quotation comes from Burkhardt and Smith et al., *The Correspondence of Charles Darwin,* abbreviated here as *Correspondence.* See *Correspondence* 6:265. Darwin's formulation of the theory of evolu-

tion by natural selection has been discussed by a large number of scholars, from which see Limoges 1970, Barrett et al. 1987, Ospovat 1981, Kohn 1989, Hodge 1985, R. Young 1985, and Browne 1983 and 1995. Much of the literature is summarised in Oldroyd 1984, Kohn 1985a, Lenoir 1987, Bowler 1989a, and Bohlin 1991.

7. Biographies of note are by Brent 1981, Bowlby 1990, Bowler 1990, and Desmond and Moore 1991. Others include West 1937, Keith 1955, Himmelfarb 1959, and Fleming 1969. How Darwin has fared in biographer's hands is discussed in part by Churchill 1982. For scientific biography as a genre see the essays in Shortland and Yeo 1996. More generally, see Epstein 1987 and 1991, and Josselson and Lieblich 1995.
8. From a large literature, see Irvine 1955, Eisley 1961, Ruse 1979, Bowler 1988 and 1989a, Greene 1991 and 1996, otherwise summarised and discussed in Kohn 1985a. For the wider implications see Oldroyd and Langham 1983, Moore 1989, Amigoni and Wallace 1995, and Numbers and Stenhouse 1999.
9. *Correspondence* 8:277.
10. The letters known to be in existence today are listed in Burkhardt and Smith et al., *Calendar of the correspondence of Charles Darwin,* abbreviated here as *Calendar.* Darwin's German correspondence is listed by Junker and Richmond 1996. See also Carroll 1976. Darwin's habits relating to his correspondence is in *Life and Letters* 1:119–21. An evaluation is given by Moore 1985b. For the role of place in a writer's work see especially Ophir and Shapin 1991, Shapin 1991, and Agar and Smith 1998, and more generally Marsh 1993.
11. *Post Office Directory of the Five Home Counties,* Kent, 1870.
12. Raverat 1952, p. 206. Darwin's accounts are in two series, his Classed Account Books and his Account Books, Down House Archives, arranged by year.
13. Hill 1862, p. 465, Perry 1992. Reflections on the impact of correspondence in society are in Chartier et al. 1997.
14. Quoted from Darwin and Wallace 1858. The original essay has not survived. See also *Correspondence* 7:517.
15. *Correspondence* 7:107.
16. Suggested in Brackman 1980.
17. McKinney 1972, pp. 153–55, states that Darwin received Wallace's essay some two weeks earlier, possibly on 3 June 1858. Brackman 1980 and Brooks 1984 agree. The latter propose that Darwin took important ideas from Wallace during the interval. The proposal has been contested in Kohn 1981 and Beddall 1988a. See also Browne 1980, Kohn 1985b, and Introduction, *Correspondence* 7, pp. xvii–xviii.
18. Gentlemanly codes of honour are summarised in Morgan 1994, St. George 1993, Curtin 1987, and Collini 1991, and in the context of science, especially Nye 1997, Barnes and Shapin 1979, and Rudwick 1985.
19. See, for example, Söderqvist 1996, in which science and the ethical life is discussed.
20. *Correspondence* 7:137.
21. *Correspondence* 7:107.
22. *Autobiography,* p. 124.
23. See especially McKinney 1972 and Kottler 1985.
24. R. W. Burkhardt 1977, Jordanova 1984, and Corsi 1988b. Evolutionary politics are discussed in A. Desmond 1989.
25. Spencer 1852 and 1857, developed more thoroughly in Spencer 1862 and Buckle 1857–61. *Vestiges* (Chambers 1844) is discussed in Yeo 1984 and J. Secord 2000.
26. Exemplary sources are Chadwick

1975, Hilton 1988, and Brooke 1991.
27. Corsi 1988a.
28. Nineteenth-century progressionist thought is examined in Eisley 1961, Greene 1996, and Bowler 1976 and 1989b, and the essays in Moore 1989. For phrenology see Cooter 1984; for spontaneous generation see Farley 1977. Scientific secularisation is dealt with in F. M. Turner 1974 and Lightman 1987. Secularisation in general is in Chadwick 1975, Herrick 1985, and Helmstadter and Lightman 1990.
29. For Grant see Jesperson 1948–49, A. Desmond 1984 and 1989; for Blyth see Geldart 1879, Eisley 1959 and 1979.
30. C. Lyell 1830–33. Discussed in the introduction to the fascimile edition by M. J. S. Rudwick, vol. 1, pp. xxix–xxxv, and in J. Secord 1997, pp. xxiii–xl.
31. Anonymous tract, *A Brief and Complete Refutation of the Anti-Scriptural Theory of Geologists,* by a Clergyman of the Church of England (London, 1853); and James Alexander Smith [J.A.S.], *Atheisms of Geology: Sir Charles Lyell, Hugh Miller etc. Confronted with the Rocks* (London, 1857). See also Miller 1857. The issue is discussed generally by Brooke 1979 and 1991, Gillespie 1979, and Moore 1986.
32. Cannon 1978 gives the classic account. For science in general see especially Yeo 1985 and 1993. For natural history see Allen 1978. A. Secord 1994 provides an important social perspective.
33. Quoted in Marchant 1916, vol. 2, pp. 227–28.
34. "A Visit to Dr. Alfred Russel Wallace," *Christian Commonwealth,* 10 December 1903, pp. 176–77. Another interview is in Rockell 1912.
35. Most of Wallace's reminiscences are in Wallace 1905, with other letters and reminiscences in Marchant 1916. Accessible biographies are George 1964, McKinney 1972, Fichman 1981, and Raby 2001.
36. Moore 1997.
37. Wallace 1905, vol. 1, p. 87. These points are discussed in R. Smith 1972 and Kottler 1974. Oppenheim 1985 gives a general account of Victorian spiritualism and Winter 1998 discusses mesmerism.
38. Cardwell 1972 and Barnes and Shapin 1977. On Robert Owen's doctrines see Harrison 1969.
39. Wallace 1905, vol. 1, pp. 88–89.
40. Wallace's scientific reading is described in Beddall 1968 and McKinney 1969. For his appreciation of Darwin's *Journal of Researches,* 1839, see Wallace 1905, vol. 1, p. 256.
41. A memoir of Bates is in Clodd 1892. See also Bates 1863 and Moon 1976. Of the three public libraries then existing in Leicester, the subscription prices, other than that of the Mechanics' Institute (now the Central Library), were probably too high for Wallace to have joined. The Mechanics' Institute library opened in 1835, four years after the institute. Its book collection was absorbed into the Free Library in 1870. I am very grateful to Bill Brock and Mr. C. Hodgson for this information. See also Patterson 1954, pp. 238–39.
42. Marchant 1916, vol. 1, pp. 24–28. Beddall 1969, Moon 1976, and especially Camerini 1996 discuss their collecting endeavours.
43. Wallace's reminiscences are in Wallace 1898, p. 137, Wallace 1905, vol. 1, pp. 254–55, 257, and McKinney 1969.
44. Stevens's obituary, *The Entomologist* (1899) 32:264. For the market economy of natural history specimens see Camerini 1997 and Larsen 1996.
45. For Wallace as a collector see George 1979 and Camerini 1996. Wallace estimated the value of his specimens

in Wallace 1905, vol. 2, p. 377. The profit from Bates's collections was £800; see Clodd 1892, p. lxiv.
46. Wallace 1853 and more generally Stepan 2001, especially pp. 57–84.
47. DAR 205.3:156–57; *Correspondence* 5:182.
48. St. John 1994, p. 274. See also Baring Gould 1909.
49. Wallace 1905, vol. 1, p. 344. Wallace's remarks were written up for publication in *Chambers' Journal* (1856) 5:325–27 and *Annals and Magazine of Natural History* (1856) 17:386–90.
50. Wallace 1855. *Correspondence* 5:519, 521n1, 6:91n10.
51. *Correspondence* 6:514–15.
52. Camerini 1994.
53. Wallace 1905, vol. 1, pp. 361–62.
54. Eventually published as the second part of Darwin and Wallace 1858. McKinney 1966 and 1972 show that Wallace was not entirely accurate in signing the essay "Ternate" since it was principally written on the island Gilolo. Furthermore if Wallace meant that it was completed and posted from Ternate, as seems reasonable, he was inaccurate in dating it February. McKinney's research indicates that the essay was principally written during February and completed in March. See also Brooks 1984.
55. Linnean Society 1908, pp. 5–11.
56. Moore 1997. For the Malthusian context see R. Young 1985, James 1979, and Dolan 2000. See Browne 1983 and 1992 for biogeographical considerations.
57. C. Darwin 1845, pp. 166–67, 365.
58. *Correspondence* 7:116. Hardy 1993 examines epidemic disease.
59. Other cases involving Lyell in priority issues are given in Silliman 1995 and Bynum 1984, the latter contested by Wilson 1996. A classic study of scientific priority is in Merton 1973.
60. *Correspondence* 7:118, 279.
61. Linnean Society 1908, p. 77. On the Linnean Society, see Gage and Stearn 1988.
62. Hardy 1993, p. 56. It is possible that Henrietta inadvertently infected Charles Waring Darwin, since people with streptococcal sore throats can harbour the causative agent for scarlet fever (a haemolytic streptococcus).
63. *Correspondence* 7:121. Darwin's memorial is in DAR 210.13, transcribed in Appendix V, *Correspondence* 7:521.
64. *Emma Darwin* 2:162. The baby may have been slightly affected by mercury poisoning, which can lead to developmental disorders and speech delays. Mercury was a common component of Victorian medicines.
65. Jalland 1996.
66. *Correspondence* 7:121.
67. *Correspondence* 7:121–22.
68. D. Porter 1993 and Dupree 1988.
69. *Correspondence* 6:445–50.
70. The draft is in DAR 6, transcribed in Appendix III, *Correspondence* 7:507–11.
71. Mayr 1991. See also Browne 1980.
72. *Origin* 112.
73. Linnean Society 1908, p. 90, and *Correspondence* 7:123–24.
74. *Correspondence* 7:127, 129.
75. Linnean Society 1908, p. 83. Attendance lists, although recorded by the secretary, were for various reasons often incomplete. It is possible several other fellows and guests were present. I thank Gina Douglas, Linnean Society, for this information. Samuel Stevens was the only Stevens to be a fellow at that time, elected 3 December 1850. See also Moody 1971.
76. Hooker's reminiscences are in Linnean Society 1908, pp. 12–16, and differ in several details from the letters transcribed in *Correspondence* 7:117–24. See also *Life and Letters* 2:125–26.
77. George Bentham to Francis Darwin, 30 May 1882, DAR 140.

78. Bell 1859.
79. Bell 1859, pp. xiii–iv.
80. Moody 1971.

CHAPTER 2:
"MY ABOMINABLE VOLUME"

1. *Correspondence* 7:128.
2. Neither letter has survived to the present day.
3. Not received until October 1858, Wallace 1905, vol. 1, p. 363. For the ownership of scientific theories see Merton 1973, Becher 1989.
4. *Correspondence* 7:166. These absent letters have led Brooks 1984 and Brackman 1980 to propose some sleight of hand.
5. *Correspondence* 9:373.
6. See *Emma Darwin* 2:210.
7. Jalland 1996, which includes an account of Darwin's attitude to the death of close family members. Bowlby 1990, pp. 457–66, discusses the impact of the death of Darwin's mother.
8. *Correspondence* 7:127, 130.
9. *Correspondence* 7:165.
10. Published 20 August 1858. Darwin probably received his copies in early October; see *Correspondence* 7:168.
11. Cohen 1985 and England 1997. See also D. Porter 1993.
12. R. Owen 1858, pp. lxxv, lxxxv, xc–xcii.
13. Sloan 1992. For a general account of Owen see Rupke 1994.
14. Watson 1847–60, vol. 4, pp. 524–25. Watson explained his progressionist views in Watson 1845.
15. See *Correspondence* 7:292. Haughton's remarks were reported in *Journal of the Geological Society of Dublin* 8 (1857–60):152. They were not repeated in Haughton 1860.
16. Wollaston 1921, pp. 118, 112, and also quoted in Cohen 1985.
17. Hooker to Gray, 21 October 1858, quoted from D. Porter 1993, pp. 32–33.
18. *Correspondence* 7:130, and Hooker 1859. D. Porter 1993 gives the background.
19. Gray 1859.
20. Dupree 1988, p. 259.
21. L. Agassiz 1859. See especially Windsor 1979 and Lurie 1960.
22. Bunbury 1891–93, Middle Life 3, p. 105.
23. Bunbury 1891–93, Middle Life 3, pp. 194–95.
24. L. Huxley 1900, vol. 1, p. 159.
25. *Correspondence* 7:198.
26. *Correspondence* 7:240.
27. *Correspondence* 7:301.
28. Burrow 1968, introduction, pp. 41–42.
29. *Life and Letters* 1:155.
30. Explored in Hyman 1962, Beer 1983 and 1985, Bulhof 1992, and Flint 1995. See also Cannon 1968, Kohn 1996, the essays in Knoepflmacher and Tennyson 1977, and Dear 1991. The modern form of scientific text that emerged during the nineteenth century is discussed in Myers 1990.
31. *Origin* 459.
32. Darwin's debt to nineteenth-century philosophy has generated a number of important studies, for example, Ellegard 1957, Ghiselin 1969, Hull 1974, and Ruse 1975.
33. *More Letters* 1:195. On making knowledge see Golinski 1998.
34. T. Porter 1986 and 1995, Gigerenzer 1989, and Kruger, Daston, and Heidelberger 1987.
35. From an abundant literature, see Hacking 1975, Gooding, Pinch and Schaffer 1989, and Pickering 1992.
36. Discussed in Yeo 1985 and Poovey 1998. A general account of the rise of facticity in the Victorian period is Cannon 1978, pp. 73–110.
37. Hodge 1977. On invitations to believe, and on trust in science, see Shapin 1994.
38. *Origin* 31. Darwin's analogy between artificial and natural

selection is discussed in Ruse 1975, Evans 1984, Cornell 1984, R. Young 1985, Weingart 1995.

39. *Correspondence* 7:274, 277.
40. R. Young 1985, also Maasen 1995 and Bowler 1995.
41. *Origin* 75.
42. *Origin* 63. See also Schwartz 1974.
43. *Origin* 84.
44. *Notebooks,* E 48.
45. *Correspondence* 7:265.
46. Beer 1986, Shapin and Barnes 1979.
47. Bowler 1974, Amigoni and Wallace 1995.
48. Spencer 1864, ch. 12, sec. 165. See also Peel 1971, Haines 1991, and Weingart 1995.
49. *Origin* 171, 188.
50. *Natural selection* 250.
51. Hodge 1977 compares Darwin's writings over the decade. For mankind see Herbert 1974–77 and H. Gruber 1974.
52. Bajema 1988 and Cook 1990.
53. From a wide literature on the origins and metaphorical nature of Darwin's principle of divergence, see Schweber 1980, Browne 1980, Kohn 1981 and 1985b, Limoges 1971 and 1994, Tammone 1995. Tree metaphors are discussed in H. Gruber 1987.
54. *Correspondence* 6:236. On Chambers generally, see Yeo 1984 and J. Secord 2000. For Darwin's reaction to Chambers, see Browne 1995, pp. 457–69.
55. *Origin* 484, 485–86.
56. *Origin* 490.
57. *Correspondence* 7:238. Darwin at Malvern is discussed in Browne 1990. The history of hydropathy is most conveniently found in Metcalfe 1906 and E. S. Turner 1967.
58. Pevsner and Nairn 1962. The house was previously called Compton Hall. See *Post Office Directory of the Six Counties,* Surrey, 1859, p. 1222. William Temple named it after the other Moor Park in Hertfordshire because Temple so admired the gardens. The house contained some excellent plasterwork, especially in the central stairwell.
59. Henrietta Darwin, DAR 246.
60. *Correspondence* 7:416.
61. Lane 1857.
62. *Correspondence* 7:385.
63. DAR 140(3):75–76.
64. Lane 1882.
65. *Correspondence* 7:81.
66. *Origin* 220.
67. *Correspondence* 6:178.
68. *Autobiography,* p. 93. On Darwin's theism see Brown 1986, Colp 1987, Kohn 1989, and Moore 1989.
69. Rowell 1974, McDannell and Lang 1988, Berman 1988, and Wheeler 1990. Some discussions of the wives of noted sceptics are in Rose 1983, Healey 1986, and Jalland 1996.
70. *Correspondence* 7:84.
71. DAR 140(3):31.
72. Greist 1970, p. 5, and Finkelstein 1993. Betham-Edwards 1919, pp. 150–55, gives a character sketch of Mudie. Darwin's subscription rate is taken from an entry in his Classed Account Books, Down House Archives, 13 May 1875.
73. Sutherland 1976, 156. Developments in nineteenth-century publishing are surveyed in Altick 1957, Feather 1988, and Chartier 1995, readerships in Cipolla 1969 and Eco 1979.
74. Darwin recorded these titles in his Reading Notebooks, DAR 119, 128 (transcribed and identified in *Correspondence* 4:434–573), but stopped around 1860. On his reading in general see Beer 1985. His library is catalogued in Rutherford 1908.
75. *Life and Letters* 1:124–25.
76. DAR 140(3):44.
77. Wedgwood and Wedgwood 1980, pp. 260–61.
78. Mulock 1864.
79. On hysteria see Bynum 1985, Oppenheim 1991, Gilman et al. 1993, and Micale 1995.
80. Horstman 1985, pp. 94–95, citing Law Reports.
81. *Life and Letters* 1:141–42.

82. *Correspondence* 7:249.
83. *Origin* 221.
84. DAR 112(B):48.
85. Murray 1919 and Paston 1932.
86. Shortland and Yeo 1996, introduction, p. 23.
87. Haynes 1916, p. 233, and Paston 1932, p. 174.
88. Elwin 1902, vol. 1, p. 352.
89. Elwin to John Murray, 3 May 1859, John Murray Archives, London. I am grateful to John Murray for allowing me access to the archives. Also printed in *Correspondence* 7:288–90. The quote comes from p. 289.
90. *Autobiography,* p. 137.
91. DAR 219.1:33.
92. *Correspondence* 7:303.
93. *Correspondence* 7:296.
94. L. Huxley 1918, vol. 1, p. 496.
95. Huxley Papers, Imperial College Archives, 16 April 1859. Dawson 1946, p. 72.
96. DAR 219.1:26.
97. DAR 210.6, reprinted in *Correspondence* 7:264.
98. *Correspondence* 9:29.
99. *Life and Letters* 2:140. See also *Correspondence* 7:196–97.
100. *Entomologist's Weekly Intelligencer* 6 (1859):99. Reprinted in *Correspondence* 7:310.
101. Lyell to Huxley 17 June 1859, Imperial College Archives, Huxley Papers 6:20, partly reprinted in Wilson 1970, 262, 314nn77, 78. Lyell used the expression again in 1863, this time to Darwin; see *Correspondence* 11:231.
102. Longford 1964, 286. Lyell's address is in C. Lyell 1859.
103. *Athenaeum,* 24 September 1859, 404. On the *Athenaeum*'s place in Victorian publishing see Marchand 1971.
104. *Correspondence* 7:337.
105. Yeo 1985 and 1993 discuss the iconic status of Whewell and Bacon in the nineteenth century.
106. *Correspondence* 7:324, 328.
107. *Correspondence* 7:331.

CHAPTER 3:
PUBLISH AND BE DAMNED

1. *Correspondence* 7:222. For the *Origin*'s publishing history see Freeman 1977 and Peckham 1959, especially pp. 775–85, which includes transcripts of Murray's accounts. The *Publisher's Circular,* 1 December 1859, p. 603, announced the price and that the book was released during the previous two-week period.
2. *Correspondence* 7:365, 366. Darwin's copy arrived in Ilkley on 3 November 1859.
3. The rhetoric of correspondence is touched on in Chartier et al. 1997 and Chartier 1995. Chase and Levenson 2000 discusses the interplay between private and public, a contrast to Huxley's rhetoric as described in Jensen 1989 and A. Desmond 1997. Some of the contemporary impact of letters as a form of communication can be seen in the rise of the epistolary novel, as in Kauffman 1992 and Earle 1999.
4. Richard Freeman saw a personally inscribed copy of the second edition (published January 1860) at Sotheby's London; see Freeman 1977, p. 78. Darwin's lists of presentation copies of his works are in DAR 210.18. His list for the first edition is transcribed in Appendix III, *Correspondence* 8:554–70.
5. *Correspondence* 7:369.
6. Contemporary elite networks and "core-sets" in science are discussed in Collins 1981, Rudwick 1985, and Morrell and Thackray 1981. See also Haskell 1984 and Brake et al. 1990. More generally see L. Stone and Stone 1986.
7. *Correspondence* 8:315.
8. DAR 210.18, *Correspondence* 8:554–70.
9. I thank Frank James for this information. John Murray Archive, London, Faraday to Murray, 2 December 1859.

10. *Athenaeum,* 19 November 1859, pp. 653, 660.
11. *Correspondence* 7:387.
12. Murray 1909, p. 540.
13. It is not possible to be completely precise. Peckham 1959, p. 775, gives figures from Murray's ledgers as follows: 1,250 printed, minus 5 for registration at Stationers' Hall, 12 free copies to the author, 41 presented as reviews, leaving a total of 1,192 for sale. Freeman 1977 follows these figures. Neither had access to the list that Darwin prepared showing the 80 or so copies also purchased by him at a reduced rate for presentation. Subtracting this list (93 inclusive of Darwin's free ones), the Stationers' Hall copies, and the review copies leaves a probable total of 1,111 available for sale to the trade.
14. Topham 1992. The sale is reported in *Athenaeum,* 26 November 1859, p. 706, and *Publisher's Circular,* 1 December 1859, pp. 599–600.
15. Greist 1970. Finkelstein 1993 uncovers Mudie's financial collusion with three London publishing houses.
16. Betham-Edwards 1919.
17. Unpublished lecture, Janet Browne, "Science and Medicine in Mudie's Circulating Library," University of London, February 2000. On science reading in the period see Brock 1996.
18. Undated letter to Mr. and Mrs. Mudie, Illinois University Library, *Calendar* 13829.
19. *Correspondence* 7:395.
20. *Correspondence* 7:339, 340.
21. Wilson 1970. See also Bartholomew 1973.
22. K. M. Lyell 1890, vol. 2, p. 329.
23. *Correspondence* 7:409.
24. *Correspondence* 7:359 and 9:188.
25. *Correspondence* 7:383, 426.
26. Arbuckle 1983, p. 186.
27. *Life and Letters* 2:197, repeated in L. Huxley 1900, vol. 1, p. 170. Huxley's response more generally is discussed in Bartholomew 1975, Paradis 1978, Di Gregorio 1984, and A. Desmond 1994.
28. *Correspondence* 7:391.
29. *Correspondence* 7:398.
30. *Correspondence* 7:396.
31. *Emma Darwin* 2:187.
32. See McDannell and Lang 1988, Rowell 1974, and Wheeler 1990.
33. *Correspondence* 8:134.
34. The original letter is not extant. The quotation is taken from a letter from FitzRoy to William Hepworth Dixon, printed in *Correspondence* 7:414n3.
35. *Times,* 1 December 1859, p. 8.
36. *Correspondence* 7:413.
37. *Correspondence* 7:380. For Kingsley, see Colloms 1975 and Brock 1996.
38. The remark is printed in *Origin,* 2d ed., p. 481. Its course through later editions is mapped in Peckham 1959, p. 748.
39. Peckham 1959, p. 753.
40. Murray 1909, p. 542.
41. Rupke 1994, pp. 12–105, and Appendix VI, *Correspondence* 7, Memorials presented to the British government. Stearn 1981 describes the founding of the Natural History Museum in South Kensington, London. Gunther 1975 gives an idea of the work of natural history keepers in the British Museum.
42. See particularly A. Desmond 1982 and 1994.
43. *Origin* 310.
44. *Correspondence* 7:421–23.
45. *Origin* 184.
46. *Times,* 26 December 1859, p. 8; *Correspondence* 7:458.
47. From a wide range of literature see Feather 1988, Eco 1979, and Sutherland 1976. Meadows 1980, Sheets-Pyenson 1985, Brock and Meadows 1998, J. Secord 2000, and Frasca-Spada and Jardine 2000 deal with the rise in science publishing.
48. Various elements of this revolution in reading patterns are discussed in Altick 1957, Ellegard 1957, and Feather 1988. Widening literacy rates are in Cipolla 1969.

49. For newspapers see A. Smith 1979, for magazines see A. Sullivan 1984, and for periodicals see Shattock and Wolff 1982 and Vann and Van Arsdel 1994. Medical journals are discussed in Bynum, Lock, and Porter 1992. General reviews of Darwin's major works are itemised and discussed in Ellegard 1990, the scientific reviews in Hull 1973. Houghton 1966–89 is the standard source for leading Victorian periodicals.
50. J. Secord 2000 gives *Vestiges'* publication figures. Even though Darwin's theories were discussed in a greater number and wider variety of organs than *Vestiges*, it appears that until the 1880s there were more actual copies of *Vestiges* in circulation than of the *Origin*. I thank John van Wyhe for an illuminating chart comparing the relative numbers of copies of *Vestiges* with George Coombe's *Constitution of Man*, and Darwin's *Origin* from 1835 to 1895. Coombe easily outstripped the other two.
51. A comprehensive list is in Ellegard 1990.
52. Bevington 1941, pp. 238–88. Thomas Rymer Jones occasionally did natural history for the *Saturday Review*. Even Huxley could not squeeze the name out of the editor, John Douglas Cook. The *Wellesley Index* (Houghton 1966–89) lists most of the names but not this.
53. Darwin's pamphlet collection is held in the Darwin Archive, Cambridge University Library, indexed by P. Vorzimmer. A fresh listing is due to appear by Di Gregorio and Gill; see Di Gregorio 1990. The newspaper clippings are in DAR 226.
54. "List of reviews of Origin of Species," DAR 262.
55. L. Huxley 1900, vol. 1, p. 363.
56. Imperial College Archives, Huxley Papers, 31 December 1859, 22. Also in Caudill 1994.
57. Bartholomew 1975.
58. *Correspondence* 8:117n11. Darwin refers to T. H. Huxley 1860a.
59. T. H. Huxley 1860b.
60. *Correspondence* 7:432.
61. T. H. Huxley 1860b, p. 556.
62. Barton 1983 and A. Desmond 1994.
63. L. Huxley 1900, vol. 2, p. 114.
64. Quoted from Hull 1973, p. 223.
65. *Correspondence* 7:423. See also Herschel's printed opinion in Herschel 1861, p. 12.
66. *Life and Letters* 2:261n.
67. Lyell to Darwin, 1–2 May 1856, *Correspondence* 6:89.
68. Wollaston 1860. Quoted from Hull 1973, p. 140.
69. *Correspondence* 8:444.
70. *Origin* 285.
71. John Phillips, Presidential Address, *Proceedings of the Geological Society of London*, 1860, pp. xxxvi, xlix–l.
72. *Correspondence* 8:495.
73. Hooker thought the review was by the Rev. Richard Whatley; see L. Huxley 1918, vol. 1, pp. 512, 515. For Sedgwick see Clark and Hughes 1890, vol. 2, p. 360.
74. Clark and Hughes 1890, vol. 2, pp. 360–61.
75. Cambridge University examination papers, Geology, March 1860.
76. Matthew's claim was stated in *Gardeners' Chronicle*, 7 April 1860, 312–13, relating to his book Matthew 1831. Darwin's response was also printed in *Gardeners' Chronicle*, 21 April 1860, 362–63. See *Correspondence* 8:154–55, 156, and Dempster 1996.
77. *Life and Letters* 3:41.
78. Freke 1860. See *Correspondence* 9:10.
79. *Correspondence* 8:189.
80. Harvey's letters were dated August and October 1860; *Correspondence* 8:322–32, 415–421. See also L. Huxley 1918, vol. 1, pp. 515–20.
81. *Punch*, 10 November 1859, p. 182. Humour and satire in the history of science is discussed in Rudwick

1975, Browne 1992, and Paradis 1997.

82. R. Owen 1860. Darwin's annotated copy is in the Darwin Archive.
83. *Correspondence* 8:154.
84. Rupke 1994 makes the point that the ultimate success of Darwinism encourages historians to be over-eager to accept Darwin's point of view in this affair. Owen's son ignored the quarrel in R. S. Owen 1894.
85. *Essays and reviews* was published 15–31 March 1860, confirmed by the *Publisher's Circular,* 2 April 1860, p. 166. See Parker 1860. A general historical analysis is in Ellis 1980 and Altholz 1994.
86. Altholz 1994, p. 68.
87. Discussed in Ellis 1980 and Altholz 1994.
88. Elwin to Murray, 3 December 1859, John Murray Archives, London. I am grateful to John Murray for allowing me access to the archives. See Shattock 1989 for the politics of the *Edinburgh Review* and the *Quarterly.*
89. Meacham 1970.
90. Eventually published as Wilberforce 1861. The author is identified as Wilberforce in Houghton 1966–89, vol. 1, p. 743. Elwin had previously argued with Wilberforce on doctrinal matters, so this commission was something of a rapprochement; see Elwin 1902, vol. 1, pp. 189–92.
91. Parker 1860, p. 139. See particularly Corsi 1988a for Powell. Wilberforce was involved in a response issued in Goulburn 1862.
92. Wilberforce 1860, p. 258.
93. Wilberforce 1860, pp. 235, 239.
94. Helpful accounts of nineteenth-century secularisation are to be found in Chadwick 1975 and Brooke 1991. See also Moore 1979, Lightman 1987, and Lindberg and Numbers 1986.
95. Morrell and Thackray 1981.
96. *Report of the British Association for the Advancement of Science,* Oxford 1860, Committee lists. See also Morrell and Thackray 1981.
97. Brock and Curtois 1977, p. 708, and Acland and Ruskin 1859. I am grateful to Colin Hughes for allowing me to consult his work on the museum.
98. Meacham 1970. Temple's sermon was on "the present relations of science to religion"; Temple 1860.
99. *Correspondence* 8:200–1.
100. The occasion is described in Clark and Hughes 1890, vol. 2, pp. 361–62, and in a letter to Hooker (Henslow's son-in-law) printed in L. Huxley 1918, vol. 1, pp. 512–14. On Victorian Cambridge see Winstanley 1947.
101. Henslow's Class list, Cambridge University Archives, O.XIV.261. There were eighty-four students in 1860.
102. *Cambridge Herald and Huntingdonshire Gazette,* 19 May 1860.
103. Published later on, with a few expansions, in Phillips 1860.
104. Hooker to Huxley, September 1859, Huxley Papers, Imperial College London.
105. Quoted from Jensen 1988, p. 164. See also Lucas 1979 and Gilley 1981. Blinderman 1971 and C. Gross 1993a and 1993b discuss the hippocampus minor.
106. Draper 1860, later expanded into Draper 1864. Draper's earlier writings on these subjects touched on the same issues, although his clearest statement of conflict between science and religion was in Draper 1875. See Fleming 1950. Historical accounts of the theme of conflict between science and religion are in Moore 1979 and Brooke 1991.
107. Crichton-Browne 1930, p. 188.
108. Chambers 1860. His review of the *Origin* was favourable; see Chambers 1859.
109. Mellersh 1968, p. 274.
110. The meeting is reported in *Athenaeum,* 14 July 1860, pp.

64–65. See also Wilberforce 1860, p. 239.

111. Tuckwell 1900, pp. 51–54. On 4 July 1860, Lyell reported that Wilberforce referred to Huxley's grandfather and grandmother. K. M. Lyell 1881, vol. 2, p. 335.
112. Jenson 1988, p. 168.
113. Cohen 1985, p. 598.
114. *Athenaeum,* 14 July 1860, p. 65.
115. Stoney's letter is in DAR 106/7:36. The Bible-waving incident is possibly apocryphal. The first printed account of it seems to be in West 1937, p. 252, probably using Stoney's letter as an uncited source. The incident is not mentioned in Mellersh 1968. See also letter from J. V. Carus, 15 November 1866, DAR 161.
116. Recollections were sent to Francis Darwin after Darwin's death; DAR 106/7 (ser. 2):22, 30. There are further recollections in DAR 112(A): 1–114.
117. Gilley 1981.
118. Discussed in F. M. Turner 1993. See also Haskell 1984.
119. *Correspondence* 8:277, 280.

CHAPTER 4: FOUR MUSKETEERS

1. *Correspondence* 8:305.
2. *Vanity Fair,* by "Ape" (Carlo Pelligrini). Wilberforce (Statesmen No. 25) "Not a brawler"; and Huxley (Men of the Day No. 19) "A great medicine-man among the inquiring redskins."
3. *Life and Letters* 2:325n. Original in DAR 112 (ser. 2):88.
4. L. Huxley 1900, vol. 1, pp. 180–89, gives an account of the meeting gathered from recollections, but several of these are repeated verbatim from other biographies; see Browne 1978. Erasmus Darwin's undated remark comes from his letters in the Wedgwood/Mosely archive, Wedgwood Archive Collection, Keele University, and is cited by courtesy of the Trustees of the Wedgwood Museum, Barlaston.
5. Wilberforce 1860, p. 264.
6. *Correspondence* 8:293.
7. *Correspondence* 8:281–82, 285.
8. *Correspondence* 8:403, in which Darwin said, "I am utterly ashamed & groan over my handwriting."
9. *Correspondence* 8:274.
10. *Correspondence* 8:294.
11. *Correspondence* 8:277. See also *Correspondence* 9:368, in which he told Asa Gray, "I care more for your & Hooker's opinion than for that of all the rest of the world, & for Lyell's on geological points."
12. Lewes 1860, p. 603.
13. Morrell and Thackray 1981, L. Stone and Stone 1986. See also R. Young 1985, pp. 23–55, on the comon intellectual context of Victorian England. Altick 1974, Briggs 1965, and Curtin 1987 examine Victorian class and manners.
14. Cooter and Pumfrey 1994 and Poovey 1995 discuss the theme in historical context. A. Secord 1994 illuminates science in the pub. A. Desmond 1989 and J. Secord 2000 reveal other evolutionary interests in Victorian England.
15. Quoted from Ellegard 1990, p. 41.
16. Annan 1955, J. Gross 1969. See also Rudwick 1985 for the geological elite.
17. Bartholomew 1974, Bynum 1984. On geological images and representations of geological time see Rudwick 1992.
18. The imperial role of the Hookers at Kew is analysed in Brockway 1979, Drayton 2000, Hobhouse 1985, and Bellon 2001. For biographical studies of Hooker see Allan 1967 and R. Desmond 1998 and 1999. Bellon 2001 examines Hooker's scientific status.
19. Dupree 1988, pp. 174–232, passim.
20. Victorius 1932. These are described in Freeman 1977, items 377, 378. The Appleton edition, item 379,

called "Revised edition" was almost certainly issued to keep sales buoyant while the firm negotiated with Gray. For authorship and copyright legislation see Barnes 1974, Saunders 1992, and Feather 1994.

21. Freeman 1977, item 380.
22. Freeman 1977 lists all of Appleton's editions of the *Origin* and Darwin's other titles. On Darwinism in America more generally see Russett 1976 and Numbers 1998.
23. *Origin*, historical preface, reprinted in Appendix IV, *Correspondence* 8:571–83.
24. Dupree 1988, p. 267.
25. In a letter from Asa Gray to Hooker, reprinted in *Life and Letters* 2:268. See also L. Agassiz 1860.
26. Most conveniently found in Gray 1861. The American response to the religious issues raised by Darwin's work is in Moore 1979, Lindberg and Numbers 1986, Livingstone 1987, and J. Roberts 1988. The response of James Dwight Dana, also an opponent, was of interest to Darwin. See Sanford 1965.
27. *Correspondence* 8:298.
28. *Correspondence* 8:350.
29. *Correspondence* 8:405.
30. *Life and Letters* 3:113. Huxley's scientific life is documented in Paradis 1978, Di Gregorio 1984, and A. Desmond 1994 and 1997.
31. *Correspondence* 8:366.
32. On the hippocampus debate see C. Smith 1992, C. Gross 1993a and 1993b, and Wilson 1996.
33. L. Huxley 1900, vol. 1, p. 210.
34. Huxley's aims for the *Natural History Review* are given in A. Desmond 1994, pp. 284, 289–90, 295. For Darwin's response, see *Correspondence* 8:294–95. Huxley continued to protest that it was "not a party journal"; ibid. p. 527.
35. *Correspondence* 9:1.
36. L. Huxley 1900, vol. 1, p. 190.
37. *Correspondence* 8:232.
38. *Correspondence* 8:220.
39. Wallace 1905, vol. 1, p. 374.
40. Farley and Geison 1974. Bronn is discussed in Junker 1991. Bronn's review of the *Origin of Species* is translated in Hull 1973.
41. For Darwin and Darwinism in Germany see Kelly 1981, Corsi and Weindling 1985, and Montgomery 1988. For metaphors see Weingart 1995. For translations in science see Rupke 2000. More generally, see also Jordan and Patten 1995 and Topham 1998.
42. Belloc 1941, p. 12.
43. *Correspondence* 8:64, 71.
44. J. Harvey 1997.
45. J. Harvey 1997. I have used Dr. Harvey's translations from the French with grateful acknowledgement.
46. *Correspondence* 10:398–400.
47. Darwinism in France is discussed by Conry 1974, Corsi and Wendling 1985, Stebbins 1988, J. Harvey 1997, and Gayon 1998.
48. J. Harvey 1997, pp. 80–121.
49. *Correspondence* 9:200. Something of Julia Wedgwood's life is given in Wedgwood and Wedgwood 1980 and Curle 1937.
50. *Emma Darwin* 2:193.
51. *Correspondence* 8:395.
52. *Correspondence* 8:451.
53. *Life and Letters* 1:137, *Emma Darwin* 2:180.
54. *Correspondence* 9:9.
55. *Correspondence* 9:21.
56. *Correspondence* 8:491.
57. Bonney 1919, p. 154. The text of the paper is not extant, although many of Darwin's notes are in DAR 54. See Appendix IV, *Correspondence* 9:405–6 for an account of the meeting.
58. *Correspondence* 9:33–34.
59. Quoted from A. Desmond 1994, p. 294.
60. *Macmillan's Magazine* 3 (1861):336 and Jenyns 1862, pp. 212–13.
61. Clark and Hughes 1890, vol. 2, p. 371; see also Bunbury 1891–93, Middle Life 3, p. 264.
62. *Correspondence* 9:98–99.

63. Gray 1861. See also *Correspondence* 8:388. Darwin's list of presentation copies of Gray's pamphlet is transcribed in Appendix III, *Correspondence* 9:393–404.
64. DAR 262.11 and *Journal,* p. 15.
65. Addressed in Ritvo 1987.
66. Vaucaire 1930, Mandelstam 1994, and McCook 1996.
67. Du Chaillu 1861a. His book was published as Du Chaillu 1861b.
68. *Punch,* 18 May 1861, p. 206. See A. Desmond 1994, p. 296.
69. *Blackwood's Edinburgh Magazine,* 89 (1861):614–17. On satire see Browne 1992 and Paradis 1997.
70. Paradis 1997. On scientific caricature see Rudwick 1973 and Browne 2001.
71. *Athenaeum,* 13 April 1861, p. 498.
72. *Correspondence* 9:100.
73. *Athenaeum,* 21 September 1861, pp. 372–73, and 28 September 1861, p. 408. The gorilla wars are in McCook 1996. Blinderman 1971 deals with the aftermath of the Oxford debate.
74. Kingsley 1862. For accounts of Kingsley see Colloms 1975 and Brock 1996.
75. Charles Kingsley, *The Water Babies,* London, 1863, pp. 156–57. Alexander Macmillan's remark is quoted from Colloms 1975, p. 256.
76. The first illustrator of Kingsley's *Water Babies* was J. Noel Paton. By 1886 Sambourne was acknowledged as Britain's greatest living caricaturist, the chief illustrator of *Punch.*
77. L. Huxley 1900, vol. 2, pp. 435–39.
78. *Blackwood's Magazine* 89 (1861):166.

## CHAPTER 5: EYES AMONG THE LEAVES

1. *Correspondence* 9:269.
2. Note on the general aspect, Down House MS 4.3. See also *More Letters* 1:33–36.
3. *Correspondence* 11:266. I am grateful to Janet Bell Garber for allowing me to see her study of Darwin's experimental work, J. B. Garber, "Charles Darwin as a Laboratory Director," Ph.D. dissertation, University of California Los Angeles, 1989.
4. Darwin's experimental work in these decades is discussed in Bowler 1974, Reinberger and McLaughlin 1984, J. Secord 1981 and 1985, J. Harvey 1997b, and Winther 2000. Darwin's connection with William Tegetmeier is documented in E. Richardson 1916 and the Victorian breeding world in Ritvo 1987. From a large body of literature on experimental science see Shapin 1988, Gooding, Pinch, and Schaffer 1989, and Gayon and Zallen 1998. Darwin's microscopes are described in Burnett 1992.
5. *Life and Letters* 1:129, 130.
6. Sprengel 1793. Darwin read this on Robert Brown's advice in 1841; *Autobiography,* pp. 127, 128.
7. Shteir 1996. See also Allen 1980. Women in science are generally discussed in Abir-Am and Outram 1987, Pycior, Slack, and Abir-Am 1996, and Kohlstedt and Longino 1997.
8. Nevill 1919, p. 56. R. Desmond 1977 states that she kept a notable garden. For botanical women in general see Shteir 1996.
9. John Horwood appears in the 1871 census at that address, aged forty-seven.
10. Duthie 1988.
11. Reinikka 1972.
12. *Orchids* 79.
13. *Correspondence* 10:331.
14. Ghiselin 1969, Gillespie 1979. On natural theology see Brooke 1991.
15. *Correspondence* 8:496. See also M. Roberts 1997.
16. *Correspondence* 8:106.
17. *Correspondence* 8:224.
18. *Correspondence* 8:258.
19. *Correspondence* 8:275.
20. *Autobiography,* pp. 92–93.

21. *Correspondence* 9:178.
22. *Correspondence* 10:59.
23. *Correspondence* 9:196.
24. *Correspondence* 9:205.
25. *Correspondence* 9:216.
26. Hughes 1993. On the Victorian family see especially Wohl 1978, Davidoff and Hall 1987, and M. J. Peterson 1989. Chase and Levenson 2000 examine privacy.
27. *Correspondence* 10:80.
28. *Correspondence* 10:92.
29. DAR 219.1:49.
30. DAR 219.1:57.
31. *Correspondence* 10:148.
32. *Correspondence* 10:75.
33. See J. Secord 2000.
34. *Athenaeum,* 27 July 1861, p. 116, referring to Miller 1861.
35. Grant 1861. See also Jesperson 1948–49. The dedicatory letter is reprinted in *Correspondence* 9:127–28.
36. A. Desmond 1989, 389–97.
37. Spencer 1904, vol. 2, p. 50.
38. Paul 1988.
39. *Correspondence* 10:438.
40. *Autobiography,* pp. 108–9; see also *Life and Letters,* 3:193–94.
41. A jibe first used by Huxley against Henry Bastian, L. Huxley 1900, vol. 1, p. 332.
42. Fawcett had reviewed *Origin* favourably in *Macmillan's Magazine,* Fawcett 1860. For Fawcett and Darwin's correspondence on the matter see *Correspondence* 9:204. The quotation comes from Mill's letter to Alexander Bain, 11 April 1860, transcribed in ibid., p. 205n3. See also the discussion of Mill in the preface of Hull 1973.
43. Mill 1862, vol. 2, p. 18n.
44. Stephen 1924, p. 75. See also Goldman 1989.
45. McKenzie 1981, T. Porter 1986 and 1995, Kruger et al. 1987, and Power 1996.
46. See particularly Schweber 1980, Schabas 1990, Mirowski 1994, and the essays in Cohen 1994, especially Limoges 1994 and Bowler 1995.
47. Discussed generally in Barnes and Shapin 1979, R. Young 1985, Cohen 1994, and Golinski 1998. These social movements swiftly intermeshed with the competitive ethos in business and economic concerns; see for example Hofstadter 1945, Wyllie 1959, Gasman 1971, and G. Jones 1980 on social Darwinism.
48. Marx to Engels, 18 June 1862. The translation is taken from Marx 1985, p. 381, and reads slightly differently from some other sources. For Marx's views on Darwin see Pancaldi 1994.
49. Marx 1985, pp. 234, 246.
50. Marx 1985, p. 543.
51. Stevenson 1932, p. 95. On Tennyson in general see Martin 1980. See also Irvine 1959, Henkin 1963, Wolff 1977, P. Morton 1984, and Levine 1988 for the impact of scientific imagery on creative literature.
52. Stevenson 1932, p. 96.
53. Tennyson *In Memoriam* 1860, given in Stevenson 1932, p. 98.
54. Stevenson 1932, p. 107.
55. Curle 1937.
56. Stevenson 1932, p. 180–81n.
57. Quoted from Stebbins 1988, p. 159.
58. The remarks occur in the fourth and fifth essays, *Cornhill Magazine* 1 (1860):441–47, 598–607.
59. Haight 1954–78, vol. 3, p. 214.
60. Especially Beer 1985, Shuttleworth 1984. See also Paradis and Postlewait 1981.
61. Müller 1864, p. 357.
62. Müller 1901, vol. 2, pp. 11, 17. Language and evolutionary ideas are examined in Beer 1989, Harris 1996, and in regard to Schleicher, Alter 1999. General remarks are in Burke and Porter 1987.
63. *Correspondence* 10:505.
64. DAR 219.1:64.
65. Uglow 1993, pp. 560–61.
66. Walls 1995, pp. 121, 194, 275.
67. M. Conway 1904, vol. 1, pp. 249–50.
68. Finney 1993, pp. 98–99, 169n55.
69. Sheets-Pyenson 1988 has much on

McCoy and Dawson; see especially pp. 30–31 for anti-Darwinian feeling. MacLeod 1982b deals with geographical factors in the structure of science.

70. Fulford 1968, p. 99. See also *Life and Letters* 3:32, where Lyell reported to Darwin in 1865 that he had "an animated conversation on Darwinism with the Princess Royal."
71. *Bromley Record,* 1 February 1861, p. 12. I thank Randal Keynes for making this available.
72. *Orchids* 283–84.
73. *Correspondence* 9:309.
74. *Correspondence* 9:279.
75. *Journal of the Proceedings of the Linnean Society of London* 7 (1862–63):xv.
76. DAR 112(B):100.
77. *Literary Churchman,* 16 July 1862.
78. Kingsley 1874, p. xxvii, quoted from *Correspondence* 10:634.
79. Gruber and Thackray 1992, p. 17.
80. DAR 219.1:66.
81. *Correspondence* 10:537.
82. DAR 219.1:9.
83. *Correspondence* 10:641.
84. *Correspondence* 10:219.

CHAPTER 6: BATTLE OF THE BOOKS

1. Wallace 1905, vol. 1, p. 384. The birds of paradise are in Scherren 1905, p. 134.
2. Wallace 1905, vol. 1, p. 385.
3. *Correspondence* 10:217.
4. Wallace's copy of the *Origin of Species* 1859 is in the Keynes Collection, Cambridge University Library. Beddall 1988b recounts its history and transcribes Wallace's annotations.
5. Wallace 1905, vol. 2, p. 1.
6. Wallace 1905, vol. 1, p. 417.
7. Camerini 1994.
8. From A. R. Wallace, 4 December [1869], DAR 106/7 (ser. 2):88–89. Wallace often expressed his views on their relationship, for example in Wallace 1903.
9. Darwin's post-*Origin* experimental years are rarely examined by historians. See, however, Vorzimmer 1963 and 1972, Olby 1963, Geison 1969, Rheinberger and McLoughlin 1984, Bartley 1992, and J. Secord 1985. On general topics in plant experimentation see H. Roberts 1929 and Olby 1985.
10. Discussed in Stauffer 1975, pp. 33–34.
11. *Variation* is analysed in Ghiselin 1969, pp. 160–86.
12. *Correspondence* 8:230.
13. *Variation* 2:5–10, 13–15, 22 passim.
14. On social insects and evolutionary theory, see R. Richards 1981, Prete 1990, and J. Clark 1997.
15. *Correspondence* 10:387–88.
16. *Correspondence* 10:193.
17. DAR 112(B):220.
18. *Variation* 1:137, 213 (1875 ed.).
19. *Correspondence* 11:154.
20. *Correspondence* 12:337.
21. Keynes 1943, p. 35.
22. DAR 219.1:56.
23. Pritchard 1896, p. 60. See also Moore 1977 on the education of Darwin's sons.
24. DAR 219.1:69.
25. F. Darwin 1920, pp. 125–26.
26. *Emma Darwin* 2:164.
27. *Correspondence* 10:505.
28. *Correspondence* 10:487.
29. DAR 117:8.
30. *Correspondence* 10:355.
31. *Correspondence* 10:596.
32. *Correspondence* 10:625.
33. *Correspondence* 10:300.
34. *Journal of Horticulture* 4 (1863):93. I thank Perry O'Donovan for this information. See *Correspondence* 10:114–15 and Appendix V, *Correspondence* 11:729–40.
35. *Athenaeum,* March 1863, p. 417; 25 April 1863, pp. 554–55. Quoted from Barrett 1977, vol. 2, p. 78. Also reprinted in Appendix VII, *Correspondence* 11:754–68.
36. *Correspondence* 11:278.
37. *Correspondence* 11:393.
38. *Correspondence* 9:163.

39. *Correspondence* 10:104.
40. *Correspondence* 12:47–48.
41. Down House MS, Classified Account Books.
42. *Correspondence* 10:331.
43. *Correspondence* 11:333.
44. *Correspondence* 12:318–19. See also Brogan 1975.
45. *Calendar* 4467, properly dated 19 April 1865, and *Calendar,* 16 April 1866. Gray Herbarium, Harvard University.
46. Bowler 1983 and 1988.
47. Quoted from Lurie 1960, p. 310. See also pp. 312–13.
48. Moore 1991. See also Peckham 1959 and Bowler 1985.
49. *Correspondence* 8:299.
50. C. Lyell 1863, p. xi. He had already used the same expression, although with a slightly different twist, in *Elements of Geology,* 1851, p. 17. For the rise of "human antiquity" as a category of thought see Grayson 1983, Reader 1988, Bowler 1986, and Van Riper 1993.
51. Compare the first edition of Figuier, published in 1863 and illustrated by Eduard Riou, and the second edition, published 1867. Discussed in Moser 1998, especially plates 5.7 and 5.9. See also Rudwick 1976 and 1989.
52. R. Owen to J. Murray, 21 [no month] 1866, John Murray Archives.
53. *Correspondence* 11:181, 173.
54. *Correspondence* 11:173–74. Darwin refers to a letter to Lyell, 4 February, in which he says, "I have just received the great book.—Very sincere thanks for it. . . . I have turned over pages on species & am very much pleased to see you hit on many of the points which seem to me most important & not generally touched on by others. I have read last chapt. with very great interest." Ibid., p. 114.
55. *Correspondence* 11:223.
56. *Correspondence* 11:217–18.
57. Bynum 1984, Wilson 1996.
58. *Correspondence* 10:611.
59. L. Huxley 1900, vol. 1, p. 264n, and *Correspondence* 11:176–77.
60. T. H. Huxley 1863, pp. 109, 111.
61. Lyell quoted from Bibby 1959, p. 92. Falconer quoted from *Correspondence* 11:179.
62. Denison 1865 and Pattison 1863, p. 17.
63. *Correspondence* 11:148.
64. *Correspondence* 11:177.
65. *Public Opinion,* 2 May 1863, pp. 497–98, also issued separately as a pamphlet. What is probably Darwin's copy is in DAR 221.4:143. Reprinted in Appendix VIII, *Correspondence* 11:769–75.
66. The attribution to George Pycroft is discussed in *Correspondence* 10:770.
67. See the memoir of Bates in Clodd 1892.
68. *Correspondence* 10:54–55.
69. On Romanes, see Marchant 1916, vol. 2, pp. 36–38. For Huxley, see Wallace 1905, vol. 1, p. 36.
70. Raby 2001, p. 167.
71. On mimicry, see especially Blaisdell 1982. Darwin's review is reprinted in *Collected Papers* 2:87–92.
72. Bates 1863, p. 261.
73. *Correspondence* 11:326.
74. L. Darwin 1929, p. 121.
75. *Correspondence* 10:417, 515.
76. *Correspondence* 11:595–96, 603–4, 607–8. On Goodsir, see Jacyna 1983. On Busk, see Cook 1997.
77. *Correspondence* 11:438.
78. Colp 1977, p. 205n55, gives Lane's final address as Harley Street, the medical centre of London.
79. Erasmus Darwin to Fanny Wedgwood, 11 October 1863, Wedgwood/Mosely archive, Wedgwood Archive Collection, Keele University.
80. *Correspondence* 11:423, 438.
81. DAR 112(A):79–82.
82. *Correspondence* 11:620. The Darwins stayed in Villa Nuova, Malvern Wells.
83. *Correspondence* 11:640.
84. *Correspondence* 11:644–45, 646.

CHAPTER 7: INVALID

1. Emma Darwin's diary, DAR 242.
2. Medical notes supplied by Darwin to Dr. John Chapman, 16 May 1865, University of Virginia Library, transcribed in Colp 1977, p. 83. By "rocking" Darwin means the motion of horse-drawn carriages and railway trains.
3. Browne 1998. Colp 1977 reviews the extensive literature, updated in Colp 1998. Some of the differing diagnoses can be found in Kempf 1918, Hubble 1953, Good 1954, Adler 1959, Foster 1965, Winslow 1971, Pickering 1974, Bowlby 1990, F. Smith 1990 and 1992, and D. Young 1997.
4. Bernstein 1984. Colp 1998 reopens the case for Chagas' disease.
5. Darwin to Chapman, 7 June 1865, University of Virginia Library.
6. DAR 112(B):93.
7. "A visit to Darwin's village," *Evening News*, 12 February 1909, p. 4.
8. Most of these remarks are drawn from Francis Darwin's manuscript recollections, DAR 140(3).
9. *Life and Letters* 1:122.
10. DAR 140(3):15.
11. *Correspondence* 11:501.
12. *Correspondence* 12:37, 57.
13. L. Darwin 1929, p. 120.
14. The possibility of nervous disorder is presented most notably by Pickering 1974 and Colp 1977. See also Bowlby 1990 and Desmond and Moore 1990.
15. Discussed in Browne 1998. See also Bynum 1997.
16. DAR 112(B):51. See also Graham 1984 and Bailin 1994.
17. DAR 112(B):35, 49. See also Ehrenreich and English 1973.
18. *Emma Darwin* 2:331. The social politics of illness are analysed in Sicherman 1977, Haley 1978, and Bailin 1994. For the "nervous" patient, see Bynum 1982 and 1985, Berrios 1985, and Oppenheim 1991. Hysteria is analysed as a category in Micale 1995 and Gilman et al. 1993. Model case studies are in Bynum and Neve 1985 and Wiltshire 1992.
19. *Correspondence* 11:602–3.
20. On the medical profession and its relationships with patients see especially M. J. Peterson 1978, R. Porter 1985, and Digby 1994.
21. *Correspondence* 11:690.
22. *Correspondence* 11:695.
23. To W. E. Darwin, 22 June 1866, DAR 185. I thank Duncan Porter for these references.
24. *Correspondence* 11:506. On contemporary imagery of women as sickroom nurses, see Bailin 1994 and Winter 1995. Women acting as amanuenses and the way letters can generate power structures are discussed in Goldsmith 1989. On women's role more generally see Vicinus 1972 and Kohlstedt and Longino 1997. On women's biographies see Wagner-Martin 1994.
25. *Correspondence* 11:616.
26. DAR 219.1:80.
27. *Correspondence* 11:689.
28. DAR 219.9:15.
29. *Correspondence* 12:29.
30. *Correspondence* 12:311, 312.
31. DAR 219.1:78.
32. *Correspondence* 12:387.
33. *Correspondence* 12:212.
34. *Correspondence* 12:31.
35. Bartholomew 1976 and MacLeod 1971. See also *Life and Letters* 3:27–28, *Correspondence* 11:662, and *Royal Society Minutes of Council* 3(1858–69):197. A full account is given in Appendix IV, *Correspondence* 12:509–27. A general assessment of the Royal Society in this period is in M. B. Hall 1984.
36. *Correspondence* 11:662.
37. *Correspondence* 11:669.
38. *Proceedings of the Royal Society of London* 13(1863–64):508.
39. Bartholomew 1976, pp. 215–17.
40. *Correspondence* 12:423.
41. *Emma Darwin* 2:204.

42. Jenson 1970, p. 64. On the X Club generally, see MacLeod 1969, Barton 1990 and 1998, and A. Desmond 2001.
43. Jenson 1970, p. 63.
44. MacLeod 1969 and Roos 1981. See also the centenary edition, *Nature* 1989.
45. *Correspondence* 12:18.
46. *Correspondence* 12:384. Wallace was awarded the Royal (Gold) Medal in 1868.
47. I thank Gudrun Richardson of the Royal Society for checking nominations between 1860 and 1893.
48. MacLeod 1969, p. 312.
49. Snell and Ell 2001.
50. Lubbock's role is discussed in Hutchinson 1914, vol. 1, pp. 57–58. See also Brock and MacLeod 1976.
51. *Correspondence* 12:35. See Colenso 1862–79. Colenso is discussed by Hinchcliff 1964 and Rowse 1989.
52. Bradford 1996 and Jenkins 1996. Disraeli was caricatured in *Punch*, 10 December 1864, 239. The quotation comes from the caption.
53. Lorimer 1978, Stepan 1982, and E. Richards 1989. More generally see Stocking 1987.
54. On racial anthropology in general, see Gillespie 1977, E. Richards 1989, Edwards 1992, J. Harvey 1993, and Alter 1999.
55. Lorimer 1978.
56. Stepan 1982 and Lorimer 1978. See also Vogt 1864.
57. Wallace 1864. See *Correspondence* 12:204. Darwin's letter to Wallace on the subject is in ibid., 216–17.
58. Wallace 1864, p. clxv. See also R. Smith 1972 and Durant 1979.
59. Lubbock 1865, Van Riper 1993, and Bynum 1984.
60. Leopold 1980.
61. *Life and Letters* 3:52–53. Darwin criticised his son William for easy assumptions about the issue; ibid., p. 53. The Eyre affair is examined in Semmel 1962.
62. L. Huxley 1900, vol. 1, p. 278.
63. From Asa Gray, 15 May 1865, DAR 165.
64. To Asa Gray, 15 August [1865], Gray Herbarium, Harvard University.
65. From a number of studies on the translation of science see especially Glick 1988a and, more recently, Rupke 2000. The development of specialised language and jargon is touched on in Burke and Porter 1995.
66. Altick 1957, Ellegard 1957, and Eco 1979.
67. *Life and Letters* 3:88.
68. Scudo and Acanfora 1985, Vucinich 1988, and Todes 1989.
69. Vucinich 1988.
70. *Correspondence* 12:265–68 and Junker and Richmond 1996, p. 19.
71. Haeckel 1866. Haeckel's work and Darwinism in Germany are discussed in Gasman 1971, Kelly 1981, Corsi and Weindling 1985, pp. 685–98, and Montgomery 1988. More generally see Nyhart 1995.
72. Müller 1864, quoted from the English translation 1869, p. 114.
73. *More Letters* 1:312.
74. See L. Clark 1984 for the cool response.
75. Much of the French reaction is given in J. Harvey 1997. See also Conry 1974 and J. Harvey 1983.
76. *Life and Letters* 3:118.
77. Quatrefages to Darwin, 29 March 1869, DAR 175.
78. To Hooker, 8 March [1869], DAR 94:116–17.
79. Núñez 1977 and Glick 1988b.
80. Pancaldi 1991.
81. Described in Corsi and Weindling 1985.
82. *Emma Darwin* 2:207.
83. *Emma Darwin* 2:224.
84. Emma to Fanny Allen, late June–early July 1865, Wedgwood/Mosely archive, Wedgwood Archive Collection, Keele University, 422.
85. Haight 1940, p. 237. For the tradition of entrepreneurial medicine in

which Chapman was engaged see R. Porter 1989.

86. Quoted from Colp 1977, p. 83.
87. Chapman 1864.
88. H. B. Jones 1850. On Bence Jones's medical system see Coley 1973 and M. Stone 1998. Francis Darwin's recollection is in DAR 140(3):13.
89. Mellersh 1968. Burton 1986 deals with FitzRoy's meteorological researches, especially FitzRoy 1862. See Anderson 1987 and Gates 1988 for Victorian suicide.
90. Mellersh 1968, pp. 243–46. Much of Darwin's knowledge of these developments came through his old *Beagle* friend Bartholomew Sulivan; see Sulivan 1896.
91. Burton 1986.
92. To Hooker, 4 May [1865], DAR 115:268, Mellersh 1968, p. 283.
93. Colp 1985, p. 390.
94. Wedgwood/Mosely archive, Wedgwood Archive Collection, Keele University, 422. See also *Life and Letters* 3:36.
95. *More Letters* 2:156–57.
96. *Emma Darwin* 2:209.
97. J. C. Wedgwood 1909, p. 17. Emma Darwin's views on Langton are in DAR 219.9.81.
98. See for example Briggs 1981.
99. Royal Society of London Archives, Soirée Bills, 1861–72; Posters, 1863–72. I am grateful to M. B. Hall and the Royal Society's archivists for their help on this matter.
100. Soirée Bills, Royal Society of London Archives, 1861–72
101. DAR 219.9:42. It may have been that the pair of silver candlesticks mentioned in Darwin's will as "presented to me by the Royal Society," were given on this occasion.
102. Haeckel 1866, in which he discussed Darwin's theories throughout, especially vol. 2, pt. 2, pp. 163–294. See Darwin to Ernst Haeckel, 18 August [1866], Haeckel Haus, Friedrich-Schiller-Universitat, Jena. For Haeckel's voice see DAR 219.9.202.
103. Quoted from Desmond and Moore 1991, p. 539.
104. *More Letters* 2:350; see also DAR 219.9:202.
105. From Hooker, 16 January 1866, DAR 102:53–55.
106. L. Huxley 1918, pp. 98–105. See also Beer 1996, which discusses this story as a "parable."
107. DAR 219.1:84.
108. Walford 1868, an updated version of Reeve 1863–67. On fame see Braudy 1986 and Goffman 1990.
109. *More Letters* 1:264. The craze for *cartes de visite* is discussed in Wynter 1862 and Darrah 1981. See also Browne 1998. On portrait photography more generally see Prescott 1985 and G. Clarke 1994. On representation and self-representation see Gilman 1976, Tagg 1988, Yeo 1988, Shortland 1996, Green-Lewis 1996, Armstrong 1998, Homans 1999, and Jordanova 2000.
110. Marchant 1916, vol. 1, p. 248.
111. Burke 2001.
112. To Hooker, 25 November [1867], DAR 94:37–38.
113. Woolner 1917.
114. To Hooker, 26 November [1868], DAR 94:98–101.
115. Erasmus Darwin to Emma Darwin, DAR 105(ser. 2):121. Darwin's bust was exhibited by Woolner at the Royal Academy in 1870. It is currently displayed at Down House.

## CHAPTER 8: THE BURDEN OF HEREDITY

1. Useful sources are Vorzimmer 1963, Olby 1963, Geison 1969, Bartley 1992, and Winther 2000. On commercial animal breeding see Russell 1986.
2. *Variation* 2:358.
3. *Notebooks* B2, 4.
4. Hodge 1985. See also Farley 1982 on nineteenth-century notions of animal and plant reproduction.
5. Victorian manners and social

connections are touched on in Briggs 1965, Davidoff 1973, Curtin 1987, Collini 1991, and Cannadine 1999.

6. Gay 1984–88, Davidoff and Hall 1987, and L. Stone and Stone 1986. See also M. J. Peterson 1989 and Wedgwood and Wedgwood 1980 for family connectivity in the period.
7. A. Richardson 1998a and 1998b. See also Bender 1996.
8. Economic themes in science are explored in Mirowski 1994, Power 1996, and Poovey 1998.
9. *Variation* 2:404.
10. *Notebooks* 690.
11. *Variation* 2:175.
12. *Variation* 2:143.
13. Jenkin 1867. See especially Morris 1994 and Cookson and Hempstead 2000.
14. Quoted from Hull 1973, pp. 315, 316.
15. Cookson and Hempstead 2000, pp. 165–68, although the authors claim that Darwin did not adjust his argument to accommodate Jenkin's comments.
16. *Correspondence* 11:223. On Lamarck see R. W. Burkhardt 1977, Jordanova 1984, and Corsi 1988b.
17. A. Desmond 1989, Desmond and Moore 1990.
18. *Variation* 2:387.
19. In 1842 Darwin served on the British Association committee that drew up rules for zoological nomenclature; see *Correspondence,* vol. 2. On classification and evolutionary theory see Crowson 1958, Ritvo 1995, and McOuat 2001.
20. To George Darwin, 27 May [1867], DAR 210.1.1.
21. To H. W. Bates, 22 February 1868, Houghton Library, Harvard University; to Charles Lyell, 19 March 1868, American Philosophical Society, Philadelphia.
22. *More Letters* 2:371.
23. To T. H. Farrer, 29 October [1868], Linnean Society of London.
24. To Hooker, 21 March [1867], DAR 94:13.
25. The inscribed copy to an "opponent" was listed for sale at Sotheby's 11 December 1992. Presentation lists are in DAR 210.18. It seems that Darwin usually sent a copy of his books to Camilla Ludwig, as if she too were a member of the family. A number of items, mostly offprints of significant reviews of Darwin's works, are listed in Sotheby's sale as above, described as "from the collection of C. Ludwig, Leipzig."
26. Freeman 1977, p. 122. See Scudo and Acanfora 1985.
27. *Variation* 2:321. Julia Pastrana is discussed in Howard 1977 and Bondeson 1997.
28. To Hooker, 17 November [1867], DAR 94:35–36. Hooker's reaction to pangenesis is in L. Huxley 1918, vol. 2, pp. 109–10. The remark to Huxley is in *More Letters* 1:287.
29. Stamhuis et al. 1999. See also De Vries 1909.
30. From Hooker, 26 February 1868, DAR 102:200–3; DAR 94:67.
31. To Hooker, 19 November [1869], DAR 94:159–61, partly in *Life and Letters* 3:110.
32. Waller 2001. The standard biography is Forrest 1974.
33. Means 1876.
34. Waller 2001.
35. Galton 1908, pp. 287–88.
36. Galton 1869, p. 210.
37. Galton 1908, p. 290.
38. DAR 219.9.80.
39. Pearson 1914–30, vol. 2, p. 160. The full correspondence between Darwin and Galton on pangenesis is printed in ibid., pp. 156–77.
40. Bynum 1991. Pearson 1913–40 provides an authoritative account of Galton's laws of heredity, vol. 2, pp. 298–309. See also Forrest 1974, pp. 187–206.
41. *Life and Letters* 3:120.
42. To John Tyndall, 8 September 1870, Royal Institution of Great Britain.
43. Gayon and Zallen 1998.
44. *Variation* 2:430.
45. *Variation* 2:426–27.

46. F. Darwin 1916, p. ix.
47. DAR 140(4):55–56.
48. DAR 219.1:85.
49. *Life and Letters* 1:135. See also DAR 219.1.17.
50. F. Darwin 1916, p. xii.
51. DAR 219.1:17.
52. *Emma Darwin* 2:216.
53. DAR 219.8.6. See also 219.9.56. George Darwin's letter to his brother Leonard is in DAR 219.6.2.
54. *Emma Darwin* 2:218.
55. Cobbe 1904, pp. 487–88.
56. T. H. Huxley 1869.
57. L. Huxley 1900, vol. 1, p. 300.
58. *More Letters* 1:313.
59. They rented Redoubt House, Freshwater; see Hinton 1992, p. 18.
60. *Life and Letters* 3:92, 102, and *Emma Darwin* 2:220–22.
61. E. Agassiz 1868, vol. 2, p. 666. Longfellow was invited to a garden party at Tennyson's house at which more than forty guests attended. Thwaite 1996, p. 443.
62. *Emma Darwin* 2:220. Neither of these photographs has been located in catalogues of J. M. Cameron's works.
63. *Emma Darwin* 2:222.
64. See Page 1983 for interviews and recollections of Tennyson, many of them engineered by Cameron.
65. Hoge 1981. See Thwaite 1996 for Emily Tennyson, especially p. 459 on Darwin's theories.
66. Allingham 1907, pp. 184, 185.
67. Hopkinson 1986 and Hinton 1992, p. 33. See also Cameron 1893 and Weaver 1984. For portrait photography in Victorian England see Prescott 1985, Weaver 1989, G. Clarke 1994, and Jordanova 2000. More generally see Fyfe and Law 1988 and Cowling 1989.
68. *Photographic News* 8 (3 June 1864):266; *British Journal of Photography* 11 (1864):261.
69. Cameron 1893, p. 9.
70. Hopkinson 1986, p. 68.
71. *Correspondence* 12:240. The attribution to William Darwin is in a letter from Darwin to Asa Gray, ibid., pp. 212, 214n. Hooker refers to J. R. Herbert's fresco of Moses on Mount Sinai in the Moses Room, Houses of Parliament, completed in 1864.
72. *Correspondence* 12:271.
73. Something of the history of beards is given in Asser 1966 and Cooper 1971. Darwin's beard is discussed in Browne 1998.
74. James Hannay, "The Beard," *Westminster Review* 62 (1854):48–67, in which Hannay states that the beard was a symbol of "revolution, democracy and dissatisfaction with existing institutions. . . . only a few travellers, artists, men of letters and philosophers wear it" (p. 49).
75. Colp 1985. Berg 1951 summarizes the psychoanalytic view.
76. *Descent of Man* 2:317–23, 372, 379–80. Gender relations in Darwin's evolutionary biology are examined in E. Richards 1983 and 1989, Russett 1989, and Jann 1994. See also Mangen and Walvin 1987, D. Roberts 1978, and Shortland 1996. Masculinity is discussed in Tosh 1999.
77. The Royal Society copy, on the original Colnagi mount, with blind stamp, is the only copy I have seen with this mechanically reproduced text at the bottom.
78. *Life and Letters* 3:102 and Classed Account Books, 19 August 1868, Down House Archives.
79. From Hooker, 30 August 1868, DAR 102:229–32.
80. *More Letters* 2:376–77.
81. *Emma Darwin* 2:215.
82. To Hooker, 2 July [1870], DAR 94:175–76.
83. *Life and Letters* 3:75.
84. A copy is in DAR 53(i)ser. B:2.
85. Freeman 1977, pp. 120–22. Printed in 1868 but also distributed in manuscript during 1867. It is reproduced in facsimile in Freeman and Gautry 1975.

86. Browne 1985a.
87. *More Letters* 1:287.
88. *Life and Letters* 2:91.
89. Wallace 1905, vol. 2, p. 386. See McKinney 1972 and, more generally on sexual selection, Cronin 1991. For Argyll, see Campbell 1867 and 1869 and Gillespie 1977.
90. To Charles Kingsley, 10 June [1867], DAR 96:28–29, 32.
91. *Dublin University Magazine* 74 (1869):589, quoted from Ellegard 1990, p. 308.
92. L. Huxley 1918, vol. 2, p. 121.
93. Marchant 1916, vol. 1, p. 221.
94. Dupree 1988, p. 338.
95. The rise of naturalism is discussed in R. Young 1970, Chadwick 1975, Royle 1974, F. M. Turner 1974, and Lightman 1987. See also Paradis and Postlewait 1981, Helmstadter and Lightman 1990, and Lightman 1997.
96. L. Huxley 1900, vol. 1, pp. 319–20. See also A. Desmond 1997, pp. 374–75. For the use of "agnostic" before Huxley see Blinderman 1995.
97. K. M. Lyell 1881, vol. 2, p. 341. See also Dupree 1988, pp. 299–301.
98. DAR 219.9.60.
99. *More Letters* 1:312
100. *Life and Letters* 3:107.
101. For example, *Variation* 2:224.
102. *More Letters* 1:267–68.
103. Wallace's copy of Darwin 1859 is in Cambridge University Library, Keynes collection; see especially pp. 82–86. Beddell 1988 transcribes Wallace's annotations.
104. Paul 1988. See also Peel 1971 and Haines 1991.
105. Spencer 1864–67, vol. 1, pp. 444–45.
106. Touched on in Depew and Weber 1995.
107. *More Letters* 1:261.
108. Spencer 1864–67, vol. 1, pp. 444–45.
109. Marchant 1916, vol. 1, p. 242.
110. Toulmin and Goodfield 1965, Gould 1987, and Burchfield 1990. See also Geikie 1874.
111. Wallace 1905, vol. 2, p. 39.
112. *Life and Letters* 3:121.
113. A. R. Wallace to Charles Kingsley, 7 May 1869, Knox College Library, Galesburg.
114. DAR 219.8.9. Darwin and Wallace's interactions with Blyth are discussed variously in Eisley 1959, Beddall 1973, and Schwartz 1974. A general account of Blyth's life is in Geldart 1879.
115. Marchant 1916, vol. 1, p. 235.
116. Wallace 1869b. Darwin's remark is in Marchant 1916, vol. 1, p. 241.
117. K. M. Lyell 1881, vol. 2, p. 441.
118. Marchant 1916, vol. 1, p. 244. Wallace's spiritualism is discussed in R. Smith 1972. See also Wallace 1866 and 1875.
119. Quoted from Ellegard 1990, p. 309.
120. Oppenheim 1985. On mesmerism see Winter 1998.
121. Wallace 1905, vol. 2, p. 280.
122. Wallace 1905, vol. 2, p. 286.
123. Marchant 1916, vol. 1, p. 251.
124. *Life and Letters* 3:112.

CHAPTER 9: SON OF A MONKEY

1. *Correspondence* 6:515.
2. From W. W. Reade, [?January–April 1870], DAR 176.
3. Woolner 1917. It is not entirely clear when Darwin saw the statue. It was exhibited at the Royal Academy galleries in 1866.
4. To T. Woolner, 10 March [1869?], Bodleian Library, Oxford. See *Descent* 1:22.
5. *Descent* 1:33.
6. To T. H. Farrer, 13 [May 1870], Linnean Society of London.
7. *Life and Letters* 3:129. William Farr's visit to Down House is in DAR 219.9.90.
8. *Nature*, 28 April 1870.
9. G. Darwin 1875, p. 22.
10. *Descent* 2:403.
11. Barton 1990 and 1998, A. Desmond 2001.
12. *More Letters* 3:135.

13. Mivart 1871, p. 60. Darwin's copy is in Darwin Library, Cambridge University Library.
14. Mivart's life and work are discussed in J. Gruber 1960. Quotation from ibid., p. 52.
15. *More Letters* 3:144–45.
16. From Mivart, 10 January 1872, DAR 171.
17. *Life and Letters* 3:106.
18. Cobbe 1904, pp. 485–88. The quotation is from p. 487.
19. The adjusted letter is in the *Echo*, 25 August 1870, p. 2. See *Emma Darwin* 2:432 and DAR 219.9.91.
20. Cobbe 1894, vol. 2, p. 126.
21. Marchant 1916, vol. 1, p. 251.
22. Calculated from Ellegard 1990.
23. Unpublished lecture, Janet Browne, "Science and Medicine in Mudie's Circulating library," University of London, February 2000.
24. To David Forbes, 31 July [1870], American Philosophical Society, Philadelphia.
25. Classed Accounts, Down House Archives.
26. *Descent* 1:206.
27. Searby 1997, vol. 3, pp. 203–33, and MacLeod 1982a.
28. Keynes 1943, p. 4.
29. *Emma Darwin* 2:253.
30. Raverat 1952, p. 146.
31. *Autobiography*, p. 97.
32. Cambridge University Examination papers, 1871–72.
33. Stebbins 1988. See also letter from J.L.A. Quatrefages, 29 March 1869, DAR 175.
34. George Darwin to Henrietta Darwin, April 1869, Wedgwood/Mosely archive, Wedgwood Archive Collection, Keele University, 473.
35. Pearson 1914–30, vol. 2, p. 176.
36. Emma Darwin to Fanny Allen, Wedgwood/Mosely archive, Wedgwood Archive Collection, Keele University.
37. *Life and Letters* 3:124. See also the letter from Sedgwick, 30 May 1870, *More Letters* 2:236–37.
38. Emma Darwin to Fanny Allen, undated letter, Wedgwood/Mosely archive, Wedgwood Archive Collection, Keele University.
39. Woolner 1917, p. 284.
40. See Jones and Gladstone 1998.
41. Atlay 1903, p. 348. See also Ward 1965.
42. A. Desmond 1997. Royle 1984 gives a sense of how dangerous Huxley's views might have seemed to conservative Oxford dons. See also Lightman 1987.
43. L. Huxley 1900, vol. 1, p. 331.
44. To B. J. Sulivan, 30 June 1870, *Calendar* 7256.
45. Cronin 1991. See particularly Jann 1994.
46. *Descent* 1:35.
47. *Descent* 1:62.
48. Taub 1993, Alter 1999, pp. 73–79, 100. See also *Descent* 1:53.
49. Alter 1999, 111–26.
50. *Nature*, 6 January 1870, p. 257, referring to Schleicher 1863, translated by Bikkers in 1869.
51. *Descent* 1:56.
52. Emma Darwin's letter is in DAR 219.9.72. Julia Wedgwood's critique is in DAR 139.12:17.
53. *Descent* 1:73.
54. *Descent* 1:71.
55. *Descent* 2:404–5.
56. *Descent* 1:180.
57. *Descent* 1:205.
58. From W. D. Fox, 18 [November 1870], DAR 164.
59. *Descent* 2:368–69.
60. *Descent* 2:327–28. Jann 1994 analyses the masculine assumptions underpinning Darwin's sexual biology. See also E. Richards 1983.
61. To Asa Gray, 15 March [1870], Gray Herbarium, Harvard University.
62. Marchant 1916, vol. 1, p. 247.
63. W. Elwin to John Murray, 21 September 1870, John Murray Archives. Printed in part in Paston 1932, pp. 230–32.
64. University College London, Archives, Pearson papers 613:3. For female assistance in scientific matters see M. J. Peterson 1989.

65. DAR 112:144.
66. To Henrietta Darwin, [March? 1870], British Library, Add 58373.
67. To Henrietta Darwin, [March–June 1870], DAR 185.
68. *Emma Darwin* 2:241.
69. DAR 219.9.72. See *Descent* 1:77–78.
70. From John Murray, 28 September [1870], DAR 171.
71. From John Murray, 1 July 1870, DAR 171.
72. Marchant 1916, vol. 1, p. 253.
73. From John Murray, 28 September [1870], DAR 171.
74. Freeman 1977.
75. *Times,* 8 April 1871, quoted from Ellegard 1990, p. 300.
76. *Edinburgh Review,* July 1871, 195–235.
77. *World,* 19 March 1871.
78. Quotations drawn from reviews collected in DAR 129. See also Ellegard 1990, pp. 293–331.
79. DAR 140.4:19.
80. Reviews in DAR 129, pp. 75, 104.
81. *Blackwood's Magazine,* April 1871.
82. F. O. Morris 1875, p. 15. Darwin's copy is in DAR 139.12:1.
83. Marchant 1916, vol. 2, p. 32.
84. *Life and Letters* 3:134, 136.
85. Marchant 1916, vol. 1, p. 127.
86. Wallace 1905, vol. 2, p. 48.
87. *Life and Letters* 3:133.
88. *Fraser's Magazine* n.s. 5 (1872):409–21.
89. Stephen 1924, especially pp. 88–89.
90. *Life and Letters* 3:139.
91. To John Murray, 3 June [1871], John Murray Archives.
92. *Life and Letters* 3:149. See also *More Letters* 1:333.
93. G. Darwin 1873. See L. Huxley 1900, vol. 1, pp. 425–26. Numerous letters were exchanged through December 1874 and January 1875; see *Calendar* 9759 to 9823, passim. An account is given in A. Desmond 1997, pp. 71–72.
94. J. Gruber 1960, pp. 112–14.
95. Marchant 1916, vol. 1, p. 269.
96. Emma Darwin to Fanny Allen, undated letter, Wedgwood/Mosely archive, Wedgwood Archive Collection, Keele University.
97. DAR 210.2.
98. Henrietta Darwin to George Darwin, June 1871, DAR 245:298.
99. Hudson 1972, p. 298.
100. Litchfield 1910, pp. 124–25. See also Harrison 1954.
101. To W. D. Fox, 16 July 1872, American Philosophical Society, Philadelphia.
102. DAR 245:45. See also Emma Darwin's letter to Henrietta Litchfield, in which she said that "sickness marries one," DAR 219.9.95.
103. *Emma Darwin* 2:247.
104. *More Letters* 2:100–1.
105. To William Ogle, 17 November [1870], Down House MS 5:4.
106. To W. E. Darwin, 11 February [1872], British Museum (Natural History). For Darwin's work on expression see Browne 1985a, Montgomery 1985, and Paul Ekman's edition of *Expression of the Emotions,* C. Darwin 1872.
107. *Expression* 94 and Litchfield 1910, pp. 216–17.
108. M. Vaughan-Williams to Henrietta Darwin, undated, DAR 180. See also DAR 219.9.84.
109. *Life and Letters* 1:114. Emma Darwin's letter is in DAR 219.9.85.
110. British Library Add MS 58373.
111. DAR 219.9.14.
112. *Expression* 62.
113. Classed Account books, Down House Archives.
114. See especially Lynch 1985, Lynch and Woolgar 1988, Tagg 1988, Tucker 1997, and Armstrong 1998. More generally on questions of representation, see Edwards 1992, Gilman 1995, and Green-Lewis 1996.
115. I am grateful to Phillip Prodger for bringing these views to my attention. See also Prodger 1998.
116. I thank Edward Wakeling for confirming this attribution.
117. Discussed in Bondeson 1997.

118. Prodger 1998.
119. Browne 1985a and 1985b. See also Hartley 2001.
120. DAR 53(1):40.
121. Eckman, introduction, *Expression* xxi–xxxvi.
122. DAR 143. For Crichton-Browne's system of psychiatry see Neve and Turner 1995. For photography and expressions of the insane see Gilman 1976 and Browne 1985b, and more generally on images of insanity Gilman 1995. For the continuing traditions of physiognomy see Cowling 1989, Edwards 1992, and Hartley 2001.
123. J. E. Jenkins, *Ginx's Baby: His Birth and Other Misfortunes* (London 1870), running to thirty-six editions by 1876. The edition illustrated by Frederick Barnard was published in 1876.
124. E. Jones 1973. See also Prichard 1988.
125. Prodger 1998.
126. Classed Account Books, Down House Archives.
127. E. Jones 1973, p. 102. The volume also reproduces one half of the composite *Ginx's Baby,* p. 107
128. DAR 53(i), ser. C:95.
129. Marchant 1916, vol. 1, p. 272.
130. Montgomery 1985.

CHAPTER 10: DARWIN IN THE DRAWING ROOM

1. Edward Aveling, *Students' Magazine of Science and Art,* 2 September 1878.
2. Wallace 1883, p. 420. On "Darwinism" as a category see Moore 1991. Several of Darwin's closest friends went on to publish books of essays carrying in their title variants of the words "Darwinism" or "Darwinina." See T. H. Huxley 1893, Gray 1876, Romanes 1892, and Wallace 1889.
3. Discussed, for example, in Altick 1978, Jardine, Secord, and Spary 1996, and Yanni 1999. The rise of a consumer society is examined in McKendrick, Brewer, and Plumb 1982 and T. Richards 1990. See particularly P. Anderson 1991 on the printed image and transformation of popular culture, and Ashton 1991 and 1996 on Lewes and Eliot respectively.
4. Several versions are in existence, some bronze. One is illustrated on the cover of Kew Books Newsletter 6, 1976. A more convenient illustration is in Service 2000. A "monkey with Darwin's head" was listed in the Darwin Centenary Exhibition, Cambridge 1909; see Freeman 1978, p. 98.
5. Howard 1977. For monstrosities as part of biological thought see E. Richards 1994.
6. The Wedgwood ware, "The Darwinian Theory," is illustrated in Reilly 1989, vol. 2, p. 130, and was produced by Émile Lessore in 1862. I doubt whether it was ever available in mass-produced form. I am grateful to the Royal Photographic Society and Ms. Debbie Ireland for helping me with questions about Cameron's copyright photographs. R. D. Wood kindly provided access to his list of Mrs. Cameron's Copyright Registrations, from which this information is taken.
7. I thank Teresa Termet Huget for this information.
8. Bingham 1994. I am extremely grateful to A. Walker Bingham for providing me with information about this firm and its advertisements. See also Holcombe 1979, 367–77.
9. Cantor 1996 examines public images of a scientific hero. See also Fara 1997 and 2000. In general, see Tagg 1988, Edwards 1992, Green-Lewis 1996, Armstrong 1998, and Jordanova 2000. Useful case studies are in Friedman and Dorley 1985 and Rose 1999.
10. Ryan 1996 discusses some of these

metaphors of cartography and exploration.

11. *Autobiography,* p. 78.
12. The history of caricature is discussed in George 1959, Weschler 1982, Townsend 1992, and Hallett 1999. For caricature and satire in science see Rudwick 1975 and 1976, Browne 1992, and Paradis 1997.
13. From a large literature on medical caricature see Haslam 1996 and R. Porter 2001.
14. Weschler 1982 for Paris, Townsend 1992 for Berlin.
15. *Grove Dictionary of Art,* vol. 5, pp. 755–61. See also J. R. Harvey 1970.
16. "A logical refutation of Mr. Darwin's theory," *Punch,* 1 April 1871, p. 130.
17. A rough estimate is that there was one picture on every page of *Punch.* An annual volume ran to some 270 pages. See Huggett 1978.
18. From D. Thomas [1871?], DAR 178.
19. *Harper's Weekly,* 23 December 1871, p. 1209.
20. "Man is but a worm," *Punch's Almanack,* 6 December 1881.
21. Hague 1884, p. 760.
22. *Harper's Weekly,* 4 January 1879.
23. Unattributed, *Harper's Weekly,* 22 February 1873, p. 160.
24. W. Smith 1967.
25. M. Conway 1904, vol. 2, p. 357, discusses the prints, although it confuses one with the other. The attribution to Holyoake is made in W. Smith 1967, pp. xiii–xvi, which reprints the second version as endpapers. I am grateful to Jim Moore for his help on this issue.
26. Darwin's copies are in DAR 141:10 and 11.
27. The first version certainly ran to three impressions, but each batch was probably rather few in number. Since one copy hung in the City Temple, the pivotal institution for revivalist preaching run by Joseph Parker, it would be fair to say that the intended audience was probably highly nonconformist. A copy marked "Third Impression" was offered for sale at Sotheby's, 11 December 1992. See W. Smith for the Temple copy.
28. *Globe,* 7 November 1872.
29. Allingham 1907, p. 185.
30. Groeben 1982, pp. 93–94n33.
31. Groeben 1982.
32. Maitland 1906, p. 301. For Romanes see Schwartz 1995.
33. Family enterprises are examined in Davidoff and Hall 1987 and M. J. Peterson 1989. The relationship between spectacle and family intimacy is examined in Chase and Levenson 2000.
34. *Emma Darwin* 2:278.
35. Illustrated in Desmond and Moore 1990.
36. Allingham 1907, p. 239.
37. M. Conway 1904, vol. 1, pp. 358–59. See also Wilson and MacArthur 1934, vol. 6, p. 328.
38. Leeds University, Samuel Smiles Correspondence, SS/A/1,72. I am grateful to Anne Secord for this quotation.
39. *Emma Darwin* 2:260.
40. *Emma Darwin* 2:278–79.
41. Emma Darwin to Fanny Allen, undated letter, Wedgwood/Mosely archive, Wedgwood Archive Collection, Keele University.
42. *Emma Darwin* 2:264.
43. Stecher 1961, p. 247.
44. *Dictionary of National Biography,* 22 vols. (Oxford: Oxford University Press, 1882–1912), supplement 1901–1911, ed. S. Lee, vol. 1, entry on Leslie Stephen.
45. Maitland 1906, p. 300. See also Annan 1984, pp. 97–98.
46. Sotheby catalogue, 11 December 1992, item 143.
47. M. Conway 1904, vol. 2, p. 173.
48. M. Conway 1904, vol. 2, pp. 324–26.
49. Haight 1954–78, vol. 9, pp. 87–88.
50. From Adolph Reuter, 11 January 1870, DAR 176.
51. Moore 1985b.
52. DAR 140(3):30.

53. Classed Account Books, Down House Archives. Darwin entered the charges for parcels separately.
54. *Life and Letters* 1:112–3,121.
55. To R. L. Tait, 13 January 1880, Shrewsbury School. See also Shepherd 1982.
56. *More Letters* 2:443.
57. DAR 140 (3):31.
58. From Thomas Burgess, 26 March 1875, DAR 106/7(ser. 4):15–16.
59. From F. E. Abbot, 18 July 1872, DAR 159.
60. F. Abbot, "Darwin's Theory of Conscience and Its Relation to Scientific Ethics," *Index*, 12 March 1874. Letter to Abbot, 30 March 1874, Harvard University Archives, quoted from *Calendar* 9377.
61. *Life and Letters* 3:18n but no attribution except 1871. The actual citation is letter to J. D. Hooker, 1 February [1871], DAR 94:188–89.
62. Groeben 1982, p. 29.
63. O'Brien 1970, Farley 1977, and especially Strick 1999. Nineteenth-century microscopical and cellular investigations are discussed in Churchill 1979.
64. Huxley to Spencer, see Marchant 1916, vol. 2, p. 239.
65. *Life and Letters* 3:168, 169.
66. Hudson 1972, p. 315.
67. L. Huxley 1900, vol. 1, pp. 366–67.
68. Colp 1985, p. 385n6. Galton's correspondence with Candolle on the issue is in Pearson 1913–40, vol. 2, pp. 134–49. For contemporary views on character see Galton 1865, Candolle 1873, and Smiles 1859, 1871, and 1887. More generally on character and biography see the essays in Shortland and Yeo 1996, especially Cantor 1996 and Söderqvist 1996.
69. To Hyacinth Hooker, 31 January 1877, Case Western Reserve University.
70. Published in *Das Ausland*, 2 April 1870. See also Preyer 1891.
71. To William Preyer, 15 May 1870, DAR 147.
72. Newpaper clipping in DAR 226.2:5.
73. *Life and Letters* 3:229.
74. *Life and Letters* 3:179. Galton's aims are discussed in Pearson 1913–40, vol. 2, pp. 178–79.
75. *Life and Letters* 3:177–79.
76. *Life and Letters* 1:149. I am grateful to Ann Dally for the citation from Trollope. The words are spoken by old Giles Hoggett, who exhorts Mr. Crawley as to the best course in life, and turns the tide of the plot. See also Colp 1998, p. 223n50.
77. *Emma Darwin* 2:259.
78. DAR 140(3).
79. *Emma Darwin* 2:355, 376. See Bermingham 1986 for landscape and the English rustic tradition, and Paradis 1985, Krasner 1992, and Worster 1985 for Darwin's ecological vision.
80. Healey 1986. See also Keynes 2001.
81. See letter to John Lubbock, 23 February 1874, DAR 97(ser. C):44–45.
82. DAR 210.15.
83. To J. J. Weir, 18 September [1872], Countway Library, Harvard University.
84. From J. S. Craig, 4 November 1872, DAR 96:112.
85. Another count makes the total eighty-four. Most of the certificates, formal invitations, and scrolls are in DAR 229 and 230.
86. Feuer 1975, Colp 1982, and H. Gruber 1961. The volume is at Down House.
87. Pancaldi 1994, p. 265.
88. DAR 219.1.92. On spiritualism see Oppenheim 1985.
89. Imperial College Archives, Huxley Papers 28:221–23 and supporting correspondence dated 1874.
90. Haight 1968, p. 469; Lewes's diary quoted from Baker 1995, vol. 1, p. 223.
91. *Life and Letters* 3:187.
92. See L. Huxley 1900, vol. 1, pp. 421–22, for Huxley's description. Accounts of the séance are in DAR 154:124–28 and DAR 119.1:116.

93. To G. H. Darwin, 2 [April 1875], DAR 210.1.2.
94. Oppenheim 1985. Milner 1996 discusses the Slade case.

CHAPTER 11: ENGLAND'S GREEN AND PLEASANT LAND

1. Freeman 1977, p. 87.
2. *Life and Letters* 3:195.
3. L. Huxley 1918, vol. 2, pp. 151–53. For Darwin's experimental botany see Allan 1977 and F. Darwin 1899 and 1909. Sachs 1890 and A. G. Morton 1981 provide a survey account of later-nineteenth-century developments. Much of Darwin's work involved investigations based on Gärtner 1849 and Kölreuter 1761–66.
4. Bynum 1994.
5. F. Darwin 1909, p. 390.
6. Hooker 1874. On the fly-trap see Nelson 1990.
7. Allan 1977, p. 243, and *Life and Letters* 3:324.
8. *Life and Letters* 3:327–28.
9. DAR 219.1.89.
10. *Life and Letters* 3:289.
11. *Life and Letters* 3:276n.
12. *More Letters* 2:419.
13. *Life and Letters* 3:290.
14. *Life and Letters* 1:147.
15. DAR 219.11:30.
16. *Emma Darwin* 2:265.
17. *Life and Letters* 3:279.
18. DAR 251.
19. DAR 112:105; also partly in *Life and Letters* 1:117.
20. *Life and Letters* 3:197.
21. DAR 93:413. The proposals are in DAR 93:414.
22. Thoroughly documented in French 1975 and Rupke 1987. See also Cobbe 1904 and Hutton 1989. The rise of physiology as a research discipline is discussed particularly in Geison 1978 and Bynum 1994. See also Butler 1988.
23. *British Medical Journal* 2 (1874): 741–54, 828, where Magnan is wrongly identified as Eugene Magnan.
24. *Life and Letters* 3:199–201.
25. *Descent* 1:40.
26. *Bromley Record*, 1 September 1863, p. 168; also in Barrett 1977, vol. 2, pp. 83–84. See *Emma Darwin* 2:200–1 for the text of Emma's second letter.
27. *Life and Letters* 3:202–3. Virchow is discussed in Ackerknecht 1953.
28. *Emma Darwin* 2:274.
29. DAR 219.8.29.
30. *Life and Letters* 1:125.
31. DAR 112 (ser. B):3d.
32. *Emma Darwin* 2:273. No portrait or sketch of Emma Darwin by Ouless has been located.
33. DAR 219.1:95, 96.
34. To Hooker, 30 March 1875, DAR 93:382.
35. DAR 219.1.103.
36. DAR 219.1.91.
37. From Ernst von Hesse-Wartegg, 20 September 1875, DAR 166.
38. De Beer 1983, p. 110.
39. DAR 245:55.
40. *Autobiography*, p. 21. A perceptive account of Darwin's autobiography is given in Colp 1985. See also Rosenberg 1989 and the introduction to Neve 2002.
41. DAR 112:117. See Graber and Miles 1988.
42. Broughton 1999.
43. To J. V. Carus, 17 July 1879, Deutsche Staatsbibliotek, Berlin.
44. Useful analyses of the autobiographical genre are in Cockshut 1984, L. Peterson 1986, Henderson 1989, Machann 1994, and J. Conway 1998.
45. *Autobiography*, p. 97.
46. Machann 1994.
47. *Autobiography*, p. 138.
48. *Emma Darwin*, 2nd ed., 1915, vol. 2, p. 170.
49. Fleming 1960. Something of the same wish continues among historians; see for example Kohn 1966.
50. *Autobiography*, pp. 22, 45.
51. Colp 1985, L. Peterson 1986, and

Rosenberg 1989 for comments on the organising devices that can be found in Darwin's *Autobiography.* From a large literature on self-fashioning, see Gagnier 1991, Shortland 1996, and the essays in Lawrence and Shapin 1998.

52. Poovey 1989, Nye 1997, and Broughton 1999 deal with some of these issues. For masculinity in Victorian life, see Mangen and Walvin 1987 and Tosh 1999.
53. See especially Colp 1985 on these omissions.
54. *Autobiography,* p. 120.
55. J. Conway 1998.
56. L. Huxley 1900, vol. 2, p. 39. See also p. 113.
57. *Autobiography,* p. 125.
58. *Autobiography,* p. 87.
59. Deconversion and epiphany are discussed in Barbour 1994 and Barros 1998. See also Henderson 1989.
60. Emma Darwin to N. A. von Mengden, 8 April 1879, quoted from Junker and Richmond 1996, p. 154.
61. DAR 139.12:17
62. *Emma Darwin* 2:190.
63. *Autobiography,* pp. 93–94n. Emma Darwin's letters to William Darwin, the eldest son, on this question are in DAR 219.1. See also Colp 1985 and Moore 1994.
64. *Life and Letters* 1:26–107. The religious section as printed in *Life and Letters* ought to have begun on p. 69. Some of these letters are in DAR 112(A):19–27 and 210.8:43. See Barlow 1959.
65. *Autobiography,* p. 94.
66. *Emma Darwin* 2:278.
67. Cattermole 1987.
68. DAR 219.1.93.
69. F. Darwin 1877. See Darwin to F. J. Cohn, 8 August 1877, DAR 143.
70. G. Darwin 1875, p. 41.
71. DAR 210.6.
72. Jalland 1996, pp. 343–50.
73. *Life and Letters* 1:135.
74. To J.P.M. Weald, 30 July 1870, American Philosophical Society, Philadelphia.
75. Ackernecht 1953, p. 200.
76. Prospekt, *Kosmos* 1 (1877):1–3. More generally see Gasman 1971, Kelly 1981, and Weingart 1995.
77. From John Brigg, 6 July 1877, DAR 160.
78. DAR 219.1.86. Darwin's letters of thanks are in *Life and Letters* 3:225–27.
79. Colp 1983. Emma Darwin's diary has Sunday 11 March 1877 as the date.
80. Matthew 1986, vol. 9, p. 199. See also Gladstone to his wife, ibid., p. 221. For Gladstone generally, see Bassett 1936 and Checkland 1971. The visit is discussed in Colp 1983.
81. To C. E. Norton, 16 March 1877, Houghton Library, Harvard University.
82. Gladstone's interest in colour perception arose from his work in classical history. He thought the vagueness of the terminology for colour in ancient Greece meant that these peoples were at a lower stage of development and incapable of refined colour discrimination. Darwin discussed similar points in relation to the heightened natural perceptions of primitives. See Kuklick 1994.
83. To W. E. Gladstone, 4 August [1879], British Library.
84. C. Darwin 1877, reprinted in H. Gruber 1974, pp. 464–74.
85. H. Gruber 1974, pp. 224–29. George Darwin made good use of these infant observations in after-dinner speeches. When elected president of the Cambridge Philosophical Society he read out extracts pertaining to himself. *Christ's College Magazine,* Michaelmas term 1914.
86. Browne 1989, Schiebinger 1991.
87. Chandrasekhar 1981.
88. From Charles Bradlaugh, 5 June 1877, DAR 160.
89. To Charles Bradlaugh, 6 June

[1877], DAR 202, partly printed in Bradlaugh 1895, vol. 2, p. 24, where there is a full account of the trial. See also Manvell 1976 and Arnstein 1965. For Victorian views on contraception see McLaren 1978 and Hall and Porter 1995.

90. To ? Truelove, 1 July 1878, American Philosophical Society, Philadelphia. The elder Truelove was being tried for obscenity after publishing a cheap edition of Robert Dale Owen's *Moral Physiology, or, a Plain and Brief Treatise on the Population Question.*
91. Letter to G. A. Gaskell, *More Letters* 2:50. See also Weikart 1995.

CHAPTER 12: HOME IS THE SAILOR

1. *Earthworms* 313. On Darwin's investigations see Satchell 1983 and Elliott 1995.
2. To Francis Darwin, 10 June 1877, DAR 211, in which Darwin refers to these bundles of notes.
3. *Emma Darwin* 2:281.
4. *More Letters* 2:214. More generally on animal intelligence, see Boakes 1984 and R. Richards 1987.
5. *Emma Darwin* 2:282.
6. *Emma Darwin* 2:285–86. According to Emma, Darwin at first wished to decline the degree because of the "bother" of visiting Cambridge for the award. DAR 219.9.146.
7. *Cambridge Chronicle,* 24 November 1877, p. 4.
8. *Emma Darwin* 2:285–86.
9. Quoted from Brent 1981, p. 499.
10. Stirling 1926, p. 101.
11. *Emma Darwin* 2:321.
12. Atkins 1974, Moore 1985a.
13. Annual Audit, Bromley Savings Bank, *Bromley Record,* 1 February 1863. For Darwin's correspondence with John Innes on some of these matters see Stecher 1961, p. 216.
14. Wedgwood and Wedgwood 1980, p. 345.
15. "A Visit to Darwin's Village," *Evening News,* 12 February 1909.
16. Russell Gebbett 1977, pp. 29–41. The quotation is from p. 31. See Gosden 1961 on Friendly Societies and Gosden 1973 on other voluntary associations.
17. Stecher 1961, p. 242.
18. Freeman 1977, p. 157.
19. *Life and Letters* 2:289.
20. Stecher 1961, p. 233.
21. DAR 140(4):64.
22. Stecher 1961, p. 219. On the duties of Victorian clergymen, see Colloms 1977 and Haig 1984.
23. "A Visit to Darwin's Village," *Evening News,* 12 February 1909.
24. "A Visit to Darwin's Village," *Evening News,* 12 February 1909. See also Atkins 1974, p. 48.
25. DAR 219.1:110. Bell demonstrated his first commercially available telephone at the Centennial Exposition in Philadelphia in 1876.
26. F. Darwin 1920, p. 58.
27. Nash 1921, p. 27. See Colp 1977, 228n30. Victorian servants are discussed in Horn 1995. See also Girouard 1978 and Mitchell 1996. A standard source on the working classes in Victorian Britain is E. Thompson 1963. Atkins 1974, pp. 72–77, describes Darwin's household.
28. Atkins 1974, p. 74.
29. Atkins 1974, p. 73. The model was given to Karl Pearson and displayed for several years at UCL before being returned to Down House.
30. Jordan 1922, vol. 1, p. 273. Darwin's dislike of remonstrating with employees was displayed in his dealings with a footman called Duberry; see DAR 219.9.8 and 12; and Lettington's "idleness," DAR 219.9.59 and 210.
31. John Lubbock, speaking at the Darwin-Wallace Centenary celebrations, Linnean Society 1908, pp. 57–58.
32. B. Darwin 1933, p. 22.
33. DAR 140(4):77–78.

34. Paul Harvey, *Oxford Companion to English Literature,* 4th ed. (1967), entry for Trollope.
35. Anderson and Cottrell 1974.
36. Jackson 1999. The interplay between Darwin's principles and economic activity are discussed concisely in Wyllie 1959.
37. Down House MS 11.5
38. Atkins 1974, pp. 95–100, and Down House MS 11.14
39. See D. Roberts 1978 and Tosh 1999.
40. DAR 140(4):49.
41. Papers in DAR 210.23.
42. A. Desmond 1997, pp. 104, 132–33.
43. DAR 219.1.147 and 219.9.270. See also *Emma Darwin* 2:321.
44. Chadarevian 1996. On self-recording instruments in physiology see Chadarevian 1993. See also Lynch 1985 and Lynch and Woolgar 1988. Darwin's microscopes are described in Burnett 1992.
45. C. Darwin 1880, pp. 571, 573.
46. A description of a typical experiment is given in *More Letters* 2:415.
47. J. Secord 1986.
48. On the history of experiment see Shapin 1988 and Gooding, Pinch, and Schaffer 1989. Darwin's experimental work is discussed in Rheinberger and McLaughlin 1984.
49. Ophir and Shapin 1991, Smith and Agar 1998, and Forgan 1996, p. 453.
50. *Life and Letters* 1:146.
51. "A Visit to Darwin's Village," *Evening News,* 12 February 1909, p. 4.
52. K. Timiriazev, "At Darwin's in Down," *Russkie Vedomosti* (Moscow), 1909, vol. XLVI, Nos. 24–25. See also Vuccinich 1988 and Atkins 1974, pp. 85–86.
53. Obituary notice, *Nature* (*Botany*), 1882, extracted from Candolle 1882.
54. DAR 219.1:122.
55. Raverat 1952, p. 185.
56. *Emma Darwin* 2:298.
57. Most of these papers are in DAR 210.29. See Freeman 1984.
58. Proof sheets of *Erasmus Darwin* (Krause 1879) are in DAR 210.11:45.
59. Krause 1879, pp. 48, 68. See also Appendix 1, Barlow 1958, pp. 149–66. On the rhetorical effects of distinguishing between "head" and "hand" see Ophir and Shapin 1976.
60. Bowler 1983 and 1988. For Butler see H. F. Jones 1920, Greenacre 1963, and Appendix 2, Barlow 1988, pp. 167–219.
61. *Athenaeum,* 31 January 1880.
62. DAR 210.11:46.
63. DAR 219.1.134. I thank Jim Paradis for discussions about Butler.
64. Francis Darwin's papers on the affair are in DAR 139(11).
65. From Anthony Rich, 9 February 1881, DAR 176.
66. Barlow 1958, pp. 167–73.
67. *Life and Letters* 1:129, *Emma Darwin* 2:299. See also DAR 219.1.125.
68. *Life and Letters* 1:125.
69. Letter to Victor Marshall, 14 September 1879, American Philosophical Society, Philadelphia.
70. DAR 149(3):11.
71. *Emma Darwin* 2:302. See also DAR 219.1.131 and 219.9.218.
72. Chadarevian 1996.
73. DAR 262.10:1–27. The stone currently embedded in the lawn at Down House is apparently a recreation of an experiment originally carried out in 1929 under Horace Darwin's supervision. Horace reported on the movements in H. Darwin 1900.
74. Groeben 1982.
75. Groeben 1982, p. 63.
76. *Life and Letters* 3:352–53.
77. L. Huxley 1918, vol. 2, pp. 237–39.
78. Darwin's papers relating to this scheme are in DAR 196.3.
79. Colp 1992.
80. Wallace 1905, vol. 2, p. 378.
81. Marchant 1916, vol. 1, pp. 314–15.
82. Nash 1921, pp. 132–35.
83. Feuer 1975.

84. Aveling 1883, p. 4. There is a copy in DAR 139(12). The veneration of Darwin's absent presence has been discussed in part in Browne 1998. See also Leder 1990.
85. *Life and Letters* 1:143, *Emma Darwin* 2:315.
86. To Wallace, 12 July 1881, British Library.
87. Frithiof Holmgren, professor of physiology at Uppsala University, *Times*, 18 April 1881.
88. To William Darwin, 4 August [1881], DAR 210.6.
89. Sakula 1982. The painting was by Archibald Preston Tilt, and hangs in the Wellcome Trust, London.
90. Prescott 1985. See also Jordanova 2000.
91. Marian Collier's drawing is in the National Portrait Gallery, London.
92. DAR 219.1.144. In another letter, Emma Darwin mentions that in addition to Jackson a second understudy was used; DAR 219.9.199.
93. I am very grateful to Trudy Prescott Nuding for this information and for allowing me access to her unpublished research.
94. *Autobiography*, p. 43. See Carlyle 1881, vol. 2, pp. 207–8.
95. Murray 1919, p. 18.
96. From Robert Cooke [John Murray], 5 November 1881, DAR 171.
97. To Francis Darwin, 9 November [1881], DAR 211.
98. *Index*, 22 December 1881.
99. F. Darwin 1920, p. 57.
100. *Emma Darwin* 2:294, and DAR 219.9.188.
101. *Emma Darwin*, 2nd ed., 1915, vol. 2, p. 395.
102. L. Darwin 1929, pp. 119–20.
103. DAR 140(3):36, and letter to Asa Gray, 28 January 1876, Gray Herbarium, Harvard University.
104. DAR 140(3):12.
105. *Emma Darwin* 2:254.
106. On Darwin's purported deathbed conversion see Moore 1994. Emma Darwin's diary records the figures "3 1/2," presumably the time of Darwin's death. See also *Life and Letters* 3:358. I am grateful to the Darwin Correspondence Project for letting me see a copy of Probate of Darwin's will, dated 4 June 1882. His estate was valued at £146,911. The will was witnessed by William Jackson, the butler.
107. Emma Darwin's notes on Darwin's death, *Emma Darwin* 2:328–29.
108. Jordan 1922, vol. 1, p. 273.
109. Matthew 1990, p. 244.
110. Raverat 1952, p. 176.
111. Quoted from Moore 1982, pp. 110, 111. See A. R. Hall 1966 for scientists buried in Westminster Abbey.

# *Bibliography*


Abir Am, Pnina, and Dorinda Outram, eds. 1987. *Uneasy careers and intimate lives: women in science, 1789–1987*. New Brunswick, N.J.: Rutgers University Press.

Ackerknecht, Erwin H. 1953. *Rudolf Virchow: doctor, statesman, anthropologist*. Madison: University of Wisconsin Press.

Acland, Henry, and John Ruskin. 1859. *The Oxford Museum*. London.

Adler, S. W. 1959. Darwin's illness. *Nature* 184:1102–3.

Agassiz, Elizabeth C., ed. 1885. *Louis Agassiz: his life and correspondence*. 2 vols. London.

Agassiz, Louis. 1859. Essay on classification. In Louis Agassiz, *Contributions to the natural history of the United States*. 3 vols. Boston, 1857–62. Vol. 1.

———. 1860. On the Origin of Species. *American Journal of Science and Arts* 30:142–54. Also in *Annals and Magazine of Natural History* 6:219–32.

Allan, Mea. 1967. *The Hookers of Kew, 1785–1911*. London: Michael Joseph.

———. 1977. *Darwin and his flowers: the key to natural selection*. London: Faber & Faber.

Allen, David Elliston. 1978. *The naturalist in Britain: a social history*. Harmondsworth: Penguin Books.

———. 1980. The women members of the Botanical Society of London, 1836–56. *British Journal for the History of Science* 13:240–54.

———. 1996. Tastes and crazes. In N. Jardine, J. A. Secord, and E. Spary, eds. *Cultures of natural history*, 394–407. Cambridge: Cambridge University Press.

Allingham, William. 1907. *A diary*. Ed. Helen Allingham and D. Radford. London: Macmillan.

Alter, Stephen G. 1999. *Darwinism and the linguistic image: language, race, and natural theology in the nineteenth century*. Baltimore: Johns Hopkins University Press.

Altholz, Josef L. 1994. *Anatomy of a controversy: the debate over "Essays and Reviews," 1860–1864*. Aldershot: Scholar Press.

Altick, Richard D. 1957. *The English common reader: a social history of the mass reading public, 1800–1900*. Chicago: University of Chicago Press.

———. 1974. *Victorian people and ideas*. London: Dent.

———. 1978. *The shows of London*. Cambridge, Mass.: Harvard University Press.

Amigoni, David, and Jeff Wallace, eds. 1995. *Charles Darwin's "The Origin of Species": new interdisciplinary essays.* Manchester: Manchester University Press.

Anderson, B. L., and P. L. Cottrell. 1974. *Money and banking in England: the development of the banking system, 1694–1914.* Newton Abbot: David & Charles.

Anderson, Olive. 1987. *Suicide in Victorian and Edwardian England.* Oxford: Clarendon Press.

Anderson, Patricia. 1991. *The printed image and the transformation of popular culture, 1790–1860.* Oxford: Clarendon Press.

Annan, Noel G. 1955. The intellectual aristocracy. In J. H. Plumb, ed., *Studies in social history: a tribute to G. M. Trevelyan,* 241–87. London: Longmans, Green.

———. 1984. *Leslie Stephen: the godless Victorian.* London: Weidenfeld & Nicolson.

Anon. 1853. *A brief and complete refutation of the anti-scriptural theory of geologists.* By a Clergyman of the Church of England. London.

Anon. 1857. *Atheisms of Geology. Sir C. Lyell, Hugh Miller, etc. confronted with the rocks.* By J.A.S. London.

Appleman, Philip, William A. Madden, and Michael Wolff. 1959. *1859: entering an age of crisis.* Bloomington: Indiana University Press.

Arbuckle, Elisabeth W. S., ed. 1983. *Harriet Martineau's letters to Fanny Wedgwood.* Stanford, Calif.: Stanford University Press.

Armstrong, Carol. 1998. *Scenes in a library: reading the photograph in the book.* Cambridge, Mass: MIT Press.

Arnstein, Walter L. 1965. *The Bradlaugh case: a study in late Victorian opinion and politics.* Oxford: Clarendon Press.

Ashton, Rosemary. 1991. *G. H. Lewes: A life.* Oxford: Clarendon Press.

———. 1996. *George Eliot. A life.* London: Allen Lane.

Asser, Joyce. 1966. *Historic hairdressing.* London: Sir Isaac Pitman.

Atkins, Hedley. 1974. *Down, the home of the Darwins: the story of a house and the people who lived there.* London: Royal College of Surgeons of England.

Atlay, J. B. 1903. *Sir Henry Wentworth Acland . . . a memoir.* London: Smith & Elder.

*Autobiography:* see Barlow, Nora, ed., 1958.

Aveling, Edward. 1883. *The religious views of Charles Darwin.* London: Freethought Publishing.

Bailin, Miriam. 1994. *The sickroom in Victorian fiction: the art of being ill.* Cambridge: Cambridge University Press.

Bajema, Carl J. 1988. Charles Darwin on man in the first edition of the *Origin of Species. Journal of the History of Biology* 21:403–10

Baker, William, ed. 1995. *The letters of George Henry Lewes.* 2 vols. Victoria, B.C.: University of Victoria Press.

Barbour, John D. 1994. *Versions of deconversions: autobiography and the loss of faith.* Charlottesville: University Press of Virginia.

Baring-Gould, S., and C. A. Bampfylde. 1909. *A history of Sarawak under its two white rajahs, 1839–1908.* London: H. Sotheran.

Barlow, Nora, ed. 1958. *The autobiography of Charles Darwin, 1809–1882, with original omissions restored.* London: Collins.

Barnes, Barry, and Steven Shapin. 1977. Science, nature and control: interpreting Mechanics' Institutes. *Social Studies of Science* 7:31–74.

Barnes, James J. 1974. *Authors, publishers and politicians. The quest for an Anglo-American copyright agreement.* London: Routledge.

Barrett, Paul H., ed. 1977. *The collected papers of Charles Darwin.* 2 vols. Chicago: University of Chicago Press.

Barrett, Paul H., et al., eds. 1987. *Charles Darwin's notebooks, 1836–1844: geology, transmutation of species, metaphysical enquiries.* Cambridge: Cambridge University Press.

Barros, Carolyn. 1998. *Autobiography: narrative of transformation.* Ann Arbor: University of Michigan Press.

Bartholomew, Michael. 1973. Lyell and evolution: an account of Lyell's response to the prospect of an evolutionary ancestry for man. *British Journal for the History of Science* 6:261–303.

———. 1975. Huxley's defence of Darwin. *Annals of Science* 32:525–35.

———. 1976. The award of the Copley medal to Charles Darwin. *Notes and Records of the Royal Society* 30:209–18.

Bartley, Mary M. 1992. Darwin and domestication: studies on inheritance. *Journal of the History of Biology* 25:307–33.

Barton, Ruth. 1983. Evolution: the Whitworth gun in Huxley's war for the liberation of science from theology. In David Oldroyd and Ian Langham, eds. *The wider domain of evolutionary thought,* 261–87. Dordrecht: Reidel.

———. 1990. "An influential set of chaps": the X Club and Royal Society politics, 1864–1885. *British Journal for the History of Science* 23:58–81.

———. 1998. "Huxley, Lubbock and Half a Dozen Others": professionals and gentlemen in the formation of the X Club, 1851–1864. *Isis* 89:410–44.

Bassett, A. T., ed. 1936. *Gladstone to his wife.* London: Methuen.

Bates, Henry W. 1863. *The naturalist on the River Amazons: a record of adventures, habits of animals, sketches of Brazilian and Indian life, and aspects of nature under the Equator during eleven years of travel.* 2 vols. London.

Becher, Tony. 1989. *Academic tribes and territories: intellectual enquiry and the cultures of disciplines.* Milton Keynes: Open University Press.

Beddall, Barbara G. 1968. Wallace, Darwin and the theory of natural selection: a study in the development of ideas and attitudes. *Journal of the History of Biology* 1:261–323.

———, ed. 1969. *Wallace and Bates in the tropics: an introduction to the theory of natural selection.* New York: Macmillan.

———. 1973. "Notes for Mr. Darwin." Letters to Charles Darwin from Edward Blyth at Calcutta: a study in the process of discovery. *Journal of the History of Biology* 6:69–95.

———. 1988a. Darwin and divergence: the Wallace connection. *Journal of the History of Biology* 21:1–68.

———. 1988b. Wallace's annotated copy of Darwin's *Origin of Species. Journal of the History of Biology* 21:265–89.

Beer, Gillian. 1983. *Darwin's plots: evolutionary narrative in Darwin, George Eliot and nineteenth-century fiction.* London and Boston: Routledge & Kegan Paul.

———. 1985. Darwin's reading and the fictions of development. In David Kohn, ed., *The Darwinian heritage,* 543–88. Princeton, N.J.: Princeton University Press in association with Nova Pacifica.

———. 1986. The face of nature: anthropomorphic elements in the language of the *Origin of Species.* In Ludmilla Jordanova, ed., *Languages of nature: critical essays on science and literature,* 207–43. London: Free Association Books.

———. 1989. Darwin and the growth of language theory. In John Christie and Sally Shuttleworth, eds., *Nature Transfigured: Science and literature, 1700–1900,* 152–70. Manchester: Manchester University Press.

———. 1996. *Open fields: science in cultural encounter.* Oxford: Clarendon Press.

Belloc, Marie A. 1941. *"I, too, have lived in Arcadia." A record of love and childhood.* London: Macmillan.

Bellon, Richard. 2001. Joseph Dalton Hooker's ideals for a professional man of science. *Journal of the History of Biology* 34:51–82.

Bender, Bert. 1996. *Descent of love: Darwin and the theory of sexual selection in American fiction,*

*1871–1926*. Philadelphia: University of Pennsylvania Press.
Bentham, George. 1863. Presidential address. *Proceedings of the Linnean Society of London* 4 (1859–64): xi–xxix.
Berg, Charles. 1951. *The unconscious significance of hair*. London: Allen & Unwin.
Berman, David. 1988. *A history of atheism in Britain: from Hobbes to Russell*. London: Croom Helm.
Bermingham, Ann. 1986. *Landscape and ideology: the English rustic tradition, 1740–1860*. Berkeley: University of California Press.
Bernstein, Ralph B. 1984. Darwin's illness: Chagas disease resurgens. *Journal of the Royal Society of Medicine* 77:608–9.
Berrios, Germen E. 1985. Obsessional disorders during the nineteenth century: terminological and classificatory issues. In W. F. Bynum, R. Porter, and M. Shepherd, eds. *The anatomy of madness: essays in the history of psychiatry*, vol. 1, *People and Ideas*, 166–87. London: Tavistock Publications.
Betham-Edwards, M. 1919. *Mid-Victorian memories*. London: John Murray.
Bevington, M. M. 1941. *The "Saturday Review," 1855–1868: representative educated opinion in Victorian England*. New York: Columbia University Press.
Bibby, Cyril. 1959. *T. H. Huxley: scientist, humanist and educator*. London: Watts.
Bingham, A. Walker. 1994. *The snake-oil syndrome: patent medicine and advertising*. Hanover, Mass.: Christopher Publishing House.
Blaisdell, Muriel. 1982. Natural theology and nature's disguises. *Journal of the History of Biology* 15:163–89.
Blinderman, Charles S. 1970. The great bone case. *Perspectives in Biology and Medicine* 14:370–93.
———. 1995. The descent of words. *Language Quarterly* 33:224–41.
Boakes, Robert. 1984. *From Darwin to behaviourism: psychology and the minds of animals*. Cambridge: Cambridge University Press.
Bohlin, Ingemar. 1991. Robert M. Young and Darwin historiography. *Social Studies of Science* 21:597–648.
Bondeson, Jan. 1997. *A cabinet of medical curiosities*. New York: Tauris.
Bonney, Thomas G. 1919. *Annals of the Philosophical Club of the Royal Society*. London: Macmillan.
Bowlby, John. 1990. *Charles Darwin: a biography*. London: Hutchinson.
Bowler, Peter J. 1974. Darwin's changing concepts of variation. *Journal of the History of Medicine and the Allied Sciences* 29:196–212.
———. 1975. The changing meaning of "evolution." *Journal of the History of Ideas* 36:95–114.
———. 1976. *Fossils and progress: paleontology and the idea of progressive evolution in the nineteenth century*. New York: Science History Publications.
———. 1983. *The eclipse of Darwinism: anti-Darwinian evolution theories in the decades around 1900*. Baltimore: Johns Hopkins University Press.
———. 1985. Scientific attitudes to Darwinism in Britain and America. In David Kohn, ed., *The Darwinian heritage*, 641–81. Princeton, N.J.: Princeton University Press in association with Nova Pacifica.
———. 1986. *Theories of human evolution: a century of debate, 1844–1944*. Baltimore: Johns Hopkins University Press.
———. 1988. *The non-Darwinian revolution: reinterpreting a historical myth*. Baltimore: Johns Hopkins University Press.
———. 1989a. *Evolution: the history of an idea*. Berkeley: University of California Press.
———. 1989b. *The invention of progress: the Victorians and the past*. Oxford: Basil Blackwell.

———. 1990. *Charles Darwin: the man and his influence.* Cambridge: Cambridge University Press.

———. 1995. Social metaphors in evolutionary biology. In S. Maasen, E. Mendelsohn, and P. Weingart, eds., *Biology as society, society as biology: metaphors,* 107–26. Dordrecht: Kluwer.

Brackman, Arnold C. 1980. *A delicate arrangement: the strange case of Charles Darwin and Alfred Russel Wallace.* New York: Times Books.

Bradford, Sarah. 1996. *Disraeli.* London: Phoenix Giant.

Bradlaugh, Hypatia B. 1894. *Charles Bradlaugh: a record of his life and work.* 2nd ed. 2 vols. London: T. Fisher Unwin.

Brake, L., A. Jones, and L. Madden, eds. 1990. *Investigating Victorian journalism.* Basingstoke: Macmillan.

Braudy, Leo. 1986. *The frenzy of renown: a history of fame.* Oxford: Oxford University Press.

Brent, Peter. 1981. *Charles Darwin: a man of enlarged curiosity.* London: Heinemann.

Briggs, Asa. 1959. *The age of improvement.* London: Longmans.

———. 1965. *Victorian people: a reassessment of persons and themes, 1851–67.* Harmondsworth: Penguin Books.

———. 1981. Prince Albert and the arts and sciences. In J.A.S. Phillips, ed., *Prince Albert and the Victorian age.* Cambridge: Cambridge University Press.

———. 1990. *Victorian things.* Harmondsworth: Penguin Books.

Brock, William H. 1980. The development of commercial science journals in Victorian England. In A. J. Meadows, ed., *The development of science publishing in Europe,* 95–122. Amsterdam: Elzevier.

———. 1996. *Glaucus:* Kingsley and the seaside naturalists. In William H. Brock, *Science for all: studies in the history of Victorian science and education,* 25–36. Aldershot, Hampshire: Variorum.

Brock, William H., and Roy M. MacLeod. 1976. The scientists' declaration: reflexions on science and belief in the wake of *Essays and Reviews,* 1864–5. *British Journal for the History of Science* 9:39–66.

Brock, William H., and A. J. Meadows. 1998. *The lamp of learning: Taylor & Francis and the development of science publishing.* 2nd ed. London: Taylor & Francis.

Brockway, Lucille H. 1979. *Science and colonial expansion: the role of the British Royal Botanic Gardens.* Studies in Social Discontinuity. New York: Academic Press.

Brogan, Hugh, ed. 1975. *The American Civil War: extracts from The Times, 1860–1865.* London: Times Books.

Brooke, John H. 1979. The natural theology of the geologists: some theological strata. In L. Jordanova and R. Porter, eds., *Images of the earth: essays in the history of the environmental sciences,* 39–64. Chalfont St Giles: British Society for the History of Science.

———. 1991. *Science and religion: some historical perspectives.* Cambridge: Cambridge University Press.

Brooks, John Langdon. 1984. *Just before the Origin: Alfred Russel Wallace's theory of evolution.* New York: Columbia University Press.

Broughton, Trev Lynn. 1999. *Men of letters, writing lives: masculinity and literary auto/biography in the late Victorian period.* London: Routledge.

Brown, Frank B. 1986. The evolution of Darwin's theism. *Journal of the History of Biology* 19:1–45. Also published as *The evolution of Darwin's religious views.* Macon, Ga.: Mercer University Press.

Browne, Janet. 1978. The Charles Darwin-Joseph Hooker correspondence: an analysis of manuscript resources and their use in biography. *Journal of the Society for the Bibliography of Natural History* 8:351–66.

———. 1980. Darwin's botanical arithmetic and the "principle of divergence," 1854–1858. *Journal of the History of Biology* 13:53–89.

———. 1983. *The secular ark: studies in the history of biogeography.* New Haven, Conn.: Yale University Press.

———. 1985a. Darwin and the expression of the emotions. In David Kohn, ed., *The Darwinian heritage,* 307–26. Princeton, N.J.: Princeton University Press in association with Nova Pacifica.

———. 1985b. Darwin and the face of madness. In W. F. Bynum, R. Porter, and M. Shepherd, eds., *The anatomy of madness: essays in the history of psychiatry,* vol. 1, *People and ideas,* 151–65. London: Tavistock Publications.

———. 1989. Botany for gentlemen: Erasmus Darwin and the *Loves of the Plants. Isis* 80:593–621.

———. 1990. Spas and sensibilities: Darwin at Malvern. In W. F. Bynum and Roy Porter, eds., *The medical history of spas and waters,* Medical History Supplement 5, 102–13.

———. 1992a. Squibs and snobs: science in humorous British undergraduate magazines around 1830. *History of Science* 30:165–97.

———. 1992b. A science of empire: British biogeography before Darwin. *Review d'histoire des Sciences* 4:453–75.

———. 1995. *Charles Darwin: voyaging.* New York: Knopf.

———. 1998. "I could have retched all night": Charles Darwin and his body. In Christopher Lawrence and Steven Shapin, eds., *Science incarnate: historical embodiments of natural knowledge,* 240–87. Chicago: University of Chicago Press.

———. 2001. Darwin in caricature: a study in the popularisation and dissemination of evolution. *Proceedings of the American Philosophical Society* 145:496–509.

Buckle, Henry Thomas. 1857–61. *History of civilization in England.* 2 vols. London.

Bulhof, Ilse. 1988. The Netherlands. In Thomas Glick, ed., *The comparative reception of Darwinism,* 269–306. Chicago: University of Chicago Press.

———. 1992. *The language of science: a study of the relationship between literature and science in the perspective of a hermeneutical ontology with a case study of Darwin's "The Origin of Species."* Leiden: E. J. Brill.

Bunbury, Frances J., ed. 1891–93. *Memorials of Sir C.J.F. Bunbury, Bart.* Middle Life, vols. 1–3; Later Life, vols. 1–5. Mildenhall.

Burchfield, Joe D. 1990. *Lord Kelvin and the age of the earth.* With a new afterword. Chicago: University of Chicago Press.

Burke, Peter. 2001. *Eyewitnessing: the uses of images as historical evidence.* London: Reaktion Books.

Burke, Peter, and Roy Porter, eds. 1987. *The social history of language.* Cambridge: Cambridge University Press.

———. 1995. *Languages and jargons: contributions to a social history of language.* Cambridge: Polity Press.

Burkhardt, Frederick H., Sydney Smith, et al., eds. 1983–2001. *The correspondence of Charles Darwin.* Vols. 1–12 (1821–64). Cambridge: Cambridge University Press.

———. 1994. *Calendar of the correspondence of Charles Darwin.* Rev ed. Cambridge: Cambridge University Press.

Burkhardt, Richard W. 1977. *The spirit of system: Lamarck and evolutionary biology.* Cambridge, Mass.: Harvard University Press.

Burn, William L. 1964. *The age of equipoise: a study of the mid-Victorian generation.* London: George Allen & Unwin.

Burnett, W.A.S. 1992. Darwin's microscopes. *Microscopy* 36:604–27.

Burrow, John W. 1966. *Evolution and society.* Cambridge: Cambridge University Press.

———, ed. 1968. Introduction. In *The Origin of species by Charles Darwin.* Reprint edition. Harmondsworth: Pelican Classics, 11–48.

Burton, James. 1986. Robert FitzRoy and the early history of the Meteorological Office. *British Journal for the History of Science* 19:147–76.

Butler, Stella. 1988. Centers and peripheries: the development of British physiology, 1870–1914. *Journal of the History of Biology* 21:473–500.

Bynum, W. F. 1983. Darwin and the doctors: evolution, diathesis and germs in nineteenth century Britain. *Gesnerus* 40:43–53.

———. 1984. Charles Lyell's *Antiquity of man* and its critics. *Journal of the History of Biology* 17:153–87.

———. 1985. The nervous patient in eighteenth and nineteenth century Britain: the psychiatric origins of British neurology. In W. F. Bynum, R. Porter, and M. Shepherd, eds., *The anatomy of madness: essays in the history of psychiatry*, vol. 1, *People and ideas*, 88–102. London: Tavistock Publications.

———. 1991. The historical Galton. In Milo Keynes, ed., *Sir Francis Galton: the legacy of his ideas*, 33–44. Basingstoke: Macmillan.

———. 1994. *Science and the practice of medicine in the nineteenth century.* New Haven, Conn.: Yale University Press.

———, ed. 1997. *Gastroenterology in Britain: historical essays.* London: Wellcome Institute for the History of Medicine.

Bynum, W. F., S. Lock, and R. Porter, eds. 1992. *Medical journals and medical knowledge: historical essays.* London: Routledge.

Bynum, W. F., and Michael Neve. 1985. Hamlet on the couch. In W. F. Bynum, R. Porter, and M. Shepherd, eds. *The anatomy of madness: essays in the history of psychiatry*, vol. 1, *People and ideas*, 289–304. London: Tavistock Publications.

Bynum, W. F., and R. S. Porter, eds. 1991. *Living and dying in London.* Medical History Supplement 11. London: Wellcome Institute for the History of Medicine.

*Calendar:* see Burkhardt, Frederick H., Sydney Smith, et al., eds., 1994.

Camerini, Jane. 1994. Evolution, biogeography and maps: an early history of Wallace's line. In R. M. MacLeod and P. Rehbock, eds., *Darwin's laboratory: evolutionary theory and natural history in the Pacific*, 70–109. Honolulu: University of Hawaii Press.

———. 1996. Wallace in the field. In H. Kuklick and R. E. Kohler, eds., *Science in the field. Osiris* 11:44–65.

———. 1997. Remains of the day: early Victorians in the field. In B. Lightman, ed., *Victorian Science in Context*, 354–77. Chicago: University of Chicago Press.

Cameron, Henry H. H. 1893. *Alfred, Lord Tennyson and his friends.* With 25 portraits by Julia Cameron. London.

Campbell, George Douglas, 8th Duke of Argyll. 1867. *The reign of law.* London.

———. 1869. *Primeval man: an examination of some recent speculations.* London.

Candolle, Alphonse de. 1873. *Histoire des sciences et des savants depuis deux siècles.* Geneva.

———. 1882. Darwin considère au point de vue des causes de son succès. *Archives des Sciences Physiques et Naturelles* 7:481–95.

Cannadine, David. 1999. *The rise and fall of class in Britain.* New York: Columbia University Press.

Cannon, Susan F. [W. F.] 1964. Scientists and Broad churchmen: an early Victorian intellectual network. *Journal of British Studies* 4:65–88.

———. 1978. *Science in culture: the early Victorian period.* New York: Science History Publications.

Cannon, W. F. 1968. Darwin's vision in *On the Origin of Species.* In G. Levine and W. Madden, eds., *The Art of Victorian Prose*, 154–76. Oxford: Oxford University Press.

Cantor, Geoffrey. 1996. The scientist as hero: public images of Michael Faraday. In M. Shortland and R. Yeo, eds., *Telling lives in science*, 171–93.

Cambridge: Cambridge University Press.

Cardwell, D.S.L. 1972. *The organization of science in England.* 2nd ed. London: Heinemann.

Carlyle, Thomas. 1881. *Reminiscences.* Edited by James Anthony Froude. 2 vols. London.

Carroll, P. Thomas. 1976. *An annotated calendar of the letters of Charles Darwin in the American Philosophical Society.* Wilmington: Scholarly Resources Inc.

Cattermole, M.J.G. 1987. *Horace Darwin's shop: a history of the Cambridge Instrument Company, 1878–1968.* Bristol: Hilger.

Caudill, Edward. 1994. The bishop-eaters: the publicity campaign for Darwin and *On the Origin of Species. Journal of the History of Ideas* 55:441–60.

Chadarevian, Soraya de. 1993. Graphical method and discipline: self-recording instruments in nineteenth-century physiology. *Studies in the History and Philosophy of Science* 24:267–91.

———. 1996. Laboratory science versus country-house experiments: the controversy between Julius Sachs and Charles Darwin. *British Journal for the History of Science* 29:17–41

Chadwick, Owen. 1975. *The secularization of the European mind in the nineteenth century.* Cambridge: Cambridge University Press.

Chambers, Robert. 1844. *Vestiges of the natural history of creation.* Edited with an introduction by J. A. Secord. Chicago: University of Chicago Press, 1994.

———. 1859. Review of the *Origin of Species. Chambers's Journal* 12:388–90. Reprinted in James A. Secord, *Robert Chambers, Vestiges of the natural history of creation,* 208–10. Facsimile ed. Chicago: University of Chicago Press, 1994.

———. 1860. *Vestiges of the natural history of creation.* 11th ed. London.

Chandrasekhar, Sripati. 1981. *A dirty filthy book: the writings of Charles Knowlton and Annie Besant on reproductive physiology and birth control and an account of the Bradlaugh-Besant trial.* Berkeley: University of California Press.

Chapman, John. 1864. *Sea sickness: its nature and treatment.* London.

———. 1873. *Neuralgia and kindred diseases of the nervous system.* London.

Chartier, Roger. 1995. *Forms and meanings: texts, performances and audiences from codex to computer.* Philadelphia: University of Pennsylvania Press.

Chartier, Roger, Alain Boureau, and Cecile Dauphin. 1997. *Correspondence: models of letter writing from the Middle Ages to the nineteenth century.* Translated by Christopher Woodall. Cambridge: Polity Press.

Chase, Karen, and Michael Levenson. 2000. *The spectacle of intimacy: a public life for the Victorian family.* Princeton, N.J.: Princeton University Press.

Checkland, Sydney G. 1971. *The Gladstones: a family biography, 1764–1851.* Cambridge: Cambridge University Press.

Churchill, Frederick B. 1979. Sex and the single organism: biological theories of sexuality in mid-nineteenth century. *Studies in the History of Biology* 3:139–77.

———. 1982. Darwin and the historian. In R. J. Berry, ed., *Charles Darwin: a commemoration,* 45–68. London: Linnean Society of London.

Cipolla, Carlo M. 1969. *Literacy and development in the West.* Harmondsworth: Penguin Books.

Clark, J. W., and T. M. Hughes, eds. 1890. *The life and letters of the reverend Adam Sedgwick.* 2 vols. London.

Clark, John F. M. 1997. "The ants were duly visited": making sense of John Lubbock, scientific naturalism and the senses of social insects. *British Journal for the History of Science* 30:151–76.

Clark, Linda L. 1984. *Social Darwinism in France.* University: University of Alabama Press.

Clarke, Graham, ed. 1994. *The portrait in photography.* London: Reaktion Books.

Clodd, Edward, ed. 1892. *The naturalist on the river Amazons . . . by H. W. Bates.* With a memoir of the author by E. Clodd. London.

Cobbe, Frances Power. 1872. *Darwinism in morals and other essays.* London: Williams and Norgate.

———. 1894. *Life of Frances Power Cobbe by herself.* 2 vols. London.

———. 1904. *Life of Frances Power Cobbe as told by herself.* Posthumous ed. London: Swan Sonneschein.

Cockshut, A.O.J. 1984. *The art of autobiography in nineteenth and twentieth century England.* New Haven, Conn.: Yale University Press.

Cohen, I. Bernard. 1985. Three notes on the reception of Darwin's ideas on natural selection (Henry Baker, Alfred Newton, Samuel Wilberforce). In David Kohn, ed., *The Darwinian heritage,* 589–607. Princeton, N.J.: Princeton University Press in association with Nova Pacifica.

———, ed. 1994. *The natural sciences and the social sciences: some critical and historical perspectives.* Dordrecht: Kluwer.

Colenso, John W. 1862–79. *The Pentateuch and Book of Joshua critically examined.* 5 vols. London.

Coley, N. G. 1973. Henry Bence Jones M.D. F.R.S. (1813–1873). *Notes and Records of the Royal Society* 28:31–56.

Collini, Stefan. 1991. *Public moralists: political thought and intellectual life in Britain, 1850–1930.* Oxford: Clarendon Press.

Collins, H. M. 1981. The place of the "core set" in modern science. *History of Science* 19:6–19.

Colloms, Brenda. 1975. *Charles Kingsley: the lion of Eversley.* London: Constable.

———. 1977. *Victorian country parsons.* London: Constable.

Colp, Ralph. 1977. *To be an invalid: the illness of Charles Darwin.* Chicago: University of Chicago Press.

———. 1982. The myth of the Darwin-Marx letter. *History of Political Economy* 14:461–82.

———. 1983. Notes on William Gladstone, Karl Marx, Charles Darwin, Kliment Timiriazev, and the "Eastern Question" of 1876–78. *Journal of the History of Medicine and Allied Sciences* 38:178–85.

———. 1985. Notes on Charles Darwin's autobiography. *Journal of the History of Biology* 18:357–401.

———. 1987. Charles Darwin's "insufferable grief." *Free Associations* 9:6–44.

———. 1989. Charles Darwin's past and future biographies. *History of Science* 27:167–97.

———. 1992. "I will gladly do my best": how Charles Darwin obtained a civil list pension for Alfred Russel Wallace. *Isis* 83:3–26

———. 1998. *To be an Invalid* redux. *Journal of the History of Biology* 31:211–40.

Conry, Yvette. 1974. *L'introduction du darwinisme en France au XIXe siècle.* Paris: Vrin.

Conway, Jill K. 1998. *When memory speaks: reflections on autobiography.* New York: Knopf.

Conway, Moncure Daniel. 1904. *Autobiography: memories and experiences.* 2 vols. London: Cassell.

Cook, Gordon C. 1997. George Busk F.R.S. (1807–1886), nineteenth century polymath: surgeon, parasitologist, zoologist and palaeontologist. *Journal of Medical Biography* 5:88–101

Cooke, Kathy J. 1990. Darwin on man in the *Origin of Species:* an addendum to the Bajema-Bowler debate. *Journal of the History of Biology* 23:517–512.

Cookson, Gillian, and Colin Hempstead. 2000. *A Victorian scientist and engineer: Fleeming Jenkin and the birth of electrical engineering.* Aldershot, Surrey: Ashgate.

Cooper, Wendy. 1971. *Hair: sex, society, symbolism.* London: Aldus Books.

Cooter, Roger. 1984. *The cultural meaning of popular science: phrenology and the organisation of consent in nineteenth-century Britain*. Cambridge: Cambridge University Press.

Cooter, Roger, and Stephen Pumfrey. 1994. Separate spheres and public places: reflections on the history of science, popularization and science in popular culture. *History of Science* 32:237–67.

Cornell, John F. 1984. Analogy and technology in Darwin's vision of nature. *Journal of the History of Biology* 17:303–44.

*Correspondence:* see Burkhardt, Frederick H., Sydney Smith, et al., eds., 1983–99.

Corsi, Pietro. 1988a. *Science and religion: Baden Powell and the Anglican debate, 1820–1860*. Cambridge: Cambridge University Press.

———. 1988b. *The age of Lamarck: evolutionary theory in France, 1790–1830*. Translated by Jonathan Mandelbaum. Berkeley: University of California Press.

Corsi, Pietro, and Paul Weindling. 1985. Darwinism in Germany, France and Italy. In David Kohn, ed., *The Darwinian heritage*, 683–729. Princeton, N.J.: Princeton University Press in association with Nova Pacifica.

Cowling, Margaret. 1989. *The artist as anthropologist: the representation of type and character in Victorian art*. Cambridge: Cambridge University Press.

Crichton-Browne, James. 1930. *What the doctor thought*. London: E. Benn.

Cronin, Helena. 1991. *The ant and the peacock: altruism and sexual selection from Darwin to today*. Cambridge: Cambridge University Press.

Crowson, R. A. 1958. Darwin and classification. In S. A. Barnett, ed., *A century of Darwin*, 102–29. Cambridge: Cambridge University Press.

Curle, Richard, ed. 1937. *Robert Browning and Julia Wedgwood: a broken friendship as revealed in their letters*. London: John Murray and Jonathan Cape.

Curtin, Michael. 1987. *Propriety and position: a study of Victorian manners*. New York: Garland.

Darrah, William C. 1981. *Cartes de visite in nineteenth century photography*. Gettysburg, Pa.: W. C. Darrah.

Darwin, Bernard. 1928. *Green memories*. London: Hodder & Stoughton.

———. 1955. *The world that Fred made: an autobiography*. London: Chatto & Windus.

Darwin, Charles R. 1845. *Journal of researches*. 2nd ed. London. Reprinted as *The voyage of the Beagle*, edited by H. G. Cannon. London: J. M. Dent, 1959.

———. 1859. *On the origin of species by means of natural selection, or the preservation of favoured races in the struggle for life*. London. Facsimile edition with an introduction by Ernst Mayr. Cambridge, Mass.: Harvard University Press, 1964.

———. 1868. *The variation of animals and plants under domestication*, 2 vols. London. Facsimile edition with new foreword by Harriet Ritvo. Baltimore: Johns Hopkins University Press. 1998.

———. 1871. *The descent of man and selection in relation to sex*. 2 vols. Facsimile ed. with an introduction by John T. Bonner and R. M. May. Princeton, N.J.: Princeton University Press, 1981.

———. 1872. *The expression of the emotions in man and animals*. Reprint ed. with an introduction, afterword, and commentaries by Paul Ekman. London: HarperCollins, 1998.

———. 1875. *Insectivorous plants*. London.

———. 1876. *The effects of cross and self fertilisation in the vegetable kingdom*. London.

———. 1877a. A biographical sketch of an infant. *Mind: Quarterly Review of Psychology and Philosophy* 2:285–94.

———. 1877b. *The different forms of flowers on plants of the same species*. London.

———. 1877c. *The various contrivances by which orchids are fertilised by insects*. 2nd ed. Revised with a new foreword by Michael Ghiselin.

Chicago: University of Chicago Press, 1984.
———. 1880. *The power of movement in plants.* Assisted by Francis Darwin. London.
———. 1881. *The formation of vegetable mould, through the action of worms, with observations on their habits.* London.
———. *Erasmus Darwin:* see Krause, Ernst, 1879.
Darwin, Charles, and Alfred Russel Wallace. 1858. On the tendency of species to form varieties; and on the perpetuation of varieties and species by natural means of selection. *Journal of the Proceedings of the Linnean Society of London (Zoology)* 3:53–62. Reprinted in Linnean Society 1908, 87–107.
Darwin, Francis. 1877. On the protrusion of protoplasmic filaments from the glandular hairs of the common teasel (*Dipsacus sylvestris*). *Quarterly Journal of Microscopical Science* 17:169–74, 245–72.
———, ed. 1887. *The life and letters of Charles Darwin.* 3 vols. London.
———. 1899. The botanical work of Darwin. *Annals of Botany,* pp. ix–xix.
———. 1909. Darwin's work on the movement of plants. In A. C. Seward, ed. 1909. *Darwin and modern science,* 385–400. Cambridge: Cambridge University Press.
———. 1912. FitzRoy and Darwin, 1831–36. *Nature* 88:547–48.
———. 1916. Memoir of Sir George Darwin by his brother Sir Francis Darwin. In G. H. Darwin, *Scientific papers* 5:ix–xxiii. 5 vols. Cambridge: Cambridge University Press.
———. 1920a. *Springtime and other essays.* London: John Murray.
———. 1920b. *The story of a childhood.* Privately printed. Edinburgh: Oliver & Boyd.
Darwin, Francis, and A. C. Seward, eds. 1903. *More letters of Charles Darwin: a record of his work in a series of hitherto unpublished letters.* 2 vols. London: John Murray.
Darwin, George H. 1873. On beneficial restrictions to liberty of marriage. *Contemporary Review* 22:412–26.
———. 1875. Marriages between first cousins in England and their effects. *Fortnightly Review* 28:22–41.
Darwin, Horace. 1900. On the small vertical movements of a stone laid on the ground. *Proceedings of the Royal Society of London* 68:253–61.
Darwin, Leonard. 1929. Memories of Down House. *Nineteenth Century* 106:118–23.
Davidoff, Leonore. 1973. *The best circles: society, etiquette and the season.* London: Croom Helm.
Davidoff, Leonore, and Catherine Hall. 1987. *Family fortunes: men and women of the English middle classes, 1780–1850.* London: Hutchinson.
Dawson, Albert. 1903. A visit to Alfred Russel Wallace. *Christian Commonwealth,* 10 December 1903, 176–78.
Dawson, Warren. 1946. *The Huxley papers: a descriptive catalogue of the correspondence, manuscripts and miscellaneous papers of the Rt. Hon. Thomas Henry Huxley . . . preserved in the Imperial College of Science and Technology.* London: Macmillan.
Dear, Peter, ed. 1991. *The literary structure of scientific argument.* Philadelphia: University of Pennsylvania Press.
De Beer, Gavin, ed. 1959. Darwin's Journal. *Bulletin of the British Museum (Natural History) Historical Series* 2:1–21.
———, ed. 1983. *Autobiographies: Charles Darwin. Thomas Henry Huxley.* Oxford: Oxford University Press.
Dempster, W. J. 1996. *Natural selection and Patrick Matthew: evolutionary concepts in the nineteenth century.* Rev. ed. Edinburgh: Pentland Press.
Denison, William. 1865. *An attempt to approximate to the antiquity of man by induction from well established facts.* Madras.
Depew, David J., and Bruce H. Weber. 1995. *Darwinism evolving: systems dynamics and the genealogy of*

*natural selection*. Cambridge, Mass.: MIT Press.

*Descent of Man:* see Darwin, Charles, 1871.

Desmond, Adrian J. 1982. *Archetypes and ancestors: palaeontology in Victorian London, 1850–1875*. London: Blond & Briggs.

———. 1984. Robert E. Grant's later views on organic development: the Swiney lectures on "Palaeozoology," 1853–1857. *Archives of Natural History* 11:395–413.

———. 1989. *The politics of evolution: morphology, medicine and reform in radical London*. Chicago: University of Chicago Press.

———. 1994. *Huxley: the devil's disciple*. London: Michael Joseph.

———. 1997. *Huxley: evolution's high priest*. London: Michael Joseph.

———. 2001. Redefining the X axis: "professionals," "amateurs," and the making of mid-Victorian biology. *Journal of the History of Biology* 34:3–50.

Desmond, Adrian J., and James R. Moore. 1991. *Darwin*. London: Michael Joseph.

Desmond, Ray. 1977. *Dictionary of British and Irish botanists and horticulturists: including plant collectors and botanical artists*. 3rd ed. London: Taylor & Francis.

———. 1998. *Kew: the history of the Royal Botanic Gardens, Kew*. London: Harvill.

———. 1999. *Sir Joseph Dalton Hooker: traveller and plant collector*. Woodbridge, Suffolk: Antique Collectors' Club Ltd. with the Royal Botanic Gardens, Kew.

De Vries, Hugo. 1909. Variation. In A. C. Seward, *Darwin and modern science*, 66–84. Cambridge: Cambridge University Press.

Di Gregorio, Mario A. 1984. *T. H. Huxley's place in natural science*. New Haven, Conn.: Yale University Press.

———, ed. 1990. *Charles Darwin's marginalia*. With the assistance of Nick Gill. New York: Garland.

Digby, Anne. 1994. *Making a medical living: doctors and patients in the English market for medicine, 1720–1911*. Cambridge: Cambridge University Press.

Dolan, Brian, ed. 2000. *Malthus, medicine and morality: "Malthusianism" after 1798*. Amsterdam: Rodophi.

Draper, John William. 1860. On the intellectual development of Europe, considered with reference to the views of Mr. Darwin and others, that the progression of organisms is determined by law. *Report of the 30th meeting of the British Association for the Advancement of Science held at Oxford, July 1860*, 115–16. London, 1861.

———. 1864. *History of the intellectual development of Europe*. 2 vols. London.

———. 1872. *History of the conflict between religion and science*. London.

Drayton, Richard. 2000. *Nature's government: science, imperial Britain, and the "improvement" of the world*. New Haven, Conn.: Yale University Press.

Du Chaillu, Paul B. 1861a. The geographical features and natural history of a hitherto unexplored region of Western Africa. *Proceedings of the Royal Geographical Society*, 25 February 1861, 108–12.

———. 1861b. *Explorations and adventures in equatorial Africa; with accounts of the manners and customs of the people, and of the chase of the gorilla . . .* London.

Dupree, A. Hunter. 1988. *Asa Gray: American botanist, friend of Darwin*. Reprint ed. Baltimore: Johns Hopkins University Press.

Durant, John. 1979. Scientific naturalism and social reform in the thought of Alfred Russel Wallace. *British Journal for the History of Science* 12:31–58.

———, ed. 1985. *Darwinism and divinity: essays on evolution and religious belief*. Oxford: Blackwell.

Duthie, Ruth. 1988. *Florists' flowers and societies*. Princes Risborough: Shire Publications Ltd.

Earle, Rebecca, ed. 1999. *Epistolary selves: letters and letter writers 1600–1945*. Aldershot: Ashgate.

Eco, Umberto. 1979. *The role of the reader: explorations in the semiotics of texts*. Bloomington: Indiana University Press.

Edwards, Elizabeth, ed. 1992. *Anthropology and photography, 1860–1920*. New Haven, Conn.: Yale University Press.

Ehrenreich, Barbara, and Deidre English. 1973. *Complaints and disorders: the sexual politics of sickness*. London: Writers and Readers Publishing Cooperative.

Eisley, Loren. 1959. Charles Darwin, Edward Blyth and the theory of natural selection. *Proceedings of the American Philosophical Society* 103:94–114.

———. 1961. *Darwin's century: evolution and the men who discovered it*. New York: Doubleday.

———. 1979. *Darwin and the mysterious Mr. X: new light on the evolutionists*. London, Toronto, Melbourne: J. M. Dent.

Ellegard, Alvar. 1957a. The Darwinian revolution and nineteenth-century philosophies of science. *Journal of the History of Ideas* 18:362–93.

———. 1957b. The readership of the periodical press in mid-Victorian Britain. *Acta Universitatis Gothoburgensis* 63:3.

———. 1990. *Darwin and the general reader: the reception of Darwin's theory of evolution in the British periodical press, 1859–1872*. With a new foreword by D. L. Hull. Chicago: University of Chicago Press.

Elliott, Charles. 1995. Darwin and the earthworms. In *The transplanted gardener*, 117–23. London: Viking.

Ellis, Ieuan. 1980. *Seven against Christ: a study of "Essays and Reviews."* Leiden: Brill.

Elwin, Warwick, ed. 1902. *Some XVIII century men of letters: biographical essays by the Rev. Whitwell Elwin, some time editor of the Quarterly Review, with a memoir.* 2 vols. London: John Murray.

*Emma Darwin:* see Litchfield, H. E., ed., 1904.

England, Richard. 1997. Natural selection before the *Origin:* public reactions of some naturalists to the Darwin-Wallace papers (Thomas Boyd, Arthur Hussey, and Henry Baker Tristram). *Journal of the History of Biology* 30:267–90.

Epstein, W. H. 1987. *Recognising biography.* Philadelphia: University of Pennsylvania Press.

———. 1991. *Contesting the subject: essays in the postmodern theory and practice of biography and biographical criticism.* West Lafayette, Ind.: Purdue University Press.

Evans, L. T. 1984. Darwin's use of the analogy between artificial and natural selection. *Journal of the History of Biology* 17:113–40

*Expression:* see Darwin, Charles, 1872.

Fara, Patricia. 1997. The Royal Society's portrait of Joseph Banks. *Notes and Records of the Royal Society* 51:199–210

———. 2000. Faces of genius: images of Newton in eighteenth-century England. In Geoffrey Cubitt and Allen Warren, eds., *Heroic reputations and exemplary lives,* 57–81. Manchester: Manchester University Press.

Farley, John. 1977. *The spontaneous generation controversy from Descartes to Oparin.* Baltimore: Johns Hopkins University Press.

———. 1982. *Gametes and spores: ideas about sexual reproduction, 1750–1914.* Baltimore: Johns Hopkins University Press.

Farley, John, and Gerald Geison. 1974. Science, politics and spontaneous generation in nineteenth-century France: the Pasteur-Pouchet debate. *Bulletin of the History of Medicine* 48:161–98.

Fawcett, Henry. 1860. A popular exposition of Mr. Darwin on the *Origin of Species. Macmillan's Magazine* 3:81–92.

Feather, John. 1988. *A history of British publishing*. London: Croom Helm.

———. 1994. *Publishing, piracy and politics: an historical study of copyright in Britain*. London: Mansell.

Feuer, Lewis. 1975. Is the Darwin-Marx correspondence authentic? *Annals of Science* 32:1–12.

Fichman, Martin. 1981. *Alfred Russel Wallace*. Boston: Twayne.

Finkelstein, David. 1993. "The secret": British publishers and Mudie's struggle for economic survival, 1861–64. *Publishing History* 34:21–50.

Finney, Colin. 1993. *Paradise revealed: natural history in nineteenth-century Australia*. Melbourne: Museum of Victoria.

FitzRoy, Robert. 1862. An explanation of the Meteorological Telegraphy and its basis, now under trial at the Board of Trade. *Notices of the Proceedings at the Meetings of the Members of the Royal Institution of Great Britain* 3:444–56.

Fleming, Donald. 1950. *John William Draper and the religion of science*. Philadelphia: University of Pennsylvania Press.

———. 1960. Charles Darwin, the anaesthetic man. *Victorian Studies* 4:219–36

Flint, Kate. 1995. Origins, species and *Great Expectations*. In David Amigoni and Jeff Wallace, eds., *Charles Darwin's "The Origin of Species": new interdisciplinary essays*, 152–73. Manchester: Manchester University Press.

Forrest, D. W. 1974. *Francis Galton: the life and work of a Victorian genius*. London: Paul Elek.

Foster, W. D. 1965. A contribution to the problem of Darwin's ill-health. *Bulletin of the History of Medicine* 39:476–78.

Fox, Robert. 1997. The University museum and Oxford science, 1850–1880. In M. G. Brock and M. C. Curtois, eds., *The history of the University of Oxford*. vol. 6, *Nineteenth century Oxford*, 641–91. Oxford: Oxford University Press.

Franklin, Allan. 1990. *Experiment, right or wrong*. Cambridge: Cambridge University Press.

Frasca-Spada, Marina, and Nick Jardine, eds. 2000. *Books and the sciences in history*. Cambridge: Cambridge University Press.

Freeman, Richard B. 1977. *The works of Charles Darwin: an annotated bibliographical handlist*. 2nd ed. Folkestone: Dawson.

———. 1978. *Charles Darwin: a companion*. Folkestone: Dawson.

———. 1984. *Darwin pedigrees*. London: Privately printed.

Freeman, Richard B., and Peter J. Gautrey. 1975. Charles Darwin's queries about expression. *Bulletin of the British Museum (Natural History) Historical Series* 4:205–19.

Freke, Henry. 1860. *Observations upon Mr. Darwin's recently published work—"On the origin of species by natural selection."* Dublin: Privately printed.

———. 1861. *On the origin of species by means of organic affinity*. Dublin.

French, Richard. 1975. *Antivivisection and medical science in Victorian society*. Princeton, N.J.: Princeton University Press.

Friedman, A. J., and C. C. Dorley. 1985. *Einstein as myth and muse*. Cambridge: Cambridge University Press.

Fulford, Roger, ed. 1968. *Dearest mama: letters between Queen Victoria and the crown princess of Prussia, 1861–1864*. London: Evans Bros.

Fyfe, Gordon, and John Law, eds. 1988. *Picturing power: visual depiction and social relations*. Sociological Review Monographs, 35. London: Routledge.

Gage, Andrew T., and William T. Stearn. 1988. *A bicentenary history of the Linnean Society of London*. London: Academic Press.

Gagnier, Regenia. 1991. *Subjectivities: a history of self-representation in Britain, 1832–1920*. Oxford: Oxford University Press.

Galton, Francis. 1865. Hereditary talent and character. *Macmillan's Magazine* 12:157–66, 318–27.

———. 1869. *Hereditary genius: an inquiry into its laws and consequences.* Facsimile edition with an introduction by H. J. Eysenck. London: Julian Friedmann Publishers, 1978.

———. 1908. *Memories of my life.* London: Methuen.

Gärtner, Karl F. von. 1849. *Versuche und beobachtungen über die Bastarderzeugung im Pflanzenreich.* Stuttgart.

Gasman, D. 1971. *The scientific origins of National Socialism: social Darwinism in Ernst Haeckel and the German Monist League.* London: Macdonald.

Gates, Barbara T. 1988. *Victorian suicides: mad crimes and sad histories.* Princeton, N.J.: Princeton University Press.

Gay, Peter. 1984–98. *The Bourgeois Experience: Victoria to Freud.* 5 vols. New York: Oxford University Press.

Gayon, Jean. 1998. *Darwinism's struggle for survival.* Cambridge: Cambridge University Press.

Gayon, Jean, and Doris Zallen. 1998. The role of the Vilmorin Company in the promotion and diffusion of the experimental science of heredity in France, 1840–1920. *Journal of the History of Biology* 31:241–62.

Geikie, James. 1874. *The great ice age and its relation to the antiquity of man.* London.

Geison, Gerald L. 1969. Darwin and heredity: the evolution of his hypothesis of pangenesis. *Journal of the History of Medicine* 24:375–411.

———. 1978. *Michael Foster and the Cambridge school of physiology: the scientific enterprise in late Victorian society.* Princeton, N.J.: Princeton University Press.

Geldart, H. D. 1879. Notes on the life and writings of Edward Blythe. *Transactions of the Norfolk and Norwich Naturalists Society* 3:38–46.

George, M. D. 1959. *English political caricature: a study of opinion and propaganda.* Oxford: Clarendon Press.

George, Wilma. 1964. *Biologist philosopher. A study of the life and writings of Alfred Russel Wallace.* London: Abelard Schuman.

———. 1979. Alfred Wallace, the gentle trader: collecting in Amazonia and the Malay Archipelago 1848–1862. *Journal of the Society for the Bibliography of Natural History* 9:503–14.

Ghiselin, Michael T. 1969. *The triumph of the Darwinian method.* Berkeley: University of California Press.

Gigerenzer, Gerd, et al. 1989. *The empire of chance: how probability changed science and everyday life.* Cambridge: Cambridge University Press.

Gillespie, Neal C. 1979. *Charles Darwin and the problem of creation.* Chicago: University of Chicago Press.

Gillespie, Neil R. 1977. The Duke of Argyll, evolutionary anthropology, and the art of scientific controversy. *Isis* 68:40–54.

Gilley, S. 1981. The Huxley-Wilberforce debate: a reconsideration. In Keith Robbins, ed., *Religion and Humanism,* 325–40. Studies in Church History 17. Oxford: Basil Blackwell for the Ecclesiastical History Society.

Gilman, Sander, ed. 1976. *The face of madness: Hugh W. Diamond and the origin of psychiatric photography.* New York: Brunner/Mazel.

———. 1995. *Health and illness: images of difference.* London: Reaktion Books.

Gilman, Sander, Helen King, Roy Porter, George Rousseau, and Elaine Showalter. 1993. *Hysteria Beyond Freud.* Berkeley: University of California Press.

Girouard, Mark. 1978. *Life in the English country house.* New Haven, Conn.: Yale University Press.

Glick, Thomas F., ed. 1988a. *The comparative reception of Darwinism.* Reprinted with a new preface. Chicago: University of Chicago Press.

———. 1988b. Spain. In Thomas Glick, ed., *The comparative reception of Darwinism,* 307–45. Chicago: University of Chicago Press.

Goffman, Erving, 1990. *The presentation of self in everyday life.* 2nd ed. London: Penguin Books.

Goldman, Lawrence, ed. 1989. *The blind Victorian: Henry Fawcett and British Liberalism*. Cambridge: Cambridge University Press.

Goldsmith, Elizabeth C., ed. 1989. *Writing the female voice: essays on epistolary literature*. Boston: Northeastern University Press.

Golinski, Jan. 1998. *Making natural knowledge*. Cambridge: Cambridge University Press.

Good, Rankine. 1954. The life of the shawl. *Lancet* pt.i:106–7.

Gooding, D., T. Pinch, and S. Schaffer, eds. 1989. *The uses of experiment: studies in the natural sciences*. Cambridge: Cambridge University Press.

Gosden, Peter H.J.H. 1961. *The friendly societies in England, 1815–1875*. Manchester: Manchester University Press.

———. 1973. *Self-help: voluntary associations in the nineteenth century*. London: Batsford.

Goulburn, J.E.M., et al. 1862. *Replies to Essays and Reviews. With a preface by the Lord Bishop of Oxford and letters from the Radcliffe Observer (R. Main) and the Reader in Geology in the University of Oxford (J. Phillips). With a note by Professor Owen*. 2nd ed. Oxford.

Gould, Stephen J. 1987. *Time's arrow, time's cycle: myth and metaphor in the discovery of geological time*. Cambridge, Mass.: Harvard University Press.

Graber, Robert Bates, and Lynate P. Miles. 1988. In defence of Darwin's father. *History of Science* 26:97–102.

Graham, Hilary. 1984. *Women, health and the family*. Brighton: Harvester Press.

Grant, Robert E. 1861. *Tabular view of the primary divisions of the animal kingdom*. London.

Gray, Asa. 1859. Diagnostic characters of new species of phaenogamous plants, collected in Japan by Charles Wright, botanist of the U.S. North Pacific Exploring Expedition. *Memoirs of the American Academy of Arts and Sciences* ns. 6 (1857–9):377–452.

———. 1861. *Natural selection not inconsistent with natural theology: a free examination of Darwin's treatise on the Origin of Species and of its American reviewers*. London.

———. 1876. *Darwiniana: essays and reviews pertaining to Darwinism*. New York.

Grayson, Donald K. 1983. *The establishment of human antiquity*. New York: Academic Press.

Greenacre, Phyllis. 1963. *The quest for the father: a study of the Darwin-Butler controversy, as a contribution to the understanding of the creative individual*. New York: International Universities Press.

Greene, John C. 1991. *Science, ideology and world view: essays in the history of evolutionary ideas*. Berkeley: University of California Press.

———. 1996. *The death of Adam: evolution and its impact on Western thought*. Rev ed. Iowa: Iowa State University Press.

Green-Lewis, Jennifer. 1996. *Framing the Victorians: photography and the culture of realism*. Ithaca, N.Y.: Cornell University Press.

Griest, G. L. 1970. *Mudie's circulating library and the Victorian novel*. Newton Abbot: David & Charles.

Groeben, Christiane, ed. 1982. *Charles Darwin, 1809–1882, Anton Dohrn, 1840–1909: correspondence*. Naples: Macchiaroli.

Gross, Charles G. 1993a. Hippocampus minor and man's place in nature: a case study in the social construction of neuroanatomy. *Hippocampus* 3:403–15.

———. 1993b. Huxley versus Owen: the hippocampus minor and evolution. *Trends in Neuroscience* 16:493–98.

Gross, John. 1969. *The rise and fall of the man of letters: aspects of English literary life since 1800*. London: Weidenfeld & Nicolson.

Gruber, H. E. 1961. Darwin and *Das Kapital*. *Isis* 52:582–83.

———. 1974. Darwin on man: a psychological study of creativity. New York: Dutton.

———. 1987. Darwin's "Tree of Nature" and other images of wide scope. In Judith Weschler, ed., *On aesthetics in science,* 121–40. Cambridge, Mass.: MIT Press.

Gruber, Jacob W. 1960. *A conscience in conflict: the life of St. George Jackson Mivart.* New York: Columbia University Press for Temple University Publications.

Gruber, J. W., and Thackray, J. C. 1992. *Richard Owen commemoration: three studies.* London: Natural History Museum Publications.

Gunther, Albert E. 1975. *A century of zoology at the British Museum through the lives of two keepers, 1815–1914.* Folkestone, Kent: Dawson & Sons.

Hacking, Ian. 1975. *The emergence of probability: a philosophical study of early ideas about probability, induction and statistical inference.* Cambridge: Cambridge University Press.

Haeckel, Ernst. 1866. *Generelle morphologie der organism.* 2 vols. Berlin.

Hague, James. 1884. A reminiscence of Mr. Darwin. *Harper's New Monthly Magazine* 69:759–63.

Haig, Alan. 1984. *The Victorian clergy.* London: Croom Helm.

Haight, Gordon S. 1940. *George Eliot and John Chapman, with Chapman's diaries.* New Haven, Conn.: Yale University Press.

———, ed. 1954–78. *The George Eliot letters.* 9 vols. New Haven, Conn.: Yale University Press.

———. 1968. *George Eliot: a biography.* Oxford: Oxford University Press.

Haines, Valerie A. 1991. Spencer, Darwin and the question of reciprocal influence. *Journal of the History of Biology* 24:409–31.

Haley, Bruce, 1978. *The healthy body and Victorian culture.* Cambridge, Mass.: Harvard University Press.

Hall, A. Rupert. 1966. *The Abbey scientists.* London: Nicolson.

Hall, Lesley, and Roy Porter. 1995. *The facts of life: the creation of sexual knowledge in Britain, 1650–1950.* New Haven, Conn.: Yale University Press.

Hall, Marie Boas. 1984. *All scientists now: the Royal Society in the nineteenth century.* Cambridge: Cambridge University Press.

Hallett, Mark. 1999. *The spectacle of difference: graphic satire in the age of Hogarth.* New Haven, Conn.: Yale University Press.

Hardy, Anne. 1993. *The epidemic streets: infectious disease and the rise of preventative medicine, 1856–1900.* Oxford: Clarendon Press.

Harris, Roy, ed. 1996. *The origin of language.* Bristol: Thoemmes Press.

Harrison, John F. C. 1954. *A history of the Working Men's College, 1854–1954.* London: Routledge & Kegan Paul.

———. 1969. *Robert Owen and the Owenites in Britain and America: the quest for the new moral world.* London: Routledge.

Harte, N., and R. Quinault, eds. 1996. *Land and society in Britain, 1700–1914. Essays in honour of F. M. L. Thompson.* Manchester: Manchester University Press.

Hartley, Lucy. 2001. *Physiognomy and the meaning of expression in nineteenth-century culture.* Cambridge: Cambridge University Press

Harvey, John R. 1970. *Victorian novelists and their illustrators.* London: Sidgwick & Jackson.

Harvey, Joy. 1983. Evolutionism transformed: positivists and materialists in the Société d'Anthropologie de Paris from the Second Empire to Third Republic. In David Oldroyd and Ian Langham, eds., *The wider domain of evolutionary thought,* 289–310. Dordrecht: Reidel.

———. 1993. Types and races: the politics of colonialism and anthropology in the nineteenth century. In A. Lafuente, A. Elena, and M. L. Ortega, eds., *Mundializacion de la ciencia y cultura nacional,* 527–37. Madrid: Doce Calles.

———. 1997a. *"Almost a man of genius": Clemence Royer, feminism, and nineteenth-century science.* New Brunswick, N.J.: Rutgers University Press.

———. 1997b. Les esprits fertiles et le problème de la sterilité experimentale: Charles Darwin et le Muséum d'histoire naturelle. In Claude Blanckaert et al., eds., *Le Muséum au premier siècle de son histoire,* 341–61. Paris: Muséum d'histoire naturelle, Archives.

Haskell, Thomas L., ed. 1984. *The authority of experts: studies in history and theory.* Bloomington: Indiana University Press.

Haslam, Fiona. 1996. *From Hogarth to Richardson: medicine in art in eighteenth-century Britain,* Liverpool: Liverpool University Press.

Haughton, Samuel. 1860. Biogenesis. *Natural History Review* 7:23–32.

Haynes, E.S.P. 1916. Master George Pollack. *Cornhill Magazine* ns. 41:232–37.

Healey, Edna. 1986. *Wives of fame: Mary Livingstone, Jenny Marx, Emma Darwin.* London: Sidgwick & Jackson.

Helmstadter, Richard, and Bernard Lightman, eds. 1990. *Victorian faith in crisis: essays in continuity and change in nineteenth-century religious belief.* Basingstoke: Macmillan.

Henderson, Heather. 1989. *The Victorian self: autobiography and biblical narrative.* Ithaca, N.Y.: Cornell University Press.

Henkin, Leo. 1963. *Darwinism in the English novel 1860–1910: the impact of evolution on Victorian fiction.* New York: Russell & Russell.

Herbert, Sandra. 1974–77. The place of man in the development of Darwin's theory of transmutation. *Journal of the History of Biology* 7:217–58; 10:155–227.

Herrick, Jim. 1985. *Against the faith: essays on deists, skeptics and atheists.* Buffalo: Prometheus Books.

Herschel, J. F. 1861. *Physical Geography. From the Encyclopaedia Britannica.* Edinburgh.

Hill, Mathew. 1862. On the Post-office. *Notices of the Proceedings of the Royal Institution of Great Britain* 3:457–66.

Hilton, Boyd. 1988. *The age of atonement: the influence of Evangelicalism on social and economic thought, 1795–1865.* Oxford: Clarendon Press.

Himmelfarb, Gertrude. 1959. *Darwin and the Darwinian revolution.* London: Chatto & Windus.

Hinchcliff, Peter B. 1964. *John William Colenso, Bishop of Natal.* London: Nelson.

Hinton, Brian. 1992. *Immortal faces: Julia Margaret Cameron on the Isle of Wight.* Newport, Hants: Isle of Wight County Press.

Hobhouse, Henry. 1985. *Seeds of change: five plants that transformed mankind.* London: Sidgwick & Jackson.

Hobsbawn, Eric J. 1975. *The age of capital, 1848–1875.* London: Abacus.

Hodge, M.J.S. 1977. The structure and strategy of Darwin's "long argument." *British Journal for the History of Science* 10:237–45.

———. 1985. Darwin as a lifelong generation theorist. In David Kohn, ed., *The Darwinian heritage,* 207–43. Princeton, N.J.: Princeton University Press in association with Nova Pacifica.

Hofstadter, Richard. 1945. *Social Darwinism in American thought, 1860–1915.* Philadelphia: University of Pennsylvania Press.

Hoge, James, ed. 1981. *Lady Tennyson's journal.* Charlottesville: University Press of Virginia.

Holcombe, Henry W. 1979. *Patent medicine tax stamps: a history of the firms using United States private die proprietary medicine tax stamps.* Lawrence, Mass.: Quarterman Publications.

Homans, Margaret. 1999. *Royal representations: Queen Victoria and British culture, 1837–1876.* Chicago: University of Chicago Press.

Hooker, Joseph Dalton. 1859. On the flora of Australia . . . an introductory

essay to the *Flora Tasmaniae*. In *Flora Tasmaniae. Pt 3 of the Botany of the Antarctic voyage of H.M. Discovery Ships Erebus and Terror, in the years 1839 to 1843 under the command of Captain Sir James Clark Ross*. 2 vols. London, 1855–60.
———. 1874. The carnivorous habits of plants. *Report of the meeting of the British Association for the Advancement of Science held at Belfast, 1874*, 102–16. London, 1875. Reprinted in *Nature* 10 (1874):366–72.
Hopkinson, Amanda. 1986. *Julia Margaret Cameron*. London: Virago.
Hoppen, K. Theodore. 1998. *The mid-Victorian generation, 1846–1886*. The New Oxford History of England. Oxford: Oxford University Press.
Horn, Pamela. 1995. *The rise and fall of the Victorian servant*. London: Alan Sutton.
Horstman, Allen. 1985. *Victorian divorce*. London: Croom Helm.
Houghton, Walter E. 1957. *The Victorian frame of mind, 1830–1870*. New Haven, Conn.: Yale University Press.
———, ed. 1966–89. *The Wellesley index to Victorian periodicals, 1824–1900*. 5 vols. Toronto: University of Toronto Press.
Howard, Martin. 1977. *Victorian grotesque: an illustrated excursion into medical curiosities, freaks and abnormalities principally of the Victorian age*. London: Jupiter Books.
Howarth, O.J.R., and E. K. Howarth. 1933. *A history of Darwin's parish, Downe, Kent*. Southampton: Russell & Co.
Hubble, Douglas. 1953. The life of the shawl. *Lancet* pt.ii:1351–54.
Hudson, Derek. 1972. *Munby: man of two worlds: the life and diaries of Arthur J. Munby, 1828–1910*. London: John Murray.
Huggett, Frank E. 1978. *Victorian England as seen by "Punch."* London: Sidgwick & Jackson.
Hughes, Kathryn. 1993. *The Victorian governess*. London: Hambledon Press.
Hull, David, ed. 1973. *Darwin and his critics: the reception of Darwin's theory of evolution by the scientific community*. Chicago: University of Chicago Press.
———. 1974. *The philosophy of the biological sciences*. Englewood Cliffs, N.J.: Prentice-Hall.
Hutchinson, Horace G., ed. 1914. *Life of Sir John Lubbock, Lord Avebury*. 2 vols. London: Macmillan.
Hutton, Richard H. 1989. *A Victorian spectator: uncollected writings*. Edited with an introduction by Robert Tener and Malcolm Woodfield. Bristol: Bristol Press.
Huxley, Leonard, ed. 1900. *The life and letters of Thomas Henry Huxley*. 2 vols. London: Macmillan.
———, ed. 1918. *Life and letters of Sir Joseph Dalton Hooker*. 2 vols. London: John Murray.
Huxley, Thomas Henry. 1859a. On the persistent types of animal life. *Notices of the Proceedings of the Royal Institution of Great Britain* 3 (1858–62):90–93.
———. 1859b. Time and life: Mr. Darwin's *Origin of Species*. *Macmillan's Magazine* 1:142–48.
———. 1860a. On species and races, and their origin. *Notices of the Proceedings of the Royal Institution of Great Britain* 3 (1858–62):195–200.
———. 1860b. Darwin on the origin of species. *Westminster Review* ns. 17:541–70.
———. 1863. *Evidence as to Man's Place in Nature*. London.
———. 1869. On the physical basis of life. Delivered in Edinburgh 1868. *Fortnightly Review* 11:129–45.
———. 1893. *Darwinina: essays*. London.
Hyman, Stanley E. 1962. *The tangled bank. Darwin, Marx, Frazer and Freud as imaginative writers*. New York: Atheneum.
Irvine, William. 1955. *Apes, angels, and Victorians: the story of Darwin, Huxley, and evolution*. New York, London, Toronto: McGraw-Hill.

Jackson, Alan A. 1999. *London's local railways*. 2nd ed. Harrow Weald, Middlesex: Capital Transport Publishing.

Jacyna, L. S. 1983. John Goodsir and the making of cellular reality. *Journal of the History of Biology* 16:75–99.

Jalland, Pat. 1996. *Death in the Victorian family*. Oxford: Oxford University Press.

James, Patricia. 1979. *"Population" Malthus: his life and times*. London: Routledge & Kegan Paul.

Jann, Rosemary. 1994. Darwin and the anthropologists: sexual selection and its discontents. *Victorian Studies* 37:287–306.

Jardine, N., J. A. Secord, and E. Spary, eds. 1996. *Cultures of natural history*. Cambridge: Cambridge University Press.

Jardine, Nicholas. 1991. *Scenes of inquiry*. Oxford: Clarendon Press.

Jenkin, Fleeming. 1867. The *Origin of Species*. *North British Review* 46:277–318.

Jenkins, Terence A. 1996. *Disraeli and Victorian Conservatism*. London: Macmillan.

Jenson, J. Vernon. 1970. The X Club: fraternity of Victorian scientists. *British Journal for the History of Science* 5:63–72.

———. 1988. Return to the Wilberforce-Huxley debate. *British Journal for the History of Science* 21:161–179.

———. 1989. *Thomas Henry Huxley: communicating for science*. Newark: University of Delaware Press.

Jenyns [Blomefield], Leonard. 1862. *Memoir of the Rev. John Stevens Henslow, M.A.* Cambridge.

Jesperson, P. H. 1948–49. Charles Darwin and Dr. Grant. *Lychnos*, pp. 159–67.

Jones, Edgar Y. 1973. *Father of art photography: O. J. Rejlander, 1813–1875*. Newton Abbot: David & Charles.

Jones, Greta. 1980. *Social Darwinism and English thought: the interaction between biological and social theory*. Sussex: Harvester Press.

Jones, H. F. 1920. *Samuel Butler, author of Erewhon (1835–1902): a memoir*. 2 vols. London: Macmillan.

Jones, Henry Bence. 1850. *On animal chemistry in its application to stomach and renal diseases*. London: John Churchill.

Jones, Jo Elwyn, and J. Francis Gladstone. 1998. *The Alice companion: a guide to Lewis Carroll's Alice books*. Basingstoke: Macmillan.

Jordan, David Starr. 1922. *The days of a man: being memories of a naturalist, teacher and minor prophet of democracy*. 2 vols. London: George Harrap.

Jordan, John, and Robert Patten, eds. 1995. *Literature in the marketplace*. Cambridge: Cambridge University Press.

Jordanova, Ludmilla. 1984. *Lamarck*. Oxford: Oxford University Press.

———. 2000. *Defining features: scientific and medical portraits, 1660–2000*. London: Reaktion Books.

Josselson, Ruthellen, and Amia Lieblich, eds. 1995. *Interpreting experience: the narrative study of lives*. London: Sage.

*Journal*: see De Beer, Gavin, ed., 1959.

*Journal of researches*: see Darwin, Charles, 1845.

Junker, Thomas. 1991. Heinrich Georg Bronn und Die Entestehung der Arten. *Sudhoffs Archiv* 75:180–208.

Junker, Thomas, and Marsha Richmond, eds. 1996. *Charles Darwin's correspondence with German naturalists*. Marburg an der Lahn: Basilisken-Presse.

Kauffman, Linda S. 1992. *Special delivery: epistolary modes in modern fiction*. Chicago: University of Chicago Press.

Keith, Arthur. 1927. Darwin's home. In *Man's origin*, 32–40. London: Watts.

———. 1955. *Darwin revalued*. London: Watts.

Kelly, Alfred. 1981. *The descent of Darwin: the popularisation of Darwinism in Germany, 1860–1914*. Chapel Hill: University of North Carolina Press.

Kempf, Edward. 1918. Charles Darwin—the affective sources of his inspiration and anxiety neurosis. *Psychoanalytic Review* 5:151–92.

Keynes, Margaret. 1943. *Leonard Darwin.* Cambridge: Cambridge University Press.

Keynes, Randal. 2001. *Annie's box: Charles Darwin, his daughter and human evolution.* London: Fourth Estate.

Kingsley, Charles. 1862. *Speech of Lord Dundreary in Section D . . . on the great hippocampus question.* Cambridge: Privately printed.

———. 1874. *Westminster sermons.* London.

Knoepflmacher, U. C., and G. B., Tennyson, eds. 1977. *Nature and the Victorian imagination.* Berkeley: University of California Press.

Kohlstedt, Sally G., and Helen H. Longino, eds. 1997. *Women, gender and science: new directions. Osiris,* vol. 12.

Kohn, David. 1981. On the origin of the principle of divergence. *Science* 213:1105–8.

———, ed. 1985a. *The Darwinian heritage.* Princeton, N.J: Princeton University Press in association with Nova Pacifica.

———. 1985b. Darwin's principle of divergence as internal dialogue. In David Kohn, ed., *The Darwinian heritage,* 245–57. Princeton, N.J.: Princeton University Press in association with Nova Pacifica.

———. 1989. Darwin's ambiguity: the secularization of biological meaning. *British Journal for the History of Science* 22:215–39.

———. 1996. The aesthetic construction of Darwin's theory. In Alfred I. Tauber, ed., *The elusive synthesis: aesthetics and science,* 13–48. Dordrecht: Kluwer.

Kölreuter, Joseph Gottlieb. 1761–66. *Vorlaufige Nachricht von einigen das Geschlecht der Pflanzen betreffenden Versuchen und Beobachtungen.* Leipzig.

Kottler, Malcolm J. 1974. Wallace, the origin of man and spiritualism. *Isis* 65:145–92.

———. 1985. Charles Darwin and Alfred Russel Wallace: two decades of debate over natural selection. In David Kohn, ed., *The Darwinian heritage,* 367–432. Princeton, N.J.: Princeton University Press in association with Nova Pacifica.

Krasner, James. 1992. *The entangled eye: visual perception and the representation of nature in post-Darwinian narrative.* Oxford: Oxford University Press.

Krause, Ernst L. 1879. *Erasmus Darwin.* Translated from the German by W. S. Dallas. With a preliminary notice by C. Darwin. London.

Kruger, Lorenz, Lorraine Daston, and Michael Heidelberger, eds. 1987. *The probabilistic revolution.* 2 vols. Cambridge, Mass.: MIT Press.

Kuklick, Henrika. 1994. The color blue: from research in the Torres Strait to an ecology of human behaviour. In R. M. MacLeod and P. Rehbock, eds., *Darwin's laboratory. Evolutionary theory and natural history in the Pacific,* 339–67. Honolulu: University of Hawaii Press.

Lane, Edward. 1857. *Hydrotherapy: or, the natural system of medical treatment. An explanatory essay.* London.

———. 1882. *Letter read by Dr. B. W. Richardson F.R.S. at his lecture on Chas. Darwin F.R.S. in St. George's Hall, Langham Place, October 22nd., 1882.* Privately printed and published.

Larsen, Anne. 1996. Equipment for the field. In N. Jardine, J. A. Secord, and E. Spary, eds., *Cultures of natural history,* 358–77. Cambridge: Cambridge University Press.

Lawrence, Christopher, and Steven Shapin, eds. 1998. *Science incarnate: historical embodiments of natural knowledge.* Chicago: University of Chicago Press.

Leder, Drew. 1990. *The absent body.* Chicago: University of Chicago Press.

Lenoir, Timothy. 1987. The Darwin industry. *Journal of the History of Biology* 20:115–30.

Leopold, Joan. 1980. *Culture in comparative and evolutionary perspective:*

*E. B. Tylor and the making of Primitive Culture*. Berlin: Dietrich Reimer Verlag.

Levine, George. 1988. *Darwin and the novelists: patterns of science in Victorian fiction*. Cambridge, Mass.: Harvard University Press.

Lewes, George H. 1860. Studies in animal life [chapter 4]. *Cornhill Magazine* 1:438–47.

*Life and Letters:* see Darwin, Francis, ed., 1883.

Lightman, Bernard. 1987. *The origins of agnosticism. Victorian belief and the limits of knowledge*. Baltimore: Johns Hopkins University Press.

———, ed. 1997. *Victorian science in context*. Chicago: University of Chicago Press.

Limoges, Camille. 1970. *La sélection naturelle: étude sur la première constitution d'un concept (1837–1859)*. Paris: Presses Universitaires de France.

———. 1971. Darwin, Milne-Edwards et le principe de divergence. *Actes du XIle congrès internationale d'histoire de science* 8:111–15.

———. 1994. Milne-Edwards, Darwin, Durkheim and the division of labour: a case study in reciprocal conceptual exchanges between the social and the natural sciences. In I. B. Cohen, ed., *The natural sciences and the social sciences: some critical and historical perspectives*, 317–43. Dordrecht: Kluwer.

Lindberg, David, and Ronald L. Numbers, eds. 1986. *God and nature: historical essays on the encounter between Christianity and science*. Berkeley: University of California Press.

Linnean Society of London. 1908. *The Darwin-Wallace celebration held on Thursday 1st July 1908*. London: Linnean Society of London.

Litchfield, Henrietta E., ed. 1904. *Emma Darwin, wife of Charles Darwin: a century of family letters*. 2 vols. Cambridge: Privately printed.

———. 1910. *Richard Buckley Litchfield: a memoir written for his friends*. Cambridge: Privately printed.

Livingstone, David N. 1987. *Darwin's forgotten defenders: the encounter between evangelical theology and evolutionary thought*. Edinburgh: Scottish Academic Press.

Longford, Elizabeth. 1964. *Victoria R.I.* London: Weidenfeld & Nicolson.

Lorimer, Douglas A. 1978. *Colour, class and the Victorians: English attitudes to the Negro in the mid-nineteenth century*. Leicester: Leicester University Press.

Lubbock, John. 1865. *Pre-historic times as illustrated by ancient remains, and the manners and customs of modern savages*. London.

———. 1870. *The origin of civilisation and the primitive condition of man. Mental and social condition of savages*. London.

Lucas, J. R. 1979. Wilberforce and Huxley: a legendary encounter. *Historical Journal* 22:313–30.

Lurie, Edward. 1960. *Louis Agassiz: a life in science*. Chicago: University of Chicago Press.

Lyell, Charles. 1830–33. *Principles of geology, being an attempt to explain the former changes of the earth's surface, by reference to causes now in operation*. Facsimile edition with an introduction by M.J.S. Rudwick. 3 vols. Chicago: University of Chicago Press, 1991.

———. 1838. *Elements of geology*. London.

———. 1859. On the occurrence of works of human art in post-pliocene deposits. *Report of the British Association for the Advancement of Science held at Aberdeen, 1858* 29:93–95.

———. 1863. *The geological evidences of the antiquity of man, with remarks on theories of the origin of species by variation*. London.

Lyell, Katherine M., ed. 1881. *Life, letters, and journals of Sir Charles Lyell*. 2 vols. London.

Lynch, Michael. 1985. Discipline and the material form of images: an analysis of scientific visibility. *Social Studies of Science* 15:37–66.

Lynch, Michael, and Steve Woolgar, eds. 1988. *Representation in scientific practice*. Dordrecht: Kluwer.

Maasen, Sabine. 1995. Who is afraid of metaphors? In S. Maasen, E. Mendelsohn, and P. Weingart, eds., *Biology as society, society as biology: metaphors*, 11–35. Dordrecht: Kluwer.

Machann, Clinton. 1994. *The genre of autobiography in Victorian literature*. Ann Arbor: University of Michigan Press.

MacKenzie, Donald A. 1981. *Statistics in Britain, 1865–1930: the social construction of scientific knowledge*. Edinburgh: Edinburgh University Press.

MacLeod, Roy M. 1969a. The genesis of *Nature*. *Nature* 224:423–61.

———. 1969b. The X Club: a social network of science in late Victorian England. *Notes and Records of the Royal Society* 24:305–22.

———. 1971. Of medals and men: a reward system in Victorian science. *Notes and Records of the Royal Society* 26:81–105.

———. 1982a. Breaking the circle of the sciences: the natural sciences tripos and the "examination revolution." In *Days of judgement: science, examinations and the organisation of knowledge in late Victorian England*, 189–212. Driffield: Nafferton.

———. 1982b. On visiting the "moving metropolis": reflections on the architecture of imperial science. *Historical Records of Australian Science* 5:1–16.

Maitland, Frederic William, ed. 1906. *The life and letters of Leslie Stephen*. London: Duckworth.

Mandelstam, Joel. 1994. Du Chaillu's stuffed gorillas and the savants from the British Museum. *Notes and Records of the Royal Society* 48:227–45

Mangen, James, and James Walvin, eds. 1987. *Manliness and morality: middle-class masculinity in Britain and America, 1800–1940*. Manchester: Manchester University Press.

Manvell, Roger. 1976. *The trial of Annie Besant and Charles Bradlaugh*. London: Elek/Pemberton.

Marchand, Leslie A. 1971. *The Athenaeum: a mirror of Victorian culture*. Chapel Hill: University of North Carolina Press.

Marchant, James, ed. 1916. *Alfred Russel Wallace: letters and reminiscences*. 2 vols. London: Cassell.

Marsh, Kate, ed., 1993. *Writers and their houses*. London: Hamish Hamilton.

Martin, Robert B. 1980. *Tennyson. The unquiet heart*. Oxford: Clarendon Press.

Marx, Karl. 1985. *Letters*. vol. 41 in *Collected works*. 46 vols. London: Lawrence & Wishart, 1975–92.

Matthew, H. Colin G. 1986. *The Gladstone Diaries*. Vol. 9, January 1875–December 1880. Oxford: Clarendon Press.

———. 1990. *The Gladstone Diaries*. Vol. 10, January 1881–June 1883. Oxford: Clarendon Press.

Matthew, Patrick. 1831. *On naval timber and arboriculture*. London.

Mayr, Ernst. 1991. *One long argument: Charles Darwin and the genesis of modern evolutionary thought*. London: Allen Lane.

McCook, Stuart. 1996. "It may be truth, but it is not evidence": Paul Du Chaillu and the legitimation of evidence in the field sciences. *Osiris* 11:177–97.

McDannell, Colleen, and Bernhard Lang. 1988. *Heaven: a history*. New Haven, Conn.: Yale University Press.

McKendrick, Neil, John Brewer, and J. H. Plumb. 1982. *The birth of a consumer society: the commercialization of eighteenth-century England*. London: Europa Publications.

McKinney, H. Lewis. 1966. Alfred Russel Wallace and the discovery of natural selection. *Journal of the History of Medicine and Allied Sciences* 21:333–57.

———. 1969. Wallace's earliest observations on evolution: 28 December 1845. *Isis* 60:370–73.

———. 1972. *Wallace and natural selection.* New Haven, Conn.: Yale University Press.

McLaren, Angus. 1978. *Birth control in nineteenth century England.* London: Croom Helm.

McOuat, Gordon. 2001. Cataloguing power: delineating "competent naturalists" and the meaning of species in the British Museum. *British Journal for the History of Science* 34:1–28.

Meacham, Standish. 1970. *Lord Bishop: the life of Samuel Wilberforce, 1805–1873.* Cambridge, Mass.: Harvard University Press.

Meadows, A. J., ed. 1980. *Development of science publishing in Europe.* Amsterdam: Elsevier Science Publishers.

Means, John O., ed. 1876. *The prayer gauge debate. By Prof. Tyndall, F. Galton and others, against Dr. Littledale, President McCosh, the Duke of Argyll, Canon Liddon, and "The Spectator."* Boston.

Mellersh, H.E.L. 1968. *FitzRoy of the Beagle.* London: Rupert Hart-Davis.

Merton, Robert. 1973. Priorities in scientific discovery. In R. Merton, *The sociology of science: theoretical and empirical investigations,* 286–324. Chicago: University of Chicago Press.

Metcalfe, Richard. 1906. *The rise and progress of hydropathy in England and Scotland.* London: Simpkin, Marshall, Hamilton, Kent & Co.

Micale, Mark. 1995. *Approaching hysteria: disease and its interpretation.* Princeton, N.J.: Princeton University Press.

Mill, John Stuart. 1862. *A system of logic, ratiocinative and inductive: being a connected view of the principles of evidence and the methods of scientific investigation.* 5th ed. 2 vols. London.

Miller, Hugh. 1857. *The testimony of the rocks; or geology in its bearings on the two theologies, natural and revealed.* Edinburgh.

———. 1861. *Footprints of the Creator.* New ed. by Mrs. Miller with a memoir by Louis Agassiz. Edinburgh.

Milner, Richard. 1996. Charles Darwin and associates, ghostbusters. *Scientific American* 275:96–101.

Mirowski, Philip, ed. 1994. *Natural images in economic thought.* Cambridge: Cambridge University Press.

Mitchell, Sally. 1996. *Daily life in Victorian England.* Westport, Conn.: Greenwood Press.

Mivart, St. George. 1871. *On the genesis of species.* London.

Montgomery, William M. 1985. Charles Darwin's thought on expressive mechanisms in evolution. In Gail Zivin, ed., *The development of expressive behaviour: biology-environment interactions,* 27–50. New York: Academic Press.

———. 1988. Germany. In Thomas Glick, ed., *The comparative reception of Darwinism,* 81–116. Chicago: University of Chicago Press.

Moody, J.W.T. 1971. The reading of the Darwin and Wallace papers: an historical "non-event." *Journal of the Society for the Bibliography of Natural History* 5:474–76.

Moon, H. P. 1976. *Henry Walter Bates, F.R.S., 1825–93, explorer, scientist and Darwinian.* Leicester: Leicestershire Museums.

Moore, James R. 1977. On the education of Darwin's sons: the correspondence between Charles Darwin and the Reverend G. V. Reed, 1857–1864. *Notes and Records of the Royal Society* 32:51–70.

———. 1979. *The post-Darwinian controversies. A study of the Protestant struggle to come to terms with Darwin in Great Britain and America, 1870–1900.* Cambridge: Cambridge University Press.

———. 1982. Charles Darwin lies in Westminster Abbey. In R. J. Berry, ed., *Charles Darwin: a commemoration,* 97–113. London: Linnean Society of London.

———. 1985a. Darwin of Down: the evolutionist as squarson-naturalist. In David Kohn, ed., *The Darwinian heritage,* 435–81. Princeton, N.J.:

Princeton University Press in association with Nova Pacifica.

———. 1985b. Darwin's genesis and revelations. *Isis* 76:570–80.

———. 1986. Geologists and interpreters of Genesis in the nineteenth century. In David C. Lindberg and Ron L. Numbers, eds., *God and nature: historical essays on the encounter between Christianity and science,* 322–50. Berkeley: University of California Press.

———, ed. 1989a. *History, humanity and evolution: essays for John C. Greene.* Cambridge: Cambridge University Press.

———. 1989b. Of love and death: why Darwin "gave up Christianity." In James Moore, ed., *History, humanity and evolution: essays for John C. Greene,* 195–229. Cambridge: Cambridge University Press.

———. 1991. Deconstructing Darwinism: the politics of evolution in the 1860s. *Journal of the History of Biology* 24:353–408.

———. 1994. *The Darwin legend.* Grand Rapids, Mich.: Baker Books.

———. 1997. Wallace's Malthusian moment: the common context revisited. In B. Lightman, ed., *Victorian Science in Context,* 290–311. Chicago: University of Chicago Press.

*More letters:* see Darwin, Francis, and A. C. Seward, eds., 1903.

Morgan, M. 1994. *Manners, morals and class in England, 1774–1858.* Basingstoke: Macmillan.

Morrell, Jack B., and Arnold Thackray. 1981. *Gentlemen of science: early years of the British Association for the Advancement of Science.* Oxford: Oxford University Press.

Morris, Francis Orpen. 1875. *All the articles of the Darwin faith.* London.

Morris, Susan S. 1994. Fleeming Jenkin and the *Origin of Species:* a reassessment. *British Journal for the History of Science* 27:313–43.

Morton, A. G. 1981. *History of botanical science.* London: Academic Press.

Morton, Peter. 1984. *The vital science: biology and the literary imagination, 1860–1900.* London: George Allen & Unwin.

Moser, Stephanie. 1998. *Ancestral images: the iconography of human origins.* Ithaca, N.Y.: Cornell University Press.

Müller, Frederick Max. 1864. *Lectures on the science of language delivered at the Royal Institution in April, May and June 1861.* London.

———. 1901. *My autobiography: a fragment.* London: Longmans.

Müller, Johann F. T. (known as Fritz). 1864. *Für Darwin.* Leipzig. Translated by William Dallas as *Facts and arguments for Darwin.* London: John Murray, 1869.

Mulock, Dinah [Craik]. 1864. *The Water Cure.* London.

Murray, John. 1909. Darwin and his publisher. *Science Progress* 3 (1908–9):537–42.

———. 1919. *John Murray III, 1808–1892: a brief memoir.* London: John Murray.

Myers, Greg. 1990. *Writing biology: texts in the social construction of scientific knowledge.* Madison: University of Wisconsin Press.

Nash, L. A. 1921. Some memories of Charles Darwin. *Overland Monthly* (May 1921) 77:26–29.

Nash, Wallis. 1919. *A lawyer's life on two continents.* Boston: Richard G. Badger.

*Natural selection:* see Stauffer, R. C., ed., 1975.

*Nature.* 1989. Centenary edition. 224:423–61.

Nelson, E. Charles. 1990. *Aphrodite's mousetrap: a biography of Venus's flytrap, with facsimiles of an original pamphlet and the manuscripts of John Ellis, F.R.S.* Aberystwyth: Boethius Press.

Neve, Michael. 1993. Charles Darwin: Down House, Downe, Kent. In Kate Marsh, ed., *Writers and their houses,* 151–58. London: Hamish Hamilton.

———, ed. 2002. *Charles Darwin's Autobiography.* London: Penguin Books.

Neve, Michael, and Trevor Turner. 1995. What the doctor thought and did: Sir

James Crichton-Browne (1840–1938). *Medical History* 39:399–432.

Nevill, Ralph H. 1919. *Life and letters of Lady Dorothy Nevill*. London: Methuen.

*Notebooks:* see Barrett, Paul H., et al., eds., 1987.

Numbers, Ronald L. 1998. *Darwinism comes to America*. Cambridge, Mass.: Harvard University Press.

Numbers, Ronald L., and John Stenhouse, eds. 1999. *Disseminating Darwinism: the role of place, race, religion, and gender.* Cambridge: Cambridge University Press.

Núñez, Diego. 1977. *El darwinismo en España*. Madrid: Editorial Castalia.

Nye, Robert. 1997. Medicine and science as masculine "fields of honour." In Sally Gregory Kohlstedt and Helen E. Longino, eds., *Women, gender, and science. Osiris* 12:60–79.

Nyhart, Lynn. 1995. *Biology takes form: animal morphology and the German universities, 1800–1900*. Chicago: University of Chicago Press.

O'Brien, Charles. 1970. *Eozoon Canadense,* "the dawn animal of Canada." *Isis* 61:206–23.

Olby, R. C. 1963. Charles Darwin's manuscript of pangenesis. *British Journal of the History of Science* 1:251–63.

———. 1985. *Origins of Mendelism*. 2nd ed. Chicago: University of Chicago Press.

Oldroyd, David. 1984. How did Darwin arrive at his theory? The secondary literature to 1982. *History of Science* 22:325–74.

Oldroyd, David, and Ian Langham, eds. 1983. *The wider domain of evolutionary thought*. Dordrecht: Reidel.

Ophir, Adi, and Steven Shapin. 1991. The place of knowledge: a methodological survey. *Science in Context* 4:3–21.

Oppenheim, Janet. 1985. *The other world: spiritualism and psychical research in England, 1850–1914*. Cambridge: Cambridge University Press.

———. 1991. *"Shattered Nerves": doctors, patients, and depression in Victorian England*. New York: Oxford University Press.

*Origin:* see Darwin, Charles, 1859.

Ospovat, Dov. 1981. *The development of Darwin's theory: natural history, natural theology, and natural selection, 1838–1859*. Cambridge: Cambridge University Press.

Owen, Richard. 1858. Address. *Report of the 28th meeting of the British Association for the Advancement of Science held at Leeds, September 1858,* xlix–cx. London, 1859.

———. 1860. Darwin on the origin of species. *Edinburgh Review* 111:487–532.

Owen, Richard S., ed. 1894. *The life of Richard Owen*. 2 vols. London.

Page, Norman, ed. 1983. *Tennyson: interviews and recollections*. Basingstoke: Macmillan.

Pancaldi, Giuliano. 1991. *Darwin in Italy: science across cultural frontiers*. Rev ed. Translated by R. B. Morelli. Bloomington: Indiana University Press.

———. 1994. The technology of nature: Marx's thoughts on Darwin. In I. B. Cohen, ed., *The natural sciences and the social sciences*, 257–74. Dordrecht: Kluwer.

Paradis, James G. 1978. *T. H. Huxley: man's place in nature*. Lincoln: University of Nebraska Press.

———. 1981. Darwin and landscape. In James G. Paradis and Thomas Postlewait, eds., *Victorian science and Victorian values: literary perspectives*, 85–110. New York: New York Academy of Sciences.

———. 1997. Science and satire in Victorian culture. In Bernard Lightman, ed., *Victorian science in context*, 143–75. Chicago: University of Chicago Press.

Paradis, James G., and Thomas Postlewait, eds. 1981. *Victorian science and Victorian values: literary perspectives*. New York: New York Academy of Sciences.

Parker, J., ed. 1860. *Essays and Reviews. By F. Temple, R. Williams, B. Powell,*

*H. B. Wilson, C. W. Goodwin, M. Pattison, B. Jowett.* London.

Paston, George. 1932. *At John Murray's: records of a literary circle, 1843–92.* London: John Murray.

Patterson, Alfred T. 1954. *Radical Leicester: a history of Leicester 1780–1850.* Leicester: University of Leicester Press.

Pattison, S. R. 1863. "The Antiquity of Man": an examination of Sir Charles Lyell's recent work. London.

Paul, Diane B. 1988. The selection of the "Survival of the Fittest." *Journal of the History of Biology* 21:411–24.

Pearson, Karl, ed. 1914–30. *The life, letters, and labours of Francis Galton.* 4 vols. Cambridge: Cambridge University Press.

Peckham, Morse, ed. 1959a. *The "Origin of Species" by Charles Darwin. A variorum text.* Philadelphia: University of Pennsylvania Press.

———. 1959b. Darwinism and Darwinisticism. *Victorian Studies* 4:19–40.

Peel, John D. Y. 1971. *Herbert Spencer: the evolution of a sociologist.* London: Heinemann Educational.

Perkin, Harold. 1969. *The origins of modern English society, 1780–1880.* London: Routledge & Kegan Paul.

Perry, C. R. 1992. *The Victorian post office: the growth of a bureaucracy.* Woodbridge, Suffolk: Boydell Press for the Royal Historical Society.

Peterson, Linda H. 1986. *Victorian autobiography: the tradition of self-interpretation.* New Haven, Conn.: Yale University Press.

Peterson, M. Jeanne. 1978. *The medical profession in mid-Victorian London.* Berkeley: University of California Press.

———. 1989. *Family, love and work in the lives of Victorian gentlewomen.* Bloomington: Indiana University Press.

Pevsner, Nikolaus, and Ian Nairn. 1962. *The buildings of England. Surrey.* Harmondsworth: Penguin Books.

Phillips, John. 1860. *Life on the earth, its origin and succession.* Cambridge.

Pickering, Andrew, ed. 1992. *Science as practice and culture.* Chicago: University of Chicago Press.

Pickering, George W. 1974. *Creative malady: illness in the lives and minds of Charles Darwin, Florence Nightingale, Mary Baker Eddy, Sigmund Freud, Marcel Proust and Elizabeth Barrett Browning.* London: George Allen & Unwin.

Poovey, Mary. 1995. *Making a social body: British cultural formation, 1830–1864.* Chicago: University of Chicago Press.

———. 1998. *A history of the modern fact: problems of knowledge in the sciences of wealth and society.* Chicago: University of Chicago Press.

Porter, Duncan M. 1993. On the road to the *Origin* with Darwin, Hooker, and Gray. *Journal of the History of Biology* 26:1–38

Porter, Roy, ed. 1985. *Patients and practitioners.* Cambridge: Cambridge University Press.

———. 1989. *Health for sale: quackery in England, 1650–1850.* Manchester: Manchester University Press.

———. 2001. *Bodies politic: disease, death and the doctors in Britain, 1650–1914.* London: Reaktion Books.

Porter, Theodore. 1986. *The rise of statistical thinking, 1820–1900.* Princeton, N.J.: Princeton University Press.

———. 1995. *Trust in numbers: the pursuit of objectivity in science and public life.* Princeton, N.J.: Princeton University Press.

Power, Michael. 1996. *Accounting and science: natural inquiry and commercial reason.* Cambridge: Cambridge University Press.

Prescott, Gertrude [Nuding]. 1985. Fame and photography: portrait publications in Great Britain, 1856–1900. Ph.D. thesis, University of Texas, Austin.

Prete, Frederick R. 1990. The conundrum of the honey bees: one impediment to the publication of Darwin's theory. *Journal of the History of Biology* 23:271–90.

Preyer, William. 1891. Briefe von Darwin. *Deutsche Rundschau.* 67:356–90.

Pritchard, Charles. 1886. *Annals of our school life addressed to the "old boys" of the Clapham Grammar School.* Oxford: Privately printed.

Pritchard, Michael. 1988. Commercial photographers in nineteenth century Britain. *History of Photography* 11:213–15.

Prodger, Philip. 1998. Photography and *The expression of the emotions.* In Paul Ekman, ed., *Charles Darwin. The expression of the emotions in man and animals,* 399–410. London: HarperCollins.

Pycior, Helena, Nancy Slack, and Pnina Abir Am, eds. 1996. *Creative couples in the sciences.* New Brunswick, N.J.: Rutgers University Press.

Raby, Peter. 2001. *Alfred Russel Wallace: a life.* London: Chatto & Windus.

Raverat, Gwen. 1952. *Period piece: a Cambridge childhood.* London: Faber & Faber.

Reader, John. 1988. *Missing links: the hunt for earliest man.* Harmondsworth: Penguin Books.

Reeve, Lovell Augustus. 1863–67. *Portraits of men of eminence in literature, science and art.* With biographical memoirs by Edward Walford. 6 vols. London: Lovell Reeve.

Reilly, Robin. 1989. *Wedgwood.* 2 vols. London: Macmillan, Stockton Press.

Reinikka, Merle A. 1972. *A history of the orchid.* Coral Gables, Fla.: University of Miami Press.

Rheinberger, Hans-Jorg, and Peter McLaughlin. 1984. Darwin's experimental natural history. *Journal of the History of Biology* 17:345–68.

Richards, Evelleen. 1983. Darwin and the descent of women. In David Oldroyd and Ian Langham, eds., *The wider domain of evolutionary thought,* 57–111. Dordrecht: D. Reidel.

———. 1989. Huxley and women's place in science: the "woman question" and the control of Victorian anthropology. In J. R. Moore, ed., *History, humanity and evolution: essays for J. C. Greene,* 253–84. Cambridge: Cambridge University Press.

———. 1994. A political anatomy of monsters, hopeful and otherwise: teratogeny, transcendentalism and evolutionary theorizing. *Isis* 85:377–411.

Richards, Robert J. 1981. Instinct and intelligence in British natural theology: some contributions to Darwin's theory of the evolution of behavior. *Journal of the History of Biology* 14:193–230.

———. 1987. *Darwin and the emergence of evolutionary theories of mind and behaviour.* Chicago: University of Chicago Press.

Richards, Thomas. 1990. *The commodity culture of Victorian England: advertising and spectacle, 1851–1914.* Stanford, Calif.: Stanford University Press.

Richardson, Angelique. 1998a. "How I mismated myself for love of you!": the biologisation of romance in Hardy's *A group of noble dames. Thomas Hardy Journal* 14 (2):59–76.

———. 1998b. "Some science underlies all art": the dramatization of sexual selection and racial biology in Thomas Hardy's *A pair of blue eyes* and *The well-beloved. Journal of Victorian Culture* 3:302–38.

Richardson, Edmund W. 1916. *A veteran naturalist: being the life and work of W. B. Tegetmeier.* London: Witherby.

Ritvo, Harriet. 1987. *The animal estate: the English and other creatures in the Victorian age.* Cambridge, Mass.: Harvard University Press.

———. 1995. Classification and continuity in the *Origin of Species.* In David Amigoni and Jeff Wallace, eds., *Charles Darwin's "The Origin of Species": new interdisciplinary essays,* 47–67. Manchester: Manchester University Press.

Roberts, David. 1978. The paterfamilias of the Victorian governing classes. In Anthony Wohl, ed. *The Victorian Family,* 59–81. London: Croom Helm.

Roberts, H. F. 1929. *Plant hybridization before Mendel.* Princeton, N.J.: Princeton University Press.

Roberts, Jon H. 1988. *Darwinism and the divine in America: Protestant intellectuals and organic evolution, 1859–1900*. Madison: University of Wisconsin Press.

Roberts, Michael B. 1997. Darwin's doubts about design: the Darwin-Gray correspondence of 1860. *Science and Christian Belief* 9:113–27.

Rockell, F. 1912. The last of the great Victorians: a special interview with Alfred Russel Wallace. *Millgate Monthly* 7, pt. 2:657–63.

Romanes, George. 1892. *Darwin and after Darwin*. London.

Roos, David A. 1981. The aims and intentions of *Nature*. In James Paradis and Thomas Postlewait, eds., *Victorian science and Victorian values*, 159–80. New York: New York Academy of Sciences.

Rose, Jacqueline. 1999. The cult of celebrity. *New Formations* 36:9–20.

Rose, Phyllis. 1983. *Parallel lives: five Victorian marriages*. New York: Alfred A. Knopf.

Rosenberg, J. D. 1989. Mr. Darwin collects himself. In L. S. Lockridge, J. Maynard, and D. D. Stone, eds., *Nineteenth-century lives: essays presented to Jerome Hamilton Buckley*, 82–111. Cambridge: Cambridge University Press.

Rowell, Geoffrey. 1974. *Hell and the Victorians: a study of the nineteenth-century theological controversies concerning eternal punishment and the future life*. Oxford: Clarendon Press.

Rowse, A. L. 1989. *The controversial Colensos*. Truro: Cornish Publications.

Royle, E. 1974. *Victorian infidels: the origins of the British secularist movement, 1791–1866*. Manchester: Manchester University Press.

Rudwick, M.J.S. 1975. Caricature as a source for the history of science: De la Beche's anti-Lyellian sketches of 1831. *Isis* 66:534–60.

———. 1976. The emergence of a visual language for geological science, 1760–1840. *History of Science* 14:149–95.

———. 1985. *The great Devonian controversy: the shaping of scientific knowledge among gentlemanly specialists*. Chicago: University of Chicago Press.

———. 1992. *Scenes from deep time: early pictorial representations of the prehistoric world*. Chicago: University of Chicago Press.

Rupke, Nicolaas, ed. 1987. *Vivisection in historical perspective*. London: Croom Helm.

———. 1994. *Richard Owen, Victorian naturalist*. New Haven, Conn.: Yale University Press.

———. 2000. Translation studies in the history of science: the example of *Vestiges*. *British Journal for the History of Science* 33:209–22.

Ruse, Michael. 1975a. Charles Darwin and artificial selection. *Journal of the History of Ideas* 36:339–50.

———. 1975b. Darwin's debt to philosophy: an examination of the influence of the philosophical ideas of J.F.W. Herschel and W. Whewell on the development of Charles Darwin's theory of evolution. *Studies in History and Philosophy of Science* 6:159–81.

———. 1979. *The Darwinian revolution: science red in tooth and claw*. Chicago: University of Chicago Press.

Russell, Nicholas. 1986. *Like engend'ring like: heredity and animal breeding in early modern England*. Cambridge: Cambridge University Press.

Russell-Gebbett, Jean. 1977. *Henslow of Hitcham: botanist, educationalist and clergyman*. Lavenham, Suffolk: Terence Dalton.

Russett, Cynthia Eagle. 1976. *Darwin in America: the intellectual response, 1865–1912*. San Francisco: W. H. Freeman.

———. 1989. *Sexual science: the Victorian construction of womanhood*. Cambridge, Mass.: Harvard University Press.

Rutherford, H. W. 1908. *Catalogue of the library of Charles Darwin now in the Botany School, Cambridge*. Cambridge: Cambridge University Press.

Ryan, Simon. 1996. *The cartographic eye: how explorers saw Australia*. Cambridge: Cambridge University Press.

Sachs, Julius von. 1890. *History of Botany, 1530–1860*. Translated by H.E.F. Garnsey, additions by I. B. Balfour. Oxford: Clarendon Press.

Sakula, Alex. 1982. Baroness Burdett-Coutts' garden party: the International Medical Congress, London, 1881. *Medical History* 26:183–90.

Sanford, William F. 1965. Dana and Darwinism. *Journal of the History of Ideas* 26:531–46.

Satchell, J. E., ed. 1983. *Earthworm ecology: from Darwin to vermiculture*. London: Chapman & Hall.

Saunders, David. 1992. *Authorship and copyright*. London: Routledge.

Schabas, Margaret. 1990. Ricardo naturalised: Lyell and Darwin on the economy of nature. In D. Moggridge, ed., *Perspectives on the History of Economic Thought*. London: Edward Elgar.

Scherren, Henry. 1905. *The Zoological Society of London: a sketch of its foundation and development*. London: Cassell.

Schiebinger, Londa. 1991. The private life of plants: sexual politics in Carl Linnaeus and Erasmus Darwin. In Marina Benjamin, ed., *Science and sensibility: gender and scientific enquiry, 1780–1945*, 121–43. Oxford: Blackwell.

Schleicher, August. 1863. *Die Darwinische Theorie und die Sprachwissenschaft*. Weimar. Translated by Alexander Bikkers as *Darwinism tested by the science of language*. London, 1869.

Schwartz, Joel S. 1974. Charles Darwin's debt to Malthus and Edward Blyth. *Journal of the History of Biology* 7:301–18.

———. 1995. George John Romanes's defence of Darwinism: the correspondence of Charles Darwin and his chief disciple. *Journal of the History of Biology* 28:281–316.

Schweber, S. 1980. Darwin and the political economists: divergence of character. *Journal of the History of Biology* 13:195–289.

Scudo, Francesco, and Michele Acanfora. 1985. Darwin and Russian evolutionary biology. In David Kohn, ed., *The Darwinian heritage*, 731–52. Princeton, N.J.: Princeton University Press in association with Nova Pacifica.

Searby, Peter. 1997. *A history of the University of Cambridge*. Vol. 3, 1750–1870. Cambridge: Cambridge University Press.

Secord, Anne. 1994a. Corresponding interests: artisans and gentleman in nineteenth-century natural history. *British Journal for the History of Science* 27:383–408.

———. 1994b. Science in the pub: artisan botanists in early nineteenth century Lancashire. *History of Science* 32:269–315.

Secord, James A. 1981. Nature's fancy: Charles Darwin and the breeding of pigeons. *Isis* 72:163–86.

———. 1985. Darwin and the breeders: a social history. In David Kohn, ed., *The Darwinian heritage*, 519–42. Princeton, N.J.: Princeton University Press in association with Nova Pacifica.

———. 1986. The Geological Survey of Great Britain as a research school, 1839–1855. *History of Science* 24:223–75.

———, ed. 1997. *Charles Lyell, Principles of Geology*. Edited with an introduction. London: Penguin Books.

———. 2000. *Victorian sensation: the extraordinary publication, reception, and secret authorship of "Vestiges of the Natural History of Creation."* Chicago: University of Chicago Press.

Sedgwick, Adam. 1860. Objections to Mr. Darwin's theory of the origin of species. *Spectator*, 24 March 1860, pp. 285–86; 7 April 1860, pp. 334–35.

Semmel, B. 1962. *The Governor Eyre controversy*. London: Macgibbon & Kee.

Service, Robert. 2000. *Lenin: a biography*, London: Macmillan.

Seward, A. C., ed. 1909. *Darwin and modern science: essays in commemoration of the centenary of the birth of Charles Darwin and of the fiftieth anniversary of the publication of the "Origin of Species."* Cambridge: Cambridge University Press.

Shapin, Steven. 1988. The house of experiment in 17th-century England. *Isis* 79:373–404.

———. 1991. "The mind is its own place": science and solitude in seventeenth-century England. *Science in Context* 4:191–218.

———. 1994. *A social history of truth: civility and science in seventeenth-century England.* Chicago: University of Chicago Press.

Shapin, Steven, and Barry Barnes. 1976. Head and hand: rhetorical resources in British pedagogical writings, 1770–1850. *Oxford Review of Education* 2:231–54.

———. 1979. Darwin and social Darwinism: purity and history. In B. Barnes and S. Shapin, eds., *Natural order: historical studies of scientific culture,* 125–42. Beverly Hills: Sage Publications.

Shattock, Joanne. 1989. *Politics and reviewers: the "Edinburgh" and the "Quarterly" in the early Victorian age.* Leicester: Leicester University Press.

Shattock, Joanne, and Michael Wolff, eds. 1982. *The Victorian periodical press.* Leicester: Leicester University Press.

Sheets-Pyenson, Susan. 1985. Popular science periodicals in Paris and London: the emergence of a low scientific culture, 1820–1875. *Annals of Science* 42:549–72.

———. 1988. *Cathedrals of science: the development of colonial natural history museums during the late nineteenth-century.* Montreal: McGill–Queen's University Press.

Shepherd, J. A. 1982. Lawson Tait—disciple of Charles Darwin. *British Medical Journal* 284:1386–87.

Shortland, Michael, ed. 1996. Bonneted mechanic and narrative hero: the self-modelling of Hugh Miller. In M. Shortland, ed., *Hugh Miller and the controversies of Victorian science,* 14–86. Oxford: Oxford University Press.

Shortland, Michael, and Richard Yeo, eds. 1996. *Telling lives in science: essays on scientific biography.* Cambridge: Cambridge University Press.

Shteir, Ann B. 1996. *Cultivating women, cultivating science: Flora's daughters and botany in England 1760–1860.* Baltimore: Johns Hopkins University Press.

Shuttleworth, Sally. 1984. *George Eliot and nineteenth-century science: the make-believe of a beginning.* Cambridge: Cambridge University Press.

Sicherman, Barbara. 1977. The uses of diagnosis: doctors, patients, and neurasthenia. *Journal of the History of Medicine* 32:33–54.

Silliman, Robert H. 1995. The Hamlet affair: Charles Lyell and the North Americans. *Isis* 86:541–61.

Sloan, Phillip R., ed. 1992. *The Hunterian lectures in comparative anatomy, May and June 1837.* By Richard Owen. London: Natural History Museum Publications.

Smiles, Samuel. 1859. *Self help, with illustrations of character and conduct.* London.

———. 1871. *Character.* London.

———. 1887. *Life and labour, or characteristics of men of industry, culture and genius.* London.

Smith, Anthony. 1979. *The newspaper: an international history.* London: Thames & Hudson.

Smith, C.U.M. 1992. The hippopotamus test: a controversy in nineteenth-century brain science. *Cogito* 1:69–74.

Smith, Crosbie, and Jon Agar, eds. 1998. *Making space for science: territorial themes in the shaping of knowledge.* Basingstoke: Macmillan.

Smith, Fabienne. 1990. Charles Darwin's ill health. *Journal of the History of Biology* 23:443–59.

———. 1992. Charles Darwin's health problems: the allergy hypothesis.

Journal of the History of Biology 25:285–306.

Smith, Roger. 1972. Alfred Russel Wallace: philosophy of nature and man. *British Journal for the History of Science* 6:177–99.

Smith, Warren Sylvester. 1967. *The London heretics, 1870–1914*. London: Constable.

Snell, K.D.M., and Paul S. Ell. 2001. *Rival Jerusalems: the geography of Victorian religion*. Cambridge: Cambridge University Press.

Söderqvist, Thomas. 1996. Existential projects and existential choices: science biography as an edifying genre. In M. Shortland and R. Yeo, eds., *Telling lives: essays on scientific biography*, 45–84. Cambridge: Cambridge University Press.

Spencer, Herbert. 1852. The development hypothesis. *The Leader*, 20 March.

———. 1857. Progress: its law and cause. *Westminster Review* 11:445–85.

———. 1862. *System of synthetic philosophy: first principles*. London.

———. 1864–67. *Principles of biology*. 2 vols. London.

———. 1904. *An autobiography*. 2 vols. London: Williams & Norgate.

Sprengel, Christian Konrad. 1793. *Das entdeckte Geheimniss der Natur im Bau und in der Befruchtung der Blumen*. Berlin.

St. George, Andrew. 1993. *The descent of manners: etiquette, rules and the Victorians*. London: Chatto & Windus.

St. John, Spenser. 1994. *The life of Sir James Brooke, rajah of Sarawak: from his personal papers and correspondence*. Reprinted with an introduction by R.H.W. Reece. Oxford: Oxford University Press.

Stamhuis, Ida, Onno Meijer, and Erik Zevenhuizen. 1999. Hugo De Vries on heredity, 1889–1903: statistics, Mendelian laws, pangenes, mutations. *Isis* 90:238–67.

Stauffer, Robert C., ed. 1975. *Charles Darwin's Natural Selection, being the second part of his big species book written from 1856 to 1858*. Cambridge: Cambridge University Press.

Stearn, William T. 1981. *The natural history museum at South Kensington: a history of the British Museum (Natural History) 1753–1980*. London: Heinemann in association with the British Museum (Natural History).

Stebbins, Robert E. 1988. France. In Thomas Glick, ed., *The comparative reception of Darwinism*, 117–63. Chicago: University of Chicago Press.

Stecher, Robert M. 1961. The Darwin-Innes letters: the correspondence of an evolutionist with his vicar, 1848–1884. *Annals of Science* 17:201–58.

Stepan, Nancy. 1982. *The idea of race in science: Great Britain, 1800–1960*. London: Macmillan in association with St. Anthony's College, Oxford.

———. 2001. *Picturing tropical nature*. London: Reaktion Books.

Stephen, Leslie. 1885. *Life of Henry Fawcett*. London.

———. 1924. *Some early impressions*. London: Hogarth Press.

Stevenson, Lionel. 1932. *Darwin among the poets*. Chicago: University of Chicago Press.

Stirling, A.M.W., ed. 1926. *The Richmond papers from the correspondence and manuscripts of George Richmond, R.A., and his son Sir William Richmond, R.A., K.C.B.* London: William Heinemann.

Stocking, George. 1987. *Victorian anthropology*. London: Macmillan.

Stone, L., and J.C.F. Stone. 1986. *An open elite? England 1540–1880*. Abridged ed. Oxford: Clarendon Press.

Stone, Marvin J. 1998. Henry Bence Jones and his protein. *Journal of Medical Biography* 6:53–57

Strick, James. 1999. Darwinism and the origin of life: the role of H. C. Bastian in the British spontaneous generation debates, 1868–1873. *Journal of the History of Biology* 32:51–92

Sulivan, Henry N., ed. 1896. *Life and letters of the late Admiral Sir*

*Bartholomew James Sulivan, K.C.B., 1810–1890.* London.

Sullivan, Alvin, ed. 1984. *British literary magazines.* Vol. 3, *The Victorian and Edwardian age, 1837–1913.* Westport, Conn.: Greenwood Press.

Sutherland, J. A. 1976. *Victorian novelists and publishers.* Chicago: University of Chicago Press.

Tagg, John. 1988. *The burden of representation: essays on photographies and histories.* Amherst: University of Massachusetts Press.

Tammone, William. 1995. Competition, the division of labor and Darwin's principle of divergence. *Journal of the History of Biology* 28:109–31.

Taub, Liba. 1993. Evolutionary ideas and "empirical" methods: the analogy between language and species in works by Lyell and Schleicher. *British Journal for the History of Science* 26:171–93.

Temple, Frederick. 1860. *The present relations of science to religion: a sermon preached on Act Sunday, July 1, 1860, before the University of Oxford, during the meeting of the British Association.* Oxford and London.

Thompson, E. P. 1963. *The making of the English working class.* London: Victor Gollancz.

Thompson, F.M.L. 1988. *The rise of respectable society: a social history of Victorian Britain, 1830–1900.* London: Fontana Press.

Thwaite, Ann. 1996. *Emily Tennyson: the poet's wife.* London: Faber & Faber.

Todes, Daniel P. 1989. *Darwin without Malthus: the struggle for existence in Russian evolutionary thought.* Oxford: Oxford University Press.

Topham, Jonathan. 1992. Science and popular education in the 1830s: the role of the Bridgewater Treatises. *British Journal for the History of Science* 25:397–430.

———. 1998. Beyond the "common context": the production and reading of the Bridgewater Treatises. *Isis* 89:233–62.

Tosh, John. 1999. *A man's place: masculinity and the middle-class home in Victorian England.* New Haven, Conn.: Yale University Press.

Toulmin, Stephen, and June Goodfield. 1965. *The discovery of time.* Chicago: University of Chicago Press.

Townsend, Mary L. 1992. *Forbidden laughter: popular humour and the limits of repression in nineteenth-century Prussia.* Ann Arbor: University of Michigan Press.

Tucker, Jennifer. 1997. Photography as witness, detective, and impostor: visual representation in Victorian science. In B. Lightman, ed., *Victorian Science in Context,* 378–408. Chicago: University of Chicago Press.

Tuckwell, William. 1900. *Reminiscences of Oxford.* London: Cassell.

Tuke, Margaret. 1939. *A history of Bedford College for Women, 1849–1937.* London: Oxford University Press.

Turner, Ernest S. 1967. *Taking the cure.* London: Michael Joseph.

Turner, Frank M. 1974. *Between science and religion.* New Haven, Conn.: Yale University Press.

———. 1993. *Contesting cultural authority: essays in Victorian intellectual life.* Cambridge: Cambridge University Press.

Uglow, Jenny. 1993. *Elizabeth Gaskell: a habit of stories.* London: Faber & Faber.

Van Riper, A. Bowdoin. 1993. *Men among the mammoths: Victorian science and the discovery of human prehistory.* Chicago: University of Chicago Press.

Vann, J. D., and R. T. Van Arsdel, eds. 1994. *Victorian periodicals and Victorian society.* Toronto: University of Toronto Press.

Vaucaire, Michel. 1930. *Paul Du Chaillu: gorilla hunter.* Translated by E. P. Watts. London: Harper & Brothers.

*Vestiges:* see Chambers, Robert, 1860.

Vicinus, Martha, ed. 1972. *Suffer and be still: women in the Victorian age.* Bloomington: Indiana University Press.

Victorius, P. B. 1932. Editions of Darwin. *The Colophon: A Book Collector's Quarterly* 9.

Vogt, Carl. 1864. *Lectures on man: his place in creation and in the history of the earth*. London.

Vorzimmer, Peter J. 1963. Charles Darwin and blending inheritance. *Isis* 54:371–90.

———. 1972. *Charles Darwin. The years of controversy: the "Origin of Species" and its critics, 1859–82*. London: University of London Press.

Vucinich, Alexander. 1988. *Darwin in Russian thought*. Berkeley: University of California Press.

Wagner-Martin, Linda. 1994. *Telling women's lives: the new biography*. New Brunswick, N.J.: Rutgers University Press.

Walford, Edward. 1868. *Representative men in literature, science, and art: the photographic portraits from life by Ernest Edwards*. London: A. W. Bennett.

Wallace, A. R. 1853. *A narrative of travels on the Amazon and Rio Negro*. London.

———. 1855. On the law which has regulated the introduction of new species. *Annals and Magazine of Natural History* 2nd ser. 16:184–96.

———. 1864. The origin of the human races and the antiquity of man deduced from the theory of natural selection. *Journal of the Anthropological Society of London* (previously *Anthropological Review*) 2:clviii–clxx.

———. 1866. *The scientific aspect of the supernatural: indicating the desirableness of an experimental inquiry by men of science into the alleged powers of clairvoyants and mediums*. London.

———. 1869a. *The Malay archipelago: the land of the orang-utan, and the bird of paradise: a narrative of travel, with studies of man and nature*. 2 vols. London.

———. 1869b. Sir Charles Lyell on geological climates and the *Origin of Species*. *Quarterly Review* 126:359–94.

———. 1870. *Contributions to the theory of natural selection: a series of essays*. London.

———. 1875. *On miracles and modern spiritualism: three essays*. London and Glasgow.

———. 1876. *The geographical distribution of animals, with a study of the relations of living and extinct faunas as elucidating the past changes of the earth's surface*. 2 vols. London.

———. 1880. *Island life; or, the phenomena and causes of insular faunas and floras, including a revision and attempts at solution of the problem of geological climates*. London.

———. 1883. The debt of science to Darwin. *Century Magazine* n.s. 3:420–32.

———. 1889. *Darwinism: an exposition of the theory of natural selection with some of its applications*. London: Macmillan.

———. 1898. *The wonderful century: its successes and its failures*. London.

———. 1903. My relations with Darwin. *Black and White*, 17 January 1903, 78–79.

———. 1905. *My life: a record of events and opinions*. 2 vols. London: Chapman & Hall.

Waller, John C. 2001. Gentlemanly men of science: Sir Francis Galton and the professionalisation of the British life sciences. *Journal of the History of Biology* 34:83–114.

Wallich, George C. 1870. *Eminent men of the day: photographs by G. C. Wallich*. London.

Walls, Laura Dassow. 1995. *Seeing new worlds: Henry David Thoreau and nineteenth-century natural science*. Madison: University of Wisconsin Press.

Ward, William R. 1965. *Victorian Oxford*. London: Frank Cass.

Watson, Hewett Cottrell. 1845. On the theory of "progressive development" applied in the explanation of the origin and transmutation of species. *Phytologist* 2:108–13, 140–47.

———. 1847–60. *Cybele Britannica; or British plants, and their geographical relations*. 4 vols. and supplement. London.

Weaver, Michael, 1984. *Julia Margaret Cameron, 1815–1879*. London: Herbert Press.

———. 1989. Julia Margaret Cameron: the stamp of divinity. In M. Weaver, ed., *British photography in the nineteenth century: the fine art tradition*. Cambridge: Cambridge University Press.

Wedgwood, Barbara, and Hensleigh Wedgwood. 1980. *The Wedgwood circle, 1730–1897: four generations of a family and their friends*. London: Studio Vista.

Wedgwood, Josiah C. 1940. *Memoirs of a fighting life*. London: Hutchinson.

Weikart, Richard. 1995. A recently discovered Darwin letter on social Darwinism. *Isis* 86:609–11.

Weingart, Peter. 1995. Struggle for existence: selection and retention of a metaphor. In S. Maasen, E. Mendelsohn, and P. Weingart, eds., *Biology as society, society as biology: metaphors*, 127–51. Dordrecht: Kluwer.

Weschler, Judith. 1982. *A human comedy: physiognomy and caricature in nineteenth century Paris*. London: Thames & Hudson.

West, Geoffrey. 1937. *Charles Darwin, the fragmentary man*. London: G. Routledge & Sons.

Wheeler, Michael. 1990. *Death and the future life in Victorian literature and theology*. Cambridge: Cambridge University Press.

Wilberforce, Samuel. 1860. Darwin's *Origin of Species*. *Quarterly Review* 108:225–64.

———. 1861. Review of *Essays and reviews*. *Quarterly Review* 109:248–305.

Wilson, D. A., and D. A. MacArthur. 1934. *Life of Thomas Carlyle*. Vol. 6, *Carlyle in old age*. London: Kegan Paul.

Wilson, Leonard G. 1996a. Brixham Cave and Sir Charles Lyell's *The Antiquity of Man:* the roots of Hugh Falconer's attack on Lyell. *Archives of Natural History* 23:79–97.

———. 1996b. The gorilla and the question of human origins: the brain controversy. *Journal of the History of Medicine and Allied Sciences* 51:184–207.

Wiltshire, John. 1992. *Jane Austen and the body: "the picture of health."* Cambridge: Cambridge University Press.

Winslow, John H. 1971. *Darwin's Victorian malady: evidence for its medically induced origin*. Philadelphia: American Philosophical Society.

Winsor, Mary P. 1969. *Starfish, jellyfish and the order of life: issues in nineteenth-century science*. New Haven, Conn.: Yale University Press.

Winstanley, D. A. 1947. *Later Victorian Cambridge*. Cambridge: Cambridge University Press.

Winter, Alison. 1995. Harriet Martineau and the reform of the invalid in Victorian England. *Historical Journal* 38:597–616.

———. 1998. *Mesmerized: power of mind in Victorian England*. Chicago: University of Chicago Press.

Winther, Rasmus G. 2000. Darwin on variation and heredity. *Journal of the History of Biology* 33:425–55.

Wohl, Anthony, ed. 1978. *The Victorian family: structure and stresses*. London: Croom Helm.

Wolff, Robert L. 1977. *Gains and losses: novels of faith and doubt in Victorian England*. London: John Murray.

Wollaston, A.F.R. 1921. *Life of Albert Newton*. London: John Murray.

Wollaston, Thomas Vernon. 1860. Review of *On the Origin of Species*. *Annals and Magazine of Natural History* 3rd ser. 5:132–43.

Woolner, Amy. 1917. *Thomas Woolner, R.A., sculptor and poet. His life in letters*. London: Chapman & Hall.

Worster, Donald. 1985. *Nature's economy: a history of ecological ideas*. Cambridge: Cambridge University Press.

Wright, Chauncey. 1871. *Darwinism: being an examination of Mr. St. George Mivart's "Genesis of species."*

Reprinted from the *North American Review.* London.

Wyllie, Irvin G. 1959. Social Darwinism and the businessman. *Proceedings of the American Philosophical Society* 103:629–35.

Wynter, Andrew. 1862. Cartes de visite. *Once a Week* 6:1134–37; also in *Journal of the Photographic Society of London* 7:375–77.

Yanni, Carla. 1999. *Nature's museums: Victorian science and the architecture of display.* Baltimore: Johns Hopkins University Press.

Yeo, Richard. 1984. Science and intellectual authority in mid-nineteenth century Britain: Robert Chambers and *Vestiges of the Natural History of Creation. Victorian Studies* 28:5–31.

———. 1985. An idol of the market place: Baconianism in nineteenth century Britain. *History of Science* 23:251–98.

———. 1993. *Defining Science: William Whewell, natural knowledge and public debate in early Victorian Britain.* Cambridge: Cambridge University Press.

Young, D.A.B. 1997. Darwin's illness and systemic lupus erythematosus. *Notes and Records of the Royal Society* 51:77–86.

Young, Robert M. 1970. *Victorian crisis of faith: six lectures.* Edited by Anthony Symondson. London: SPCK.

———. 1985. *Darwin's metaphor: nature's place in Victorian culture.* Cambridge: Cambridge University Press.

# *Acknowledgements*

While this book has been in the making I have been helped by a large number of people, and it is a genuine pleasure to be able to thank them here. First and foremost the Wellcome Trust generously maintained my research throughout, and Bridget Ogilvie and Michael Dexter, past and present directors, have taken a special interest in all things Darwinian. More immediately, my colleagues in the Wellcome Trust Centre for the History of Medicine at University College London (formerly the Wellcome Institute for the History of Medicine) have provided wonderful support over a long period, in turn intellectual, professional, and personal, all that anyone could wish for from friends. Throughout, I have particularly appreciated the encouragement offered by Bill Bynum and Michael Neve, as well as advice from Hal Cook, Anne Hardy, Chris Lawrence, Vivian Nutton, the late Roy Porter, Cornelius O'Boyle, Tilli Tansey, and Andrew Wear. There could be no better place to work, no better interactions with colleagues. Sharon Messenger and Caroline Essex provided timely research assistance, and I thank them, and Gita Tailor and Sally Scovell, for their invaluable help. Alex Goldbloom kindly researched *Punch* on my behalf. The staff of the Wellcome Library for the History and Understanding of Medicine generously offered their expertise, time, and patience over a long period, and I gratefully acknowledge the assistance of Eric Freeman, David Pearson, Wendy Fish, Sue Gold, Richard Aspin, Julia Sheppard, Lesley Hall, John Symons, William Schupbach, and other members of the library staff. I am also very grateful to Catherine Draycott and Chris Carter for their help with photographic requests.

The Provost and Fellows of King's College Cambridge made me very welcome during a year's Visiting Senior Research Fellowship in 1996–97, and I extend warmest thanks to Pat Bateson, John Barber, Hal Dixon, Peter Jones, Rob Foley, Ian Patterson, and George Pattison. Spending the year in Cambridge, so close to the archives, and in such a community, was a great privilege that I appreciated enormously. During the same year I served on an English Heritage panel for the restoration of Down House, which was both illuminating and fun, and encouraged me rethink the way I wanted to present Darwin in this biography. In this regard I particularly thank Angela Darwin and Stephen and Randal Keynes, whose zest for research was infectious. In more recent months, Adrian Desmond, Jim Moore, and I have written about Darwin as a trio for the *New Dictionary of National Biography.* I thank them for their friendship and biographical generosity.

My largest debt lies with Cambridge University Library, where Peter Fox, the University Librarian, and Patrick Zutshi, Adam Perkins, and Godfrey Waller do so much to help Darwin scholars. Their assistance has truly been invaluable. Nor could I have completed this work without the support of friends on the Darwin Correspondence Project, both past and present, especially Fred and Anne Burkhardt, Duncan Porter, and the late Sidney Smith. Their help is very warmly acknowledged, and it is a real pleasure to thank them, and the other members of the project, in public. The volumes of *Correspondence* and the finding aids they have produced make the Darwin collections in Cambridge one of the world's finest historical resources. Alan Crowden of Cambridge University Press has been very supportive, and I am grateful for permission from the Syndics of Cambridge University Press to cite materials published in the *Correspondence of Charles Darwin.* The Syndics of Cambridge University Library similarly granted permission to quote from unpublished manuscripts and other resources in their collections. William Huxley Darwin kindly gave permission to cite Darwin materials. Richard Darwin Keynes generously allowed me to use Emma Darwin's diaries and other family manuscripts on loan to Cambridge University Library.

I also thank the archivists and librarians of those other institutions that hold related materials, especially the Trustees of the Wedgwood Museum, Barlaston, Stoke-on-Trent; John Murray Ltd.; the American Philosophical Society; the British Library; Die Deutsche Staatsbibliothek; English Heritage; the Gray Herbarium of Harvard University; Das Haeckel Haus, Jena; Leeds University Library; Case Western Reserve University; the Natural History Museum, London; the Royal Institution of Great Britain; the Royal Society of London; and the University of Virginia Library.

The British Library, the Master and Fellows of Christ's College Cambridge; English Heritage; the Gray Herbarium of Harvard University; Imperial College of Science, Technology and Medicine; the Mary Evans Picture Library; Missouri Botanical Garden Library; the Royal College of Surgeons of England; the Syndics of Cambridge University Library; the Library of University College London; and the Wellcome Library for the History and Understanding of Medicine kindly gave permission to publish images from their collections. A. Walker Bingham and Randal Keynes very generously made items available from their private collections.

And personally, it gives me special pleasure to thank Bill Bynum, Michael Neve, Joy Harvey, and Thomas Söderqvist for reading the manuscript. Their comments showed me ways in which to integrate the details of Darwin's life with his work that have made all the difference and are deeply appreciated. Juan Bacigalupo, Anne Barrett, Gillian Beer, Peter Bowler, Nicholas Browne, Joe

Cain, Jane Camerini, John Clark, Ann Dally, Angela Darwin, Sheila Dean, Adrian Desmond, Mario di Gregorio, Gina Douglas, Patricia Fara, Paul Farber, Janet Garber, Nick Gill, Rupert and Marie Hall, Jonathan Hodge, Rob Iliffe, Frank James, Ludmilla Jordanova, Milo Keynes, David Kohn, Ed Larsen, Ernst Mayr, Richard Milner, Jim Moore, Jack Morrell, Solene Morris, Virginia Murray, Trudy Prescott Nuding, Caroline Overy, Alison Pearn, Jim Paradis, Robin Reilly, Marsha Richmond, Harriet Ritvo, Anne and Jim Secord, Crosbie Smith, Jon Topham, Hugh Torrens, Edward Wakeling, Andrew Warwick, R. D. Wood, and the late John Thackray helped me in many different forms, as have countless wide-ranging discussions with other friends, colleagues, and students over the years. I have been very lucky in being able to attend so many conferences and seminars in different disciplines where Darwin's works and thoughts, and other issues of a historical and biographical nature, have generated such lively debate. It is clear that there are many ways in which Darwin can be interpreted, many different Darwins, and a rich variety of authoritative preexisting literature, all adding materially to a highly stimulating field on which I have very gratefully drawn. Last but by no means least, Kit and Evie Browne have been the hub of my family life.

Attentive readers of this volume will see that I have a special regard for the unsung heroes of book production who prop the author up. I would particularly like to thank Will Sulkin of Jonathan Cape, Sam Elworthy of Princeton University Press, and the Knopf editorial team who made my life so much smoother: Ken Schneider, Ted Johnson, Ellen Feldman, Avery Fluck, Anthea Lingeman, Justin Salvas, and Carol Carson. Charles Elliott has been a marvellously supportive editor over what has turned out to be a much longer haul than he ever expected.

# Index